Information Security and Cryptography

More information about this series at http://www.springer.com/series/4752

Hans Delfs • Helmut Knebl

Introduction to Cryptography

Principles and Applications

Third Edition

Springer

Hans Delfs
Department of Computer Science
Technische Hochschule Nürnberg
   Georg Simon Ohm
Nürnberg, Germany

Helmut Knebl
Department of Computer Science
Technische Hochschule Nürnberg
   Georg Simon Ohm
Nürnberg, Germany

ISSN 1619-7100   ISSN 2197-845X   (electronic)
Information Security and Cryptography
ISBN 978-3-662-47973-5   ISBN 978-3-662-47974-2   (eBook)
DOI 10.1007/978-3-662-47974-2

Library of Congress Control Number: 2015949232

Springer Heidelberg New York Dordrecht London

Printed on acid-free paper

Springer-Verlag GmbH Berlin Heidelberg is part of Springer Science+Business Media (www.springer.com)

Preface to the Third, Extended Edition

The third edition is a substantive extension. This is reflected in a considerable increase in pages. A number of new topics have been included. They
supplement and extend the material covered in the previous editions. The
major extensions and enhancements follow:

- We added a description of the new SHA-3 algorithm for cryptographic hash
  functions, Keccak, and the underlying sponge construction (Section 2.2.2).
- As a further important example of homomorphic encryption algorithms,
  Paillier encryption is introduced in Section 3.6.2.
- ElGamal encryption in a prime-order subgroup of a finite field is studied
  in detail in Section 3.4.4.
- An introduction to elliptic curve cryptosystems including Diffie–Hellman
  key exchange, ElGamal encryption and the elliptic curve digital signature
  standard ECDSA is given in Section 3.7.
- The basics of plane curves and elliptic curves that are necessary to understand elliptic curve cryptography are explained in Appendix A.11.
- Additional concepts of interactive proof systems, such as the special honest-
  verifier zero-knowledge property and the OR-combination of Σ-proofs, are
  addressed in Section 4.5.
- The exposition of cryptographic protocols for electronic elections and Internet voting has been substantially extended (Sections 4.6 and 4.7).
- Decryption and re-encryption mix nets and shuffles as the basic tools to
  anonymize communication are discussed in Section 4.6. A complete zero-
  knowledge proof for the correctness of a re-encryption shuffle is given.
  Such proofs are indispensable in critical applications of mix nets, such as
  electronic voting systems. The correct operations of the mix net must be
  publicly verifiable, without compromising the confidentiality of individual
  votes.
- Receipt-free and coercion-resistant elections are studied in Section 4.7. Various techniques that are useful in the design of cryptographic protocols are
  explained here, such as designated-verifier proofs, diverted proofs, untappable channels and plaintext equivalence tests.
- Unconditionally secure cryptographic schemes are not based on the practical infeasibility of a computational task. Their security is proven by using

methods from information theory. Unconditionally secure schemes are now addressed in an extra chapter, Chapter 10.

- Unconditional security is not achievable without an unconditionally secure authentication of the communication partners. Suitable message authentication codes are constructed from almost universal classes of hash functions in Section 10.3.
- Privacy amplification is one of the essential techniques to derive an unconditionally secure shared secret key from information that is publicly exchanged. We give a proof of the Privacy Amplification Theorem and explain the basic properties of Rényi entropy.
- In Section 10.5, we present an introduction to quantum cryptography. We describe, in detail, the famous BB84 protocol of Bennett and Brassard for quantum key distribution and prove its security, assuming intercept-and-resend attacks.

The description of cryptographic hash functions has been moved to Chapter 2, which is now headlined "Symmetric-Key Cryptography". Moreover, errors and inaccuracies have been corrected, and in some places, the exposition has been clarified.

Some exercises have been added. Answers to all exercises are provided on the webpage www.in.th-nuernberg.de/DelfsKnebl/Cryptography.

We thank our readers and our students for their comments and hints. Again, we are greatly indebted to our colleague Patricia Shiroma-Brockmann for proof-reading the English copy of the new and revised chapters. Finally, we would like to thank Ronan Nugent at Springer for his support.

Nürnberg, July 2015   Hans Delfs, Helmut Knebl

Preface to the Second, Extended Edition

New topics have been included in the second edition. They reflect recent progress in the field of cryptography and supplement the material covered in the first edition. Major extensions and enhancements are the following.

- A complete description of the Advanced Encryption Standard AES is given in Chapter 2.1 on symmetric encryption.
- In Appendix A, there is a new section on polynomials and finite fields. There we offer a basic explanation of finite fields, which is necessary to understand the AES.
- The description of cryptographic hash functions in Chapter 3 has been extended. It now also includes, for example, the HMAC construction of message authentication codes.[1]
- Bleichenbacher's 1-Million-Chosen-Ciphertext Attack against schemes that implement the RSA encryption standard PKCS#1 is discussed in detail in Chapter 3. This attack proves that adaptively-chosen-ciphertext attacks can be a real danger in practice.
- In Chapter 9 on provably secure encryption we have added typical security proofs for public-key encryption schemes that resist adaptively-chosen-ciphertext attacks. Two prominent examples are studied – Boneh's simple-OAEP, or SAEP for short, and Cramer–Shoup's public-key encryption.
- Security proofs in the random oracle model are now included. Full-domain-hash RSA signatures and SAEP serve as examples.

Furthermore, the text has been updated and clarified at various points. Errors and inaccuracies have been corrected.

We thank our readers and our students for their comments and hints, and we are indebted to our colleague Patricia Shiroma-Brockmann and Ronan Nugent at Springer for proof-reading the English copy of the new and revised chapters.

Nürnberg, December 2006   Hans Delfs, Helmut Knebl

---

[1] In the 3rd edition, the description of hash functions has been moved to Chapter 2.

Preface

The rapid growth of electronic communication means that issues in information security are of increasing practical importance. Messages exchanged over worldwide publicly accessible computer networks must be kept confidential and protected against manipulation. Electronic business requires digital signatures that are valid in law, and secure payment protocols. Modern cryptography provides solutions to all these problems.

This book originates from courses given for students in computer science at the Georg-Simon-Ohm University of Applied Sciences, Nürnberg. It is intended as a course on cryptography for advanced undergraduate and graduate students in computer science, mathematics and electrical engineering.

In its first part (Chapters 1–4), it covers – at an undergraduate level – the key concepts from symmetric and asymmetric encryption, digital signatures and cryptographic protocols, including, for example, identification schemes, electronic elections and digital cash. The focus is on asymmetric cryptography and the underlying modular algebra. Since we avoid probability theory in the first part, we necessarily have to work with informal definitions of, for example, one-way functions and collision-resistant hash functions.

It is the goal of the second part (Chapters 5–11) to show, using probability theory, how basic notions like the security of cryptographic schemes and the one-way property of functions can be made precise, and which assumptions guarantee the security of public-key cryptographic schemes such as RSA. More advanced topics, like the bit security of one-way functions, computationally perfect pseudorandom generators and the close relation between the randomness and security of cryptographic schemes, are addressed. Typical examples of provably secure encryption and signature schemes and their security proofs are given.

Though particular attention is given to the mathematical foundations and, in the second part, precise definitions, no special background in mathematics is presumed. An introductory course typically taught for beginning students in mathematics and computer science is sufficient. The reader should be familiar with the elementary notions of algebra, such as groups, rings and fields, and, in the second part, with the basics of probability theory. Appendix A contains an exposition of the results from algebra and number theory necessary for an understanding of the cryptographic methods. It includes proofs and covers, for example, basics like Euclid's algorithm and the Chinese Remainder Theorem, but also more advanced material like Legendre and Jacobi symbols and probabilistic prime number tests. The concepts and results from probability and information theory that are applied in the second part of the book are given in full in Appendix B. To keep the mathematics easy, we do not address elliptic curve cryptography.[2] We illustrate the key concepts of public-key cryptography by the classical examples like RSA in the quotient rings $\mathbb{Z}_n$ of the integers $\mathbb{Z}$.

The book starts with an introduction to classical symmetric encryption in Chapter 2.[3] The principles of public-key cryptography and their use for encryption and digital signatures are discussed in detail in Chapter 3. The famous and widely used RSA, ElGamal's methods and the digital signature standard, Rabin's encryption and signature schemes serve as the outstanding examples. The underlying one-way functions – modular exponentiation, modular powers and modular squaring – are used throughout the book, also in the second part.

Chapter 4 presents typical cryptographic protocols, including key exchange, identification and commitment schemes, electronic cash and electronic elections.

The following chapters focus on a precise definition of the key concepts and the security of public-key cryptography. Attacks are modeled by probabilistic polynomial algorithms (Chapter 5). One-way functions as the basic building blocks and the security assumptions underlying modern public-key cryptography are studied in Chapter 6. In particular, the bit security of the RSA function, the discrete logarithm and the Rabin function is analyzed in detail (Chapter 7). The close relation between one-way functions and computationally perfect pseudorandom generators meeting the needs of cryptography is explained in Chapter 8. Provable security properties of encryption schemes are the central topic of Chapter 9. It is clarified that randomness is the key to security. We start with the classical notions of provable security originating from Shannon's work on information theory. Typical examples of more recent results on the security of public-key encryption schemes are given, taking into account the computational complexity of attacking algorithms. A short introduction to cryptosystems, whose security can be proven by information-theoretic methods without any assumptions on the hardness of computational problems ("unconditional security approach"), supplements the section. Finally, we discuss in Chapter 11 the levels of security of digital signatures and give examples of signature schemes, whose security can be proven solely under standard assumptions like the factoring assumption, including a typical security proof.

---

[2] Elliptic curve cryptography has been included in the 3rd edition.

[3] Chapter 2 now covers symmetric encryption and cryptographic hash functions.

Each chapter (except Chapter 1) closes with a collection of exercises. Answers to the exercises are provided on the webpage for this book: www.in.th-nuernberg.de/DelfsKnebl/Cryptography.

We thank our colleagues and students for pointing out errors and suggesting improvements. In particular, we express our thanks to Jörg Schwenk, Harald Stieber and Rainer Weber. We are grateful to Jimmy Upton for his comments and suggestions, and we are very much indebted to Patricia Shiroma-Brockmann for proof-reading the English copy. Finally, we would like to thank Alfred Hofmann at Springer-Verlag for his support during the writing and publication of this book.

Nürnberg, December 2001   Hans Delfs, Helmut Knebl

Contents

1. **Introduction** 1
   1.1 Encryption and Secrecy 1
   1.2 The Objectives of Cryptography........................... 2
   1.3 Attacks 4
   1.4 Cryptographic Protocols 5
   1.5 Provable Security.. 6

2. **Symmetric-Key Cryptography** 11
   2.1 Symmetric-Key Encryption 11
       2.1.1 Stream Ciphers 12
       2.1.2 Block Ciphers.................................... 15
       2.1.3 DES 16
       2.1.4 AES 19
       2.1.5 Modes of Operation 25
   2.2 Cryptographic Hash Functions 30
       2.2.1 Security Requirements for Hash Functions 30
       2.2.2 Construction of Hash Functions 32
       2.2.3 Data Integrity and Message Authentication 42
       2.2.4 Hash Functions as Random Functions 44

3. **Public-Key Cryptography** 49
   3.1 The Concept of Public-Key Cryptography 49
   3.2 Modular Arithmetic 51
       3.2.1 The Integers 51
       3.2.2 The Integers Modulo n 53
   3.3 RSA 58
       3.3.1 Key Generation and Encryption 58
       3.3.2 Attacks Against RSA Encryption 62
       3.3.3 Probabilistic RSA Encryption 67
       3.3.4 Digital Signatures — The Basic Scheme 70
       3.3.5 Signatures with Hash Functions 71
   3.4 The Discrete Logarithm 77
       3.4.1 ElGamal Encryption 77
       3.4.2 ElGamal Signatures 78

          3.4.3   Digital Signature Algorithm 80

          3.4.4   ElGamal Encryption in a Prime-Order Subgroup 82

   3.5   Modular Squaring 85

          3.5.1   Rabin's Encryption 85

          3.5.2   Rabin's Signature Scheme 86

   3.6   Homomorphic Encryption Algorithms 87

          3.6.1   ElGamal Encryption 87

          3.6.2   Paillier Encryption 88

          3.6.3   Re-encryption of Ciphertexts 89

   3.7   Elliptic Curve Cryptography 90

          3.7.1   Selecting the Curve and the Base Point 93

          3.7.2   Diffie–Hellman Key Exchange 98

          3.7.3   ElGamal Encryption 100

          3.7.4   Elliptic Curve Digital Signature Algorithm 102

4.   **Cryptographic Protocols**.................................. 107

   4.1   Key Exchange and Entity Authentication.................. 107

          4.1.1   Kerberos 108

          4.1.2   Diffie–Hellman Key Agreement 111

          4.1.3   Key Exchange and Mutual Authentication 112

          4.1.4   Station-to-Station Protocol 114

          4.1.5   Public-Key Management Techniques 115

   4.2   Identification Schemes 117

          4.2.1   Interactive Proof Systems 117

          4.2.2   Simplified Fiat–Shamir Identification Scheme 119

          4.2.3   Zero-Knowledge 121

          4.2.4   Fiat–Shamir Identification Scheme................... 123

          4.2.5   Fiat–Shamir Signature Scheme 125

   4.3   Commitment Schemes................................... 126

          4.3.1   A Commitment Scheme Based on Quadratic Residues . 127

          4.3.2   A Commitment Scheme Based on Discrete Logarithms 128

          4.3.3   Homomorphic Commitments....................... 129

   4.4   Secret Sharing 130

   4.5   Verifiable Electronic Elections........................... 133

          4.5.1   A Multi-authority Election Scheme 135

          4.5.2   Proofs of Knowledge 138

          4.5.3   Non-interactive Proofs of Knowledge................. 142

          4.5.4   Extension to Multi-way Elections.................... 143

          4.5.5   Eliminating the Trusted Center 144

   4.6   Mix Nets and Shuffles 146

          4.6.1   Decryption Mix Nets 147

          4.6.2   Re-encryption Mix Nets 150

          4.6.3   Proving Knowledge of the Plaintext................. 153

          4.6.4   Zero-Knowledge Proofs of Shuffles 154

   4.7   Receipt-Free and Coercion-Resistant Elections 168

        4.7.1   Receipt-Freeness by Randomized Re-encryption 169

        4.7.2   A Coercion-Resistant Protocol 176

   4.8   Digital Cash 184

        4.8.1   Blindly Issued Proofs 186

        4.8.2   A Fair Electronic Cash System 192

        4.8.3   Underlying Problems 197

5. Probabilistic Algorithms 203

   5.1   Coin-Tossing Algorithms 203

   5.2   Monte Carlo and Las Vegas Algorithms 208

6. One-Way Functions and the Basic Assumptions 215

   6.1   A Notation for Probabilities 216

   6.2   Discrete Exponential Function 217

   6.3   Uniform Sampling Algorithms 223

   6.4   Modular Powers 226

   6.5   Modular Squaring 229

   6.6   Quadratic Residuosity Property 230

   6.7   Formal Definition of One-Way Functions 231

   6.8   Hard-Core Predicates 235

7. Bit Security of One-Way Functions 243

   7.1   Bit Security of the Exp Family 243

   7.2   Bit Security of the RSA Family 250

   7.3   Bit Security of the Square Family 258

8. One-Way Functions and Pseudorandomness 267

   8.1   Computationally Perfect Pseudorandom Bit Generators 267

   8.2   Yao's Theorem 275

9. Provably Secure Encryption 283

   9.1   Classical Information-Theoretic Security 284

   9.2   Perfect Secrecy and Probabilistic Attacks 288

   9.3   Public-Key One-Time Pads 292

   9.4   Passive Eavesdroppers 294

   9.5   Chosen-Ciphertext Attacks 301

        9.5.1   A Security Proof in the Random Oracle Model 304

        9.5.2   Security Under Standard Assumptions 313

10. Unconditional Security of Cryptosystems 321

   10.1 The Bounded Storage Model 322

   10.2 The Noisy Channel Model 332

   10.3 Unconditionally Secure Message Authentication 333

       10.3.1 Almost Universal Classes of Hash Functions 333

       10.3.2 Message Authentication with Universal Hash Families . 335

          10.3.3 Authenticating Multiple Messages 336
    10.4 Collision Entropy and Privacy Amplification 337
          10.4.1 Rényi Entropy 338
          10.4.2 Privacy Amplification 340
          10.4.3 Extraction of a Secret Key 341
    10.5 Quantum Key Distribution 343
          10.5.1 Quantum Bits and Quantum Measurements 344
          10.5.2 The BB84 Protocol 350
          10.5.3 Estimation of the Error Rate 353
          10.5.4 Intercept-and-Resend Attacks 354
          10.5.5 Information Reconciliation......................... 362
          10.5.6 Exchanging a Secure Key – An Example 367
          10.5.7 General Attacks and Security Proofs 368

11. Provably Secure Digital Signatures 373
    11.1 Attacks and Levels of Security 373
    11.2 Claw-Free Pairs and Collision-Resistant Hash Functions 376
    11.3 Authentication-Tree-Based Signatures 379
    11.4 A State-Free Signature Scheme 381

A. Algebra and Number Theory 397
    A.1 The Integers 397
    A.2 Residues 403
    A.3 The Chinese Remainder Theorem...................... 407
    A.4 Primitive Roots and the Discrete Logarithm 409
    A.5 Polynomials and Finite Fields 413
          A.5.1 The Ring of Polynomials 413
          A.5.2 Residue Class Rings 415
          A.5.3 Finite Fields 417
    A.6 Solving Quadratic Equations in Binary Fields 419
    A.7 Quadratic Residues 421
    A.8 Modular Square Roots 426
    A.9 The Group $\mathbb{Z}_{n^2}^*$ 430
    A.10 Primes and Primality Tests 432
    A.11 Elliptic Curves 437
          A.11.1 Plane Curves 438
          A.11.2 Normal Forms of Elliptic Curves 446
          A.11.3 Point Addition on Elliptic Curves 449
          A.11.4 Group Order and Group Structure of Elliptic Curves.. 455

B. Probabilities and Information Theory.................... 459
    B.1 Finite Probability Spaces and Random Variables 459
    B.2 Some Useful and Important Inequalities 467
    B.3 The Weak Law of Large Numbers 470
    B.4 Distance Measures 472

B.5   Basic Concepts of Information Theory 476

References 483

Index 501

Notation

		Page
$\mathbb{N}$	natural numbers: $\{1, 2, 3, \ldots\}$	51
$\mathbb{Z}$	integers	51
$\mathbb{Q}$	rational numbers	
$\mathbb{R}$	real numbers	
$\mathbb{C}$	complex numbers	
$\ln(x)$	natural logarithm of a real $x > 0$	
$\log(x)$	base-10 logarithm of a real $x > 0$	
$\log_2(x)$	base-2 logarithm of a real $x > 0$	
$\log_g(x)$	discrete base-g logarithm of $x \in \mathbb{Z}_p^*$	
$a \mid b$	$a \in \mathbb{Z}$ divides $b \in \mathbb{Z}$	397
$\lvert x \rvert$	absolute value of $x \in \mathbb{R}$	
$\lvert x \rvert$	length of a bit string x	
$\lvert x \rvert$	binary length of $x \in \mathbb{N}$	
$\lvert M \rvert$	number of elements in a set M	404
M^*	set of words $m_1 m_2 \ldots m_l, l \geq 0$, over M	
$\{0, 1\}^*$	set of bit strings of arbitrary length	
1^k	constant bit string $11 \ldots 1$ of length k	226
$a \oplus b$	bitwise XOR of bit strings $a, b \in \{0, 1\}^l$	13
$a \Vert b$	concatenation of strings a and b	
$g \circ f$	composition of maps: $g \circ f(x) = g(f(x))$	
id_X	identity map: $\mathrm{id}_X(x) = x$ for all $x \in X$	
f^{-1}	inverse of a bijective map f	
x^{-1}	inverse of a unit x in a ring	404
$\mathbb{Z}_n$	residue class ring modulo n	403
$\mathbb{Z}_n^*$	units in $\mathbb{Z}_n$	404
$a \operatorname{div} n$	integer quotient of a and n	398
$a \bmod n$	remainder of a modulo n	398, 414
$a \equiv b \bmod n$	a congruent b modulo n	403, 415

		Page
$\gcd(a,b)$	greatest common divisor of integers	397
$\operatorname{lcm}(a,b)$	least common multiple of integers	397
$\varphi(n)$	Euler phi function	405
$\mathbb{F}_q, \operatorname{GF}(q)$	finite field with q elements	417
$E(\mathbb{F}_q)$	$\mathbb{F}_q$-rational points of an elliptic curve	90, 438
$\operatorname{ord}(x)$	order of an element x in a group	409
QR_n	quadratic residues modulo n	422
QNR_n	quadratic non-residues modulo n	422
$\left(\frac{x}{n}\right)$	Legendre or Jacobi symbol	422, 423
J_n^{+1}	units in $\mathbb{Z}_n$ with Jacobi symbol 1	424
$[a,b]$	interval $a \le x \le b$ in $\mathbb{R}$	
$\lfloor x \rfloor$	greatest integer $\le x$	401
$\lceil x \rceil$	smallest integer $\ge x$	401
$O(n)$	Big-O notation	402
Primes_k	set of primes of binary length k	225
P or $P(X)$	positive polynomial	209
$\operatorname{prob}(\mathcal{E})$	probability of an event $\mathcal{E}$	459
$\operatorname{prob}(x)$	probability of an element $x \in X$	459
$\operatorname{prob}(\mathcal{E},\mathcal{F})$	probability of $\mathcal{E}$ AND $\mathcal{F}$	459
$\operatorname{prob}(\mathcal{E}\,|\,\mathcal{F})$	conditional probability of $\mathcal{E}$ assuming $\mathcal{F}$	460
$\operatorname{prob}(y\,|\,x)$	conditional probability of y assuming x	462
$\operatorname{E}(R)$	expected value of a random variable R	463
$X \bowtie W$	join of a set X with $W = (W_x)_{x \in X}$	465
XW	joint probability space	462, 465
$x \stackrel{p_X}{\leftarrow} X$	x randomly selected according to p_X	216, 466
$x \leftarrow X$	x randomly selected from X	216, 466
$x \stackrel{u}{\leftarrow} X$	x uniformly selected from X	216, 466
$x \leftarrow X, y \leftarrow Y_x$	first x, then y randomly selected	216, 467
$\operatorname{prob}(\ldots : x \leftarrow X)$	probability of $\ldots$ for randomly chosen x	216, 466
$x \leftarrow S$	x randomly generated by the random variable S	466
$y \leftarrow A(x)$	y randomly generated by A on input x	217
$\operatorname{dist}(p,\tilde{p})$	statistical distance between distributions	472
$\operatorname{H}(X)$	uncertainty (or entropy) of X	477
$\operatorname{H}(X|Y)$	conditional uncertainty (entropy)	478
$\operatorname{I}(X;Y)$	mutual information	478
$\operatorname{H}_2(X)$	Rényi entropy of order 2 of X	329
$\operatorname{H}_2(X|Y)$	conditional Rényi entropy of order 2	338

1. Introduction

Cryptography is the science of keeping secrets secret. Assume a sender, referred to here and in what follows as *Alice* (as is commonly used), wants to send a message m to a receiver, referred to as *Bob*. She uses an insecure communication channel. For example, the channel could be a computer network or a telephone line. There is a problem if the message contains confidential information. The message could be intercepted and read by an eavesdropper. Or, even worse, the adversary, as usual referred to here as *Eve*, might be able to modify the message during transmission in such a way that the legitimate recipient Bob does not detect the manipulation.

One objective of cryptography is to provide methods for preventing such attacks. Other objectives are discussed in Section 1.2.

1.1 Encryption and Secrecy

The fundamental and classical task of cryptography is to provide *confidentiality* by *encryption methods*. The message to be transmitted – it can be some text, numerical data, an executable program or any other kind of information – is called the *plaintext*. Alice *encrypts* the plaintext m and obtains the *ciphertext* c. The ciphertext c is transmitted to Bob. Bob turns the ciphertext back into the plaintext by *decryption*. To *decrypt*, Bob needs some secret information, a secret *decryption key*.[1] The adversary Eve may still intercept the ciphertext. However, the encryption should guarantee secrecy and prevent her from deriving any information about the plaintext from the observed ciphertext.

Encryption is very old. For example, *Caesar's shift cipher*[2] was introduced more than 2000 years ago. Every encryption method provides an encryption algorithm E and a decryption algorithm D. In classical encryption schemes, both algorithms depend on the same secret key k. This key k is used for both encryption and decryption. These encryption methods are therefore called

---

[1] Sometimes the terms *encipher* and *decipher* are used instead of "encrypt" and "decrypt".

[2] Each plaintext character is replaced by the character 3 to the right modulo 26, i.e., a is replaced by d, b by e, ... , x by a, y by b and z by c.

symmetric. For example, in Caesar's cipher the secret key is the offset 3 of the shift. We have

$$D(k, E(k, m)) = m \text{ for each plaintext } m.$$

Symmetric encryption and the important examples DES (Data Encryption Standard) and AES (Advanced Encryption Standard) are discussed in Section 2.1.

In 1976, Diffie and Hellman published their famous paper, *"New Directions in Cryptography"* ([DifHel76]). There they introduced the revolutionary concept of *public-key cryptography*. They provided a solution to the long-standing problem of key exchange and pointed the way to digital signatures. The *public-key encryption* methods (comprehensively studied in Chapter 3) are *asymmetric*. Each recipient of messages has a personal key $k = (pk, sk)$, consisting of two parts: pk is the encryption key and is made public, and sk is the decryption key and is kept secret. If Alice wants to send a message m to Bob, she encrypts m by use of Bob's publicly known encryption key pk. Bob decrypts the ciphertext by use of his decryption key sk, which is known only to him. We have

$$D(sk, E(pk, m)) = m.$$

Mathematically speaking, public-key encryption is a so-called *one-way function* with a *trapdoor*. Anyone can easily encrypt a plaintext using the public key pk, but the other direction is difficult. It is practically impossible to deduce the plaintext from the ciphertext, without knowing the secret key sk (which is called the trapdoor information).

Public-key encryption methods require more complex computations and are less efficient than classical symmetric methods. Thus symmetric methods are used for the encryption of large amounts of data. Before applying symmetric encryption, Alice and Bob have to agree on a key. To keep this key secret, they need a secure communication channel. It is common practice to use public-key encryption for this purpose.

1.2 The Objectives of Cryptography

Providing confidentiality is not the only objective of cryptography. Cryptography is also used to provide solutions for other problems:

1. *Data integrity.* The receiver of a message should be able to check whether the message was modified during transmission, either accidentally or deliberately. No one should be able to substitute a false message for the original message, or for parts of it.
2. *Authentication.* The receiver of a message should be able to verify its origin. No one should be able to send a message to Bob and pretend to be Alice (*data origin authentication*). When initiating a communication,

Alice and Bob should be able to identify each other (*entity authentication*).

3. *Non-repudiation*. The sender should not be able to later deny that she sent a message.

If messages are written on paper, the medium – paper – provides a certain amount of security against manipulation. Handwritten personal signatures are intended to guarantee authentication and non-repudiation. If electronic media are used, the medium itself provides no security at all, since it is easy to replace some bytes in a message during its transmission over a computer network, and it is particularly easy if the network is publicly accessible, like the Internet.

So, while encryption has a long history,[3] the need for techniques providing data integrity and authentication has resulted from the rapidly increasing significance of electronic communication.

There are symmetric as well as public-key methods to ensure the integrity of messages. Classical symmetric methods require a secret key k that is shared by sender and receiver. The message m is augmented by a *message authentication code* (MAC). The code is generated by an algorithm and depends on the secret key. The augmented message $(m, MAC(k, m))$ is protected against modifications. The receiver may test the integrity of an incoming message $(m, \overline{m})$ by checking whether

$$MAC(k, m) = \overline{m}.$$

Message authentication codes may be implemented by keyed hash functions (see Chapter 2).

Digital signatures require public-key methods (see Chapter 3 for examples and details). As with classical handwritten signatures, they are intended to provide authentication and non-repudiation. Note that non-repudiation is an indispensable feature if digital signatures are used to sign contracts. Digital signatures depend on the secret key of the signer – they can be generated only by that person. On the other hand, anyone can check whether a signature is valid, by applying a publicly known verification algorithm *Verify*, which depends on the public key of the signer. If Alice wants to sign a message m, she applies an algorithm *Sign* with her secret key sk and gets the signature $Sign(sk, m)$. Bob receives a signature s for message m, and may then check the signature by testing whether

$$Verify(pk, s, m) = ok,$$

with Alice's public key pk.

It is common not to sign the message itself, but to apply a *cryptographic hash function* (see Section 2.2) first and then sign the hash value. In schemes

---

[3] For the long history of cryptography, see [Kahn67].

like the famous RSA (named after its inventors: Rivest, Shamir and Adleman), the decryption algorithm is used to generate signatures and the encryption algorithm is used to verify them. This approach to digital signatures is therefore often referred to as the "hash-then-decrypt" paradigm (see Section 3.3.5 for details). More sophisticated signature schemes, like the probabilistic signature scheme (PSS), require more steps. Modifying the hash value by pseudorandom sequences turns signing into a probabilistic procedure (see Section 3.3.5).

Digital signatures depend on the message. Distinct messages yield different signatures. Thus, like classical message authentication codes, digital signatures can also be used to guarantee the integrity of messages.

1.3 Attacks

The primary goal of cryptography is to keep the plaintext secret from eavesdroppers trying to get some information about the plaintext. As discussed before, adversaries may also be active and try to modify the message. Then, cryptography is expected to guarantee the integrity of messages. Adversaries are assumed to have complete access to the communication channel.

Cryptanalysis is the science of studying attacks against cryptographic schemes. Successful attacks may, for example, recover the plaintext (or parts of the plaintext) from the ciphertext, substitute parts of the original message or forge digital signatures. Cryptography and cryptanalysis are often subsumed by the more general term *cryptology*.

A fundamental assumption in cryptanalysis was first stated by A. Kerckhoffs in the nineteenth century. It is usually referred to as *Kerckhoffs' Principle*. It states that the adversary knows all the details of the cryptosystem, including its algorithms and their implementations. According to this principle, the security of a cryptosystem must be based entirely on the secret keys.

Attacks on the secrecy of an encryption scheme try to recover plaintexts from ciphertexts or, even more drastically, to recover the secret key. The following survey is restricted to passive attacks. The adversary – as usual, we call her Eve – does not try to modify the messages. She monitors the communication channel and the end points of the channel. So not only may she intercept the ciphertext, but also (at least from time to time) she may be able to observe the encryption and decryption of messages. She has no information about the key. For example, Eve might be the operator of a bank computer. She sees incoming ciphertexts and sometimes also the corresponding plaintexts. Or she observes the outgoing plaintexts and the generated ciphertexts. Perhaps she manages to let encrypt plaintexts or decrypt ciphertexts of her own choice.

The possible attacks depend on the actual resources of the adversary Eve. They are usually classified as follows:

1. *Ciphertext-only attack*. Eve has the ability to obtain ciphertexts. This is likely to be the case in any encryption situation. Even if Eve cannot perform the more sophisticated attacks described below, one must assume that she can get access to encrypted messages. An encryption method that cannot resist a ciphertext-only attack is completely insecure.

2. *Known-plaintext attack*. Eve has the ability to obtain plaintext–ciphertext pairs. Using the information from these pairs, she attempts to decrypt a ciphertext for which she does not have the plaintext. At first glance, it might appear that such information would not ordinarily be available to an attacker. However, it very often is available. Messages may be sent in standard formats which Eve knows.

3. *Chosen-plaintext attack*. Eve has the ability to obtain ciphertexts for plaintexts of her choosing. Then she attempts to decrypt a ciphertext for which she does not have the plaintext. While again this may seem unlikely, there are many cases in which Eve can do just this. For example, she may send some interesting information to her intended victim which she is confident he will encrypt and send out. This type of attack assumes that Eve must first obtain whatever plaintext–ciphertext pairs she wants and then do her analysis, without any further interaction. This means that she only needs access to the encrypting device once.

4. *Adaptively-chosen-plaintext attack*. This is the same as the previous attack, except that now Eve may do some analysis on the plaintext–ciphertext pairs, and subsequently get more pairs. She may switch between gathering pairs and performing the analysis as often as she likes. This means that she either has lengthy access to the encrypting device or can somehow make repeated use of it.

5. *Chosen- and adaptively-chosen-ciphertext attacks*. These two attacks are similar to the above plaintext attacks. Eve can choose ciphertexts and get the corresponding plaintexts. She has access to the decryption device.

1.4 Cryptographic Protocols

Encryption and decryption algorithms, cryptographic hash functions, and *pseudorandom generators* (see Section 2.1.1, and Chapter 8) are the basic building blocks (also called cryptographic primitives) for solving problems involving secrecy, authentication or data integrity.

In many cases a single building block is not sufficient to solve the given problem: different primitives must be combined. A series of steps must be executed to accomplish a given task. Such a well-defined series of steps is called a *cryptographic protocol*. As is also common, we add another condition: we require that two or more parties are involved. We only use the term "protocol" if at least two people are required to complete the task.

As a counterexample, take a look at digital-signature schemes. A typical scheme for generating a digital signature first applies a cryptographic hash function h to the message m and then, in a second step, computes the signature by applying a public-key decryption algorithm to the hash value $h(m)$. Both steps are done by one person. Thus, we do not call it a protocol.

Some typical examples of protocols are protocols for user identification. There are many situations where the identity of a user, Alice, has to be verified. Alice wants to log in to a remote computer, for example, or to get access to an account for electronic banking. Passwords or PINs are used for this purpose. This method is not always secure. For example, anyone who observes Alice's password or PIN when transmitted might be able to impersonate her. We sketch a simple *challenge-and-response* protocol which prevents this attack (however, it is not perfect; see Section 4.2.1).

The protocol is based on a public-key signature scheme, and we assume that Alice has a key $k = (pk, sk)$ for this scheme. Now, Alice can prove her identity to Bob in the following way:

1. Bob randomly chooses a "challenge" c and sends it to Alice.
2. Alice signs c with her secret key, $s := Sign(sk, c)$, and sends the "response" s to Bob.
3. Bob accepts Alice's proof of identity if $Verify(pk, s, c) = ok$.

Only Alice can return a valid signature of the challenge c, because only she knows the secret key sk. Thus, Alice proves her identity, without showing her secret. No one can observe Alice's secret key, not even the verifier Bob.

Suppose that an eavesdropper Eve has observed the exchanged messages. Later, she wants to impersonate Alice. Since Bob selects his challenge c at random (from a huge set), the probability that he will use the same challenge twice is very small. Therefore, Eve cannot gain any advantage from her observations.

The parties in a protocol can be friends or adversaries. Protocols can be attacked. The attacks may be directed against the underlying cryptographic algorithms or against the implementation of the algorithms and protocol. There may also be attacks against a protocol itself. There may be passive attacks performed by an eavesdropper, where the only purpose is to obtain information. An adversary may also try to gain an advantage by actively manipulating the protocol. She might pretend to be someone else, substitute messages or replay old messages.

Important protocols for key exchange, electronic elections, digital cash and interactive proofs of identity are discussed in Chapter 4.

1.5 Provable Security

It is desirable to design cryptosystems that are *provably secure*. "Provably secure" means that mathematical proofs show that the cryptosystem resists certain types of attacks. Pioneering work in this field was done by C.E. Shannon. In his information theory, he developed measures for the amount of information associated with a message and the notion of perfect secrecy. A *perfectly secret* cipher perfectly resists all ciphertext-only attacks. An adversary gets no information at all about the plaintext, even if his resources in terms of computing power and time are unlimited. *Vernam's one-time pad* (see Section 2.1.1), which encrypts a message m by XORing it bitwise with a truly random bit string, is the most famous perfectly secret cipher. It even resists all the passive attacks mentioned above. This can be mathematically proven by Shannon's theory. Classical information-theoretic security is discussed in Section 9.1; an introduction to Shannon's information theory may be found in Appendix B. Unfortunately, Vernam's one-time pad and all perfectly secret ciphers are usually impractical. It is not practical in most situations to generate and handle truly random bit sequences of sufficient length as required for perfect secrecy.

More recent approaches to provable security therefore abandon the ideal of perfect secrecy and the (unrealistic) assumption of unbounded computing power. The computational complexity of algorithms is taken into account. Only attacks that might be *feasible* in practice are considered. "Feasible" means that the attack can be performed by an *efficient algorithm*. Of course, here the question about the right notion of efficiency arises. Certainly, algorithms with non-polynomial running time are inefficient. Vice versa, algorithms with polynomial running time are often considered as the efficient ones. In this book, we also adopt this notion of efficiency.

The way a cryptographic scheme is attacked might be influenced by random events. The adversary Eve might toss a coin to decide which case she tries next. Therefore, *probabilistic algorithms* are used to model attackers. Breaking an encryption system by, for example, a ciphertext-only attack means that a probabilistic algorithm with polynomial running time manages to derive information about the plaintext from the ciphertext, with some non-negligible probability. Probabilistic algorithms can toss coins, and their control flow may be at least partially directed by these random events. By using random sources, they can be implemented in practice. They must not be confused with non-deterministic algorithms. The notion of probabilistic (polynomial) algorithms and the underlying probabilistic model are discussed in Chapter 5.

The security of a public-key cryptosystem is based on the hardness of some computational problem (there is no efficient algorithm for solving the problem). For example, the secret keys of an RSA scheme could be easily figured out if computing the prime factors of a large integer were possible.[4] However, it is believed that factoring large integers is infeasible.[5] There are no mathematical proofs for the hardness of the computational problems used

---

[4] What "large" means depends on the available computing power. Today, a 1024-bit integer is considered large.

[5] It is not known whether breaking RSA is easier than factoring the modulus. See Chapters 3 and 6 for a detailed discussion.

in public-key systems. Therefore, security proofs for public-key methods are always conditional: they depend on the validity of the underlying assumption.

The assumption usually states that a certain function f is one-way; i.e., f can be computed efficiently, but it is infeasible to compute x from $f(x)$. These assumptions, as well as the notion of a one-way function, can be made very precise by the use of probabilistic polynomial algorithms. The probability of successfully inverting the function by a probabilistic polynomial algorithm is negligibly small, and "negligibly small" means that it is asymptotically less than any given polynomial bound (see Chapter 6, Definition 6.12). Important examples, like the factoring, discrete logarithm and quadratic residuosity assumptions, are included in this book (see Chapter 6).

There are analogies to the classical notions of security. Shannon's perfect secrecy has a computational analogy: *ciphertext indistinguishability* (or *semantic security*). An encryption is perfectly secret if and only if an adversary cannot distinguish between two plaintexts, even if her computing resources are unlimited: if the adversary Eve knows that a ciphertext c is the encryption of either m or m', she has no better chance than $1/2$ of choosing the right one. Ciphertext indistinguishability – also called *polynomial-time indistinguishability* – means that Eve's chance of successfully applying a probabilistic polynomial algorithm is at most negligibly greater than $1/2$ (Chapter 9, Definition 9.14).

As a typical result, it is proven in Section 9.4 that *public-key one-time pads* are ciphertext-indistinguishable. This means, for example, that the RSA public-key one-time pad is ciphertext-indistinguishable under the sole assumption that the RSA function is one way. A public-key one-time pad is similar to Vernam's one-time pad. The difference is that the message m is XORed with a pseudorandom bit sequence which is generated from a short truly random seed, by means of a one-way function.

Thus, one-way functions are not only the essential ingredients of public-key encryption and digital signatures. They also yield computationally perfect pseudorandom bit generators (Chapter 8). If f is a one-way function, it is not only impossible to compute x from $f(x)$, but also certain bits (called hard-core bits) of x are equally difficult to deduce. This feature is called the bit security of a one-way function. For example, the least-significant bit is a hard-core bit for the RSA function $x \mapsto x^e \bmod n$. Starting with a truly random seed, repeatedly applying f and taking the hard-core bit in each step, you get a pseudorandom bit sequence. These bit sequences cannot be distinguished from truly random bit sequences by an efficient algorithm, or, equivalently (Yao's Theorem, Section 8.2), it is practically impossible to predict the next bit from the previous ones. So they are really computationally perfect.

The bit security of important one-way functions is studied in detail in Chapter 7, including an in-depth analysis of the probabilities involved.

Randomness and the security of cryptographic schemes are closely related. There is no security without randomness. An encryption method provides secrecy only if the ciphertexts appear random to the adversary Eve. Vernam's one-time pad is perfectly secret, because, owing to the truly random key string k, the encrypted message $m \oplus k$ [6] is a truly random bit sequence for Eve. The public-key one-time pad is ciphertext-indistinguishable, because if Eve applies an efficient probabilistic algorithm, she cannot distinguish the pseudorandom key string and, as a consequence, the ciphertext from a truly random sequence.

Public-key one-time pads are secure against passive eavesdroppers who perform ciphertext-only attacks (see Section 1.3 above for a classification of attacks). However, active adversaries who perform adaptively-chosen-ciphertext attacks can be a real danger in practice – as demonstrated by Bleichenbacher's 1-Million-Chosen-Ciphertext Attack (Section 3.3.2). Therefore, security against such attacks is also desirable. In Section 9.5, we study two examples of public-key encryption schemes which are secure against adaptively-chosen-ciphertext attacks, and their security proofs. One of the examples, Cramer and Shoup's public-key encryption scheme, was the first practical scheme whose security proof is based solely on a standard number-theoretic assumption and a standard assumption about hash functions (collision resistance).

The ideal cryptographic hash function is a *random function*. It yields hash values which cannot be distinguished from randomly selected and uniformly distributed values. Such a random function is also called a *random oracle*. Sometimes, the security of a cryptographic scheme can be proven in the *random oracle model*. In addition to the assumed hardness of a computational problem, such a proof relies on the assumption that the hash functions used in the scheme are truly random functions. Examples of such schemes include the public-key encryption schemes OAEP (Section 3.3.3) and SAEP (Section 9.5.1), the above-mentioned signature scheme PSS and full-domain-hash RSA signatures (Section 3.3.5). We give the random-oracle proofs for SAEP and full-domain-hash signatures.

Truly random functions cannot be implemented, nor even perfectly approximated in practice. Therefore, a proof in the random oracle model can never be a complete security proof. The hash functions used in practice are constructed to be good approximations to the ideal of random functions. However, there have been surprising errors in the past (see Section 2.2).

We have distinguished different types of attacks on an encryption scheme. In a similar way, the attacks on signature schemes can be classified and different levels of security can be defined. We introduce this classification in Chapter 11 and give examples of signature schemes whose security can be proven solely under standard assumptions (like the factoring or the strong RSA assumption). No assumptions about the randomness of a hash function have to be made, in contrast to schemes like PSS. A typical security proof for the highest level of security is included. For the given signature scheme,

---

[6] $\oplus$ denotes the bitwise XOR operator; see page 13.

we show that not a single signature can be forged, even if the attacker Eve is able to obtain valid signatures from the legitimate signer for messages she has chosen adaptively.

The security proofs for public-key systems are always conditional and depend on (widely believed, but unproven) assumptions. On the other hand, Shannon's notion of perfect secrecy and, in particular, the perfect secrecy of Vernam's one-time pad are unconditional. Although perfect unconditional security is not reachable in most practical situations, there have been promising attempts to design practical cryptosystems which provably come close to perfect information-theoretic security. The proofs are based on classical information-theoretic methods and do not depend on unproven assumptions. In Chapter 10, we give an introduction to these cryptosystems and the techniques applied there. We review some results on systems whose security relies on the fact that communication channels are noisy or on the limited storage capacity of an adversary. An important approach to unconditional security is quantum cryptography. The famous BB84 protocol of Bennett and Brassard for quantum key distribution is described in detail, including a proof of its security. Essential techniques, such as unconditionally secure message authentication codes, Rényi entropy and privacy amplification, are addressed.

2. Symmetric-Key Cryptography

In this chapter, we give an introduction to basic methods of symmetric-key cryptography. At first, we consider symmetric-key encryption. We explain the notions of stream and block ciphers. The operation modes of block ciphers are studied and, as prominent examples for block ciphers, DES and AES are described. Later, we introduce cryptographic hash functions. Methods for the construction of cryptographic hash functions, such as Merkle–Damgård's method and the sponge construction, are explained in detail. As an application of hash functions, we get message authentication codes, MACs for short. MACs are the standard symmetric-key technique to guarantee the integrity and authenticity of messages.

2.1 Symmetric-Key Encryption

Symmetric-key encryption provides secrecy when two parties, say Alice and Bob, communicate. An adversary who intercepts a message should not get any significant information about its content.

To set up their secure communication channel, Alice and Bob first agree on a key k. They keep their shared key k secret. Before sending a message m to Bob, Alice encrypts m by using the encryption algorithm E and the key k. She obtains the ciphertext $c = E(k, m)$ and sends c to Bob. By using the decryption algorithm D and the same key k, Bob decrypts c to recover the plaintext $m = D(k, c)$.

We speak of symmetric encryption, because both communication partners use the same key k for encryption and decryption. The encryption and decryption algorithms E and D are publicly known. Anyone who knows the key can decrypt ciphertexts. Therefore, the key k has to be kept secret.

A basic problem in a symmetric scheme is how Alice and Bob can agree on a shared secret key k in a secure and efficient way. For this key exchange, the methods of public-key cryptography are needed, which we discuss in the subsequent chapters. There were no solutions to the key exchange problem, until the revolutionary concept of public-key cryptography was discovered 40 years ago.

We require that the plaintext m can be uniquely recovered from the ciphertext c. This means that for a fixed key k, the encryption map must be bijective. Mathematically, symmetric encryption may be considered as follows.

Definition 2.1. A *symmetric-key encryption scheme* consists of a map

$$E : K \times M \longrightarrow C$$

such that for each $k \in K$, the map

$$E_k : M \longrightarrow C, \ m \longmapsto E(k, m)$$

is invertible. The elements $m \in M$ are the *plaintexts* (also called *messages*). C is the set of *ciphertexts* or *cryptograms*, the elements $k \in K$ are the *keys*. E_k is called the *encryption function* with respect to the key k. The inverse function $D_k := E_k^{-1}$ is called the *decryption function*. It is assumed that efficient algorithms to compute E_k and D_k exist.

The key k is shared between the communication partners and kept secret. A basic security requirement for the encryption map E is that, without knowing the key k, it should be impossible to successfully execute the decryption function D_k. Important examples of symmetric-key encryption schemes – Vernam's one-time pad, DES and AES – are given below.

Among all encryption algorithms, symmetric-key encryption algorithms have the fastest implementations in hardware and software. Therefore, they are very well-suited to the encryption of large amounts of data. If Alice and Bob want to use a symmetric-key encryption scheme, they first have to exchange a secret key. For this, they have to use a secure communication channel. Public-key encryption methods, which we study in Chapter 3, are often used for this purpose. Public-key encryption schemes are less efficient and hence not suitable for large amounts of data. Thus, symmetric-key encryption and public-key encryption complement each other to provide practical cryptosystems.

We distinguish between *block ciphers* and *stream ciphers*. The encryption function of a block cipher processes plaintexts of fixed length. A stream cipher operates on streams of plaintext. Processing character by character, it encrypts plaintext strings of arbitrary length. If the plaintext length exceeds the block length of a block cipher, various *modes of operation* are used. Some of them yield stream ciphers. Thus, block ciphers may also be regarded as building blocks for stream ciphers.

2.1.1 Stream Ciphers

Definition 2.2. Let K be a set of keys and M be a set of plaintexts. In this context, the elements of M are called characters.
A stream cipher

$$E^* : K^* \times M^* \longrightarrow C^*, E^*(k, m) := c := c_1 c_2 c_3 \ldots$$

encrypts a stream $m := m_1 m_2 m_3 \ldots \in M^*$ of plaintext characters $m_i \in M$ as a stream $c := c_1 c_2 c_3 \ldots \in C^*$ of ciphertext characters $c_i \in C$ by using a *key stream* $k := k_1 k_2 k_3 \ldots \in K^*, k_i \in K$.

The plaintext stream $m = m_1 m_2 m_3 \ldots$ is encrypted character by character. For this purpose, there is an encryption map

$$E : K \times M \longrightarrow C,$$

which encrypts the single plaintext characters m_i with the corresponding key character k_i:

$$c_i = E_{k_i}(m_i) = E(k_i, m_i), i = 1, 2, \ldots .$$

Typically, the characters in M and C and the key elements in K are binary digits or bytes.

Of course, encrypting plaintext characters with E_{k_i} must be a bijective map, for every key character $k_i \in K$. Decrypting a ciphertext stream $c := c_1 c_2 c_3 \ldots$ is done character by character by applying the decryption map D with the same key stream $k = k_1 k_2 k_3 \ldots$ that was used for encryption:

$$c = c_1 c_2 c_3 \ldots \mapsto D(k, c) := D_{k_1}(c_1) D_{k_2}(c_2) D_{k_3}(c_3) \ldots .$$

A necessity for stream ciphers comes, for example, from operating systems, where input and output is done with so-called streams.

Of course, the key stream in a stream cipher has to be kept secret. It is not necessarily the secret key which is shared between the communication partners; the key stream might be generated from the shared secret key by a pseudorandom generator (see below).

Notation. In most stream ciphers, the binary exclusive-or operator XOR for bits $a, b \in \{0, 1\}$ – considered as truth values – is applied. We have a XOR $b = 1$, if $a = 0$ and $b = 1$ or $a = 1$ and $b = 0$, and a XOR $b = 0$, if $a = b = 0$ or $a = b = 1$. XORing two bits a and b means to add them modulo 2, i.e., we have a XOR $b = a + b \bmod 2$. As is common practice, we denote the XOR-operator by $\oplus$, $a \oplus b := a$ XOR b, and we use $\oplus$ also for the binary operator that bitwise XORs two bit strings. If $a = a_1 a_2 \ldots a_n$ and $b = b_1 b_2 \ldots b_n$ are bit strings, then

$$a \oplus b := (a_1 \text{ XOR } b_1)(a_2 \text{ XOR } b_2) \ldots (a_n \text{ XOR } b_n).$$

Vernam's One-Time Pad. The most famous example of a stream cipher is *Vernam's one-time pad* (see [Vernam19] and [Vernam26]). It is easy to describe. Plaintexts, keys and ciphertexts are bit strings. To encrypt a message $m := m_1 m_2 m_3 \ldots$, where $m_i \in \{0, 1\}$, a key stream $k := k_1 k_2 k_3 \ldots$, with $k_i \in \{0, 1\}$, is needed. Encryption and decryption are given by bitwise XORing with the key stream:

$$E^*(k, m) := k \oplus m \text{ and } D^*(k, c) := k \oplus c.$$

Obviously, encryption and decryption are inverses of each other. Each bit in the key stream is chosen at random and independently, and the key stream is used only for the encryption of one message m. This fact explains the name "one-time pad". If a key stream k were used twice to encrypt m and $\overline{m}$, we could derive $m \oplus \overline{m}$ from the cryptograms c and $\overline{c}$ and thus obtain information about the plaintexts, by computing $c \oplus \overline{c} = m \oplus k \oplus \overline{m} \oplus k = m \oplus \overline{m} \oplus k \oplus k = m \oplus \overline{m}$.

There are obvious disadvantages to Vernam's one-time pad. Truly random keys of the same length as the message have to be generated and securely transmitted to the recipient. There are very few situations where this is practical. Reportedly, the hotline between Washington and Moscow was encrypted with a one-time pad; the keys were transported by a trusted courier.

Nevertheless, most practical stream ciphers work as Vernam's one-time pad. The difference is that a pseudorandom key stream is taken instead of the truly random key stream. A pseudorandom key stream looks like a random key stream, but actually the bits are generated from a short (truly) random seed by a deterministic algorithm. In practice, such *pseudorandom generators* can be based on specific operation modes of block ciphers or on feedback shift registers. We study the operation modes of block ciphers in Section 2.1.5 (e.g. see the cipher and output feedback modes). Feedback shift registers can be implemented to run very fast on relatively simple hardware. This fact makes them especially attractive. More about these generators and stream ciphers can be found, for example, in [MenOorVan96]. There are also public-key stream ciphers, in which the pseudorandom key stream is generated by using public-key techniques. We discuss these pseudorandom bit generators and the resulting stream ciphers in detail in Chapters 8 and 9.

Back to Vernam's one-time pad. Its advantage is that one can prove that it is secure – an adversary observing a cryptogram does not have the slightest idea what the plaintext is. We discuss this point in the simplest case, where the message m consists of a single bit. Alice and Bob want to exchange one of the messages yes $= 1$ or no $= 0$. Previously, they exchanged the key bit k, which was the outcome of an unbiased coin toss.

First, we assume that each of the two messages yes and no is equally likely. The adversary, we call her Eve, intercepts the cryptogram c. Since the key bit is truly random, Eve can only derive that c encrypts yes or no with probability $1/2$. Thus, she has not the slightest idea which of the two is encrypted. Her only chance of making a decision is to toss a coin. She can do this, however, without seeing the cryptogram c.

If one of the two messages has a greater probability, Eve also cannot gain any advantage by intercepting the cryptogram c. Assume, for example, that the probability of a 1 is $3/4$ and the probability of a 0 is $1/4$. Then the cryptogram c encrypts 0 with probability $1/4$ and 1 with probability $3/4$, irrespective of whether $c = 0$ or $c = 1$. Thus, Eve cannot learn more from the cryptogram than she has learned a priori about the distribution of the plaintexts.

Our discussion for one-bit messages may be transferred to the general case of n-bit messages. The amount of information an attacker may obtain is made precise by information theory. The level of security we achieve with the one-time pad is called perfect secrecy (see Chapter 9 for details). Note that we have to assume that all messages have the same length n (if necessary, they are padded out). Otherwise, some information – the length of the message – would leak to the attacker.

The Vernam one-time pad not only resists a ciphertext-only attack as proven formally in Chapter 9, but it resists all the attacks defined in Chapter 1. Each cryptogram has the same probability. Eve does not learn anything, not even about the probabilities of the plaintexts, if she does not know them a priori. For each message, the key is chosen at random and independently from the previous ones. Thus, Eve cannot gain any advantage by observing plaintext–ciphertext pairs, not even if she has chosen the plaintexts adaptively.

The Vernam one-time pad ensures the confidentiality of messages, but it does not protect messages against modifications. If someone changes bits in the cryptogram and the decrypted cryptogram makes sense, the receiver will not notice it.

2.1.2 Block Ciphers

Definition 2.3. A *block cipher* is a symmetric-key encryption scheme with $M = C = \{0,1\}^n$ and key space $K = \{0,1\}^r$:

$$E : \{0,1\}^r \times \{0,1\}^n \longrightarrow \{0,1\}^n, \ (k,m) \longmapsto E(k,m).$$

Using a secret key k of binary length r, the encryption algorithm E encrypts plaintext blocks m of a fixed binary length n and the resulting ciphertext blocks $c = E(k,m)$ also have length n. n is called the *block length* of the cipher.

Typical block lengths are 64 (as in DES) or 128 (as in AES), typical key lengths are 56 (as in DES) or 128, 192 and 256 (as in AES).

Let us consider a block cipher E with block length n and key length r. There are 2^n plaintext blocks and 2^n ciphertext blocks of length n. For a fixed key k, the encryption function $E_k : m \mapsto E(k,m)$ maps $\{0,1\}^n$ bijectively to $\{0,1\}^n$ – it is a permutation[1] of $\{0,1\}^n$. Thus, to choose a key k, means to select a permutation E_k of $\{0,1\}^n$, and this permutation is then used to encrypt the plaintext blocks. The 2^r permutations E_k, with k running through the set $\{0,1\}^r$ of keys, form an almost negligibly small subset in the tremendously large set of all permutations of $\{0,1\}^n$, which consists of $2^n!$

---

[1] A map $f : D \longrightarrow D$ is called a *permutation* of D, if f is bijective.

elements. So, when we randomly choose an r-bit key k for E, then we restrict our selection of the encryption permutation to an extremely small subset.

From these considerations, we conclude that we cannot have the ideal block cipher with perfect secrecy in practice. Namely, in the preceding Section 2.1.1, we discussed a stream cipher with perfect secrecy, the Vernam one-time pad. Perfect secrecy results from a maximum amount of randomness: for each message bit, a random key bit is chosen (we will prove in Chapter 9 that less randomness in key generation destroys perfect secrecy, see Theorem 9.6). We conclude that the maximal level of security in a block cipher also requires a maximum of randomness, and this in turn means that – when choosing a key – we would have to select a random element from the set of all permutations of $\{0, 1\}^n$. Unfortunately, this turns out to be completely impractical. We could try to enumerate all permutations π of $\{0, 1\}^n$, $\pi_1, \pi_2, \pi_3, \ldots$, and then randomly select one by randomly selecting an index (this index would be the key). Since there are $2^n!$ permutations, we need $\log_2(2^n!)$-bit numbers to encode the indexes. By Stirling's approximation formula $k! \approx \sqrt{2\pi k}\, (k/e)^k$, we derive that $\log_2(2^n!) \approx (n - 1.44)2^n$. This is a huge number. For a block length n of 64 bits, we would need approximately 2^{67} bytes to store a single key. There is no storage medium with such capacity.

Thus, in a real block cipher, we have to restrict ourselves to much smaller keys and choose the encryption permutation E_k for a key k from a much smaller set of 2^r permutations, with r typically in the range of 56 to 256. Nevertheless, the designers of a block cipher try to approximate the ideal. The idea is to get an encryption function which behaves like a randomly chosen function from the very huge set of all permutations.

2.1.3 DES

The *data encryption standard* (DES), originally specified in [FIPS46 1977], was previously the most widely used symmetric-key encryption algorithm. Governments, banks and applications in commerce took the DES as the basis for secure and authentic communication.

We give a high-level description of the DES encryption and decryption functions. The DES algorithm takes 56-bit keys and 64-bit plaintext messages as inputs and outputs a 64-bit cryptogram:[2]

$$\text{DES} : \{0, 1\}^{56} \times \{0, 1\}^{64} \longrightarrow \{0, 1\}^{64}$$

If the key k is chosen, we get

$$\text{DES}_k : \{0, 1\}^{64} \longrightarrow \{0, 1\}^{64}, \quad x \longmapsto \text{DES}(k, x).$$

An encryption with DES_k consists of 16 major steps, called rounds. In each of the 16 rounds, a 48-bit round key k_i is used. The 16 round keys

---

[2] Actually, the 56 bits of the key are packed with 8 bits of parity.

$k_1, \ldots, k_{16}$ are computed from the 56-bit key k by using an algorithm which is studied in Exercise 1 at the end of this chapter.

In the definition of DES, one of the basic building blocks is a map

$$f : \{0,1\}^{48} \times \{0,1\}^{32} \longrightarrow \{0,1\}^{32},$$

which transforms a 32-bit message block x with a 48-bit round key $\tilde{k}$. f is composed of a substitution S and a permutation P:

$$f(\tilde{k}, x) = P(S(E(x) \oplus \tilde{k})).$$

The 32 message bits are extended to 48 bits, $x \mapsto E(x)$ (some of the 32 bits are used twice), and XORed with the 48-bit round key $\tilde{k}$. The resulting 48 bits are divided into eight groups of 6 bits, and each group is substituted by 4 bits. Thus, we get 32 bits which are then permuted by P. The cryptographic strength of the DES function depends on the design of f, especially on the design of the eight famous *S-boxes* which handle the eight substitutions (for details, see [FIPS46 1977]).

We define for $i = 1, \ldots, 16$

$$\phi_i : \{0,1\}^{32} \times \{0,1\}^{32} \longrightarrow \{0,1\}^{32} \times \{0,1\}^{32}, \ (x,y) \longmapsto (x \oplus f(k_i, y), y).$$

ϕ_i transforms 64-bit blocks and for this transformation, a 64-bit block is split into two 32-bit halves x and y. We have

$$\phi_i \circ \phi_i(x,y) = \phi_i(x \oplus f(k_i, y), y) = (x \oplus f(k_i, y) \oplus f(k_i, y), y) = (x,y).^3$$

Hence, ϕ_i is bijective and $\phi_i^{-1} = \phi_i.^4$ The fact that ϕ_i is bijective does not depend on any properties of f.

The DES_k function is obtained by composing $\phi_1, \ldots, \phi_{16}$ and the map

$$\mu : \{0,1\}^{32} \times \{0,1\}^{32} \longrightarrow \{0,1\}^{32} \times \{0,1\}^{32}, \ (x,y) \longmapsto (y,x),$$

which interchanges the left and the right half of a 64-bit block (x,y).

Namely,

$$DES_k : \{0,1\}^{64} \longrightarrow \{0,1\}^{64},$$

$$DES_k(x) := IP^{-1} \circ \phi_{16} \circ \mu \circ \phi_{15} \circ \ldots \mu \circ \phi_2 \circ \mu \circ \phi_1 \circ IP(x).$$

Here, IP is a publicly known permutation without cryptographic significance.

We see that a DES cryptogram is obtained by 16 encryptions of the same type using 16 different round keys that are derived from the original 56-bit key. ϕ_i is called the encryption of round i. After each round, except the last one, the left and the right half of the argument are interchanged. A block cipher that is computed by iteratively applying a round function

---

[3] $g \circ h$ denotes the composition of maps: $g \circ h(x) := g(h(x))$.
[4] As usual, if $f : D \longrightarrow R$ is a bijective map, we denote the inverse map by f^{-1}.

to the plaintext is called an *iterated cipher*. If the round function has the form of the DES round function ϕ_i, the cipher is called a *Feistel cipher*. H. Feistel developed the *Lucifer* algorithm, which was a predecessor of the DES algorithm. The idea of using an alternating sequence of permutations and substitutions to get an iterated cipher can be attributed to Shannon (see [Shannon49]).

Notation. We also write $\text{DES}_{k_1...k_{16}}$ for DES_k to indicate that the round keys, derived from k, are used in this order for encryption.

The following Proposition 2.4 means that the DES encryption function may also be used for decryption. For decryption, the round keys $k_1 \ldots k_{16}$ are supplied in reverse order.

Proposition 2.4. *For all messages $x \in \{0,1\}^{64}$*

$$\text{DES}_{k_{16}...k_1}(\text{DES}_{k_1...k_{16}}(x)) = x.$$

In other words,

$$\text{DES}_{k_{16}...k_1} \circ \text{DES}_{k_1...k_{16}} = \text{id}.$$

Proof. Since $\phi_i = \phi_i^{-1}$ (see above) and, obviously, $\mu = \mu^{-1}$, we get

$$\text{DES}_{k_{16}...k_1} \circ \text{DES}_{k_1...k_{16}}$$
$$= \text{IP}^{-1} \circ \phi_1 \circ \mu \circ \phi_2 \circ \ldots \mu \circ \phi_{16} \circ \text{IP} \circ \text{IP}^{-1} \circ \phi_{16} \circ \mu \circ \phi_{15} \circ \ldots \mu \circ \phi_1 \circ \text{IP}$$
$$= \text{id}.$$

This proves the proposition.   □

Shortly after DES was published, Diffie and Hellman criticized the short key size of 56 bits in [DifHel77]. They suggested using DES in multiple encryption mode. In triple encryption mode with three independent 56-bit keys k_1, k_2 and k_3, the cryptogram c is computed by $\text{DES}_{k_3}(\text{DES}_{k_2}(\text{DES}_{k_1}(m)))$. This can strengthen the DES because the set of DES_k functions is not a group (i.e., $\text{DES}_{k_2} \circ \text{DES}_{k_1}$ is not a DES_k function), a fact which was shown in [CamWie92]. Moreover, it is shown there that 10^{2499} is a lower bound for the size of the subgroup generated by the DES_k functions in the symmetric group. A small order of this subgroup would imply a less secure multiple encryption mode.

The DES algorithm is well-studied and a lot of cryptanalysis has been performed. Special methods like linear and differential cryptanalysis have been developed and applied to attempt to break DES. However, the best practical attack known is an exhaustive key search. Assume some plaintext–ciphertext pairs (m_i, c_i), $i = 1, \ldots, n$, are given. An exhaustive key search tries to find the key by testing $\text{DES}(k, m_i) = c_i$, $i = 1, \ldots, n$, for all possible $k \in \{0,1\}^{56}$. If such a k is found, the probability that k is really the key is very high. Special computers were proposed to perform an exhaustive key search (see [DifHel77]). Recently a specially designed supercomputer and a worldwide network of nearly 100 000 PCs on the Internet were able to find out the key after 22 hours and 15 minutes (see [RSALabs]). This effort recovered one key. This work would need to be repeated for each additional key to be recovered.

The key size and the block size of DES have become too small to resist the progress in computer technology. The U.S. National Institute of Standards and Technology (NIST) had standardized DES in the 1970s. After more than 20 "DES years" the search for a successor, the AES, was started.

2.1.4 AES

In January 1997, the National Institute of Standards and Technology started an open selection process for a new encryption standard – *the advanced encryption standard*, or *AES* for short. NIST encouraged parties worldwide to submit proposals for the new standard. The proposals were required to support a block size of at least 128 bits, and three key sizes of 128, 192 and 256 bits.

The selection process was divided into two rounds. In the first round, 15 of the submitted 21 proposals were accepted as AES candidates. The candidates were evaluated by a public discussion. The international cryptographic community was asked for comments on the proposed block ciphers. Five candidates were chosen for the second round: MARS (IBM), RC6 (RSA), Rijndael (Daemen and Rijmen), Serpent (Anderson, Biham and Knudsen) and Twofish (Counterpane). Three international "AES Candidate Conferences" were held, and in October 2000 NIST selected the *Rijndael cipher* to be the AES (see [NIST00]).

Rijndael (see [DaeRij02]) was developed by Daemen and Rijmen. It is an iterated block cipher and supports different block and key sizes. Block and key sizes of 128, 160, 192, 224 and 256 bits can be combined independently.

The only difference between Rijndael and AES is that AES supports only a subset of Rijndael's block and key sizes. The AES fixes the block length to 128 bits, and uses the three key lengths 128, 192 and 256 bits.

Besides encryption, Rijndael (like many block ciphers) is suited for other cryptographic tasks, for example, the construction of cryptographic hash functions (see Section 2.2.2) or pseudorandom bit generators (see Section 2.1.5). Rijndael can be implemented efficiently on a wide range of processors and on dedicated hardware.

Structure of Rijndael. Rijndael is an iterated block cipher. The iterations are called rounds. The number of rounds, which we denote by N_r, depends on the block length and the key length. In each round except the final round, the same round function is applied, each time with a different round key. The round function of the final round differs slightly. The round keys $key_1, \ldots, key_{N_r}$ are derived from the secret key k by using the key schedule algorithm, which we describe below.

We use the terminology of [DaeRij02] in our description of Rijndael. A byte, as usual, consists of 8 bits, and by a *word* we mean a sequence of 32 bits or, equivalently, 4 bytes.

Rijndael is byte-oriented. Input and output (plaintext block, key, ciphertext block) are considered as one-dimensional arrays of 8-bit-bytes. Both block length and key length are multiples of 32 bits. We denote by N_b the block length in bits divided by 32 and by N_k the key length in bits divided by 32. Thus, a Rijndael block consists of N_b words (or $4 \cdot N_b$ bytes), and a Rijndael key consists of N_k words (or $4 \cdot N_k$ bytes).

The following table shows the number of rounds N_r as a function of N_k and N_b:

N_k	N_b				
	4	5	6	7	8
4	10	11	12	13	14
5	11	11	12	13	14
6	12	12	12	13	14
7	13	13	13	13	14
8	14	14	14	14	14

In particular, AES with key length 128 bits (and the fixed AES block length of 128 bits) consists of 10 rounds.

The round function of Rijndael, and its steps, operate on an intermediate result, called the *state*. The *state* is a block of N_b words (or $4 \cdot N_b$ bytes). At the beginning of an encryption, the variable *state* is initialized with the plaintext block, and at the end, *state* contains the ciphertext block.

The intermediate result *state* is considered as a 4-row matrix of bytes with N_b columns. Each column contains one of the N_b words of *state*.

The following table shows the state matrix in the case of block length 192 bits. We have 6 state words. Each column of the matrix represents a state word consisting of 4 bytes.

$a_{0,0}$	$a_{0,1}$	$a_{0,2}$	$a_{0,3}$	$a_{0,4}$	$a_{0,5}$
$a_{1,0}$	$a_{1,1}$	$a_{1,2}$	$a_{1,3}$	$a_{1,4}$	$a_{1,5}$
$a_{2,0}$	$a_{2,1}$	$a_{2,2}$	$a_{2,3}$	$a_{2,4}$	$a_{2,5}$
$a_{3,0}$	$a_{3,1}$	$a_{3,2}$	$a_{3,3}$	$a_{3,4}$	$a_{3,5}$

The Rijndael Algorithm. An encryption with Rijndael consists of an initial round key addition, followed by applying the round function ($N_r - 1$)-times, and a final round with a slightly modified round function. The round function is composed of the SubBytes, ShiftRows and MixColumns steps and an addition of the round key (see next section). In the final round, the MixColumns step is omitted. A high level description of the Rijndael algorithm follows:

Algorithm 2.5.
  byteString *Rijndael*(byteString *plaintextBlock*, *key*)
  1   InitState(*plaintextBlock*, *state*)
  2   AddKey(*state*, *key*$_0$)
  3   for $i \leftarrow 1$ to $N_r - 1$ do
  4   SubBytes(*state*)
  5   ShiftRows(*state*)
  6   MixColumns(*state*)
  7   AddKey(*state*, *key*$_i$)
  8   SubBytes(*state*)
  9   ShiftRows(*state*)
  10 AddKey(*state*, *key*$_{N_r}$)
  11 return *state*;

The input and output blocks of the Rijndael algorithm are byte strings of $4 \cdot N_b$ bytes. In the beginning, the state matrix is initialized with the plaintext block. The matrix is filled column by column. The ciphertext is taken from the state matrix after the last round. Here, the matrix is read column by column.

All steps of the round function – SubBytes, ShiftRows, MixColumns, AddKey – are invertible. Therefore, decrypting with Rijndael means to apply the inverse functions of SubBytes, ShiftRows, MixColumns and AddKey, in the reverse order.

The Round function. We describe now the steps – SubBytes, ShiftRows, MixColumns and AddKey – of the round function. The Rijndael algorithm and its steps are byte-oriented. They operate on the bytes of the state matrix. In Rijndael, bytes are usually considered as elements of the finite field $\mathbb{F}_{2^8}$ with 2^8 elements, and $\mathbb{F}_{2^8}$ is constructed as an extension of the field $\mathbb{F}_2$ with 2 elements by using the irreducible polynomial $X^8 + X^4 + X^3 + X + 1$ (see Appendix A.5.3). Then adding (which is the same as bitwise XORing) and multiplying bytes means to add and multiply them as elements of the field $\mathbb{F}_{2^8}$.

The SubBytes Step. SubBytes is the only non-linear transformation of Rijndael. It substitutes the bytes of the state matrix byte by byte, by applying the function S_{RD}[5] to each element of the matrix *state*. The function S_{RD} is also called the S-box; it does not depend on the key. The same S-box is used for all byte positions. The S-box S_{RD} is composed of two maps, f and g. First f and then g is applied:

$$S_{RD}(x) = g \circ f(x) = g(f(x)) \quad (x \text{ a byte}).$$

Both maps, f and g, have a simple algebraic description.

---

[5] Rijmen and Daemen's S-box.

To understand f, we consider a byte x as an element of the finite field $\mathbb{F}_{2^8}$. Then f simply maps x to its multiplicative inverse x^{-1}:

$$f : \mathbb{F}_{2^8} \longrightarrow \mathbb{F}_{2^8}, \ x \longmapsto \begin{cases} x^{-1} & \text{if } x \neq 0, \\ 0 & \text{if } x = 0. \end{cases}$$

To understand g, we consider a byte x as a vector of 8 bits or, more precisely, as a vector of length 8 over the field $\mathbb{F}_2$ with 2 elements[6]. Then g is the $\mathbb{F}_2$-affine map

$$g : \mathbb{F}_2^8 \longrightarrow \mathbb{F}_2^8, \ x \longmapsto Ax + b,$$

composed of a linear map $x \mapsto Ax$ and a translation with vector b. The matrix A of the linear map and b are given by

$$A := \begin{pmatrix} 1 & 0 & 0 & 0 & 1 & 1 & 1 & 1 \\ 1 & 1 & 0 & 0 & 0 & 1 & 1 & 1 \\ 1 & 1 & 1 & 0 & 0 & 0 & 1 & 1 \\ 1 & 1 & 1 & 1 & 0 & 0 & 0 & 1 \\ 1 & 1 & 1 & 1 & 1 & 0 & 0 & 0 \\ 0 & 1 & 1 & 1 & 1 & 1 & 0 & 0 \\ 0 & 0 & 1 & 1 & 1 & 1 & 1 & 0 \\ 0 & 0 & 0 & 1 & 1 & 1 & 1 & 1 \end{pmatrix} \text{ and } b := \begin{pmatrix} 1 \\ 1 \\ 0 \\ 0 \\ 0 \\ 1 \\ 1 \\ 0 \end{pmatrix}.$$

The S-box $\mathrm{S_{RD}}$ operates on each of the state bytes of the state matrix independently. For a block length of 128 bits, we have:

$a_{0,0}$	$a_{0,1}$	$a_{0,2}$	$a_{0,3}$
$a_{1,0}$	$a_{1,1}$	$a_{1,2}$	$a_{1,3}$
$a_{2,0}$	$a_{2,1}$	$a_{2,2}$	$a_{2,3}$
$a_{3,0}$	$a_{3,1}$	$a_{3,2}$	$a_{3,3}$

$\longmapsto$

$\mathrm{S_{RD}}(a_{0,0})$	$\mathrm{S_{RD}}(a_{0,1})$	$\mathrm{S_{RD}}(a_{0,2})$	$\mathrm{S_{RD}}(a_{0,3})$
$\mathrm{S_{RD}}(a_{1,0})$	$\mathrm{S_{RD}}(a_{1,1})$	$\mathrm{S_{RD}}(a_{1,2})$	$\mathrm{S_{RD}}(a_{1,3})$
$\mathrm{S_{RD}}(a_{2,0})$	$\mathrm{S_{RD}}(a_{2,1})$	$\mathrm{S_{RD}}(a_{2,2})$	$\mathrm{S_{RD}}(a_{2,3})$
$\mathrm{S_{RD}}(a_{3,0})$	$\mathrm{S_{RD}}(a_{3,1})$	$\mathrm{S_{RD}}(a_{3,2})$	$\mathrm{S_{RD}}(a_{3,3})$

Both maps f and g are invertible. We even have $f = f^{-1}$. Thus the S-box $\mathrm{S_{RD}}$ is invertible and $\mathrm{S_{RD}}^{-1} = f^{-1} \circ g^{-1} = f \circ g^{-1}$.

The ShiftRows Step. The ShiftRows transformation performs a cyclic left shift of the rows of the state matrix. The offsets are different for each row and depend on the block length N_b.

N_b	1. row	2. row	3. row	4. row
4	0	1	2	3
5	0	1	2	3
6	0	1	2	3
7	0	1	2	4
8	0	1	3	4

---

[6] Recall that the field $\mathbb{F}_2$ with 2 elements consists of the residues modulo 2, i.e., $\mathbb{F}_2 = \mathbb{Z}_2 = \{0, 1\}$.

For a block length of 128 bits ($N_b = 4$), as in AES, ShiftRows is the map

a	b	c	d
e	f	g	h
i	j	k	l
m	n	o	p

$\longmapsto$

a	b	c	d
f	g	h	e
k	l	i	j
p	m	n	o

Obviously, ShiftRows is invertible. The inverse operation is obtained by cyclic right shifts with the same offsets.

The MixColumns Step. The MixColumns transformation operates on each column of the state matrix independently. We consider a column $a = (a_0, a_1, a_2, a_3)$ as a polynomial $a(X) = a_3 X^3 + a_2 X^2 + a_1 X + a_0$ of degree ≤ 3, with coefficients in $\mathbb{F}_{2^8}$.

Then MixColumns transforms a column a by multiplying it with the fixed polynomial

$$c(X) := 03\,X^3 + 01\,X^2 + 01\,X + 02$$

and taking the residue of the product modulo $X^4 + 1$:

$$a(X) \mapsto a(X) \cdot c(X) \bmod (X^4 + 1).$$

The coefficients of $c(X)$ are elements of $\mathbb{F}_{2^8}$. Hence, they are represented as bytes, and a byte is given by two hexadecimal digits, as usual.

The choice of the polynomial $c(X)$ is based on maximum distance separating (MDS) codes (see [DaeRij02]). If only one coefficient of $a(X)$ changes, then all four coefficients of $a(X) \cdot c(X) \bmod (X^4 + 1)$ change (see Exercise 3).

The transformations of MixColumns, multiplying by $c(X)$ and taking the residue modulo $X^4 + 1$, are $\mathbb{F}_{2^8}$-linear maps. Hence MixColumns is a linear map of vectors of length 4 over $\mathbb{F}_{2^8}$. It is given by the following 4×4-matrix over $\mathbb{F}_{2^8}$:

$$\begin{pmatrix} 02 & 03 & 01 & 01 \\ 01 & 02 & 03 & 01 \\ 01 & 01 & 02 & 03 \\ 03 & 01 & 01 & 02 \end{pmatrix}.$$

Again, bytes are represented by two hexadecimal digits.

MixColumns transforms each column of the state matrix independently. For a block length of 128 bits, as in AES, we get

$a_{0,0}$	$a_{0,1}$	$a_{0,2}$	$a_{0,3}$
$a_{1,0}$	$a_{1,1}$	$a_{1,2}$	$a_{1,3}$
$a_{2,0}$	$a_{2,1}$	$a_{2,2}$	$a_{2,3}$
$a_{3,0}$	$a_{3,1}$	$a_{3,2}$	$a_{3,3}$

$\longmapsto$

$b_{0,0}$	$b_{0,1}$	$b_{0,2}$	$b_{0,3}$
$b_{1,0}$	$b_{1,1}$	$b_{1,2}$	$b_{1,3}$
$b_{2,0}$	$b_{2,1}$	$b_{2,2}$	$b_{2,3}$
$b_{3,0}$	$b_{3,1}$	$b_{3,2}$	$b_{3,3}$

where

$$\begin{pmatrix} b_{0,j} \\ b_{1,j} \\ b_{2,j} \\ b_{3,j} \end{pmatrix} = \begin{pmatrix} 02\ 03\ 01\ 01 \\ 01\ 02\ 03\ 01 \\ 01\ 01\ 02\ 03 \\ 03\ 01\ 01\ 02 \end{pmatrix} \cdot \begin{pmatrix} a_{0,j} \\ a_{1,j} \\ a_{2,j} \\ a_{3,j} \end{pmatrix}, \ j = 0, 1, 2, 3.$$

The polynomial $c(X)$ is relatively prime to $X^4 + 1$. Therefore $c(X)$ is a unit modulo $X^4 + 1$. Its inverse is

$$d(X) = 0\text{B}\,X^3 + 0\text{D}\,X^2 + 09\,X + 0\text{E},$$

i.e., $c(X){\cdot}d(X) \bmod (X^4{+}1) = 1$. This implies that MixColumns is invertible. The inverse operation is to multiply each column of the state matrix by $d(X)$ modulo $X^4 + 1$.

AddKey. The operation AddKey is the only operation in Rijndael that depends on the secret key k, which is shared by the communication partners. It adds a round key to the intermediate result *state*. The round keys are derived from the secret key k by applying the key schedule algorithm, which is described in the next section. Round keys are bit strings and, as the intermediate results *state*, they have block length, i.e., each round key is a sequence of N_b words. AddKey simply bitwise XORs the *state* with the *roundkey* to get the new value of *state*:

$$(state, roundkey) \mapsto state \oplus roundkey.$$

Since we arrange *state* as a matrix, a round key is also represented as a round key matrix of bytes with 4 rows and N_b columns. Each of the N_b words of the round key yields a column. Then the corresponding entries of the state matrix and the round key matrix are bitwise XORed by AddKey to get the new state matrix. Note that bitwise XORing two bytes means to add two elements of the field $\mathbb{F}_{2^8}$.

Obviously, AddKey is invertible. It is inverse to itself. To invert it, you simply apply AddKey a second time with the same round key.

The Key Schedule. The secret key k consists of N_k 4-byte-words. The Rijndael algorithm needs a round key for each round and one round key for the initial key addition. Thus we have to generate $N_r + 1$ round keys (as before, N_r is the number of rounds). A round key consists of N_b words. If we concatenate all the round keys, we get a string of $N_b(N_r + 1)$ words. We call this string the expanded key.

The expanded key is derived from the secret key k by the key expansion procedure, which we describe below. The round keys

$$key_0, key_1, key_2, \ldots, key_{N_r}$$

are then selected from the expanded key $ExpKey$: key_0 consists of the first N_b words of $ExpKey$, key_1 consists of the next N_b words of $ExpKey$, and so on.

To explain the key expansion procedure, we use functions f_j, defined for multiples j of N_k, and a function g. All these functions map words (x_0, x_1, x_2, x_3), which each consist of 4 bytes, to words.

g simply applies the S-box $\mathrm{S_{RD}}$ (see SubBytes above) to each byte:

$$(x_0, x_1, x_2, x_3) \mapsto (\mathrm{S_{RD}}(x_0), \mathrm{S_{RD}}(x_1), \mathrm{S_{RD}}(x_2), \mathrm{S_{RD}}(x_3)).$$

For multiples j of N_k, i.e., $j \equiv 0 \bmod N_k$, we define f_j by

$$(x_0, x_1, x_2, x_3) \mapsto \left(\mathrm{S_{RD}}(x_1) \oplus RC\left[j/N_k\right], \mathrm{S_{RD}}(x_2), \mathrm{S_{RD}}(x_3), \mathrm{S_{RD}}(x_0)\right).$$

Here, so-called round constants $RC[i]$ are used. f_j first applies g, then executes a cyclic left shift and finally adds $RC[j/N_k]$ to the first byte.

The round constants $RC[i]$ are defined as follows. First, recall that in our representation, the elements of $\mathbb{F}_{2^8}$ are the residues of polynomials with coefficients in $\mathbb{F}_2$ modulo $P(X) = X^8 + X^4 + X^3 + X + 1$. Now, the round constant $RC[i] \in \mathbb{F}_{2^8}$ is defined by $RC[i] := X^{i-1} \bmod P(X)$.

Relying on the non-linear S-box $\mathrm{S_{RD}}$, the functions f_j and g are also non-linear.

We are ready to describe the key expansion. We denote by

$$ExpKey[j], \ 0 \leq j < N_b(N_r + 1),$$

the words of the expanded key. The first N_k words are initialized with the secret key k. The following words are computed recursively.

$$ExpKey[j] := \begin{cases} ExpKey[j - N_k] \oplus f_j(ExpKey[j-1]) & \text{if } j \equiv 0 \bmod N_k, \\ ExpKey[j - N_k] \oplus g(ExpKey[j-1]) & \text{if } N_k > 6 \text{ and} \\ & \qquad j \equiv 4 \bmod N_k, \\ ExpKey[j - N_k] \oplus ExpKey[j-1] & \text{else.} \end{cases}$$

$ExpKey[j]$ depends on $ExpKey[j - N_k]$, on $ExpKey[j-1]$ and on the round constant.

2.1.5 Modes of Operation

Block ciphers need some extension, because in practice most of the messages have a size that is distinct from the block length. Often the message length exceeds the block length. Modes of operation handle this problem. They were first specified in conjunction with DES, but they can be applied to any block cipher.

We consider a block cipher E with block length n. We fix a key k and, as usual, we denote the encryption function with this key k by

$$E_k : \{0,1\}^n \longrightarrow \{0,1\}^n,$$

for example, $E_k = \mathrm{DES}_k$. To encrypt a message m that is longer than n bits we apply a mode of operation: The message m is decomposed into blocks of some fixed bit length r, $m = m_1m_2 \ldots m_l$, and then these blocks are encrypted iteratively. The length r of the blocks m_i is not in all modes of operation equal to the block length n of the cipher. There are modes of operation, where r can be smaller than n, for example, the cipher feedback and the output feedback modes below. In electronic code book mode and cipher-block chaining mode, which we discuss first, the block length r is equal to the block length n of the block cipher.

If the block length r does not divide the length of our message, we have to complete the last block. The last block is padded out with some bits. After applying the decryption function, the receiver must remove the padding. Therefore, he must know how many bits were added. This can be achieved, for example, by storing the number of padded bits in the last byte of the last block.

Electronic Codebook Mode. The electronic code book mode is the straightforward mode. The encryption is deterministic – identical plaintext blocks result in identical ciphertext blocks. The encryption works like a codebook. Each block of m is encrypted independently of the other blocks. Transmission bit errors in a single ciphertext block affect the decryption only of that block.

In this mode, we have $r = n$. The electronic codebook mode is implemented by the following algorithm:

Algorithm 2.6.
    bitString $ecbEncrypt$(bitString m)
    1   divide m into $m_1 \ldots m_l$
    2   for $i \leftarrow 1$ to l do
    3       $c_i \leftarrow E_k(m_i)$.
    4   return $c_1 \ldots c_l$

For decryption, the same algorithm can be used with the decryption function E_k^{-1} in place of E_k.

If we encrypt many blocks, partial information about the plaintext is revealed. For example, an eavesdropper Eve detects whether a certain block repeatedly occurs in the sequence of plaintext blocks, or, more generally, she can figure out how often a certain plaintext block occurs. Therefore, other modes of operation are preferable.

Cipher-Block Chaining Mode. In this mode, we have $r = n$. Encryption in the cipher-block chaining mode is implemented by the following algorithm:

Algorithm 2.7.

    bitString *cbcEncrypt*(bitString m)

  1   select $c_0 \in \{0,1\}^n$ at random

  2   divide m into $m_1 \ldots m_l$

  3   for $i \leftarrow 1$ to l do

  4       $c_i \leftarrow E_k(m_i \oplus c_{i-1})$

  5   return $c_0 c_1 \ldots c_l$

Choosing the initial value c_0 at random prevents almost with certainty that the same initial value c_0 is used for more than one encryption. This is important for security. Suppose for a moment that the same c_0 is used for two messages m and m'. Then, an eavesdropper Eve can immediately detect whether the first s blocks of m and m' coincide, because in this case the first s ciphertext blocks are the same.

If a message is encrypted twice, then, with a very high probability, the initial values are different, and hence the resulting ciphertexts are distinct. The ciphertext depends on the plaintext, the key and a randomly chosen initial value. We obtain a *randomized encryption algorithm*.

Decryption in cipher-block chaining mode is implemented by the following algorithm:

Algorithm 2.8.

    bitString *cbcDecrypt*(bitString c)

  1   divide c into $c_0 c_1 \ldots c_l$

  2   for $i \leftarrow 1$ to l do

  3       $m_i \leftarrow E_k^{-1}(c_i) \oplus c_{i-1}$

  4   return $m_1 \ldots m_l$

The cryptogram $c = c_0 c_1 \ldots c_l$ has one block more than the plaintext. The initial value c_0 needs not be secret, but its integrity must be guaranteed in order to decrypt c_1 correctly.

A transmission bit error in block c_i affects the decryption of the blocks c_i and c_{i+1}. The block recovered from c_i will appear random (here we assume that even a small change in the input of a block cipher will produce a random-looking output), while the plaintext recovered from c_{i+1} has bit errors precisely where c_i did. The block c_{i+2} is decrypted correctly. The cipher-block chaining mode is self-synchronizing, even if one or more entire blocks are lost. A lost ciphertext block results in the loss of the corresponding plaintext block and errors in the next plaintext block.

In both the electronic codebook mode and cipher-block chaining mode, E_k^{-1} is applied for decryption. Hence, both modes are also applicable with public-key encryption methods, where the computation of E_k^{-1} requires the recipient's secret, while E_k can be easily computed by everyone.

Cipher Feedback Mode. Let lsb_l denote the l least significant (rightmost) bits of a bit string, msb_l the l most significant (leftmost) bits of a bit string, and let $\|$ denote the concatenation of bit strings.

In the cipher feedback mode, we have $1 \leq r \leq n$ (recall that the plaintext m is divided into blocks of length r). Let $x_1 \in \{0,1\}^n$ be a randomly chosen initial value. The cipher feedback mode is implemented by the following algorithm:

Algorithm 2.9.

bitString $cfbEnCrypt$(bitString m, x_1)
1   divide m into $m_1 \ldots m_l$
2   for $i \leftarrow 1$ to l do
3        $c_i \leftarrow m_i \oplus msb_r(E_k(x_i))$
4        $x_{i+1} \leftarrow lsb_{n-r}(x_i)\|c_i$
5   return $c_1 \ldots c_l$

We get a stream cipher in this way. The key stream is computed by using E_k, and depends on the key underlying E_k, on an initial value x_1 and on the ciphertext blocks already computed. Actually, x_{i+1} depends on the first $\lceil n/r \rceil$ members[7] of the sequence $c_i, c_{i-1}, \ldots, c_1, x_1$. The key stream is obtained in blocks of length r. The message can be processed bit by bit and messages of arbitrary length can be encrypted without padding. If one block of the key stream is consumed, the next block is computed. The initial value x_1 is transmitted to the recipient. It does not need to be secret if E_k is the encryption function of a symmetric cryptosystem (an attacker does not know the key underlying E_k). The recipient can compute $E_k(x_1)$ – hence m_1 and x_2 – from x_1 and the cryptogram c_1, then $E_k(x_2), m_2$ and x_3, and so on.

For each encryption, a new initial value x_1 is chosen at random. This prevents almost with certainty that the same initial value x_1 is used for more than one encryption. As in every stream cipher, this is important for security. If the same initial value x_1 is used for two messages m and m', then an eavesdropper Eve immediately finds out whether the first s blocks of m and m' coincide. In this case, the first s blocks of the generated key stream, and hence the first s ciphertext blocks are the same for m and m'.

A transmission bit error in block c_i affects the decryption of that block and the next $\lceil n/r \rceil$ ciphertext blocks. The block recovered from c_i has bit errors precisely where c_i did. The next $\lceil n/r \rceil$ ciphertext blocks will be decrypted into random-looking blocks (again we assume that even a small change in the input of a block cipher will produce a random-looking output). The cipher feedback mode is self-synchronizing after $\lceil n/r \rceil$ steps, even if one or more entire blocks are lost.

Output Feedback Mode. As in the cipher feedback mode, we have $1 \leq r \leq n$. Let $x_1 \in \{0,1\}^n$ be a randomly chosen initial value. The output feedback mode is implemented by the following algorithm:

---

[7] $\lceil x \rceil$ denotes the smallest integer $\geq x$.

Algorithm 2.10.
　bitString *ofbEnCrypt*(bitString m, x_1)
　1　divide m into $m_1 \ldots m_l$
　2　for $i \leftarrow 1$ to l do
　3　　　$c_i \leftarrow m_i \oplus \mathrm{msb}_r(E_k(x_i))$
　4　　　$x_{i+1} \leftarrow E_k(x_i)$
　5　return $c_1 \ldots c_l$

There are two different output feedback modes discussed in the literature. The one we introduced is considered to have better security properties and was specified in [ISO/IEC 10116]. In the output feedback mode, plaintexts of arbitrary length can be encrypted without padding. As in the cipher feedback mode, the plaintext is considered as a bit stream and each bit is XORed with a bit of a key stream. The key stream depends only on an initial value x_1 and is iteratively computed by $x_{i+1} = E_k(x_i)$. The initial value x_1 is transmitted to the recipient. It does not need to be secret if E_k is the encryption function of a symmetric cryptosystem (an attacker does not know the key underlying E_k). For decryption, the same algorithm can be used.

It is essential for security that the initial value is chosen randomly and independently from the previous ones. This prevents almost with certainty that the same initial value x_1 is used for more than one encryption. If the same initial value x_1 is used for two messages m and m', then identical key streams are generated for m and m', and an eavesdropper Eve immediately computes the difference between m and m' from the ciphertexts: $m \oplus m' = c \oplus c'$. Thus, it is strongly recommended to choose a new random initial value for each message.

A transmission bit error in block c_i only affects the decryption of that block. The block recovered from c_i has bit errors precisely where c_i did. However, the output feedback mode will not recover from a lost ciphertext block – all following ciphertext blocks will be decrypted incorrectly.

Security. We mentioned before that the electronic codebook mode has some shortcomings. The question arises as to what amount the mode of operation weakens the cryptographic strength of a block cipher. A systematic treatment of this question can be found in [BelDesJokRog97]. First, a model for the security of block ciphers is developed, the so-called *pseudorandom function model* or *pseudorandom permutation model*.

As discussed in Section 2.1.2, ideally we would like to choose the encryption function of a block cipher from the huge set of all permutations on $\{0, 1\}^n$ in a truly random way. This approach might be called the "truly random permutation model". In practice, we have to follow the "pseudorandom permutation model": the encryption function is chosen randomly, but from a much smaller family $(F_k)_{k \in K}$ of permutations on $\{0, 1\}^n$, like DES.

In [BelDesJokRog97], the security of the cipher-block chaining mode is reduced to the security of the pseudorandom family $(F_k)_{k \in K}$. Here, security of the family $(F_k)_{k \in K}$ means that no efficient algorithm is able to distinguish elements randomly chosen from $(F_k)_{k \in K}$ from elements randomly chosen from the set of all permutations. This notion of security for pseudorandom function families is analogously defined as the notion of computationally perfect pseudorandom bit generators, which will be studied in detail in Chapter 8. [BelDesJokRog97] also consider a mode of operation similar to the output feedback mode, called the XOR scheme, and its security is also reduced to the security of the underlying pseudorandom function family.

2.2 Cryptographic Hash Functions

Cryptographic hash functions such as SHA-1 or MD5 are widely used in cryptography. In digital signature schemes, messages are first hashed and the hash value $h(m)$ is signed in place of m. Hash values are used to check the integrity of public keys. Pseudorandom bit strings are generated by hash functions. When used with a secret key, cryptographic hash functions become *message authentication codes* (MACs), the preferred tool in protocols like SSL and IPSec to check the integrity of a message and to authenticate the sender.

A *hash function* is a function that takes as input an arbitrarily long string of bits (called a message) and outputs a bit string of a fixed length n. Mathematically, a hash function is a function

$$h : \{0,1\}^* \longrightarrow \{0,1\}^n, \ m \longmapsto h(m).$$

The length n of the output is typically between 128 and 512 bits[8]. Later, when discussing the birthday attack, we will see why the output lengths are in this range.

One basic requirement is that the hash values $h(m)$ are easy to compute, making both hardware and software implementations practical.

2.2.1 Security Requirements for Hash Functions

A classical application of cryptographic hash functions is the "encryption" of passwords. Rather than storing the cleartext of a user password pwd in the password file of a system, the hash value $h(pwd)$ is stored in place of the password itself. If a user enters a password, the system computes the hash value of the entered password and compares it with the stored value. This technique of non-reversible "encryption" is applied in operating systems. It prevents, for example, passwords becoming known to privileged users of the system such as administrators, provided it is not possible to compute a password pwd from its hash value $h(pwd)$. This leads to our first security requirement.

---

[8] The output lengths of MD5 and SHA-1 are 128 and 160 bits.

A cryptographic hash function must be a *one-way function*: Given a value $y \in \{0,1\}^n$, it is computationally infeasible to find an m with $h(m) = y$.

If a hash function is used in conjunction with digital signature schemes, the message is hashed first and then the hash value is signed in place of the original message. Suppose Alice signs $h(m)$ for a message m. An adversary Eve should have no chance to find a message $m' \neq m$ with $h(m') = h(m)$. Otherwise, she could pretend that Alice signed m' instead of m.

Thus, the hash function must have the property that given a message m, it is computationally infeasible to obtain a second message m' with $m \neq m'$ and $h(m) = h(m')$. This property is called the *second pre-image resistance*.

When using hash functions with digital signatures, we require an even stronger property. The legal user Alice of a signature scheme with hash function h should have no chance of finding two distinct messages m and m' with $h(m) = h(m')$. If Alice finds such messages, she could sign m and say later that she has signed m' and not m.

Such a pair (m, m') of messages, with $m \neq m'$ and $h(m) = h(m')$, is called a *collision* of h. If it is computationally infeasible to find a collision (m, m') of h, then h is called *collision resistant*.

Sometimes, collision resistant hash functions are called *collision free*, but that's misleading. The function h maps an infinite number of elements to a finite number of elements. Thus, there are lots of collisions (in fact, infinitely many). Collision resistance merely states that they cannot be found.

Proposition 2.11. *A collision-resistant hash function h is second-pre-image resistant.*

Proof. An algorithm computing second pre-images can be used to compute collisions in the following way: Choose m at random. Compute a pre-image $m' \neq m$ of $h(m)$. (m, m') is a collision of h. □

Proposition 2.11 says that collision resistance is the stronger property. Therefore, second pre-image resistance is sometimes also called *weak collision resistance*, and collision resistance is referred to as *strong collision resistance*.

Proposition 2.12. *A second-pre-image-resistant hash function is a one-way function.*

Proof. If h were not one-way, there would be a practical algorithm A that on input of a randomly chosen value v computes a message $\tilde{m}$ with $h(\tilde{m}) = v$, with a non-negligible probability. Given a random message m, attacker Eve could find, with a non-negligible probability, a second pre-image of $h(m)$ in the following way: She applies A to the hash value $h(m)$ and obtains $\tilde{m}$ with $h(\tilde{m}) = h(m)$. The probability that $\tilde{m} \neq m$ is high. □

Our definitions and the argument in the previous proof lack some precision and are not mathematically rigorous. For example, we do not explain what "computationally infeasible" and a "non-negligible probability" mean. It is possible to give precise definitions and a rigorous proof of Proposition 2.12 (see Chapter 11, Exercise 2).

Definition 2.13. A hash function is called a *cryptographic hash function* if it is collision resistant.

Sometimes, hash functions used in cryptography are referred to as *one-way hash functions*. We have seen that there is a stronger requirement, collision resistance, and the one-way property follows from it. Therefore, we prefer to speak of collision-resistant hash functions.

2.2.2 Construction of Hash Functions

There are no known examples of hash functions whose collision resistance can be proven without any assumptions. In Section 11.2, we give examples of (rather inefficient) hash functions that are provably collision resistant under standard assumptions in public-key cryptography, such as the factoring assumption. Here, we introduce methods to construct very efficient hash functions.

Merkle–Damgård's construction. Many cryptographic hash functions used in practice are obtained by the following method, known as *Merkle–Damgård's construction* or *Merkle's meta method*. The method reduces the problem of designing a collision-resistant hash function $h : \{0,1\}^* \longrightarrow \{0,1\}^n$ to the problem of constructing a collision-resistant function

$$f : \{0,1\}^{n+r} \longrightarrow \{0,1\}^n \qquad (r \in \mathbb{N}, r > 0)$$

with finite domain $\{0,1\}^{n+r}$. Such a function f is called a *compression function*. A compression function maps messages m of a fixed length $n + r$ to messages $f(m)$ of length n. We call r the *compression rate*.

We discuss Merkle–Damgård's construction. Let $f : \{0,1\}^{n+r} \longrightarrow \{0,1\}^n$ be a compression function with compression rate r. By using f, we define a hash function

$$h : \{0,1\}^* \longrightarrow \{0,1\}^n.$$

Let $m \in \{0,1\}^*$ be a message of arbitrary length. The hash function h works iteratively. To compute the hash value $h(m)$, we start with a fixed initial n-bit hash value $v = v_0$ (the same for all m). The message m is subdivided into blocks of length r. One block after the other is taken from m, concatenated with the current value v and compressed by f to get a new v. The final v is the hash value $h(m)$.

More precisely, we pad m out, i.e., we append some bits to m, to obtain a message $\tilde{m}$, whose bit length is a multiple of r. We apply the following padding method: A single 1-bit followed by as few (possibly zero) 0-bits as necessary are appended. Every message m is padded out with such a string

$100\ldots0$, even if the length of the original message m is a multiple of r. This guarantees that the padding can be removed unambiguously – the bits which are added during padding can be distinguished from the original message bits.[9]

After the padding, we decompose

$$\tilde{m} = m_1 \| \ldots \| m_k, \ m_i \in \{0,1\}^r, \ 1 \le i \le k,$$

into blocks m_i of length r.

We add one more r-bit block m_{k+1} to $\tilde{m}$ and store the original length of m (i.e., the length of m before padding it out) into this block right-aligned. The remaining bits of m_{k+1} are filled with zeros:

$$\tilde{m} = m_1 \| m_2 \| \ldots \| m_k \| m_{k+1}.$$

Starting with the initial value $v_0 \in \{0,1\}^n$, we set recursively

$$v_i := f(v_{i-1} \| m_i), \ 1 \le i \le k+1.$$

The last value of v is taken as hash value $h(m)$:

$$h(m) := v_{k+1}.$$

The last block m_{k+1} is added to prevent certain types of collisions. It might happen that we obtain $v_i = v_0$ for some i. If we had not added m_{k+1}, then (m, m') would be a collision of h, where m' is obtained from $m_{i+1} \| \ldots \| m_k$ by removing the padding string $10\ldots0$ from the last block m_k. Since m and m' have different lengths, the additional length blocks differ and prevent such collisions.[10]

Proposition 2.14. *Let f be a collision-resistant compression function. The hash function h constructed by Merkle's meta method is also collision resistant.*

Proof. The proof runs by contradiction. Assume that h is not collision resistant, i.e., that we can efficiently find a collision (m, m') of h. Let $(\tilde{m}, \tilde{m}')$ be the modified messages as above. The following algorithm efficiently computes a collision of f from $(\tilde{m}, \tilde{m}')$. This contradicts our assumption that f is collision resistant.

---

[9] There are also other padding methods which may be applied. See, for example, [RFC 3369].

[10] Sometimes, the padding and the length block are combined: the length of the original message is stored in the rightmost bits of the padding string. See, for example, SHA-1 ([RFC 3174]).

Algorithm 2.15.

collision $FindCollision(\text{bitString } \tilde{m}, \tilde{m}')$

1   $\tilde{m} = m_1 \| \dots \| m_{k+1}, \tilde{m}' = m_1' \| \dots \| m_{k'+1}'$ decomposed as above
2   $v_1, \dots, v_{k+1}, v_1', \dots, v_{k'+1}'$ constructed as above
3   if $|m| \neq |m'|$
4   then return $(v_k \| m_{k+1}, v_{k'}' \| m_{k'+1}')$
5   for $i \leftarrow 1$ to k do
6   if $v_i \neq v_i'$ and $v_{i+1} = v_{i+1}'$
7   then return $(v_i \| m_{i+1}, v_i' \| m_{i+1}')$
8   for $i \leftarrow 0$ to $k - 1$ do
9   if $m_{i+1} \neq m_{i+1}'$
10   then return $(v_i \| m_{i+1}, v_i' \| m_{i+1}')$

Note that $h(m) = v_{k+1} = v_{k'+1}' = h(m')$. If $|m| \neq |m'|$ we have $m_{k+1} \neq m_{k'+1}'$, since the length of the string is encoded in the last block. Hence $v_k \| m_{k+1} \neq v_{k'}' \| m_{k'+1}'$. We obtain a collision $(v_k \| m_{k+1}, v_{k'}' \| m_{k'+1}')$, because $f(v_k \| m_{k+1}) = h(m) = h(m') = f(v_{k'}' \| m_{k'+1}')$. On the other hand, if $|m| = |m'|$, then $k = k'$, and we are looking for an index i with $v_i \neq v_i'$ and $v_{i+1} = v_{i+1}'$. $(v_i \| m_{i+1}, v_i' \| m_{i+1}')$ is then a collision of f, because $f(v_i \| m_{i+1}) = v_{i+1} = v_{i+1}' = f(v_i' \| m_{i+1}')$. If no index with the above condition exists, we have $v_i = v_i'$, $1 \leq i \leq k + 1$. In this case, we search for an index i with $m_{i+1} \neq m_{i+1}'$. Such an index exists, because $m \neq m'$. $(v_i \| m_{i+1}, v_i' \| m_{i+1}')$ is then a collision of f, because $f(v_i \| m_{i+1}) = v_{i+1} = v_{i+1}' = f(v_i' \| m_{i+1}')$.   □

Merkle–Damgård's construction is based on a collision-resistant compression function. The result is a collision-resistant hash function h. The output length of h is equal to the output length of the compression function.

For a long time, Merkle–Damgård's construction was the preferred method for building hash functions. Recently, another construction for building hash functions was published.

The Sponge Construction. The *sponge construction* has gained attention because Keccak, the winner of the SHA-3 competition (see page 41 below), is based on it. It was proposed by Bertoni et al. [BerDaePeeAss11] and can be regarded as a generalization of Merkle–Damgård's construction.

The sponge construction uses a permutation (and not a compression function, as Merkle–Damgård's construction does) and yields a *sponge function*

$$g : \{0,1\}^* \longrightarrow \{0,1\}^n, \; m \longmapsto g(m).$$

As required for hash functions, g is able to process binary strings m of arbitrary length. The length n of the output can be varied and adapted to the needs of the application. The sponge function g processes the input m iteratively in pieces of length r, and the output $g(m)$ is also iteratively generated in pieces of the same length r. The output is computed by iteratively applying a permutation

$$f : \{0,1\}^r \times \{0,1\}^c \longrightarrow \{0,1\}^r \times \{0,1\}^c$$

to the intermediate result s, which is called the *state*. The state s consists of two components, i.e., $s = (s_1, s_2)$, where $s_1 \in \{0,1\}^r$ and $s_2 \in \{0,1\}^c$. The input–output operations of g interact only with s_1, the first component of the state. The bits of s_2 are never directly affected by these operations. Accordingly, s_1 is called the outer part and s_2 the inner part of the state. The parameter r is called the *bit rate*, and c is called the *capacity* of the sponge function. The bit rate r and the capacity c are closely related to the performance and the security level, respectively, of the sponge function.

Now we describe the construction in detail. The sponge function g iteratively applies the permutation f. One iteration is called a round. In the first k rounds, where $k = \lfloor |m|/r \rfloor + 1$, the input is processed. These rounds are called the *absorbing phase*. In the following l rounds, called the *squeezing phase*, the output is generated. The number l of squeezing rounds depends on the desired output length n: $l = \lceil n/r \rceil$.

In each absorbing round, an input block of length r is processed. Therefore, we have to decompose the input string $m \in \{0,1\}^*$ into blocks of length r. As in Merkle–Damgård's construction, we pad out m to obtain a string $\tilde{m}$ whose bit length is a multiple of r. Then, we subdivide

$$\tilde{m} = m_1 \| \ldots \| m_k, \ m_i \in \{0,1\}^r, \ 1 \le i \le k,$$

into blocks m_i of length r. At the beginning, the bits of the internal state s are initialized to zero. Then, in round i of the absorbing phase, the i-th input block m_i is XORed with the outer part s_1 of the state and the permutation f is applied to get the new state s:

$$s = f(s_1 \oplus m_i, s_2), 1 \le i \le k.$$

When all message blocks m_i have been processed, the sponge function continues with the squeezing phase. It iteratively computes output blocks $g_1, g_2, \ldots, g_l$ of length r. The first output block g_1 is just the current value of the outer part of the state, i.e., $g_1 = s_1$. The following output blocks are obtained by iteratively permuting the state s and taking the outer part s_1 of s:

$$s := f(s), g_i := s_1, 2 \le i \le l.$$

The value $g(m)$ is obtained by truncating $g_1 \| g_2 \| \ldots \| g_l$ to the desired output length n.

The sponge construction makes it possible to define functions with arbitrary input and output lengths. Therefore, sponge functions can be applied in a variety of cryptographic applications, from hash functions with long input and short output values to pseudorandom key generators for stream ciphers, where the input consists of a short seed and the output is a long key stream.

Keccak. *Keccak* is a sponge function. It has been selected by NIST to become the new SHA-3 standard for cryptographic hash functions (see page 41 below). As we have seen in the previous section, the core of the sponge construction is a permutation f that permutes the internal state of the sponge function. We now describe and discuss the permutation that is used in Keccak. We follow the Keccak reference [BerDaePeeAss11a] and denote the permutation by

$$\text{Keccak-}f : \{0,1\}^b \longrightarrow \{0,1\}^b.$$

Parameter b is called the *width*. It is equal to the sum $r + c$ of the bit rate r and the capacity c from the sponge construction (see above). The width b takes one of the seven values $b = 25 \cdot 2^l$, $0 \le l \le 6$. The Keccak proposal for SHA-3 restricts b to the largest value $b = 25 \cdot 2^6 = 1600$.

The state $\{0,1\}^b$ is considered as a fixed memory on which Keccak-f operates. The bits of the state are arranged in a 3-dimensional array Q:

$$Q[0\ldots4][0\ldots4][0\ldots w-1], \text{ where } w = 2^l, 0 \le l \le 6.$$

Q may be regarded as a cuboid with a 5×5 square as base and a height of $w = 2^l$. For a better understanding, it is useful to consider the cuboid Q as being composed of slices and lanes, and the slices in turn are composed of rows and columns:

1. A *slice* is defined by a constant z index: $Q[0\ldots4][0\ldots4][z]$.
2. A *row* is defined by constant y and z indices: $Q[0\ldots4][y][z]$.
3. A *column* is defined by a constant x and z indices: $Q[x][0\ldots4][z]$.
4. A *lane* is defined by constant x and y indices: $Q[x][y][0\ldots w-1]$.

The state Q consists of w parallel slices, each containing 5 rows and 5 columns, and of 25 lanes. The lanes are orthogonal to the slices.

Keccak-f is composed of n iterations of the permutation

$$R := \iota \circ \chi \circ \pi \circ \rho \circ \theta.$$

The iterations are called rounds. The number of rounds depends on the width $w = 2^l$ of the permutation, namely $n = 12 + 2l$. The *step mappings*

$$\iota, \chi, \pi, \rho, \theta : Q \longrightarrow Q$$

operate on Q and are defined by the following formulas.

In these formulas, bits are considered as elements of $\mathbb{Z}_2$, i.e., the addition $+$ and the multiplication $\cdot$ are modulo 2. The first and the second index of Q are taken modulo 5 and the third index modulo w. The left side of each assignment is the new state bit. It is expressed as a function of the old state bits. For x, y, z, $0 \le x, y \le 4$, $0 \le z \le w-1$,

$$\theta : Q[x][y][z] \leftarrow Q[x][y][z] + \sum_{y'=0}^{4} Q[x-1][y'][z] + \sum_{y'=0}^{4} Q[x+1][y'][z-1],$$

$\rho : Q[x][y][z] \leftarrow Q[x][y][z-t(x,y)]$ ($t(x,y)$ is a constant for each lane),

$\pi : Q[x][y][z] \leftarrow Q[x+3y][x][z],$

$\chi : Q[x][y][z] \leftarrow Q[x][y][z] + (Q[x+1][y][z] + 1) \cdot Q[x+2][y][z],$

$\iota : Q[x][y][z] \leftarrow Q[x][y][z] + RC[i][x][y][z]$ ($RC[i]$ is a round constant).

Remarks:

1. θ adds to each bit of a column $Q[x][0\ldots4][z]$ the same values, namely the parities of the columns $Q[x-1][0\ldots4][z]$ and $Q[x+1][0\ldots4][z-1]$. Therefore, each output bit of θ is the parity over 11 input bits. Conversely, each bit of Q is input for 11 output bits of θ. Thus, modifying one bit in the input of θ affects 11 bits in the output of θ, and we see that θ provides a high level of diffusion.[11] Without θ, Keccak-f would not provide significant diffusion. To show that θ is invertible requires extensive computations (see [BerDaePeeAss11a]).

2. ρ is a cyclic shift on each lane. The offset $t(x,y)$ of the shift depends on the lane. ρ ensures that bit patterns in a slice are disrupted. The inverse of ρ shifts each lane into the other direction.

3. π changes the ordering of the lanes. The transformation of the indizes $(x,y) \longmapsto (x+3y,x)$ is an invertible linear transformation M of $\mathbb{Z}_5^2$. The inverse of π is given by the inverse M^{-1}.

4. χ is the only non-linear part of R. Without χ, Keccak-f would be linear. χ works on each row independently. So, you can view it as the application of an S-box with 5-bits in- and output on the $5w$ rows of the state. χ is invertible on each row. You can see this by inspecting all 32 input values of the S-box. Each bit is input for the computation of 3 bits. χ flips a bit if and only if the next two bits have the pattern 01. Thus, changing one bit in the input affects at least one and at most two output bits.

5. The round constants $RC[i]$, $0 \le i \le n-1$, are specified in the Keccak reference. They were derived from the output of a maximum-length linear feedback shift register. The constants differ from round to round. The inverse of ι is ι itself.

6. The design of the Keccak permutation is influenced by the wide experience in the design and cryptanalysis of iterated block ciphers. There are parallels to the construction of an iterated block cipher, such as Rijndael/AES (see Section 2.1.4). Keccak-f consists of identical rounds and each round is composed of a sequence of simple step mappings. Instead of a round key a round constant $RC[i]$ is added.

7. The Keccak step mappings are defined by bit operations. An implementation is much more efficient if all bits of a lane are simultaneously processed. This can be done by mapping the lanes to CPU words. For a width

---

[11] Diffusion measures the effect on the output if one bit in the input is changed.

of 1600 bit, each lane consists of 64 bits. If processed on a 64-bit CPU, a lane can then be stored in a single register, and the results of the step mappings are computed by using the CPU commands XOR, AND, NOT and (cyclic) SHIFT on CPU words. We see that width 1600, as proposed for SHA-3, enables an efficient implementation on 64-bit CPUs, which are typically used in modern PCs.

The Birthday Attack. One of the main questions when designing a hash function $h : \{0,1\}^* \longrightarrow \{0,1\}^n$ is how large to choose the length n of the hash values. A lower bound for n is obtained by analyzing the birthday attack.

The *birthday attack* is a brute-force attack against collision resistance. The adversary Eve randomly generates messages $m_1, m_2, m_3, \ldots$. For each newly generated message m_i, Eve computes and stores the hash value $h(m_i)$ and compares it with the previous hash values. If $h(m_i)$ coincides with one of the previous hash values, $h(m_i) = h(m_j)$ for some $j < i$, Eve has found a collision (m_i, m_j)[12]. We show below that Eve can expect to find a collision after choosing about $2^{n/2}$ messages. Thus, it is necessary to choose n so large that it is impossible to calculate and store $2^{n/2}$ hash values. If $n = 128$ (as with MD5), about $2^{64} \approx 10^{20}$ messages have to be chosen for a successful attack.[13] Many people think that today a hash length of 128 bits is no longer large enough, and that 160 bits (as in SHA-1 and RIPEMD-160) should be the lower bound (also see Section 2.2.2 below).

Attacking second-pre-image resistance or the one-way property of h with brute force would mean to generate, for a given hash value $v \in \{0,1\}^n$, random messages $m_1, m_2, m_3, \ldots$ and check each time whether $h(m_i) = v$. Here we expect to find a pre-image of v after choosing 2^n messages (see Exercise 7). For $n = 128$, we need $2^{128} \approx 10^{39}$ messages. To protect against this attack, a smaller n would be sufficient.

The surprising efficiency of the birthday attack is based on the *birthday paradox*. It says that the probability of two persons in a group sharing the same birthday is greater than $1/2$, if the group is chosen at random and has more than 23 members. It is really surprising that this happens with such a small group.

Considering hash functions, the 365 days of a year correspond to the number of hash values. We assume in our discussion that the hash function $h : \{0,1\}^* \longrightarrow \{0,1\}^n$ behaves like the birthdays of people. Each of the $s = 2^n$ values has the same probability. This assumption is reasonable. It is a basic design principle that a cryptographic hash function comes close to a

---

[12] In practice, the messages are generated by a deterministic pseudorandom generator. Therefore, the messages themselves can be reconstructed and need not be stored.

[13] To store 2^{64} 16-byte hash values, you need 2^{28} TB of storage. There are memoryless variations of the birthday attack which avoid these extreme storage requirements, see [MenOorVan96].

random function, which yields random and uniformly distributed values (see Section 2.2.4).

Evaluating h, k times with independently chosen inputs, the probability that no collisions occur is

$$p = p(s,k) = \frac{1}{s^k} \prod_{i=0}^{k-1} (s - i) = \prod_{i=1}^{k-1} \left(1 - \frac{i}{s} \right).$$

We have $1 - x \leq e^{-x}$ for all real numbers x and get

$$p \leq \prod_{i=1}^{k-1} e^{-i/s} = e^{-(1/s) \sum_{i=1}^{k-1} i} = e^{-k(k-1)/2s}.$$

The probability that a collision occurs is $1 - p$, and $1 - p \geq \frac{1}{2}$ if $k \geq \frac{1}{2} \left(\sqrt{1 + 8\ln 2 \cdot s} + 1 \right) \approx 1.18 \sqrt{s}$.

For $s = 365$, we get an explanation for the original birthday paradox, since $1.18 \cdot \sqrt{s} = 22.54$.

For the hash function h, we conclude that it suffices to choose about $2^{n/2}$ many messages at random to obtain a collision with probability $> \frac{1}{2}$.

In a hash-then-decrypt digital signature scheme, where the hash value is signed in place of the message (see Section 3.3.5 below), the birthday attack might be practically implemented in the following way. Suppose that Eve and Bob want to sign a contract m_1. Later, Eve wants to say that Bob has signed a different contract m_2. Eve generates $O(2^{n/2})$ minor variations of m_1 and m_2. In many cases, for example, if m_1 includes a bitmap, Bob might not observe the slight modification of m_1. If the birthday attack is successful, Eve gets messages $\tilde{m}_1$ and $\tilde{m}_2$ with $h(\tilde{m}_1) = h(\tilde{m}_2)$. Eve lets Bob sign the contract $\tilde{m}_1$. Later, she can pretend that Bob signed $\tilde{m}_2$.

Compression Functions from Block Ciphers. We show in this section how to derive compression functions from a block cipher, such as DES or AES. From these compression functions, cryptographic hash functions can be obtained by using Merkle–Damgård's construction.

Symmetric block ciphers are widely used and well studied. Encryption is implemented by efficient algorithms (see Chapter 2.1). It seems natural to also use them for the construction of compression functions. Though no rigorous proofs exist, the hope is that a good block cipher will result in a good compression function. Let

$$E : \{0,1\}^r \times \{0,1\}^n \longrightarrow \{0,1\}^n, \ (k, x) \longmapsto E(k, x)$$

be the encryption function of a symmetric block cipher, which encrypts blocks x of bit length n with r-bit keys k.

First, we consider constructions where the bit length of the hash value is equal to the block length of the block cipher. These schemes are called *single-length MDCs*[14]. To obtain a collision-resistant compression function, the block length n of the block cipher should be at least 128 bits.

The compression function

$$f_1 : \{0,1\}^{n+r} \longrightarrow \{0,1\}^n, \ (x\|y) \longmapsto E(y,x)$$

maps bit blocks of length $n+r$ to blocks of length n. A block of length $n+r$ is split into a left block x of length n and a right block y of length r. The right block y is used as key to encrypt the left block x.

The second example of a compression function – it is the basis of the Matyas–Meyer–Oseas hash function ([MatMeyOse85]) – was included in [ISO/IEC 10118-2]. Its compression rate is n. A block of bit length $2n$ is split into two halves, x and y, each of length n. Then x is encrypted with a key $g(y)$ which is derived from y by a function $g : \{0,1\}^n \longrightarrow \{0,1\}^r$.[15] The resulting ciphertext is bitwise XORed with x:

$$f_2 : \{0,1\}^{2n} \longrightarrow \{0,1\}^n, \ (x\|y) \longmapsto E(g(y),x) \oplus x.$$

If the block length of the block cipher is less than 128, *double-length MDCs* are used. Compression functions whose output length is twice the block length can be obtained by combining two types of the above compression functions (for details, see [MenOorVan96]).

Real Hash Functions. Most cryptographic hash functions used in practice today do not rely on other cryptographic primitives such as block ciphers. They are derived from custom-designed compression functions by applying Merkle–Damgård's construction. The functions are especially designed for the purpose of hashing, with performance efficiency in mind.

In [Rivest90], Rivest proposed MD4, which is algorithm number 4 in a family of hash algorithms. MD4 was designed for software implementation on a 32-bit processor. The MD4 algorithm is not strong enough, as early attacks showed. However, the design principles of the MD4 algorithm were subsequently used in the construction of hash functions. These functions are often called the MD4 family. The family contains the most popular hash functions in use today, such as MD5, SHA-1 and RIPEMD-160. The hash values of MD5 are 128 bits long, those of RIPEMD-160 and SHA-1 160 bits. All of these hash functions are iterative hash functions; they are constructed with Merkle–Damgård's method. The compression rate of the underlying compression functions is 512 bits.

SHA-1 is included in the Secure Hash Standard FIPS 180 of NIST ([RFC 1510]; [RFC 3174]). It is an improvement of SHA-0, which turned out to have a weakness. The standard was updated in 2002 ([FIPS 180-2]). Now

---

[14] The acronym MDC is explained in Section 2.2.3 below.

[15] For example, we can take $g(y) = y$, if the block length of E is equal to the key length, or, more generally, we can compute r key bits $g(y)$ from y by using a Boolean function.

it includes additional algorithms, referred to as SHA-2, that produce 256-bit, 384-bit and 512-bit outputs.

Since no rigorous mathematical proofs for the security of these hash functions exist, there is always the chance of a surprise attack. For example, the MD5 algorithm is very popular, but there have been very successful attacks.

In 1996, Dobbertin detected collisions $(v_0\|m, v_0\|m')$ of the underlying compression function, where v_0 is a common 128-bit string and m, m' are distinct 512-bit messages ([Dobbertin96a]; [Dobbertin96]). Dobbertin's v_0 is different from the initial value that is specified for MD5 in the Merkle–Damgård iteration. Otherwise, the collision would have immediately implied a collision of MD5 (note that the same length block is appended to both messages). Already in 1993, den Boer and Bosselaers had detected collisions of MD5's compression function. Their collisions $(v_0\|m, v_0'\|m)$ were made up of distinct initial values v_0 and the same message m. Thus, they did not fit Merkle–Damgård's method and were sometimes called pseudocollisions.

Attacks by the Chinese researchers Wang, Feng, Lai and Yu showed that MD5 can no longer be considered collision-resistant. In August 2004, these researchers published collisions for the hash functions MD4, MD5, HAVAL-128 and RIPEMD-128 ([WanFenLaiYu04]). Klima published an algorithm which works for any initial value and computes collisions of MD5 on a standard PC within a minute ([Klima06]). The collision resistance of MD5 is really broken.

Moreover, in February 2005, Wang, Yin and Yu cast serious doubts on the security of SHA-1 ([WanYinYu05]). They announced that they had found an algorithm which computes collisions with 2^{69} hash operations. This is much less than the expected 2^{80} steps of the brute-force birthday attack. With current technology, 2^{69} steps are still on the far edge of feasibility. For example, the RC5-64 Challenge finished in 2002. A worldwide network of Internet users was able to figure out a 64-bit RC5 key by a brute-force search. The search took almost 5 years, and more than 300,000 users participated (see [RSALabs]; [DistributedNet]).

All of these attacks are against collision resistance, and they are relevant for digital signatures. They are not attacks against second-pre-image resistance or the one-way property. Therefore, applications like HMAC (see Section 2.2.3), whose security is based on these properties, are not yet affected. No doubt is cast today on the collision resistance of the SHA-2 algorithms, such as SHA-256, SHA-384 and SHA-512.

The progress in attacking the collision resistance of widely used hash functions suggested the development of a new, future-proof standard hash algorithm. In November 2007, the National Institute of Standards and Technology started a public selection process for a new cryptographic hash algorithm – the *Secure Hash Algorithm-3*, SHA-3 for short ([NIST13]). SHA-3 was to be selected in a public competition, just as the Advanced Encryption Standard was (see Section 2.1.4).

In response to the call for candidate algorithms, 64 proposals were submitted from researchers around the world. The competition proceeded in three rounds; 51 candidates were selected for the first round, and 14 candidates for the second round. Five candidates were chosen for the third and final round, namely BLAKE, Grøstl, JH, Keccak and Skein. The cryptographic community supported the selection process on a large scale. Cryptanalysis and performance studies were published. NIST held a candidate conference in each round to obtain public feedback. In October 2012, NIST announced Keccak as the Secure Hash Algorithm-3. Keccak was designed by a team of cryptographers from Belgium and Italy: Bertoni, Daemen, Peeters and Van Assche.

The Keccak SHA-3 proposal ([BerDaePeeAss11b]) is based on the sponge construction (see page 34), with a fixed permutation

$$f : \{0,1\}^{r+c} \longrightarrow \{0,1\}^{r+c},$$

where f is the Keccak-f permutation with the maximum width $b = r + c = 1600$ (see page 36). Four output sizes are supported: 224, 256, 384 and 512 bits. The bit rate r varies according to the output size: 1152, 1088, 832 and 576 bits, respectively. Thus, the bit rate r decreases and the capacity $c = 1600 - r$ increases with growing output size.

2.2.3 Data Integrity and Message Authentication

Modification Detection Codes. Cryptographic hash functions are also known as *message digest functions*, and the hash value $h(m)$ of a message m is called the *digest* or *fingerprint* or *thumbprint* of m. [16] The hash value $h(m)$ is indeed a "fingerprint" of m. It is a very compact representation of m, and, as an immediate consequence of second-pre-image resistance, this representation is practically unique. Since it is computationally infeasible to obtain a second message m' with $m \neq m'$ and $h(m') = h(m)$, a different hash value would result, if the message m were altered in any way.

This implies that a cryptographic hash function can be used to control the integrity of a message m. If the hash value of m is stored in a secure place, a modification of m can be detected by calculating the hash value and comparing it with the stored value. Therefore, hash functions are also called *modification detection codes* (MDCs).

Let us consider an example. If you install a new root certificate in your Internet browser, you have to make sure (among other things) that the source of the certificate is the one you think it is and that your copy of the certificate was not modified. You can do this by checking the certificate's thumbprint. For this purpose, you can get the fingerprint of the root certificate from the issuing certification authority's webpage or even on real paper by ordinary mail (certificates and certification authorities are discussed in Section 4.1.5).

---

[16] MD5 is a "message digest function".

Message Authentication Codes. A very important application of hash functions is message authentication, which means to authenticate the origin of the message. At the same time, the integrity of the message is guaranteed. If hash functions are used for message authentication, they are called *message authentication codes*, or MACs for short.

MACs are the standard symmetric technique for message authentication and integrity protection and widely used, for example, in protocols such as SSL/TLS ([RFC 5246]) and IPSec. They depend on secret keys shared between the communicating parties. In contrast to digital signatures, where only one person knows the secret key and is able to generate the signature, each of the two parties can produce the valid MAC for a message.

Formally, the secret keys k are used to parameterize hash functions. Thus, MACs are families of hash functions

$$(h_k : \{0,1\}^* \longrightarrow \{0,1\}^n)_{k \in K}.$$

MACs may be derived from block ciphers or from cryptographic hash functions. We describe two methods to obtain MACs.

The standard method to convert a cryptographic hash function into a MAC is called *HMAC*. It is published in [RFC 2104] (and [FIPS 198], [ISO/IEC 9797-2]) and can be applied to a hash function h that is derived from a compression function f by using Merkle–Damgård's method. You can take as h, for example, MD5, SHA-1 or RIPEMD-160 (see Section 2.2.2).

We have to assume that the compression rate of f and the length of the hash values are multiples of 8, so we can measure them in bytes. We denote by r the compression rate of f in bytes. The secret key k can be of any length, up to r bytes. By appending zero bytes, the key k is extended to a length of r bytes (e.g., if k is of length 20 bytes and $r = 64$, then k will be appended with 44 zero bytes 0x00).

Two fixed and distinct strings ipad and opad are defined (the 'i' and 'o' are mnemonics for inner and outer):

$$\text{ipad} := \text{the byte 0x36 repeated } r \text{ times},$$

$$\text{opad} := \text{the byte 0x5C repeated } r \text{ times}.$$

The keyed hash value HMAC of a message m is calculated as follows:

$$\text{HMAC}(k, m) := h((k \oplus \text{opad}) \| h((k \oplus \text{ipad}) \| m)).$$

The hash function h is applied twice in order to guarantee the security of the MAC. If we apply h only once and define $\text{HMAC}(k, m) := h((k \oplus \text{ipad}) \| m)$, an adversary Eve could take a valid MAC value, modify the message m and compute the valid MAC value of the modified message, without knowing the secret key. For example, Eve may take any message m' and compute the hash value v of m' by applying Merkle–Damgård's iteration with $\text{HMAC}(k, m) =$

$h((k \oplus \text{ipad}) \| m)$ as initial value v_0. Before iterating, Eve appends the padding bits and the additional length block to m'. She does not store the length of m' into the length block, but the length of $\tilde{m} \| m'$, where $\tilde{m}$ is the padded message m (including the length block for m, see Section 2.2.2). Then v is the MAC of the extended message $\tilde{m} \| m'$, and Eve has computed it without knowing the secret key k. This problem is called the *length extension problem of iterated hash functions*. Applying the hash function twice prevents the length extension attack.

MACs can also be constructed from block ciphers. The most important construction is CBC-MAC. Let E be the encryption function of a block cipher, such as DES or AES. Then, with k a secret key, the MAC value for a message m is the last ciphertext block when encrypting m, with E in the Cipher-Block Chaining Mode CBC and key k (see Section 2.1.5). We need an initialization vector IV for CBC. For encryption purposes, it is important not to use the same value twice. Here, the IV is fixed and typically set to $0 \ldots 0$. If the block length of E is n, then m is split into blocks of length n, $m_1 \| m_2 \| \ldots \| m_l$ (pad out the last block, if necessary, for example, by appending zeros), and we compute

$$c_0 := IV,$$
$$c_i := E(k, m_i \oplus c_{i-1}), \ i = 1, \ldots, l,$$
$$\text{CBC-MAC} := c_l.$$

We have to assume that all messages m are of a fixed length. Otherwise, the construction is not secure (see Exercise 8).

Sometimes, the output of the CBC-MAC function is taken only to be a part of the last block. There are various standards for CBC-MAC, for example, [FIPS 113] and [ISO/IEC 9797-1]. A comprehensive discussion of hash functions and MACs can be found in [MenOorVan96].

2.2.4 Hash Functions as Random Functions

A random function would be the perfect cryptographic hash function h. Random means that for all messages m, each of the n bits of the hash value $h(m)$ is determined by tossing a coin. Such a perfect cryptographic hash function is also called a *random oracle*[17]. Unfortunately, it is obvious that a perfect random oracle cannot be implemented. To determine only the hash values for all messages of fixed length l would require exponentially many $(n \cdot 2^l)$ coin tosses and storage of all the results, which is clearly impossible.

Nevertheless, it is a design goal to construct hash functions which approximate random functions. It should be computationally infeasible to distinguish the hash function from a truly random function. Recall that there

---

[17] Security proofs in cryptography sometimes rely on the assumption that the hash function involved is a random oracle. An example of such a proof is given in Section 3.3.5

is a similar design goal for symmetric encryption algorithms. The ciphertext should appear random to the attacker. That is the reason why we hoped in Section 2.2.2 that a good block cipher induces a good compression function for a good hash function.

If we assume that the designers of a hash function h have done a good job and h comes close to a random oracle, then we can use h as a generator of pseudorandom bits. Therefore, we often see popular cryptographic hash functions such as SHA-1 or MD5 as sources of pseudorandomness. For example, in the Transport Layer Security (TLS) protocol ([RFC 5246]), also known as Secure Socket Layer (SSL), the client and the server agree on a shared 48-bit master secret, and then they derive further key material (for example, the MAC keys, encryption keys and initialization vectors) from this master secret by using a pseudorandom function. The pseudorandom function of TLS is based on the HMAC construction, with the hash functions SHA-1 or MD5.

Exercises

1. The following algorithm computes the round keys k_i, $i = 1, \ldots, 16$, for DES from the 64-bit key k. Only 56 of the 64 bits are used and permutated. This is done by a map PC1. The result PC1(k) is divided into two halves, C_0 and D_0, of 28 bits.

   Algorithm 2.16.
   bitString *DESKeyGenerator*(bitString k)
   1   $(C_0, D_0) \leftarrow \mathrm{PC1}(k)$
   2   for $i \leftarrow 1$ to 16 do
   3       $(C_i, D_i) \leftarrow (\mathrm{LS}_i(C_{i-1}), \mathrm{LS}_i(D_{i-1}))$
   4       $k_i \leftarrow \mathrm{PC2}(C_i, D_i)$
   5   return $k_1 \ldots k_{16}$

   Here LS_i is a cyclic left shift by one position if $i = 1, 2, 9$ or 16, and by two positions otherwise. The maps

   $$\mathrm{PC1} : \{0,1\}^{64} \longrightarrow \{0,1\}^{56}, \ \mathrm{PC2} : \{0,1\}^{56} \longrightarrow \{0,1\}^{48}$$

   are defined by Table 2.1. The tables PC1 and PC2 are read line by line and describe how to get the images, i.e.,

   $$\mathrm{PC1}(x_1, \ldots, x_{64}) = (x_{57}, x_{49}, x_{41}, \ldots, x_{12}, x_4),$$
   $$\mathrm{PC2}(x_1, \ldots, x_{56}) = (x_{14}, x_{17}, x_{11}, \ldots, x_{29}, x_{32}).$$

   The bits 8, 16, 24, 32, 40, 48, 56 and 64 of k are not used. They are defined in such a way that odd parity holds for each byte of k. A key k is defined to be *weak* if $k_1 = k_2 = \ldots = k_{16}$.
   Show that exactly four weak keys exist, and determine these keys.

		PC1							PC2			
57	49	41	33	25	17	9	14	17	11	24	1	5
1	58	50	42	34	26	18	3	28	15	6	21	10
10	2	59	51	43	35	27	23	19	12	4	26	8
19	11	3	60	52	44	36	16	7	27	20	13	2
63	55	47	39	31	23	15	41	52	31	37	47	55
7	62	54	46	38	30	22	30	40	51	45	33	48
14	6	61	53	45	37	29	44	49	39	56	34	53
21	13	5	28	20	12	4	46	42	50	36	29	32

Table 2.1: The definition of PC1 and PC2.

2. In this exercise, $\overline{x}$ denotes the bitwise complement of a bit string x. Let $\mathrm{DES} : \{0,1\}^{64} \times \{0,1\}^{56} \longrightarrow \{0,1\}^{64}$ be the DES function.

   a. Show that $\mathrm{DES}(\overline{k},\overline{x}) = \overline{\mathrm{DES}(k,x)}$, for $k \in \{0,1\}^{56}, x \in \{0,1\}^{64}$.
   b. Let $(m, \mathrm{DES}_k(m))$ be a plaintext–ciphertext pair. We try to find the unknown key k by an exhaustive key search. Show that the number of encryptions we have to compute can be reduced from 2^{56} to 2^{55} if the pair $(\overline{m}, \mathrm{DES}_k(\overline{m}))$ is also known.

3. We consider AES encryption.
   a. We modify one byte in a column of the state matrix. Show that all four bytes of the column are affected if MixColumn is applied.
   b. After how many rounds are all state bytes affected if one byte in the plaintext is modified?

4. We refer to the notation in the description of the output feedback mode on page 28. The key stream in the output feedback mode is periodic: there exists an $i \in \mathbb{N}$ such that $x_i = x_1$. The lowest positive integer with this property is called the *period* of the key stream. Assume that the encryption algorithm E_k is a randomly chosen function from the set of all permutations of $\{0,1\}^n$. Show that the average period of the key stream is $2^{n-1} + 1/2$ if the initial value $x_1 \in \{0,1\}^n$ is chosen at random.

5. Let p be a large prime number such that $q := (p-1)/2$ is also a prime. Let G_q be the subgroup of order q in $\mathbb{Z}_p^*$ (see Appendix A.4). Let g and h be randomly chosen generators of G_q. We assume that it is infeasible to compute discrete logarithms in G_q. Show that

$$ f : \{0, \ldots, q-1\}^2 \longrightarrow G_q, \ (x,y) \longmapsto g^x h^y $$

can be used to obtain a collision-resistant compression function.

6. Consider the SHA-3 algorithm. The round constant $RC[0]$ of Keccak-f is defined by $RC[0][0][0][63] = 1$ and $RC[0][x][y][z] = 0$ in all other cases. Assume that the message to be hashed consists of 1600 zero bits. Describe the state after θ has been applied in the second round.

7. Let $h : \{0,1\}^* \longrightarrow \{0,1\}^n$ be a cryptographic hash function. We assume that the hash values are uniformly distributed, i.e., each value $v \in \{0,1\}^n$ has the same probability $1/2^n$. How many steps do you expect until you succeed with the brute-force attacks against the one-way property and second pre-image resistance?

8. Explain why CBC-MAC is not secure in the following cases:
   a. If the initial value IV is chosen at random (and not the 0-bit string) and appended at the end of the MAC-code.
   b. If the MAC-code consists of all blocks and not only of the last one.
   c. If variable-length messages are allowed.

3. Public-Key Cryptography

The basic idea of public-key cryptography are public keys. Each person's key is separated into two parts: a public key for encryption available to everyone and a secret key for decryption which is kept secret by the owner. In this chapter we introduce the concept of public-key cryptography. Then we discuss some of the most important examples of public-key cryptosystems, such as the RSA, ElGamal and Rabin cryptosystems. These all provide encryption and digital signatures.

3.1 The Concept of Public-Key Cryptography

Classical symmetric cryptography provides a secure communication channel to each pair of users. In order to establish such a channel, the users must agree on a common secret key. After establishing a secure communication channel, the secrecy of a message can be guaranteed. Symmetric cryptography also includes methods to detect modifications of messages and methods to verify the origin of a message. Thus, confidentiality and integrity can be accomplished using secret key techniques.

However, public key techniques have to be used for a secure distribution of secret keys, and at least some important forms of authentication and non-repudiation also require public-key methods, such as digital signatures. A digital signature should be the digital counterpart of a handwritten signature. The signature must depend on the message to be signed and a secret known only to the signer. An unbiased third party should be able to verify the signature without access to the signer's secret.

In a public-key encryption scheme, the communication partners do not share a secret key. Each user has a pair of keys: a *secret key sk* known only to him and a *public key pk* known to everyone.

Suppose Bob has such a key pair (pk, sk) and Alice wants to encrypt a message m for Bob. Like everyone else, Alice knows Bob's public key pk. She computes the ciphertext $c = E(pk, m)$ by applying the encryption function E with Bob's public key pk. As before, we denote encrypting with a fixed key pk by E_{pk}, i.e., $E_{pk}(m) := E(pk, m)$. Obviously, the encryption scheme can only be secure if it is practically infeasible to compute m' from $c = E_{pk}(m)$.

But how can Bob then recover the message m from the ciphertext c? This is where Bob's secret key is used. The encryption function E_{pk} must have the property that the pre-image m of the ciphertext $c = E_{pk}(m)$ is easy to compute using Bob's secret key sk. Since only Bob knows the secret key, he is the only one who can decrypt the message. Even Alice, who encrypted the message m, would not be able to get m from $E_{pk}(m)$ if she lost m. Of course, efficient algorithms must exist to perform encryption and decryption.

We summarize the requirements of public-key cryptography. We are looking for a family of functions $(E_{pk})_{pk \in PK}$ such that each function E_{pk} is computable by an efficient algorithm. It should be practically infeasible to compute pre-images of E_{pk}. Such families $(E_{pk})_{pk \in PK}$ are called *families of one-way functions* or *one-way functions* for short. Here, PK denotes the set of available public keys.[1] For each function E_{pk} in the family, there should be some information sk to be kept secret which enables an efficient computation of the inverse of E_{pk}. This secret information is called the *trapdoor information*. One-way functions with this property are called *trapdoor functions*.

In 1976, Diffie and Hellman published the idea of public-key cryptography in their famous paper "New Directions in Cryptography" ([DiffHel76]). They introduced a public-key method for key agreement which is in use to this day. In addition, they described how digital signatures would work, and proposed, as an open question, the search for such a function. The first public-key cryptosystem that could function as both a key agreement mechanism and as a digital signature was the RSA cryptosystem published in 1978 ([RivShaAdl78]). RSA is named after the inventors: Rivest, Shamir and Adleman. The RSA cryptosystem provides encryption and digital signatures and is the most popular and widely used public-key cryptosystem today. We shall describe the RSA cryptosystem in Section 3.3. It is based on the difficulty of factoring large numbers, which enables the construction of one-way functions with a trapdoor. Another basis for one-way functions is the difficulty of extracting discrete logarithms. These two problems from number theory are the foundations of most public-key cryptosystems used today.

Each participant in a public-key cryptosystem needs his personal key $k = (pk, sk)$, consisting of a public and a secret (also called private) part. To guarantee the security of the cryptosystem, it must be infeasible to compute the secret key sk from the public key pk, and it must be possible to randomly choose the keys k from a huge parameter space. An efficient algorithm must be available to perform this random choice. If Bob wants to participate in the cryptosystem, he randomly selects his key $k = (pk, sk)$, keeps sk secret and publishes pk. Now everyone can use pk in order to encrypt messages for Bob.

To discuss the basic idea of digital signatures, we assume that we have a family $(E_{pk})_{pk \in PK}$ of trapdoor functions and that each function E_{pk} is

---

[1] A rigorous definition of one-way functions is given in Definition 6.12.

bijective. Such a family of trapdoor permutations can be used for digital signatures. Let pk be Alice's public key. To compute the inverse E_{pk}^{-1} of E_{pk}, the secret key sk of Alice is required. So Alice is the only one who is able to do this. If Alice wants to sign a message m, she computes $E_{pk}^{-1}(m)$ and takes this value as signature s of m. Everyone can verify Alice's signature s by using Alice's public key pk and computing $E_{pk}(s)$. If $E_{pk}(s) = m$, Bob is convinced that Alice really signed m because only Alice was able to compute $E_{pk}^{-1}(m)$.

An important straightforward application of public-key cryptosystems is the distribution of session keys. A session key is a secret key used in a classical symmetric encryption scheme to encrypt the messages of a single communication session. If Alice knows Bob's public key, then she may generate a session key, encrypt it with Bob's public key and send it to Bob. Digital signatures are used to guarantee the authenticity of public keys by certification authorities. The certification authority signs the public key of each user with her secret key. The signature can be verified with the public key of the certification authority. Cryptographic protocols for user authentication and advanced cryptographic protocols, like bit commitment schemes, oblivious transfer and zero-knowledge interactive proof systems, have been developed. Today they are fundamental to Internet communication and electronic commerce.

Public-key cryptography is also important for theoretical computer science: theories of security were developed and the impact on complexity theory should be mentioned.

3.2 Modular Arithmetic

In this section, we give a brief overview of the modular arithmetic necessary to understand the cryptosystems we discuss in this chapter. Details can be found in Appendix A.

3.2.1 The Integers

Let $\mathbb{Z}$ denote the ordered set $\{\ldots, -3, -2, -1, 0, 1, 2, 3, \ldots\}$. The elements of $\mathbb{Z}$ are called integers or numbers. The integers greater than 0 are called natural numbers and are denoted by $\mathbb{N}$. The sum $n+m$ and the product $n \cdot m$ of integers are defined. Addition and multiplication satisfy the axioms of a commutative ring with a unit element. We call $\mathbb{Z}$ the ring of integers.

Addition, Multiplication and Exponentiation. Efficient algorithms exist for the addition and multiplication of numbers.[2] An efficient algorithm is an algorithm whose running time is bounded by a polynomial in the size of its

---

[2] Simplifying slightly, we only consider non-negative integers, which is sufficient for all our purposes.

input. The size of a number is the length of its binary encoding, i.e., the size of $n \in \mathbb{N}$ is equal to $\lfloor \log_2(n) \rfloor + 1$.[3] It is denoted by $|n|$. Let $a, b \in \mathbb{N}$, $a, b \leq n$, and $k := \lfloor \log_2(n) \rfloor + 1$. The number of bit operations for the computation of $a + b$ is $O(k)$, whereas for the multiplication $a \cdot b$ it is $O(k^2)$. Multiplication can be improved to $O(k \log_2(k))$ if a fast multiplication algorithm is used.

Exponentiation is also an operation that occurs often. The square-and-multiply method (Algorithm A.27) yields an efficient algorithm for the computation of a^n. It requires at most $2 \cdot |n|$ modular multiplications. We compute, for example, a^{16} by $(((a^2)^2)^2)^2$, which are four squarings, in contrast to the 15 multiplications that are necessary for the naive method. If the exponent is not a power of 2, the computation has to be modified a little. For example, $a^{14} = ((a^2 a)^2 a)^2$ is computable by three squarings and two multiplications, instead of by 13 multiplications.

Division with Remainder. If m and n are integers, $m \neq 0$, we can divide n by m with a remainder. We can write $n = q \cdot m + r$ in a unique way such that $0 \leq r < \mathrm{abs}(m)$.[4] The number q is called the *quotient* and r is called the *remainder* of the division. They are unique. Often we denote r by $a \bmod b$.

An integer m divides an integer n if n is a multiple of m, i.e., $n = mq$ for an integer q. We say, m is a *divisor* or *factor* of n. The *greatest common divisor* $\gcd(m, n)$ of numbers $m, n \neq 0$ is the largest positive integer dividing m and n. $\gcd(0, 0)$ is defined to be zero. If $\gcd(m, n) = 1$, then m is called *relatively prime to* n, or *prime to* n for short.

The *Euclidean algorithm* computes the greatest common divisor of two numbers and is one of the oldest algorithms in mathematics:

Algorithm 3.1.
   int gcd(int a, b)
  1   while $b \neq 0$ do
  2      $r \leftarrow a \bmod b$
  3      $a \leftarrow b$
  4      $b \leftarrow r$
  5   return $\mathrm{abs}(a)$

The algorithm computes $\gcd(a, b)$ for $a \neq 0$ and $b \neq 0$. It terminates because the non-negative number r decreases in each step. Note that $\gcd(a, b)$ is invariant in the while loop, because $\gcd(a, b) = \gcd(b, a \bmod b)$. In the last step, the remainder r becomes 0 and we get $\gcd(a, b) = \gcd(a, 0) = \mathrm{abs}(a)$.

Primes and Factorization. A natural number $p \neq 1$ is a *prime number*, or simply a *prime*, if 1 and p are the only divisors of p. If a number $n \in \mathbb{N}$ is not prime, it is called *composite*. Primes are essential for setting up the public-key cryptosystems we describe in this chapter. Fortunately, there are very fast algorithms (so-called probabilistic primality tests) for finding – at

---

[3] $\lfloor x \rfloor$ denotes the greatest integer less than or equal to x.
[4] $\mathrm{abs}(m)$ denotes the absolute value of m.

least with a high probability (though not with mathematical certainty) – the correct answer to the question whether a given number is a prime or not (see Section A.10). Primes are the basic building blocks for numbers. This statement is made precise by the *Fundamental Theorem of Arithmetic*.

Theorem 3.2 *(Fundamental Theorem of Arithmetic). Let $n \in \mathbb{N}, n \geq 2$. There exist pairwise distinct primes $p_1, \ldots, p_k$ and exponents $e_1, \ldots, e_k \in \mathbb{N}, e_i \geq 1, i = 1, \ldots, k,$ such that*

$$n = \prod_{i=1}^{k} p_i^{e_i}.$$

The primes $p_1, \ldots, p_k$ and exponents $e_1, \ldots, e_k$ are unique.

It is easy to multiply two numbers, but the design of an efficient algorithm for calculating the prime factors of a number is an old mathematical problem. For example, this was already studied by the famous mathematician C.F. Gauss about 200 years ago (see, e.g., [Riesel94] for details on Gauss's factoring method). However, to this day, we do not have a practical algorithm for factoring extremely large numbers.

3.2.2 The Integers Modulo n

The Residue Class Ring Modulo n. Let n be a positive integer. Let a and b be integers. Then a is *congruent to b modulo n*, written $a \equiv b \bmod n$, if a and b leave the same remainder when divided by n or, equivalently, if n divides $a - b$. We obtain an equivalence relation. The equivalence class of a is the set of all numbers congruent to a. It is denoted by $[a]$ and called the *residue class* of a modulo n. The set of residue classes $\{[a] \mid a \in \mathbb{Z}\}$ is called the set of *integers modulo n* and is denoted by $\mathbb{Z}_n$.

Each number is congruent to a unique number r in the range $0 \leq r \leq n - 1$. Therefore the numbers $0, \ldots, n - 1$ form a set of representatives of the elements of $\mathbb{Z}_n$. We call them the *natural representatives*. Sometimes we identify the residue classes with their natural representatives, i.e., we identify $\mathbb{Z}_n$ with $\{0, 1, \ldots, n - 1\}$.

The equivalence relation is compatible with addition and multiplication in $\mathbb{Z}$, i.e., if $a \equiv a' \bmod n$ and $b \equiv b' \bmod n$, then $a + b \equiv (a' + b') \bmod n$ and $a \cdot b \equiv (a' \cdot b') \bmod n$. Consequently, addition and multiplication on $\mathbb{Z}$ induce an addition and multiplication on $\mathbb{Z}_n$:

$$[a] + [b] := [a + b],$$
$$[a] \cdot [b] := [a \cdot b].$$

Addition and multiplication satisfy the axioms of a commutative ring with a unit element. We call $\mathbb{Z}_n$ the *residue class ring modulo n*.

Although we can calculate in $\mathbb{Z}_n$ as in $\mathbb{Z}$, there are some important differences. First, we do not have an ordering of the elements of $\mathbb{Z}_n$ which is compatible with addition and multiplication. For example, if we assume that we have such an ordering in $\mathbb{Z}_5$ and that $[0] < [1]$, then $[0] < [1]+[1]+[1]+[1]+[1] = [0]$, which is a contradiction. A similar calculation shows that the assumption $[1] < [0]$ also leads to a contradiction.

Another fact is that $[a] \cdot [b]$ can be $[0]$ for $[a] \neq [0]$ and $[b] \neq [0]$. For example, $[2] \cdot [3] = [0]$ in $\mathbb{Z}_6$. Such elements – $[a]$ and $[b]$ – are called *zero divisors*.

The Prime Residue Class Group Modulo n. In $\mathbb{Z}$, elements a and b satisfy $a \cdot b = 1$ if and only if both a and b are equal to 1, or both are equal to -1. We say that 1 and -1 have multiplicative inverse elements. In $\mathbb{Z}_n$, this can happen more frequently. In $\mathbb{Z}_5$, for example, every class different from $[0]$ has a multiplicative inverse element. Elements in a ring which have multiplicative inverses are called *units* and form a group under multiplication.

An element $[a]$ in $\mathbb{Z}_n$ has the multiplicative inverse element $[b]$, if $ab \equiv 1 \bmod n$ or, equivalently, n divides $1 - ab$. This means we have an equation $nm + ab = 1$, with suitable m. The equation implies that $\gcd(a, n) = 1$. On the other hand, if numbers a, n with $\gcd(a, n) = 1$ are given, an equation $nm + ab = 1$, with suitable b and m, can be derived from a and n by the extended Euclidean algorithm (Algorithm A.6). Hence, $[a]$ is a unit in $\mathbb{Z}_n$ and the inverse element is $[b]$. Thus, the elements of the group of units of $\mathbb{Z}_n$ are represented by the numbers prime to n.

$$\mathbb{Z}_n^* := \{[a] \mid 1 \leq a \leq n - 1 \text{ and } \gcd(a, n) = 1\}$$

is called the *prime residue class group modulo n*. The number of elements in $\mathbb{Z}_n^*$ (also called the order of $\mathbb{Z}_n^*$) is the number of integers in the interval $[1, n-1]$ which are prime to n. This number is denoted by $\varphi(n)$. The function φ is called the *Euler phi function* or the *Euler totient function*.

For every element a in a finite group G, we have $a^{|G|} = e$, with e being the neutral element of G.[5] This is an elementary and easy to prove feature of finite groups. Thus, we have for a number a prime to n

$$a^{\varphi(n)} \equiv 1 \bmod n.$$

This is called *Euler's Theorem* or, if n is a prime, *Fermat's Theorem*.

If $\prod_{i=1}^{k} p_i^{e_i}$ is the prime factorization of n, then the Euler phi function can be computed by the formula (see Corollary A.31)

$$\varphi(n) = n \prod_{i=1}^{k} \left(1 - \frac{1}{p_i}\right).$$

If p is a prime, then every integer in $\{1, \ldots, p - 1\}$ is prime to p. Therefore, every element in $\mathbb{Z}_p \setminus \{0\}$ is invertible and $\mathbb{Z}_p$ is a field. The group of units

---

[5] $|G|$ denotes the number of elements of G (called the *cardinality* or *order* of G).

$\mathbb{Z}_p^*$ is a cyclic group with $p - 1$ elements, i.e., $\mathbb{Z}_p^* = \{g, g^2, \ldots, g^{p-1} = [1]\}$ for some $g \in \mathbb{Z}_{p-1}$. Such a g is called a *generator* of $\mathbb{Z}_p^*$. Generators are also called *primitive elements modulo p* or *primitive roots modulo p* (see Definition A.38).

We can now introduce three functions which may be used as one-way functions and which are hence very important in cryptography.

Discrete Exponentiation. Let p denote a prime number and g be a primitive root in $\mathbb{Z}_p$.

$$\text{Exp} : \mathbb{Z}_{p-1} \longrightarrow \mathbb{Z}_p^*, \ x \longmapsto g^x$$

is called the discrete exponential function. Exp is a homomorphism from the additive group $\mathbb{Z}_{p-1}$ to the multiplicative group $\mathbb{Z}_p^*$, i.e., $\text{Exp}(x + y) = \text{Exp}(x) \cdot \text{Exp}(y)$, and Exp is bijective. In other words, Exp is an isomorphism of groups. This follows immediately from the definition of a primitive root. The inverse function

$$\text{Log} : \mathbb{Z}_p^* \longrightarrow \mathbb{Z}_{p-1}$$

is called the *discrete logarithm function*. We use the adjective "discrete" to distinguish Exp and Log for finite groups from the classical functions defined for the real numbers.

Exp is efficiently computable, for example by the square-and-multiply method (Algorithm A.27), whereas no efficient algorithm is known to exist for computing the inverse function Log for sufficiently large primes p. This statement is made precise by the discrete logarithm assumption (see Definition 6.1).

Modular Powers. Let n be the product of two distinct primes p and q, and let e be a positive integer prime to $\varphi(n)$, $e \not\equiv 1 \bmod \varphi(n)$.

$$\text{RSA}_{n,e} : \mathbb{Z}_n \longrightarrow \mathbb{Z}_n, \ x \longmapsto x^e$$

is called the *RSA function*.

RSA function values can be efficiently computed by applying the square-and-multiply algorithm (Algorithm A.27). Conversely, it is practically infeasible to compute the pre-image x from $y = x^e$, provided the factors p and q of n are not known, i.e., $\text{RSA}_{n,e}$ is a one-way function. The one-way property depends on the computational hardness of factorization: there is no feasible algorithm for computing the factors p and q from n, if the primes p, q are very large. Therefore, it is possible to make n and e public and, at the same time, to keep p and q secret.

The RSA function is bijective and, if the primes p and q are known, it can be inverted easily.

Proposition 3.3. *Let n be the product of two distinct primes, and let e be a positive integer that is prime to $\varphi(n)$, $e \not\equiv 1 \bmod \varphi(n)$. Let d be a multiplicative inverse element of e modulo $\varphi(n)$ (note that d is also prime to $\varphi(n)$ and $\text{RSA}_{n,d}$ is defined). Then*

$$\mathrm{RSA}_{n,d} \circ \mathrm{RSA}_{n,e} = \mathrm{RSA}_{n,e} \circ \mathrm{RSA}_{n,d} = \mathrm{id}_{\mathbb{Z}_n}.$$

The proposition says that the inverse of $\mathrm{RSA}_{n,e}$ is $\mathrm{RSA}_{n,d}$. Note that $\varphi(n) = (p-1)(q-1)$ (Corollary A.31). Therefore, d can be easily computed from p and q by applying the extended Euclidean algorithm.[6] We see that the one-way function $\mathrm{RSA}_{n,e}$ is a trapdoor function, with trapdoor information d.

Proposition 3.3 is an immediate consequence of the following Proposition 3.4. This is a special version of Euler's Theorem (Proposition A.26), which implies that $x^k \equiv x^{k'} \bmod n$ if $k \equiv k' \bmod \varphi(n)$ and x is prime to n. The requirement that x is prime to n is not necessary if n is the product of distinct primes.

Proposition 3.4. *Let p, q be distinct primes and let $n - pq$. Let k, k' be positive integers such that $k \equiv k' \bmod \varphi(n)$. Then*

$$x^k \equiv x^{k'} \bmod n$$

for every $x \in \mathbb{Z}$.

Proof. Let $x \in \mathbb{Z}$. If p divides x, then $x^k \equiv x^{k'} \equiv 0 \bmod p$. If p does not divide x, then $x^{p-1} \equiv 1 \bmod p$ by Fermat's Little Theorem (see page 54; Proposition A.25), and hence

$$x^k = x^{k'+l\varphi(n)} = x^{k'+l(p-1)(q-1)} = x^{k'}\left(x^{p-1}\right)^{l(q-1)} \equiv x^{k'} \bmod p.$$

In any case, p divides $x^k - x^{k'}$. In the same way we see that q divides $x^k - x^{k'}$. Hence n divides $x^k - x^{k'}$.   □

It is now easy to derive Proposition 3.3. For every $[x] \in \mathbb{Z}_n$ we have, by Proposition 3.4,

$$([x]^e)^d = [x]^{ed} = [x]^1 = [x].$$

Modular Squares. Let p and q denote distinct prime numbers and $n = pq$.

$$\mathrm{Square} : \mathbb{Z}_n \longrightarrow \mathbb{Z}_n, \ x \longmapsto x^2$$

is called the *Square function*. Each element $y \in \mathbb{Z}_n, y \neq 0$, either has 0, 2 or 4 pre-images. If both p and q are primes $\equiv 3 \bmod 4$, then -1 is not a square modulo p and modulo q, and it easily follows that Square becomes a bijective map by restricting the domain and range to the subset of squares in $\mathbb{Z}_n^*$ (for details see Section A.7). If the factors of n are known, pre-images of Square (called *square roots*) are efficiently computable (see Proposition A.68). Again, without knowing the factors p and q, computing square roots is practically impossible for p, q sufficiently large.

---

[6] Conversely, p and q can be efficiently computed from n and d; see below.

On the Difficulty of Extracting Roots and Discrete Logarithms. Let $g, k \in \mathbb{N}$, $g, k \geq 2$, and let

$$F : \mathbb{Z} \longrightarrow \mathbb{Z}$$

denote one of the maps $x \longmapsto x^2, x \longmapsto x^k$ or $x \longmapsto g^x$. Usually these maps are considered as real functions. Efficient algorithms for the computation of values and pre-images are known. These algorithms rely heavily on the ordering of the real numbers and the fact that F is, at least piecewise, monotonic. The map F is efficiently computable by integer arithmetic with a fast exponentiation algorithm (see Section 3.2.1). If we use such an algorithm to compute F, we get the following algorithm for computing a pre-image x of $y = F(x)$.

For simplicity, we restrict the domain to the positive integers. Then all three functions are monotonic and hence injective. It is easy to find integers a and b with $a \leq x \leq b$. If $F(a) \neq y$ and $F(b) \neq y$, we call $FInvers(y, a, b)$.

Algorithm 3.5.

   int $FInvers$(int y, a, b)
  1   repeat
  2       $c \leftarrow (a + b)$ div 2
  3   if $F(c) < y$
  4   then $a \leftarrow c$
  5   else $b \leftarrow c$
  6   until $F(c) = y$
  7   return c

F is efficiently computable. The repeat-until loop terminates after $O(\log_2(|b - a|))$ steps. Hence, $FInvers$ is also efficiently computable, and we see that F considered as a function from $\mathbb{Z}$ to $\mathbb{Z}$ can easily be inverted in an efficient way.

Now consider the same maps modulo n. The function F is still efficiently computable (see above or Algorithm A.27). The algorithm $FInvers$, however, does not work in modular arithmetic. The reason is that in modular arithmetic, it does not make sense to ask in line 3 whether $F(c) < y$. The ring $\mathbb{Z}_n$ has no order which is compatible with the arithmetic operations. The best we could do to adapt the algorithm above for modular arithmetic is to test all elements c in $\mathbb{Z}_n$ until we reach $F(c) = y$. However, this leads to an algorithm with exponential running time (n is exponential in the binary length $|n|$ of n).

If the factors of n are kept secret, no efficient algorithm is known today to invert RSA and Square. The same holds for Exp. It is widely believed that no efficient algorithms exist to compute the pre-images. No one could prove, however, this statement in the past. These assumptions are defined in detail in Chapter 6. They are the basis for security proofs in public-key cryptography (Chapters 9 and 11).

3.3 RSA

The RSA cryptosystem is based on facts from elementary number theory which have been known for more than 250 years. To set up an RSA cryptosystem, we have to multiply two very large primes and make their product n public. n is part of the public key, whereas the factors of n are kept secret and are used as the secret key. The basic idea is that the factors of n cannot be recovered from n. The security of RSA depends on the tremendous difficulty of factoring.

We now describe in detail how RSA works. We discuss key generation, encryption and decryption as well as digital signatures.

3.3.1 Key Generation and Encryption

Key Generation. Each user Bob of the RSA cryptosystem has his own public and secret keys. To generate his key pair, Bob proceeds in three steps:

1. He randomly chooses large distinct primes p and q, and computes $n = p \cdot q$.
2. He chooses an integer $e, 1 < e < \varphi(n)$, that is prime to $\varphi(n)$. Then he publishes (n, e) as his public key.
3. He computes d with $ed \equiv 1 \bmod \varphi(n)$. (n, d) is his secret key.

Recall that $\varphi(n) = (p-1)(q-1)$ (Corollary A.31). The numbers n, e and d are referred to as the *modulus*, *encryption* and *decryption exponents*, respectively. The decryption exponent d can be computed with the extended Euclidean algorithm (Algorithm A.6). To decrypt a ciphertext or to generate a digital signature, Bob only needs his decryption exponent d, he does not need to know the primes p and q. Nevertheless, knowing p and q can be helpful for him (for example, to speed up decryption, see below). There is an efficient algorithm which Bob can use to recover, with very high probability, the primes from n, e and d (see Exercise 4).

An adversary should not have the slightest idea what Bob's primes are. Therefore, the primes are generated randomly. This reduces the risk that the selected primes are insecure to a negligibly small magnitude. There exist insecure prime numbers p and q, where $n = pq$ can be factorized very efficiently, for example Mersenne primes (i.e., primes of the form $2^k - 1$). We will discuss these security aspects below.

To generate the primes p and q, we first choose a large number x at random. If x is even, we replace x by $x + 1$. Then we apply a probabilistic primality test to check whether x is a prime (see Section A.10). If x is not a prime number, we repeat the procedure. We expect to test $O(\ln(x))$ numbers for primality before reaching the first prime (see Corollary A.83). The method described does not produce primes with mathematical certainty (we use a probabilistic primality test), but only with an overwhelmingly high probability, which is sufficient for all practical purposes.

At the moment, it is suggested to take 1024-bit prime numbers. No one can predict for how long such numbers will be secure, because it is difficult to predict improvements in factorization algorithms and computer technology.

The number e, as part of a key, should also be generated randomly. Whether e is prime to $\varphi(n)$ is tested with Euclid's algorithm (Algorithm A.5). In practice, the number e is often not selected randomly. Sometimes, small values or numbers of a special form are chosen for efficiency reasons (see the section on speeding up encryption on page 62). A non-random choice of e does not compromise security, but it requires some care, as Wiener's algorithm (see below) and the low-encryption-exponent attack (see Section 3.3.2) show. Choosing e as a prime between $\max(p, q)$ and $\varphi(n)$ is a way to guarantee that it will be relatively prime to $\varphi(n)$.

The RSA keys are used as parameters of RSA functions. Exponents e and e' with $e \equiv e' \bmod \varphi(n)$ yield identical RSA functions: $\mathrm{RSA}_{n,e} = \mathrm{RSA}_{n,e'}$ (Proposition 3.4). Therefore, the selection of encryption exponents e can be restricted to the range $1 < e < \varphi(n)$.

To generate numbers at random, hardware random number generators and secure pseudorandom number generators can be used. A pseudorandom generator generates a sequence of digits that look like a sequence of random digits from a short truly random seed by using a deterministic algorithm. Examples of hardware generators are the "True Random Bit Generator" of a "Trusted Platform Module" (TPM) chip ([TCG]) and quantum-mechanical sources of randomness (see, e.g., the remark on page 348). There is a wide array of literature concerning the efficient and secure generation of pseudorandom numbers (see, e.g., [MenOorVan96]). We also discuss some aspects of this subject in Chapter 8.

Encryption and Decryption. Alice encrypts messages for Bob by using Bob's public key (n, e). She can encrypt messages $m \in \{0, \ldots, n-1\}$, considered as elements of $\mathbb{Z}_n$.

Alice encrypts a plaintext $m \in \mathbb{Z}_n$ by applying the RSA function

$$\mathrm{RSA}_{n,e} : \mathbb{Z}_n \longrightarrow \mathbb{Z}_n, \ m \longmapsto m^e,$$

i.e., the ciphertext c is m^e modulo n.

Bob decrypts a ciphertext c by applying the RSA function

$$\mathrm{RSA}_{n,d} : \mathbb{Z}_n \longrightarrow \mathbb{Z}_n, \ c \longmapsto c^d,$$

where (n, d) is his secret key. Bob indeed recovers the plaintext, i.e., $m = c^d$ modulo n, because the encryption and decryption functions $\mathrm{RSA}_{n,e}$ and $\mathrm{RSA}_{n,d}$ are inverse to each other (Proposition 3.3). Encryption and decryption can be implemented efficiently (Algorithm A.27).

With the basic encryption procedure, we can encrypt bit sequences up to $k := \lfloor \log_2(n) \rfloor$ bits. If our messages are longer, we may decompose them into blocks of length k and apply the scheme described in Section 3.3.3 or a suitable mode of operation, for example the electronic codebook mode or the cipher-block chaining mode (see Section 2.1.5). The cipher feedback mode and the output feedback mode are not immediately applicable. Namely, if the initial value is not kept secret everyone can decrypt the cryptogram. See Chapter 9 for an application of the output feedback mode with RSA.

Security. An adversary knowing the factors p and q of n also knows $\varphi(n) = (p-1)(q-1)$, and then derives d from the public encryption key e using the extended Euclidean algorithm (Algorithm A.6). Thus, the security of RSA depends on the difficulty of finding the factors p and q of n, if p and q are large primes. It is widely believed that it is impossible today to factor n by an efficient algorithm if p and q are sufficiently large. This fact is known as the factoring assumption (for a precise definition, see Definition 6.9). An efficient factoring algorithm would break RSA. It is not proven whether factoring is necessary to decrypt RSA ciphertexts, but it is also believed that inverting the RSA function is intractable. This statement is made precise by the RSA assumption (see Definition 6.7).

In the construction of the RSA keys, the only inputs for the computation of the exponents e and d are $\varphi(n)$ and n. Since $\varphi(n) = (p-1)(q-1)$, we have

$$p + q = n - \varphi(n) + 1 \text{ and } p - q = \sqrt{(p+q)^2 - 4n} \text{ (if } p > q).$$

Therefore, it is easy to compute the factors of n if $\varphi(n)$ is known.

The factorization of n can be reduced to an algorithm A that computes d from n and e (see Exercise 4). The resulting factoring algorithm A' is probabilistic and of the same complexity as A. (This fact was mentioned in [RivShaAdl78]). A deterministic polynomial-time algorithm that factors n on the inputs n, e and d has been published in [CorMay07].

It is an open question as to whether an efficient algorithm for factoring can be derived from an efficient algorithm inverting RSA, i.e., an efficient algorithm that on inputs n, e and x^e outputs x. A result of Boneh and Venkatesan (see [BonVen98]) provides evidence that, for a small encryption exponent e, inverting the RSA function might be easier than factoring n. They show that an efficient factoring algorithm A which uses, as a subroutine, an algorithm for computing e-th roots – called oracle for e-th roots – can be converted into an efficient factoring algorithm B which does not call the oracle. However, they use a restricted computation model for the algorithm A – only algebraic reductions are allowed. Their result says that factoring is easy, if an efficient algorithm like A exists in the restricted computational model. The result of Boneh and Venkatesan does not expose any weakness in the RSA cryptosystem.

The decryption exponent d should be not too small. For $d < n^{1/4}$, which means that the binary length of d is less than one-quarter of the length of n, a polynomial-time algorithm to compute d has been developed by Wiener ([Wiener90]). The algorithm uses the continued fraction expansion of e/n.

Boneh and Durfee presented an improved bound. Their algorithm can compute d if $d < n^{0.292}$ ([BonDur00]). It is based on techniques for finding solutions to polynomial equations by using lattices. If the encryption exponent e is chosen randomly, then the probability that d is below these bounds is negligibly small.

Efficient factoring algorithms are known for special types of primes p and q. To give these algorithms no chance, we have to avoid such primes. First we require that the absolute value $|p - q|$ is large. This prevents the following attack: we have $(p+q)^2/4 - n = (p+q)^2/4 - pq = (p-q)^2/4$. If $|p - q|$ is small, then $(p-q)^2/4$ is also small and therefore $(p+q)^2/4$ is slightly larger than n. Thus $(p+q)/2$ is slightly larger than $\sqrt{n}$ and the following factoring method could be successful:

1. Choose successive numbers $x > \sqrt{n}$ and test whether $x^2 - n$ is a square.
2. In this case, we have $x^2 - n = y^2$. Thus $x^2 - y^2 = (x - y)(x + y) = n$, and we have found a factorization of n.

This idea for factoring numbers goes back to Fermat.

To prevent other attacks on the RSA cryptosystem, the notion of *strong primes* has been defined. A prime number is called strong if the following conditions are satisfied:

1. $p - 1$ has a large prime factor, denoted by r.
2. $p + 1$ has a large prime factor.
3. $r - 1$ has a large prime factor.

What "large" means can be derived from the attacks to prevent (see below). Strong primes can be generated by Gordon's algorithm (see [Gordon84]). If used in conjunction with a probabilistic primality test, the running time of Gordon's algorithm is only about 20% more than the time needed to generate a prime factor of the RSA modulus in the way described above. Gordon's algorithm yields a prime with high probability. The size of the resulting prime p can be controlled to guarantee a large absolute value $|p - q|$.

Strong primes are intended to prevent the $p - 1$ and the $p + 1$ factoring attacks. These are efficient if $p - 1$ or $p + 1$ have only small prime factors (see, e.g., [Forster96]; [Riesel94]). Note that, with an overwhelmingly high probability, a prime p is a strong prime if it is chosen at random and large. Moreover, strong primes do not provide more protection against factoring attacks with a modern algorithm like the number-field sieve (see [Cohen95]) than non-strong primes. Thus, the notion of strong primes has lost significance.

There is another attack which should be prevented by strong primes: decryption by iterated encryption. The idea is to repeatedly apply the encryption algorithm to the cryptogram until $c = c^{e^i}$. Then $c = \left(c^{e^{i-1}}\right)^e$, and the plaintext $m = c^{e^{i-1}}$ can be recovered. Condition 1 and 3 ensure that this attack fails, since the order of c in $\mathbb{Z}_n^*$ and the order of e in $\mathbb{Z}_{\varphi(n)}^*$ are, with high probability, very large (see Exercises 6 and 7). If p and q are chosen at random and are sufficiently large, then the probability of success of a decryption-by-iterated-encryption attack is negligible (see [MenOorVan96], p. 313). Thus, to prevent this attack, there is no compelling reason for explicitly choosing strong primes, too.

Speeding up Encryption and Decryption. The modular exponentiation algorithm is especially efficient, if the exponent has many zeros in its binary encoding. For each zero we have one less multiplication. We can take advantage of this fact by choosing an encryption exponent e with many zeros in its binary encoding. The primes 3, 17 or $2^{16} + 1$ are good examples, with only two ones in their binary encoding.

The efficiency of decryption can be improved by use of the Chinese Remainder Theorem (Theorem A.30). The receiver of the message knows the factors p and q of the modulus n. Let ϕ be the isomorphism

$$\phi : \mathbb{Z}_n \longrightarrow \mathbb{Z}_p \times \mathbb{Z}_q, \ [x] \longmapsto ([x \bmod p], [x \bmod q]).$$

Compute $c^d = \phi^{-1}(\phi(c^d)) = \phi^{-1}((c \bmod p)^d, (c \bmod q)^d)$. The computation of $(c \bmod p)^d$ and $(c \bmod q)^d$ is executed in $\mathbb{Z}_p$ and $\mathbb{Z}_q$, respectively. In $\mathbb{Z}_p$ and $\mathbb{Z}_q$ we have much smaller numbers than in $\mathbb{Z}_n$. Moreover, the decryption exponent d can be replaced by $d \bmod (p-1)$ and $d \bmod (q-1)$, respectively, since $(c \bmod p)^d \bmod p = (c \bmod p)^{d \bmod (p-1)} \bmod p$ and $(c \bmod q)^d \bmod q = (c \bmod q)^{d \bmod (q-1)} \bmod q$ (by Proposition A.25, also see "Computing Modulo a Prime" on page 412).

3.3.2 Attacks Against RSA Encryption

We now describe attacks not primarily directed against the RSA algorithm itself, but against the environment in which the RSA cryptosystem is used.

The Common-Modulus Attack. Suppose two users Bob and Bridget of the RSA cryptosystem have the same modulus n. Let (n, e_1) be the public key of Bob and (n, e_2) be the public key of Bridget, and assume that e_1 and e_2 are relatively prime. Let $m \in \mathbb{Z}_n$ be a message sent to both Bob and Bridget and encrypted as $c_i = m^{e_i}$, $i = 1, 2$. The problem now is that the plaintext can be computed from c_1, c_2, e_1, e_2 and n. Since e_1 is prime to e_2, integers r and s with $re_1 + se_2 = 1$ can be derived by use of the extended Euclidean algorithm (see Algorithm A.6). Either r or s, say r, is negative. If $c_1 \notin \mathbb{Z}_n^*$, we can factor n by computing $\gcd(c_1, n)$, thereby breaking the cryptosystem. Otherwise we again apply the extended Euclidean algorithm and compute c_1^{-1}. We can recover the message m using $(c_1^{-1})^{-r} c_2^s = (m^{e_1})^r (m^{e_2})^s = m^{re_1 + se_2} = m$. Thus, the cryptosystem fails to protect a message m if it is sent to two users with common modulus whose encryption exponents are relatively prime.

With common moduli, secret keys can be recovered. If Bob and Bridget have the same modulus n, then Bob can determine Bridget's secret key.

Namely, either Bob already knows the prime factors of n or he can compute them from his encryption and decryption exponents, with a very high probability (see Exercise 4). Therefore, common moduli should be avoided in RSA cryptosystems – each user should have his own modulus. If the prime factors (and hence the modulus) are randomly chosen, as described above, then the probability that two users share the same modulus is negligibly small.

Low-Encryption-Exponent Attack. Suppose that messages will be RSA-encrypted for a number of recipients, and each recipient has the same small encryption exponent. We discuss the case of three recipients, Bob, Bridget and Bert, with public keys $(n_i, 3)$, $i = 1, 2, 3$. Of course, the moduli n_i and n_j must satisfy $\gcd(n_i, n_j) = 1$, for $i \neq j$, since otherwise factoring of n_i and n_j is possible by computing $\gcd(n_i, n_j)$. We assume that Alice sends the same message m to Bob, Bridget and Bert. The following attack is possible: let $c_1 := m^3 \bmod n_1, c_2 := m^3 \bmod n_2$ and $c_3 := m^3 \bmod n_3$. The inverse of the Chinese Remainder isomorphism (see Theorem A.30)

$$\phi : \mathbb{Z}_{n_1 n_2 n_3} \longrightarrow \mathbb{Z}_{n_1} \times \mathbb{Z}_{n_2} \times \mathbb{Z}_{n_3}$$

can be used to compute $m^3 \bmod n_1 n_2 n_3$. Since $m^3 < n_1 n_2 n_3$, we can get m by computing the ordinary cube root of m^3 in $\mathbb{Z}$.

Small-Message-Space Attack. If the number of all possible messages is small, and if these messages are known in advance, an adversary can encrypt all messages with the public key. He can decrypt an intercepted cryptogram by comparing it with the precomputed cryptograms.

A Chosen-Ciphertext Attack. In the first phase of a chosen-ciphertext attack against an encryption scheme (see Section 1.3), adversary Eve has access to the decryption device. She obtains the plaintexts for ciphertexts of her choosing. Then, in the second phase, she attempts to decrypt another ciphertext for which she did not request decryption in the first phase.

Basic RSA encryption does not resist the following chosen-ciphertext attack. Let (n, e) be Bob's public RSA key, and let c be any ciphertext, encrypted with Bob's public key.

To find the plaintext m of c, adversary Eve first chooses a random unit $r \in \mathbb{Z}_n^*$ and requests the decryption of the random-looking message

$$r^e c \bmod n.$$

She obtains the plaintext $\tilde{m} = rm \bmod n$. Then, in the second phase of the attack, Eve easily derives the plaintext m, because we have in $\mathbb{Z}_n$

$$r^{-1} \tilde{m} = r^{-1} rm = m.$$

The attack relies on the fact that the RSA function is a ring isomorphism. It might appear that the attack is artificial and only of theoretical interest. But this is not true: Bleichenbacher's 1-Million-Chosen-Ciphertext Attack, which is very practical and described in the next section, depends on a variant of it. Another setting in which the chosen-ciphertext attack against RSA encryption may work is described in Section 3.3.4, where RSA signatures are discussed.

Bleichenbacher's 1-Million-Chosen-Ciphertext Attack. The 1-Million-Chosen-Ciphertext Attack of Bleichenbacher ([Bleichenbacher98]) is an attack against PKCS#1 v1.5 ([RFC 2313]).[7] The widely used PKCS#1 is part of the Public-Key Cryptography Standards series PKCS. These are de facto standards that are developed and published by RSA Security ([RSALabs]) in conjunction with system developers worldwide. The RSA standard PKCS#1 defines mechanisms for encrypting and signing data using the RSA public-key system. We explain Bleichenbacher's attack against encryption.

Let (n, e) be a public RSA key with encryption exponent e and modulus n. The modulus n is assumed to be a k-byte integer, i.e., $256^{k-1} < n < 256^k$. PKCS#1 defines a padding format. Messages (which are assumed to be shorter than k bytes) are padded out to obtain a formatted plaintext block m consisting of k bytes, and this plaintext block is then encrypted by using the RSA function. The ciphertext is $m^e \bmod n$, as usual. The first byte of the plaintext block m is 00, and the second byte is 02 (in hexadecimal notation). Then a padding string follows. It consists of at least 8 randomly chosen bytes, different from 00. The end of the padding block is marked by the zero byte 00. Then the original message bytes are appended. After the padding, we get

$$m = 00\|02\|\text{padding string}\|00\|\text{original message}.$$

The leading 00-byte ensures that the plaintext block, when converted to an integer, is less than the modulus.

We call a k-byte message m *PKCS conforming*, if it has the above format. A message $m \in \mathbb{Z}$ is PKCS conforming, if and only if

$$2B < m < 3B - 1,$$

with $B = 256^{k-2}$.

Adversary Eve wants to decrypt a ciphertext c. The attack is an adaptively-chosen-ciphertext attack (see Section 1.3). Eve chooses ciphertexts $c_1, c_2, c_3, \ldots$, different from c, and gets information about the plaintexts from a "decryption oracle" (imagine that Eve can supply ciphertexts to the decryption device and obtain some information on the decryption results). Adaptively means that Eve can choose a ciphertext, get information about the corresponding plaintext and do some analysis. Depending on the results of her analysis, she can choose a new ciphertext, and so on. With the help of the oracle, she computes the plaintext m of the ciphertext c. If Eve does not

---

[7] PKCS#1 has been updated. The current version is Version 2.1 ([RFC 3447]).

get the full plaintexts of the ciphertexts $c_1, c_2, \ldots$, as in Bleichenbacher's attack, such an attack is also called, more precisely, a *partial chosen-ciphertext attack*.

In Bleichenbacher's attack, on input of a ciphertext, the decryption oracle answers whether the corresponding plaintext is PKCS conforming or not. There were implementations of the SSL/TLS-protocol that contained such an oracle in practice (see below).

Eve wants to decrypt a ciphertext c which is the encryption of a PKCS conforming message m. She successively constructs intervals $[a_i, b_i] \subset \mathbb{Z}, i = 0, 1, 2, \ldots$, which all contain m and become shorter in each step, usually by a factor of 2. Eve finds m as soon as the interval has become sufficiently small and contains only one integer.

Eve knows that m is PKCS conforming and hence in $[2B, 3B - 1]$. Thus, she starts with the interval $[a_0, b_0] = [2B, 3B - 1]$. In each step, she chooses integers s, computes the ciphertext

$$\tilde{c} := s^e c \bmod n$$

of $sm \bmod n$ and queries the oracle with input $\tilde{c}$. The oracle outputs whether

$$sm \bmod n$$

is PKCS conforming or not. Whenever $sm \bmod n$ is PKCS conforming, Eve can narrow the interval $[a, b]$. The choice of the multipliers s depends on the output of the previous computations. So, the ciphertexts $\tilde{c}$ are chosen adaptively.

We now describe the attack in more detail.

Let $[a_0, b_0] = [2B, 3B - 1]$. We have $m \in [a_0, b_0]$, and the length of $[a_0, b_0]$ is $B - 1$.

Step 1: Eve searches for the smallest integer $s_1 > 1$ such that $s_1 m \bmod n$ is PKCS conforming. Since $2m \geq 4B$, the residue $s_1 m \bmod n$ can be PKCS conforming, only if $s_1 m \geq n + 2B$. Therefore, Eve can start her search with $s_1 \geq \lceil (n + 2B)/(3B - 1) \rceil$.

We have $s_1 m \in [s_1 a_0, s_1 b_0]$. If $s_1 m \bmod n$ is PKCS conforming, then

$$a_0 + tn \leq s_1 m \leq b_0 + tn$$

for some $t \in \mathbb{N}$ with $s_1 a_0 \leq b_0 + tn$ and $a_0 + tn \leq s_1 b_0$.

This means that m is contained in one of the intervals

$$[a_{1,t}, b_{1,t}] := [a_0, b_0] \cap \left[(a_0 + tn)/s_1, (b_0 + tn)/s_1 \right],$$

with $\lceil (s_1 a_0 - b_0)/n \rceil \leq t \leq \lfloor (s_1 b_0 - a_0)/n \rfloor$. We call the intervals $[a_{1,t}, b_{1,t}]$ the candidate intervals of step 1. They are pairwise disjoint and have length $< B/s_1$.

Step 2: Eve searches for the smallest integer $s_2, s_2 > s_1$, such that $s_2 m \bmod n$ is PKCS conforming. With high probability (see [Bleichenbacher98]), only one of the candidate intervals $[a_{1,t}, b_{1,t}]$ of step 1 contains a message x such that $s_2 x \bmod n$ is PKCS conforming. Eve can easily find out whether an interval $[a, b]$ contains a message x such that $s_2 x \bmod n$ is PKCS conforming. By comparing the interval boundaries, she simply checks whether

$$[s_2 a, s_2 b] \cap [a_0 + rn, b_0 + rn] \neq \emptyset \text{ for some } r \text{ with } \lceil s_2 a/n \rceil \leq r \leq \lfloor s_2 b/n \rfloor .$$

By performing this check for all of the candidate intervals $[a_{1,t}, b_{1,t}]$ of step 1, Eve finds the candidate interval containing m. We denote this interval by $[a_1, b_1]$. With a high probability, we have $s_2(b_1 - a_1) < n - B$ and then, $[s_2 a_1, s_2 b_1]$ is sufficiently short to meet only one of the intervals $[a_0 + rn, b_0 + rn]$, $r \in \mathbb{N}$, say for $r = r_2$. Now, Eve knows that

$$m \in [a_2, b_2] := [a_1, b_1] \cap \left[(a_0 + r_2 n)/s_2, (b_0 + r_2 n)/s_2 \right].$$

The length of this interval is $< B/s_2$.

In the rare case that more than one of the candidate intervals of step 1 contains a message x such that $s_2 x$ is PKCS conforming, or that more than one values for r_2 exist, Eve is left with more than one interval $[a_2, b_2]$. Then, she repeats step 2, starting with the candidate intervals $[a_2, b_2]$ in place of the candidate intervals $[a_{1,t}, b_{1,t}]$ (and searching for $s_2' > s_2$ such that $s_2' m \bmod n$ is PKCS conforming).

Step 3: Step 3 is repeatedly executed, until the plaintext m is determined. Eve starts with $[a_2, b_2]$ and (r_2, s_2), which she has computed in step 2. She iteratively computes pairs (r_i, s_i) and intervals $[a_i, b_i]$ of length $\leq B/s_i$ such that $m \in [a_i, b_i]$. The numbers r_i, s_i are chosen, such that

1. $s_i \approx 2 s_{i-1}$,
2. $[s_i a_{i-1}, s_i b_{i-1}] \cap [a_0 + r_i n, b_0 + r_i n] \neq \emptyset$,
3. $s_i m$ is PKCS conforming.

The number of multipliers s that Eve has to test by querying the decryption oracle is much smaller than in steps 1 and 2. She searches for s_i only in the neighborhood of $2 s_{i-1}$. This kind of choosing s_i works, because, after step 2, the intervals $[a_i, b_i]$ are sufficiently small. The length of $[a_{i-1}, b_{i-1}]$ is $< B/s_{i-1}$ and $s_i \approx 2 s_{i-1}$. Hence, the length of the interval $[s_i a_{i-1}, s_i b_{i-1}]$ is less than $\approx 2B$, and it therefore meets at most one of the intervals $[a_0 + rn, b_0 + rn], r \in \mathbb{N}$, say for $r = r_i$. From properties 2 and 3, we conclude that $s_i m \in [a_0 + r_i n, b_0 + r_i n]$. Eve sets

$$[a_i, b_i] := [a_{i-1}, b_{i-1}] \cap \left[(a_0 + r_i n)/s_i, (b_0 + r_i n)/s_i \right].$$

Then, $[a_i, b_i]$ contains m and its length is $< B/s_i$.

The upper bound B/s_i for the length of $[a_i, b_i]$ decreases by a factor of two in each iteration ($s_i \approx 2 s_{i-1}$). Step 3 is repeated, until $[a_i, b_i]$ contains only one integer. This integer is the searched plaintext m.

The analysis in [Bleichenbacher98] shows that for a 1024-bit modulus n, the total number of ciphertexts, for which Eve queries the oracle, is typically about 2^{20}.

Chosen-ciphertext attacks were considered to be only of theoretical interest. Bleichenbacher's attack proved the contrary. It was used against a web server with the ubiquitous SSL/TLS protocol ([RFC 5246]). In the interactive key establishment phase of SSL/TLS, a secret session key is encrypted by using RSA and PKCS#1 v1.5. For a communication server it is natural to process many messages, and to report the success or failure of an operation. Some implementations of SSL/TLS reported an error to the client, when the RSA-encrypted message was not PKCS conforming. Thus, they could be used as the oracle. The adversary could anonymously attack the server, because SSL/TLS is often applied without client authentication. To prevent Bleichenbacher's attack, the implementation of SSL/TLS-servers was improved and the PKCS#1 standard was updated. Now it uses the OAEP padding scheme, which we describe below in Section 3.3.3.

Encryption schemes that are provably secure against adaptively-chosen-ciphertext attacks are studied in more detail in Section 9.5.

The predicate "PKCS conforming" reveals one bit of information about the plaintext. Bleichenbacher's attack shows that, if we were able to compute the bit "PKCS conforming" for RSA-ciphertexts, then we could easily compute complete (PKCS conforming) plaintexts from the ciphertexts. To compute the bit "PKCS conforming" from c is as difficult as to compute m from c. Such a bit is called a secure bit. The bit security of one-way functions is carefully studied in Chapter 7. There we show, for example, that the least significant bit of an RSA-encrypted message is as secure as the whole message. In particular, we develop an algorithm that inverts the RSA function given an oracle with only a small advantage on the least significant bit.

3.3.3 Probabilistic RSA Encryption

To protect against the attacks studied in the previous section, a randomized preprocessing of the plaintexts turns out to be effective. Such a preprocessing is provided by the OAEP (optimal asymmetric encryption padding) scheme. It is not only applicable with RSA encryption. It can be used with any public-key encryption scheme whose encryption function is a bijective trapdoor function f, such as the RSA function or the modular squaring function of Rabin's cryptosystem (see Section 3.5.1).

Let
$$f : D \longrightarrow D, D \subset \{0,1\}^s,$$

be the trapdoor function that is used as the public-key encryption function. The inverse f^{-1} is the decryption function and the trapdoor information is the secret key.

In our context, we have $f = \text{RSA}_{n,e}$ and $f^{-1} = \text{RSA}_{n,d}$, where (n, e) is the public key and (n, d) is the secret key of the recipient of the messages. The domain D is $\mathbb{Z}_n$, which we identify with $\{0, 1, \ldots, n-1\}$, and $s = |n|$ is the binary length of the modulus n.

In addition, a pseudorandom bit generator

$$G : \{0, 1\}^k \longrightarrow \{0, 1\}^l$$

and a cryptographic hash function

$$h : \{0, 1\}^l \longrightarrow \{0, 1\}^k$$

are used, with $s = l + k$. Given a random seed $r \in \{0, 1\}^k$ as input, G generates a pseudorandom bit sequence of length l (see Chapter 8 for more details on pseudorandom bit generators). Cryptographic hash functions were discussed in Section 2.2.

Encryption. To encrypt a message $m \in \{0, 1\}^l$, we proceed in three steps:

1. We choose a random bit string $r \in \{0, 1\}^k$.
2. We set $x = (m \oplus G(r)) \| (r \oplus h(m \oplus G(r)))$.
   (If $x \notin D$ we return to step 1.)
3. We compute $c = f(x)$.

As always, let $\|$ denote the concatenation of strings and $\oplus$ the bitwise XOR operator.

OAEP is an embedding scheme. The message m is embedded into the input x of f such that all bits of x depend on the bits of m. The length of the message m is l. Shorter messages are padded with some additional bits to get length l. The first l bits of x, namely $m \oplus G(r)$ are obtained from m by masking with the pseudorandom bits $G(r)$. The seed r is encoded in the last k bits masked with $h(m \oplus G(r))$. The encryption depends on a randomly chosen r. Therefore, the resulting encryption scheme is not deterministic – encrypting a message m twice will produce different ciphertexts.

Decryption. To decrypt a ciphertext c, we use the function f^{-1}, the same pseudorandom random bit generator G and the same hash function h as above:

1. Compute $f^{-1}(c) = a \| b$, with $|a| = l$ and $|b| = k$.
2. Set $r = h(a) \oplus b$ and get $m = a \oplus G(r)$.

To compute the plaintext m from the ciphertext $c = f(x)$, an adversary must figure out all the bits of x from $c = f(x)$. He needs the first l bits to compute $h(a)$ and the last k bits to get r. Therefore, an adversary cannot exploit any advantage from some partial knowledge of x.

The OAEP scheme is published in [BelRog94]. OAEP has great practical importance. It has been adopted in PKCS#1 v2.0, a widely used standard which is implemented by Internet browsers and used in the secure socket layer protocol (SSL/TLS, [RFC 5246]). Using OAEP prevents Bleichenbacher's attack, which we studied in the preceding section. Furthermore, OAEP is included in electronic payment protocols to encrypt credit card numbers, and it is part of the IEEE P1363 standard.

For practical purposes, it is recommended to implement the hash function h and the random bit generator G using the secure hash algorithm SHA-1 (see Section 2.2.2 below) or some other cryptographic hash algorithm which is considered secure (for details, see [BelRog94]).

If h and G are implemented with efficient hash algorithms, the time to compute h and G is negligible compared to the time to compute f and f^{-1}. Formatting with OAEP does not increase the length of the message substantially.

Using OAEP with RSA encryption no longer preserves the multiplicative structure of numbers, and it is probabilistic. This prevents the previously discussed small-message-space attack. The low encryption exponent attack against RSA is also prevented, provided that the plaintext is individually re-encoded by OAEP for each recipient before it is encrypted with RSA.

We explained the so-called basic OAEP scheme. A slight modification of the basic scheme is the following. Let k, l be as before and let k' be another parameter, with $s = l + k + k'$. We use a pseudorandom generator

$$G : \{0,1\}^k \longrightarrow \{0,1\}^{l+k'}$$

and a cryptographic hash function

$$h : \{0,1\}^{l+k'} \longrightarrow \{0,1\}^k.$$

To encrypt a message $m \in \{0,1\}^l$, we first append k' 0-bits to m, and then we encrypt the extended message as before, i.e., we randomly choose a bit string $r \in \{0,1\}^k$ and the encryption c of m is defined by

$$c = f(((m\|0^{k'}) \oplus G(r))\|(r \oplus h((m\|0^{k'}) \oplus G(r)))).$$

Here, we denote by $0^{k'}$ the constant bit string $000\ldots0$ of length k'.

In [BelRog94], the modified scheme is proven to be secure in the random oracle model against adaptively-chosen-ciphertext attacks. The proof assumes that the hash function and the pseudorandom generator used behave like truly random functions. We describe the random oracle model in Section 3.3.5.

In [Shoup01], it was observed that there is a gap in the security proof of OAEP. This does not imply that a particular instantiation of OAEP, such as OAEP with RSA, is insecure. In the same paper, it is shown that OAEP with RSA is secure for an encryption exponent of 3. In [FujOkaPoiSte01], this result is generalized to arbitrary encryption exponents. In Section 9.5.1, we describe SAEP – a simplified OAEP – and we give a security proof for SAEP in the random oracle model against adaptively-chosen-ciphertext attacks.

3.3.4 Digital Signatures — The Basic Scheme

The RSA cryptosystem may also be used for digital signatures. Let (n, e) be the public key and d be the secret decryption exponent of Alice. We first discuss the basic signature scheme: Alice can sign messages that are encoded by numbers $m \in \{0, \ldots, n-1\}$. As usual, we consider those numbers as the elements of $\mathbb{Z}_n$, and the computations are done in $\mathbb{Z}_n$.

Signing. If Alice wants to sign a message m, she uses her secret key and computes her *signature* $\sigma = m^d$ of m by applying her decryption algorithm. We call (m, σ) a *signed message*.

Verification. Assume that Bob received a signed message (m, σ) from Alice. To verify the signature, Bob uses the public key of Alice and computes σ^e. He accepts the signature if $\sigma^e = m$.

If Alice signed the message, we have $\left(m^d\right)^e = (\mathrm{RSA}_{n,e} \circ \mathrm{RSA}_{n,d})\,(m) = m$ and Bob accepts (see Proposition 3.3). However, the converse is not true. It might happen that Bob accepts a signature not produced by Alice. Suppose Eve uses Alice's public key (n, e), chooses some m, sets $\sigma' := m$, computes $m' = m^e$ and says that (m', σ') is a message signed by Alice. Everyone verifying the signature gets $\sigma'^e = m^e = m'$ and is convinced that Alice really signed the message m'. Probably, the forged signature σ' is not very useful for Eve: the message $m' = m^e$ is not likely to be meaningful in most settings, for example, if the messages are expected to belong to some natural language. But this example shows that basic RSA signatures can be *existentially forged*. This means that an adversary can forge a signature for some message, but not for a message of his choice.

Another attack uses the fact that the RSA encryption and decryption functions are *ring homomorphisms* (see Section A.3). The image of a product is the product of the images and the image of the unit element is the unit element. If Alice signed m_1 and m_2, then the signatures for $m_1 m_2$ and m_1^{-1} are $\sigma_1 \sigma_2$ and σ_1^{-1}. These signatures can easily be computed without the secret key.

There is a chosen-message attack[8] against basic RSA signatures, which is analogous to the chosen-ciphertext attack against RSA encryption (see page 63 above). Eve can successfully forge Bob's signature for a message m, if she is first supplied with a valid signature σ for $r^e m \bmod n$, where r is randomly chosen by Eve. The forged signature is $r^{-1}\sigma$.

Assume that basic RSA signatures are used in the challenge-and-response authentication protocol which we described in the introduction (see page 6). Then this is a setting in which the chosen-ciphertext attack against RSA encryption may work, as described in the following attack.

---

[8] A detailed discussion of types of attacks against digital signature schemes is given in Section 11.1.

Attack on Encryption and Signing with RSA. The attack is possible if Bob has only one public-secret RSA key pair and uses this key pair for both encryption and digital signatures. Assume that the cryptosystem is also used for mutual authentication. On request, Bob proves his identity to Eve by signing a random number, supplied by Eve, with his secret key. Eve verifies the signature of Bob with his public key. In this situation, Eve can successfully attack Bob as follows. Suppose Eve has intercepted a ciphertext c intended for Bob:

1. Eve selects $r \in \mathbb{Z}_n^*$ at random.
2. Eve computes $x = r^e c \bmod n$, where (n, e) is the public key of Bob. She sends x to Bob to get a signature x^d (d the secret key of Bob). Note that x looks like a random number to Bob.
3. Eve computes $r^{-1}x^d$, which is the plaintext of c. Thus, she succeeds in decrypting c.

We will now discuss how to overcome these difficulties. In the basic signature scheme, adding redundancy to the messages can protect.

Signing with Redundancy. If the messages to be signed belong to some natural language, it is very unlikely that the above attacks will succeed. The messages m^e and $m_1 m_2$ will rarely be meaningful. When embedding the messages into $\{0,1\}^*$, the message space is sparse. The probability that a randomly chosen bit string belongs to the message space is small. By adding redundancy to each message we can always guarantee that the message space is sparse, even if arbitrary bit strings are admissible messages. A possible redundancy function is

$$R : \{0,1\}^* \longrightarrow \{0,1\}^*, \; x \longmapsto x \| x.$$

This principle is also used in error-detection and error-correction codes. Doubling the message we can detect transmission errors if the first half of the transmitted message does not match the second half. The redundancy function R has the additional advantage that the composition of R with the RSA function no longer preserves products.

In practice, usually another approach to prevent the attacks is taken. A cryptographic hash function is first applied to the message and then the hash value is signed instead of the original message. In this way messages of arbitrary binary length can be signed and not only messages from $\{0, \ldots, n-1\}$. On the other hand, in contrast to the basic scheme, the message cannot be recovered from the signature. We discuss this approach in the next section.

3.3.5 Signatures with Hash Functions

Let (n, e) be the public RSA key and d be the secret decryption exponent of Alice. In the basic RSA signature scheme (see Section 3.3.4), Alice can sign messages that are encoded by numbers $m \in \{0, \ldots, n-1\}$. To sign m, she applies the RSA decryption algorithm and obtains the signature $\sigma = m^d \bmod n$ of m.

Typically, n is a 1024–bit number. Alice can sign a bit string m that, when interpreted as a number, is less than n. This is a text string of at most 128 ASCII-characters. Most documents are much larger, and we are not able to sign them with basic RSA. This problem, which exists in all digital signatures schemes, is commonly solved by applying a collision resistant hash function h (see Section 2.2).

Message m is first hashed, and the hash value $h(m)$ is signed in place of m. Alice's RSA signature of m is

$$\sigma = h(m)^d \bmod n.$$

To verify Alice's signature σ for message m, Bob checks whether

$$\sigma^e = h(m) \bmod n.$$

This way of generating signatures is called the *hash-then-decrypt paradigm*. This term is even used for signature schemes, where the signing algorithm is not the decryption algorithm as in RSA (see, for example, ElGamal's Signature Scheme in Section 3.4.2).

Messages with the same hash value have the same signature. Collision resistance of h is essential for non-repudiation. It prevents Alice from first signing m and pretending later that she has signed a different message m' and not m. To do this, Alice would have to generate a collision (m, m'). Collision resistance also prevents that an attacker Eve takes a signed message (m, σ) of Alice, generates another message m' with the same hash value and uses σ as a (valid) signature of Alice for m'. To protect against the latter attack, second-pre-image resistance of h would be sufficient.

The hash-then-decrypt paradigm has two major advantages. Messages of any length can be signed by applying the basic signature algorithm, and the attacks, which we discussed in Section 3.3.4, are prevented. Recall that the hash function reduces a message of arbitrary length to a short digital fingerprint of less than 100 bytes.

The schemes which we discuss now implement the hash-then-decrypt paradigm.

Full-Domain-Hash RSA signatures. We apply the hash-then-decrypt paradigm in an RSA signature scheme with public key (n, e) and secret key d and use a hash function

$$h : \{0, 1\}^* \longrightarrow \{0, \ldots, n - 1\},$$

whose values range through the full set $\{0, \ldots, n - 1\}$ rather than a smaller subset. Such a hash function h is called a *full-domain* hash function, because the image of h is the full domain $\{0, \ldots, n - 1\}$ of the RSA function[9]. The signature of a message $m \in \{0, 1\}^*$ is $h(m)^d \bmod n$.

---

[9] As often, we identify $\mathbb{Z}_n$ with $\{0, \ldots, n - 1\}$.

The hash functions that are typically used in practical RSA schemes, like SHA, MD5 or RIPEMD, are not full-domain hash functions. They produce hash values of bit length between 128 and 512 bits, whereas the typical bit length of n is 1024 or 2048.

It can be mathematically proven that full-domain-hash RSA signatures are secure in the *random oracle model* ([BelRog93]), and we will give such a proof in this section.

For this purpose, we consider an adversary F, who attacks Bob, the legitimate owner of an RSA key pair, and tries to forge at least one signature of Bob, without knowing Bob's private key d. More precisely, F is an efficiently computable algorithm that, with some probability of success, on input of Bob's public RSA key (n, e) outputs a message m together with a valid signature σ of m.

The Random Oracle Model. In this model, the hash function h is assumed to operate as a *random oracle*. This means that

1. the hash function h is a random function (as explained in Section 2.2.4), and
2. whenever the adversary F needs the hash value for a message m, it has to call the oracle h with m as input. Then it obtains the hash value $h(m)$ from the oracle.

Condition 2 means that F always calls h as a "black box" (for example, by calling it as a subroutine or by communicating with another computer program), whenever it needs a hash value, and this may appear as a trivial condition. But it includes, for example, that the adversary has no algorithm to compute the hash values by itself; it has no knowledge about the internal structure of h, and it is stateless with respect to hash values. It does not store and reuse any hash values from previous executions. The hash values $h(m)$ appear as truly random values to him.

We assume from now on that our full-domain hash function h is a random oracle. Given m, each element of $\mathbb{Z}_n$ has the same probability $1/n$ of being the hash value $h(m)$.

The security of RSA signatures relies, of course, on the RSA assumption, which states that the RSA function is a one-way function. Without knowing the secret exponent d, it is infeasible to compute e-th roots modulo n, i.e., for a randomly chosen e-th power $y = x^e \bmod n$, it is impossible to compute x from y with more than a negligible probability (see Definition 6.7 for a precise statement).

Our security proof for full-domain-hash RSA signatures is a typical one. We develop an efficient algorithm A which attacks the underlying assumption – here the RSA assumption. In our example, A tries to compute the e-th root of a randomly chosen $y \in \mathbb{Z}_n$. The algorithm A calls the forger F as a subroutine. If F is successful in its forgery, then A is successful in computing the e-th root. Now, we conclude: since it is infeasible to compute e-th roots (by the RSA assumption), F cannot be successful, i.e., it is impossible to forge signatures. By A, the security of the signature scheme is reduced to the security of the RSA trapdoor function. Therefore, such proofs are called *security proofs by reduction*.

The security of full-domain-hash signatures is guaranteed, even if forger F is supplied with valid signatures for messages m' of its choice. Of course, to be successful, F has to produce a valid signature for a message m which is different from the messages m'. F can request the signature for a message m' at any time during its attack, and it can choose the messages m' adaptively, i.e., F can analyze the signatures that it has previously obtained, and then choose the next message to be signed. F performs an *adaptively-chosen-message attack* (see Section 11.1 for a more detailed discussion of the various types of attacks against signature schemes).

In the real attack, the forger F interacts with Bob, the legitimate owner of the secret key, to obtain signatures, and with the random oracle to obtain hash values. Algorithm A is constructed to replace both, Bob and the random oracle h, in the attack. It "simulates" the signer Bob and h.

Since A has no access to the secret key, it has a problem to produce a valid signature, when F issues a signature request for message m'. Here, the random oracle model helps A. It is not the message that is signed, but its hash value. To check if a signature is valid, the forger F must know the hash value, and to get the hash value it has to ask the random oracle. Algorithm A answers in place of the oracle. If asked for the hash value of a message m', it selects $s \in \mathbb{Z}_n$ at random and supplies s^e as the hash value. Then, it can provide s as the valid signature of m'.

The forger F cannot detect that A sends manipulated hash values. The elements s, s^e and the real hash values (generated by a random oracle) are all random and uniformly distributed elements of $\mathbb{Z}_n$. This means that forger F, when interacting with A, runs in the same probabilistic setting as in the real attack. Therefore, its probability of successfully forging a signature is the same as in the real attack against Bob.

A takes as input the public key (n, e) and a random element $y \in \mathbb{Z}_n$. Let F query the hash values of the r messages $m_1, m_2, \ldots, m_r$. The structure of A is the following.

Algorithm 3.6.

```
int A(int n, e, y)
  1   choose t ∈ {1, ..., r} at random and set h_t ← y
  2   choose s_i ∈ Z_n at random and set h_i ← s_i^e, i = 1, ..., r, i ≠ t
  3   call F(n, e)
  4   if F queries the hash value of m_i, then respond with h_i
  5   if F requests the signature of m_i, i ≠ t, then respond with s_i
  6   if F requests the signature of m_t, then terminate with failure
```

7   if F requests the signature of $m', m' \neq m_i$ for $i = 1, \ldots, r$,
8   then respond with a random element of $\mathbb{Z}_n$
9   if F returns (m, s), then return s

In step 1, A tries to guess the message $m \in \{m_1, \ldots, m_r\}$, for which F will output a forged signature. F must know the hash value of m. Otherwise, the hash value $h(m)$ of m would be randomly generated independently from F's point of view. Then, the probability that a signature s, generated by F, satisfies the verification condition $s^e \bmod n = h(m)$ is $1/n$, and hence negligibly small. Thus, m is necessarily one of the messages m_i, for which F queries the hash value.

If F requests the signature of m_i, $i \neq t$, then A responds with the valid signature s_i (line 5). If F requests the signature of $m', m' \neq m_i$ for $i = 1, \ldots, r$, and F never asks for the hash value of m', then A can respond with a random value (line 7) – F is not able to check the validity of the answer.

Suppose that A guesses the right m_t in step 1 and F forges successfully. Then, F returns a valid signature s for m_t, which is an e-th root of y, i.e., $s^e = h(m_t) = y$. In this case, A returns s. It has successfully computed an e-th root of y modulo n.

The probability that A guesses the right m_t in step 1 is $1/r$. Hence, the success probability of A is $1/r \cdot \alpha$, where α is the success probability of forger F. Assume for a moment that forger F is always successful, i.e., $\alpha = 1$. By independent repetitions of A, we then get an algorithm which successfully computes e-th roots with a probability close to 1. In general, we get an algorithm to compute e-th roots with about the same success probability α as F.

We described the notion of provable security in the random oracle model by studying full-domain hash RSA signatures. The proof says that a successful forger cannot exist in the random oracle model. But since real hash functions are never perfect random oracles (see Section 2.2.4 above), our argument can never be completed to a security proof of the real signature scheme, where a real implementation of the hash function has to be used.

In Section 9.5, we will give a random-oracle proof for Boneh's SAEP encryption scheme.

Our proof requires a full-domain hash function h – it is essential that each element of $\mathbb{Z}_n$ has the same probability of being the hash value. The hash functions used in practice usually are not full-domain hash functions, as we observed above. The scheme we describe in the next section does not rely on a full-domain hash function. It provides a clever embedding of the hashed message into the domain of the signature function.

PSS. The probabilistic signature scheme (PSS) was introduced in [BelRog96]. The signature of a message depends on the message and some randomly chosen input. The resulting signature scheme is therefore probabilistic. To set up the scheme, we need the decryption function of a public-key cryptosystem like the RSA decryption function or the decryption function of Rabin's cryptosystem (see Section 3.5). More generally, it requires a trapdoor permutation

$$f : D \longrightarrow D, D \subset \{0,1\}^n ,$$

a pseudorandom bit generator

$$G : \{0,1\}^l \longrightarrow \{0,1\}^k \times \{0,1\}^{n-(l+k)}, \ w \longmapsto (G_1(w), G_2(w))$$

and a hash function

$$h : \{0,1\}^* \longrightarrow \{0,1\}^l .$$

The PSS is applicable to messages of arbitrary length. The message m cannot be recovered from the signature σ.

Signing. To sign a message $m \in \{0,1\}^*$, Alice proceeds in three steps:

1. Alice chooses $r \in \{0,1\}^k$ at random and calculates $w := h(m\|r)$.
2. She computes $G(w) = (G_1(w), G_2(w))$ and $y := w\|(G_1(w) \oplus r)\|G_2(w)$. (If $y \notin D$, she returns to step 1.)
3. The signature of m is $\sigma := f^{-1}(y)$.

As usual, $\|$ denotes the concatenation of strings and $\oplus$ the bitwise XOR operator. If Alice wants to sign message m, she concatenates a random seed r to the message and applies the hash function h to $m\|r$. Then Alice applies the generator G to the hash value w. The first part $G_1(w)$ of $G(w)$ is used to mask r; the second part of $G(w)$, $G_2(w)$, is appended to $w\|G_1(w) \oplus r$ to obtain a bit string y of appropriate length. All bits of y depend on the message m. The hope is that mapping m into the domain of f by $m \longmapsto y$ behaves like a truly random function. This assumption guarantees the security of the scheme. Finally, y is decrypted with f to get the signature. The random seed r is selected independently for each message m – signing a message twice yields distinct signatures.

Verification. To verify the signature of a signed message (m, σ), we use the same trapdoor function f, the same random bit generator G and the same hash function h as above, and proceed as follows:

1. Compute $f(\sigma)$ and decompose $f(\sigma) = w\|t\|u$, where $|w| = l, |t| = k$ and $|u| = n - (k+l)$.
2. Compute $r = t \oplus G_1(w)$.
3. We accept the signature σ if $h(m\|r) = w$ and $G_2(w) = u$; otherwise we reject it.

PSS can be proven to be secure in the random oracle model under the RSA assumption. The proof assumes that the hash functions G and h are random oracles.

For practical applications of the scheme, it is recommended to implement the hash function h and the random bit generator G with the secure hash algorithm SHA-1 or some other cryptographic hash algorithm that is considered collision resistant. Typical values of the parameters n, k and l are $n = 1024$ bits and $k = l = 128$ bits.

3.4 The Discrete Logarithm

In Section 3.3 we discussed the RSA cryptosystem. The RSA function raises an element m to the e-th power. It is a bijective function and is efficient to compute. If the factorization of n is not known, there is no efficient algorithm for computing the e-th root. There are other functions in number theory that are easy to compute but hard to invert. One of the most important is exponentiation in finite fields. Let p be a prime and g be a primitive root in $\mathbb{Z}_p^*$ (see Section A.4). The discrete exponential function

$$\text{Exp} : \mathbb{Z}_{p-1} \longrightarrow \mathbb{Z}_p^*, \ x \longmapsto g^x$$

is a one-way function. It can be efficiently computed, for example, by the square-and-multiply algorithm (Algorithm A.27). No efficient algorithm for computing the inverse function Log of Exp, i.e., for computing x from $y = g^x$, is known, and it is widely believed that no such algorithm exists. This assumption is called the *discrete logarithm assumption* (for a precise definition, see Definition 6.1).

3.4.1 ElGamal Encryption

In contrast to the RSA function, Exp is a one-way function without a trapdoor. It does not have any additional information, which makes the computation of the inverse function easy. Nevertheless, Exp is the basis of ElGamal's cryptosystem ([ElGamal84]).

Key Generation. The recipient of messages, Bob, proceeds as follows:

1. He chooses a large prime p such that $p - 1$ has a big prime factor and a primitive root $g \in \mathbb{Z}_p^*$.
2. He randomly chooses an integer x in the range $1 \leq x \leq p - 2$. The triple (p, g, x) is the secret key of Bob.
3. He computes $y = g^x$ in $\mathbb{Z}_p$. The public key of Bob is (p, g, y), and x is kept secret.

The number $p-1$ will have a large prime factor if Bob is looking for primes p of the form $2kq + 1$, where q is a large prime. Thus, Bob first chooses a large prime q. Here he proceeds in the same way as in the RSA key generation procedure (see Section 3.3.1). Then, to get p, Bob randomly generates a k of appropriate bit length and applies a probabilistic primality test to $z = 2kq+1$. He replaces k by $k+1$ until he succeeds in finding a prime. He expects to test $O(\ln z)$ numbers for primality before reaching the first prime (see Corollary A.85). Having found a prime $p = 2kq + 1$, he randomly selects elements g in $\mathbb{Z}_p^*$ and tests whether g is a primitive root. The factorization of k is required for this test (see Algorithm A.40). Thus q must be chosen to be sufficiently large such that k is small enough to be factored efficiently.

Bob has to avoid that all prime factors of $p-1$ are small. Otherwise, there is an efficient algorithm for the computation of discrete logarithms developed by Silver, Pohlig and Hellman (see [Koblitz94]).

Encryption and Decryption. Alice encrypts messages for Bob by using Bob's public key (p, g, y). She can encrypt elements $m \in \mathbb{Z}_p$. To encrypt a message $m \in \mathbb{Z}_p$, Alice chooses an integer $k, 1 \leq k \leq p - 2$, at random. The encrypted message is the following pair (c_1, c_2) of elements in $\mathbb{Z}_p$:

$$(c_1, c_2) := (g^k, y^k m).$$

The computations are done in $\mathbb{Z}_p$. By multiplying m with y^k, Alice hides the message m behind the random element y^k.

Bob decrypts a ciphertext (c_1, c_2) by using his secret key x. Since $y^k = (g^x)^k = (g^k)^x = c_1^x$, he obtains the plaintext m by multiplying c_2 with the inverse c_1^{-x} of c_1^x:

$$c_1^{-x} c_2 = y^{-k} y^k m = m.$$

Recall that $c_1^{-x} = c_1^{p-1-x}$, because $c_1^{p-1} = [1]$ (see "Computing Modulo a Prime" on page 412). Therefore, Bob can decrypt the ciphertext by raising c_1 to the $(p-1-x)$-th power, $m = c_1^{p-1-x} c_2$. Note that $p - 1 - x$ is a positive number.

The encryption algorithm is not a deterministic algorithm. The cryptogram depends on the message, the public key and on a randomly chosen number. If the random number is chosen independently for each message, it rarely happens that two plaintexts lead to the same ciphertext.

The security of the scheme depends on the following assumption: it is impossible to compute g^{xk} (and hence $g^{-xk} = (g^{xk})^{-1}$ and m) from g^x and g^k, which is called the *Diffie–Hellman problem*. An efficient algorithm to compute discrete logarithms would solve the Diffie–Hellman problem. It is unknown whether the Diffie–Hellman problem is equivalent to computing discrete logarithms, but it is believed that no efficient algorithm exists for this problem (also see Section 4.1.2).

Like basic RSA (see Section 3.3.2), ElGamal encryption is vulnerable to a chosen-ciphertext attack. Adversary Eve, who wants to decrypt a ciphertext $c = (c_1, c_2)$, with $c_1 = g^k$ and $c_2 = m y^k$, chooses random elements $\tilde{k}$ and $\tilde{m}$ and gets Bob to decrypt $\tilde{c} = (c_1 g^{\tilde{k}}, c_2 \tilde{m} y^{\tilde{k}})$. Bob sends $m\tilde{m}$, the plaintext of $\tilde{c} = (g^{k+\tilde{k}}, m\tilde{m} y^{k+\tilde{k}})$, to Eve. Eve simply divides by $\tilde{m}$ and obtains the plaintext m of c: $m = (m\tilde{m})\tilde{m}^{-1}$. Bob's suspicion is not aroused, because the plaintext $m\tilde{m}$ looks random to him.

3.4.2 ElGamal Signatures

Key Generation. To generate a key for signing, Alice proceeds as Bob in the key generation procedure above to obtain a public key (p, g, y) and a secret key (p, g, x) with $y = g^x$.

Signing. We assume that the message m to be signed is an element in $\mathbb{Z}_{p-1}$. In practice, a hash function h is used to map the messages into $\mathbb{Z}_{p-1}$. Then the hash value is signed. The signed message is produced by Alice using the following steps:

1. She selects a random integer k, $1 \leq k \leq p - 2$, with $\gcd(k, p - 1) = 1$.
2. She sets $r := g^k$ and $s := k^{-1}(m - rx) \bmod (p - 1)$.
3. (m, r, s) is the signed message.

Verification. Bob verifies the signed message (m, r, s) as follows:

1. He verifies whether $1 \leq r \leq p - 1$. If not, he rejects the signature.
2. He computes $v := g^m$ and $w := y^r r^s$, where y is Alice's public key.
3. The signature is accepted if $v = w$; otherwise it is rejected.

Proposition 3.7. *If Alice signed the message* (m, r, s), *we have* $v = w$.

Proof.

$$w = y^r r^s = (g^x)^r (g^k)^s = g^{rx} g^{kk^{-1}(m-rx)} = g^m = v.$$

Here, recall that exponents of g can be reduced modulo $(p - 1)$, since $g^{p-1} = [1]$ (see "Computing Modulo a Prime" on page 412). □

Remarks. The following observations concern the security of the system:

1. The security of the system depends on the discrete logarithm assumption. Someone who can compute discrete logarithms can get everyone's secret key and thereby break the system totally. To find an s such that $g^m = y^r r^s$, on given inputs m and r, is equivalent to the computation of discrete logarithms.

   To forge a signature for a message m, one has to find elements r and s such that $g^m = y^r r^s$. It is not known whether this problem is equivalent to the computation of discrete logarithms. However, it is also believed that no efficient algorithm for this problem exists.

2. If an adversary Eve succeeds in getting the chosen random number k for some signed message m, she can compute $rx \equiv (m - sk) \bmod (p - 1)$ and the secret key x, because with high probability $\gcd(r, p - 1) = 1$. Thus, the random number generator used to get k must be of superior quality.

3. It is absolutely necessary to choose a new random number for each message. If the same random number is used for different messages $m \neq m'$, it is possible to compute k: $s - s' \equiv (m - m')k^{-1} \bmod (p - 1)$ and hence $k \equiv (s - s')^{-1}(m - m') \bmod (p - 1)$.

4. When used without a hash function, ElGamal's signature scheme is existentially forgeable; i.e., an adversary Eve can construct a message m and a valid signature (m, r, s) for m.

   This is easily done. Let b and c be numbers such that $\gcd(c, p - 1) = 1$. Set $r = g^b y^c$, $s = -rc^{-1} \bmod (p-1)$ and $m = -rbc^{-1} \bmod (p-1)$. Then (m, r, s) satisfies $g^m = y^r r^s$. Fortunately in practice, as observed above, a hash function h is applied to the original message, and it is the hash value that is signed. Thus, to forge the signature for a real message is not so easy. The adversary Eve has to find some meaningful message $\tilde{m}$ with $h(\tilde{m}) = m$. If h is a collision-resistant hash function, her probability of accomplishing this is very low.

5. D. Bleichenbacher observed in [Bleichenbacher96] that step 1 in the verification procedure is essential. Otherwise Eve would be able to sign messages of her choice, provided she knows one valid signature (m, r, s), where m is a unit in $\mathbb{Z}_{p-1}$.

Let m' be a message of Eve's choice, $u = m'm^{-1} \bmod (p - 1)$, $s' = su \bmod (p - 1)$, $r' \in \mathbb{Z}$ such that $r' \equiv r \bmod p$ and $r' \equiv ru \bmod (p - 1)$. r' is obtained by the Chinese Remainder Theorem (see Theorem A.30). Then (m', r', s') is accepted by the verification procedure.

3.4.3 Digital Signature Algorithm

In 1991 NIST proposed a digital signature standard (DSS) (see [NIST94]). DSS was intended to become a standard digital signature method for use by government and financial organizations. The DSS contains the digital signature algorithm (DSA), which is very similar to the ElGamal algorithm.

Key Generation. The keys are generated in a similar way as in ElGamal's signature scheme. As above, a prime p, an element $g \in \mathbb{Z}_p^*$ and an exponent x are chosen. x is kept secret, whereas p, g and $y = g^x$ are published. The difference is that g is not a primitive root in $\mathbb{Z}_p^*$, but an element of order q, where q is a prime divisor of $p - 1$.[10] Moreover, the binary size of q is required to be 160 bits.

To generate a public and a secret key, Alice proceeds as follows:

1. She chooses a 160-bit prime q and a prime p such that q divides $p - 1$ (p should have the binary length $|p| = 512 + 64t, 0 \le t \le 8$). [11] She can do this in a way analogous to the key generation in ElGamal's encryption scheme. First, she selects q at random, and then she looks for a prime p in $\{2kq + 1, 2(k + 1)q + 1, 2(k + 2)q + 1, \ldots\}$, with k a randomly chosen number of appropriate size.

2. To get an element g of order q, she selects elements $h \in \mathbb{Z}_p^*$ at random until $g := h^{(p-1)/q} \neq [1]$. Then g has order q, and it generates the unique cyclic group G_q of order q in $\mathbb{Z}_p^*$.[12] Note that in G_q elements are computed modulo p and exponents are computed modulo q.[13]

3. Finally, she chooses an integer x in the range $1 \le x \le q - 1$ at random.

4. (p, q, g, x) is the secret key, and the public key is (p, q, g, y), with $y := g^x$.

---

[10] The order of g is the smallest $e \in \mathbb{N}$ with $g^e = [1]$. The order of a primitive root in $\mathbb{Z}_p^*$ is $p - 1$.

[11] With advances in technology, larger key sizes are recommended today: the current standard [FIPS 186-4] allows the following pairs of binary lengths of p and q: $(1024, 160), (2048, 224), (2048, 256), (3072, 256)$.

[12] There is a unique subgroup G_q of order q of $\mathbb{Z}_p^*$. G_q consists of the unit element and all elements $x \in \mathbb{Z}_p^*$ of order q. It is cyclic and each member except the unit element is a generator, see Lemma A.41.

[13] See "Computing Modulo a Prime" on page 412.

Signing. Messages m to be signed by DSA must be elements in $\mathbb{Z}_q$. In DSS, a hash function h is used to map real messages to elements of $\mathbb{Z}_q$. The signed message is produced by Alice using the following steps:

1. She selects a random integer k, $1 \le k \le q-1$.
2. She sets $r := (g^k \bmod p) \bmod q$ and $s := k^{-1}(m+rx) \bmod q$. If $s = 0$, she returns to step 1, but it is extremely unlikely that this occurs.
3. (m, r, s) is the signed message.

Recall the verification condition of ElGamal's signature scheme. It says that $(m, \tilde{r}, \tilde{s})$ with $\tilde{r} = g^k \bmod p$ and $\tilde{s} = k^{-1}(m - \tilde{r}x) \bmod (p-1)$ can be verified by

$$y^{\tilde{r}}\tilde{r}^{\tilde{s}} \equiv g^m \bmod p.^{14}$$

Now suppose that, as in DSA, $\tilde{s}$ is defined by use of $(m + \tilde{r}x)$, $\tilde{s} = (m + \tilde{r}x)k^{-1} \bmod (p-1)$, and not by use of $(m - \tilde{r}x)$, as in ElGamal's scheme. Then the equation remains valid if we replace the exponent $\tilde{r}$ of y by $-\tilde{r}$:

$$y^{-\tilde{r}}\tilde{r}^{\tilde{s}} \equiv g^m \bmod p. \tag{3.1}$$

In the DSA, g and hence $\tilde{r}$ and y are elements of order q in $\mathbb{Z}_p^*$. Thus, we can replace the exponents $\tilde{s}$ and $\tilde{r}$ in (3.1) by $\tilde{s} \bmod q = s$ and $\tilde{r} \bmod q = r$. So, we have the idea that a verification condition for (r, s) may be derived from (3.1) by reducing $\tilde{r}$ and $\tilde{s}$ modulo q. This is not so easy, because the exponent $\tilde{r}$ also appears as a base on the left-hand side of (3.1). The base cannot be reduced without destroying the equality. To overcome this difficulty, we first transform (3.1) to

$$\tilde{r}^s \equiv g^m y^r \bmod p.$$

Now, the idea of DSA is to remove the exponentiation on the left-hand side. This is possible because s is a unit in $\mathbb{Z}_q^*$. For $t = s^{-1} \bmod q$, we get

$$\tilde{r} \equiv (g^m y^r)^t \bmod p.$$

Now we can reduce by modulo q on both sides and obtain the verification condition of DSA:

$$r = \tilde{r} \bmod q \equiv ((g^m y^r)^t \bmod p) \bmod q.$$

A complete proof of the verification condition is given below (Proposition 3.8). Note that the exponentiations on the right-hand side are done in $\mathbb{Z}_p$.

---

[14] We write "mod p" to make clear that computations are done in $\mathbb{Z}_p$.

Verification. Bob verifies the signed message (m, r, s) as follows:

1. He verifies that $1 \leq r \leq q - 1$ and $1 \leq s \leq q - 1$; if not, then he rejects the signature.
2. He computes the inverse $t := s^{-1}$ of s modulo q and $v := ((g^m y^r)^t \bmod p) \bmod q$, where y is the public key of Alice.
3. The signature is accepted if $v = r$; otherwise it is rejected.

Proposition 3.8. *If Alice signed the message (m, r, s), we have $v = r$.*

Proof. We have

$$
\begin{aligned}
v &= ((g^m y^r)^t \bmod p) \bmod q = (g^{mt} g^{rxt} \bmod p) \bmod q \\
&= (g^{(m+rx)t \bmod q} \bmod p) \bmod q = (g^k \bmod p) \bmod q \\
&= r.
\end{aligned}
$$

Note that exponents can be reduced by modulo q, because $g^q \equiv 1 \bmod p$. $\square$

Remarks:

1. Compared with ElGamal, the DSA has the advantage that signatures are fairly short, consisting of two 160-bit numbers.
2. In DSA, most computations – in particular the exponentiations – take place in the field $\mathbb{Z}_p^*$. The security of DSA depends on the difficulty of computing discrete logarithms. So it relies on the discrete logarithm assumption. This assumption says that it is infeasible to compute the discrete logarithm x of an element $y = g^x$ randomly chosen from $\mathbb{Z}_p^*$, where p is a sufficiently large prime and g is a primitive root of $\mathbb{Z}_p^*$ (see Definition 6.1 for a precise statement). Here, as in some cryptographic protocols discussed in Chapter 4 (commitments, electronic elections and digital cash), the base g is not a primitive root (with order $p - 1$), but an element of order q, where q is a large prime divisor of $p - 1$. To get the secret key x, it would suffice to find discrete logarithms for random elements $y = g^x$ from the much smaller subgroup G_q generated by g. Thus, the security of DSA (and some protocols discussed in Chapter 4) requires the (widely believed) stronger assumption that finding discrete logarithms for elements randomly chosen from the subgroup G_q is infeasible.
3. The remarks on the security of ElGamal's signature scheme also apply to DSA and DSS.
4. In the DSS, messages are first hashed before being signed by DSA. The DSS suggests taking SHA-1 for the hash function.

3.4.4 ElGamal Encryption in a Prime-Order Subgroup

Often, ElGamal encryption, like many discrete-log-based cryptographic protocols (see Chapter 4 for examples), is applied in a DSS/DSA setting: Bob's public key (p, q, g, y) consists of primes p and q, with q dividing $p - 1$; $g \in \mathbb{Z}_p^*$

is an element of order q (and not a primitive root). His secret key is an exponent $x \in \{1, \ldots, q-1\}$, which is the discrete logarithm of the public element y: $y = g^x \in \mathbb{Z}_p^*$.

Key Generation. Bob creates his key as follows:

1. He randomly generates a prime q and a cofactor k of appropriate binary length. Then he looks for the first prime p in the sequence $\{2kq+1, 2(k+1)q+1, 2(k+2)q+1, \ldots\}$. The binary length $|q|$ of q must be sufficiently large (in the DSS, $|q| \in \{160, 224, 256\}$), and the binary length of k has to be chosen such that the binary length of p is as desired (for example, $|p| = 1024$ or $|p| = 2048$).
2. He randomly generates an element $h \in \mathbb{Z}_p^*$ and sets $g := h^{(p-1)/q}$. Since $g^q = h^{p-1} = [1]$ by Fermat's Little Theorem (Proposition A.25) and q is a prime, $\mathrm{ord}(g) = q$ or $g = [1]$. If $g = [1]$, Bob repeats the procedure and generates a new h.
3. He randomly generates his secret $x \in \{1, \ldots, q-1\}$ and sets $y := g^x$.

Remark. There is a unique subgroup G_q of $\mathbb{Z}_p^*$ of order q. G_q consists of the unit element and all elements of order q. It is cyclic, and each member except for the unit element is a generator (see Lemma A.41 and "Computing Modulo a Prime" on page 412 for a proof and details). In particular, Bob's base element g is a generator of G_q:

$$G_q = \{g^x \mid x \in \mathbb{Z}_q\} = \{[1] = g^0, g^1, g^2, \ldots, g^{q-1}\}.^{15}$$

Encryption and Decryption. Both procedures work as in the original ElGamal schema. The difference is that only messages $m \in G_q$ from the subgroup of order q can be encrypted. To encrypt a message $m \in G_q$ for Bob, Alice randomly chooses an integer $k, 1 \leq k \leq q-1$, and computes the ciphertext (c_1, c_2):

$$(c_1, c_2) := (g^k, y^k \cdot m).$$

The computations are done in $\mathbb{Z}_p$, i.e., we calculate with residues modulo p. But all of the elements g, y, m, c_1, c_2 occurring are of order q. Thus, actually, the computation takes place in the subgroup G_q.

Bob decrypts a ciphertext (c_1, c_2) by using his secret key x. As in the original ElGamal schema, $y^k = (g^x)^k = (g^k)^x = c_1^x$ and Bob obtains the plaintext $m \in G_q$ by multiplying c_2 by the inverse c_1^{-x} of c_1^x:

$$c_1^{-x} c_2 = y^{-k} y^k m = m.$$

Since c_1 has order q, we have $c_1^{-x} = c_1^{q-x}$, and Bob can also compute $m = c_1^{q-x} c_2$.

---

[15] Note that we may identify $\mathbb{Z}_q$ and the set $\{0, 1, 2, \ldots, q-1\}$ of residues modulo q.

Remark. The original ElGamal encryption is a very secure encryption scheme, but it is not perfect. It is not *ciphertext-indistinguishable* (see Exercise 3 in Chapter 9). We discuss this security feature, which is equivalent to *semantic security*, in Chapter 9. ElGamal encryption in a prime-order subgroup G_q of $\mathbb{Z}_p^*$ does not have this weakness.[16] But now we have another problem. The plaintext space is G_q. There are applications where the plaintexts can be considered as elements of G_q in a natural way. Some examples are the encryption of secret keys that are chosen as random numbers (one can take as a key a random number g^x, where $x \in \mathbb{Z}_q$ is randomly generated) and the encryption of votes in the election protocol that we will study in Section 4.5.1. But, in practice, we typically want to encrypt some data given as a bit string. If the plaintext space of an encryption scheme is a residue class group $\mathbb{Z}_n$, then all of the integers z with $0 \leq z \leq n-1$ can be encrypted, and hence also all of the bit strings with length less than $|n|$, and then, by applying a suitable mode of operation, such as CBC (see Section 2.1.5), arbitrary bit strings. What we need for a practical application of prime-order-subgroup ElGamal encryption is a way to consider the plaintext messages in a unique way as elements of G_q, i.e., we need an encoding $\mathbb{Z}_q \longrightarrow G_q$ of messages $m \in \mathbb{Z}_q$. Simply encoding $m \in \mathbb{Z}_q$ as g^m usually does not work, because decoding then means to compute the discrete logarithm of $y = g^m$, which is infeasible in general. There are research efforts to develop variants of ElGamal encryption which avoid such encodings (see, e.g., [ChePaiPoi06]).

Encoding of Messages. An efficient encoding $\mathbb{Z}_q \longrightarrow G_q$ is known in the special case where $p = 2q + 1$.[17] Let $p = 2q + 1$.[18] Then G_q is the subgroup QR_p of quadratic residues of $\mathbb{Z}_p$ (see Section A.7, Definition A.54), since QR_p is the (only) subgroup of $\mathbb{Z}_p$ of order $q = (p-1)/2$ (Proposition A.56). The prime q is odd and hence $p \equiv 3 \bmod 4$. For primes $p \equiv 3 \bmod 4$, $[-1] \in \mathbb{Z}_p$ is not a square, i.e., $[-1] \notin \mathrm{QR}_p$ (Theorem A.59). In the following description, we identify $\mathbb{Z}_p$ and $\mathbb{Z}_q$ with the sets of residues $\{0, 1, 2, \ldots, p-1\}$ and $\{0, 1, 2, \ldots, q-1\}$, respectively.

An element $m \in \{0, 1, 2, \ldots, q-1\}$ is encoded as follows:

$$m \mapsto \begin{cases} m+1 & \text{if } m+1 \in \mathrm{QR}_p, \\ -(m+1) \bmod p = p - (m+1) & \text{if } m+1 \notin \mathrm{QR}_p. \end{cases}$$

It is easy to check whether $m + 1 \in \mathrm{QR}_p$, by applying Euler's criterion (Proposition A.58): $m + 1 \in \mathrm{QR}_p$ if $(m+1)^q \equiv 1 \bmod p$ and $m + 1 \notin \mathrm{QR}_p$ if $(m+1)^q \equiv -1 \bmod p$. In the second case, $-(m+1) \bmod p$ is a square modulo p because $m+1$ is not a square and -1 is not a square. The encodings resulting

---

[16] It can be proven that ElGamal encryption in a prime-order subgroup is ciphertext-indistinguishable, assuming that the decision Diffie–Hellman problem (Section 4.8.3) cannot be solved efficiently ([TsiYun98]).

[17] This means that q is a Sophie Germain prime (see Section 11.4).

[18] In key generation, we randomly generate primes q until we have found one where $2q + 1$ is a prime.

from the first case are squares $w \in \mathbb{Z}_p^* = \{1, 2, \ldots, p-1\}$ with $1 \leq w \leq q$, and the encodings resulting from the second case are squares w with $q + 1 \leq w \leq p-1$. Conversely, we can decode an element $w \in \mathrm{QR}_p \subset \{1, 2, \ldots, p-1\}$ as follows:

$$w \mapsto \begin{cases} w - 1 & \text{if } w \leq q, \\ p - w - 1 & \text{if } w > q. \end{cases}$$

3.5 Modular Squaring

Breaking the RSA cryptosystem might be easier than solving the factoring problem. It is widely believed to be equivalent to factoring, but no proof of this assumption exists. Rabin proposed a cryptosystem whose underlying encryption algorithm is provably as difficult to break as the factorization of large numbers (see [Rabin79]).

3.5.1 Rabin's Encryption

Rabin's system is based on the modular squaring function

$$\text{Square} : \mathbb{Z}_n \longrightarrow \mathbb{Z}_n, \ m \longmapsto m^2.$$

This is a one-way function with a trapdoor, if the factoring of large numbers is assumed to be infeasible (see Section 3.2.2).

Key Generation. As in the RSA scheme, we randomly choose two large distinct primes, p and q, for Bob. The scheme works with arbitrary primes, but primes p and q such that $p, q \equiv 3 \bmod 4$ speed up the decryption algorithm. Such primes are found as in the RSA key generation procedure. We are looking for primes p and q of the form $4k + 3$ in order to get the condition $p, q \equiv 3 \bmod 4$. Then $n = pq$ is used as the public key and p and q are used as the secret key.

Encryption and Decryption. We suppose that the messages to be encrypted are elements in $\mathbb{Z}_n$. The modular squaring one-way function is used as the encryption function E:

$$E : \mathbb{Z}_n \longrightarrow \mathbb{Z}_n, \ m \longmapsto m^2.$$

To decrypt a ciphertext c, Bob has to compute the square roots of c in $\mathbb{Z}_n$. The computation of modular square roots is discussed in detail in Section A.8. For example, square roots modulo n can be efficiently computed, if and only if the factors of n can be efficiently computed. Bob can compute the square roots of c because he knows the secret key p and q.

Using the Chinese Remainder Theorem (Theorem A.30), he decomposes $\mathbb{Z}_n$:

$$\phi : \mathbb{Z}_n \longrightarrow \mathbb{Z}_p \times \mathbb{Z}_q, \ c \longmapsto (c \bmod p, c \bmod q).$$

Then he computes the square roots of c mod p and the square roots of c mod q. Combining the solutions, he gets the square roots modulo n. If p divides c (or q divides c), then the only square root modulo p (or modulo q) is 0. Otherwise, there are two distinct square roots modulo p and modulo q. Since the primes are $\equiv 3 \bmod 4$, the square roots can be computed by one modular exponentiation (see Algorithm A.67). Bob combines the square roots modulo p and modulo q by using the Chinese Remainder Theorem, and obtains four distinct square roots of c (or two roots in the rare case that p or q divides c, or the only root 0 if $c = 0$). He has to decide which of the square roots represents the plaintext. There are different approaches. If the message is written in some natural language, it should be easy to choose the right one. If the messages are unstructured, one way to solve the problem is to add additional information. The sender, Alice, might add a header to each message consisting of the Jacobian symbol $\left(\frac{m}{n}\right)$ and the sign bit b of m. The sign bit b is defined as 0 if $0 \leq m < n/2$, and 1 otherwise. Now Bob can easily determine the correct square root (see Proposition A.72).

Remark. The difficulty of computing square roots modulo n is equivalent to the difficulty of computing the factors of n (see Proposition A.70). Hence, Rabin's encryption resists ciphertext-only attacks as long as factoring is impossible. The basic scheme, however, is completely insecure against a chosen-ciphertext attack. If an adversary Eve can use the decryption algorithm as a black box, she can determine the secret key using the following attack. She selects m in the range $0 < m < n$ and computes $c = m^2 \bmod n$. Decrypting c delivers y. There is a 50% chance that $m \not\equiv \pm y \bmod n$, and in this case Eve can easily compute the prime factors of n from m and y (see Lemma A.69). If the Jacobian symbol $\left(\frac{m}{n}\right)$ is added to each message m as sketched above, Eve may choose m with $\left(\frac{m}{n}\right) = -1$, but add $+1$ to the header. Then she will be certain to get a square root y with $m \not\equiv \pm y$. Applying some proper formatting, as in the OAEP scheme, can prevent this attack.

3.5.2 Rabin's Signature Scheme

The decryption function in Rabin's cryptosystem is only applicable to quadratic residues modulo n. Therefore, the system is not immediately applicable as a digital signature scheme. Before applying the decryption function, we usually apply a hash function to the message to be signed. Here we need a collision-resistant hash function whose values are quadratic residues modulo n. Such a hash function is obtained by the following construction. Let M be the message space and let

$$h : M \times \{0,1\}^k \longrightarrow \mathbb{Z}_n, \ (m, x) \longmapsto h(m, x)$$

be a hash function. To sign a message m, Bob generates pseudorandom bits x using a pseudorandom bit generator and computes $h(m, x)$. Knowing the factors of n, he can easily test whether $z := h(m, x) \in \mathrm{QR}_n$. z is a square if and only if $z \bmod p$ and $z \bmod q$ are squares, and $z \bmod p$ is a square in $\mathbb{Z}_p$, if and only if $z^{(p-1)/2} \equiv 1 \bmod p$ (Proposition A.58). He repeatedly chooses pseudorandom bit strings x until $h(m, x)$ is a square in $\mathbb{Z}_n$. Then he computes a square root y of $h(m, x)$ (e.g., using Proposition A.68). The signed message is defined as

$$(m, x, y).$$

A signed message (m, x, y) is verified by testing

$$h(m, x) = y^2.$$

If an adversary Eve is able to make Bob sign hash values of her choice, she can figure out Bob's secret key (see above).

3.6 Homomorphic Encryption Algorithms

ElGamal encryption is probabilistic and homomorphic. Probabilistic, homomophic public-key encryption algorithms have important and interesting applications, for example, in electronic voting protocols and re-encryption mix nets (see Chapter 4). In this section, we will first discuss the homomorphic property of ElGamal encryption and then describe another important probabilistic, homomorphic encryption scheme, Paillier encryption.

3.6.1 ElGamal Encryption

Here, we consider ElGamal encryption, as described in Section 3.4.2 above. Let p be a large random prime, $g \in \mathbb{Z}_p^*$ be a primitive root, $x \in \mathbb{Z}_{p-1}$ be a random exponent and $y = g^x$, and let (pk, sk) be the ElGamal key pair with secret key $sk = (p, g, x)$ and public key $pk = (p, g, y)$. A message $m \in \mathbb{Z}_p$ is encrypted by $E_{pk}(m, \omega) = (g^\omega, y^\omega m)$, where $\omega \in \mathbb{Z}_{p-1}$ is a randomly chosen element.

ElGamal encryption is homomorphic with respect to the multiplication of plaintexts and ciphertexts and the addition of randomness elements. Namely, we have

$$E_{pk}(m \cdot m', \omega + \omega') = E_{pk}(m, \omega) \cdot E_{pk}(m', \omega'),$$

because

$$(g^\omega, y^\omega m) \cdot (g^{\omega'}, y^{\omega'} m') = (g^\omega g^{\omega'}, y^\omega m y^{\omega'} m') = (g^{\omega+\omega'}, y^{\omega+\omega'} mm').$$

Sometimes ElGamal encryption is applied in the following way (for an example, see Section 4.5.1): messages $m \in \mathbb{Z}_{p-1}$ are first encoded as g^m and then encrypted, so that $\tilde{E}_{pk}(m, \omega) := (g^\omega, y^\omega g^m)$. This variant of ElGamal encryption is a homomorphic encryption $\tilde{E}_{pk} : \mathbb{Z}_{p-1} \times \mathbb{Z}_{p-1} \longrightarrow \mathbb{Z}_p^* \times \mathbb{Z}_p^*$

with respect to the addition of plaintexts and, as before, with respect to the multiplication of ciphertexts and the addition of randomness elements. We have

$$\tilde{E}_{pk}(m + m', \omega + \omega') = \tilde{E}_{pk}(m, \omega) \cdot \tilde{E}_{pk}(m', \omega').$$

Ciphertexts that are generated in this way can only be efficiently decrypted if the plaintexts are from a sufficiently small subset $\{m_1, m_2, \ldots\}$ of $\mathbb{Z}_{p-1}$ such that m can be computed from $z = g^m$, for example by comparing z with $g^{m_1}, g^{m_2}, \ldots$.

The same arguments show that ElGamal encryption in a prime-order subgroup (Section 3.4.4) has analogous homomorphic properties. In this setting, we have another random large prime, q, which divides $p - 1$; $sk = (p, q, g, x)$; $pk = (p, q, g, y)$; and $g \in \mathbb{Z}_p^*$ is an element of order q. The domains and ranges of the encryption maps E and $\tilde{E}$ are more restricted: $\omega \subset \mathbb{Z}_q = \{0, 1, \ldots, q-1\}$ and $m \in G_q$ or $m \in \mathbb{Z}_q$, respectively. The components of the ciphertext (c_1, c_2) are elements of G_q. As always, G_q denotes the subgroup of order q of $\mathbb{Z}_p^*$.

3.6.2 Paillier Encryption

We now give another important example of a probabilistic homomorphic encryption algorithm. Paillier encryption ([Paillier99]) works in the residue class group $\mathbb{Z}_{n^2}^*$, where n is an RSA modulus. The structure of this group is studied in detail in Section A.9.

Let p, q be large random primes, $p \neq q$, and let $n := pq$. As usual, we choose p and q to be of the same bit length, $|p| = |q|$. This implies that n is prime to $\varphi(n) = (p - 1)(q - 1)$.

Let $g \in \mathbb{Z}_{n^2}^*$ be an element of order n (for example $g = [1 + n]$). The public key pk is (n, g). The primes p, q are kept secret. The secret key is $\lambda := \mathrm{lcm}(p - 1, q - 1)$, the least common multiple of $p - 1$ and $q - 1$.[19]

To encrypt a message $m \in \mathbb{Z}_n$, we randomly choose an element ω in $\mathbb{Z}_n^*$ and set

$$c = E_{pk}(m, \omega) = g^m \omega^n,$$

i.e., we use the isomorphism $E_{pk} : \mathbb{Z}_n \times \mathbb{Z}_n^* \longrightarrow \mathbb{Z}_{n^2}^*$ (Proposition A.79) as the encryption function. Since

$$E_{pk}(m + m', \omega \cdot \omega') = E_{pk}(m, \omega) \cdot E_{pk}(m', \omega'),$$

E_{pk} is homomorphic.

Ciphertexts $c \in \mathbb{Z}_{n^2}$ can be efficiently decrypted by using the secret key λ. Namely, every $c \in \mathbb{Z}_{n^2}^*$ has a unique representation $g^m \omega^n$, where $m \in \mathbb{Z}_n$ and $\omega \in \mathbb{Z}_n^*$ (Proposition A.79), and m can be efficiently computed from c by

$$m = \left[L\left(c^\lambda \bmod n^2 \right) \right] \left[L\left(g^\lambda \bmod n^2 \right) \right]^{-1},$$

---

[19] $\lambda = \lambda(n)$ is called the *Carmichael function* of n.

where $L(x) = (x - 1)/n$ (see Proposition A.80).

The security of Paillier encryption is based on the composite residuosity assumption. Let $z \in \mathbb{Z}_{n^2}^*$ be a randomly chosen element. Then z has a unique representation $z \equiv g^x r^n$, where $x \in \mathbb{Z}_n$ and $r \in \mathbb{Z}_n^*$ (Proposition A.79). The *composite residuosity assumption* says that it is infeasible to compute x from z if the secret key is not known. Computing x from z means to decrypt the Paillier ciphertext z.

The composite residuosity assumption is stronger than the RSA assumption or the factoring assumption, as the following arguments show. Let $z = g^x r^n \in \mathbb{Z}_{n^2}^*$. If you are able to solve the RSA$[n, n]$ problem, i.e., if you are able to compute n-th roots modulo n, then you can compute g^x. Namely, you first compute r, which is the n-th root of z modulo n, $r = (z \bmod n)^{1/n}$ (note that $g \equiv 1 \bmod n$ by Proposition A.77). Then $y := z r^{-n} \equiv g^x \bmod n^2$. From y, you get x by $x = [L(y)][L(g)]^{-1}$ (see Proposition A.78).

The composite residuosity assumption guarantees that Paillier encryption is a one-way function. A stronger assumption, the *decisional composite residuosity assumption*, says that it is infeasible to find out whether a randomly chosen element z from $\mathbb{Z}_{n^2}^*$ is an n-th residue, i.e., whether it belongs to the subgroup $R_n = \{w^n \mid w \in \mathbb{Z}_{n^2}^*\}$. Note that $z = g^x r^n$ is an n-th residue if and only if $x = 0$. Thus, if you can compute x from z, it is easy to distinguish between n-th residues and non-residues. The decisional composite residuosity assumption implies that Paillier encryption is ciphertext-indistinguishable (see Exercise 8 in Chapter 9)[20].

3.6.3 Re-encryption of Ciphertexts

If a probabilistic homomorphic public-key encryption algorithm E is used, such as ElGamal or Paillier encryption, it is easy to re-encrypt a ciphertext c without increasing the length of the ciphertext. Re-encryption leaves the plaintext unchanged, but the randomness is modified.

Let $c := E_{pk}(m, \omega)$ be a ciphertext generated by using the public key pk and randomness ω. To re-encrypt c, Alice randomly chooses ω', encrypts the unit element e by using ω' and multiplies c with $E_{pk}(e, \omega')$. In detail:

1. *ElGamal encryption:* Let $c = E_{pk}(m, \omega) = (g^\omega, y^\omega m)$ be an ElGamal ciphertext. Alice re-encrypts c by randomly generating $\omega' \in \mathbb{Z}_{p-1}$ (or $\omega' \in \mathbb{Z}_q$ if ElGamal encryption in the subgroup is used) and multiplying c by the encryption $E_{pk}([1], \omega') = (g^{\omega'}, y^{\omega'})$ of $[1]$:

$$c' = (g^{\omega'}, y^{\omega'}) \cdot (g^\omega, y^\omega m) = (g^{\omega + \omega'}, y^{\omega + \omega'} m) = E_{pk}(m, \omega + \omega').$$

   c' is the encryption of m obtained by using randomness $\omega + \omega'$.

---

[20] Ciphertext indistinguishability, which is equivalent to semantic security, is discussed in Chapter 9.

2. *Paillier encryption*: To re-encrypt $c = E_{pk}(m, \omega) = g^m \cdot \omega^n$, Alice randomly chooses $\omega' \in \mathbb{Z}_n^*$ and multiplies c by $E_{pk}([0], \omega') = g^0 \cdot \omega'^n = \omega'^n$:

$$c' = \omega'^n \cdot g^m \cdot \omega^n = g^m \cdot (\omega \cdot \omega')^n = E_{pk}(m, \omega\omega').$$

c' is the encryption of m obtained by using randomness $\omega \cdot \omega'$.

3.7 Elliptic Curve Cryptography

Elliptic curves that are defined over a finite field $\mathbb{F}_q$ are widely used in cryptography (see Section A.11 for a short introduction to elliptic curves). The curves are defined either over a prime field, i.e., $\mathbb{F}_q = \mathbb{Z}_p$, with $q = p$ a prime number, or over a binary field, $\mathbb{F}_q = \mathbb{F}_{2^l}$, with $q = 2^l$ a power of 2 (for some basics on finite fields $\mathbb{F}_q$, see Section A.5.3). There are many elliptic curves over $\mathbb{F}_q$. A specific elliptic curve E over $\mathbb{F}_q$ is given by the coefficients $a, b \in \mathbb{F}_q$ of its defining equation.

If $\mathbb{F}_q = \mathbb{Z}_p$ is a prime field, then E is the set of solutions of its short Weierstrass equation

$$E(\mathbb{F}_q) = E(\mathbb{Z}_p) = \{(x, y) \in \mathbb{Z}_p^2 \mid y^2 = x^3 + ay + b\},$$

where a, b are any elements of $\mathbb{Z}_p$ with $4a^3 + 27b^2 \neq 0$ (see Propositions A.107 and A.110).[21]

If $\mathbb{F}_q = \mathbb{F}_{2^l}$ is a binary field, then E is the set of solutions of its short Weierstrass equation

$$E(\mathbb{F}_q) = E(\mathbb{F}_{2^l}) = \{(x, y) \in \mathbb{F}_{2^l}^2 \mid y^2 + xy = x^3 + ax^2 + b\},$$

where a, b are any elements of $\mathbb{F}_{2^l}$ with $b \neq 0$ (Propositions A.107 and A.110).

There are elliptic curves over $\mathbb{F}_{2^l}$ with short Weierstrass equations $Y^2 + aY = X^3 + bX + c$ (Proposition A.107), but these curves are not suitable for cryptographic applications (see Section 3.7.1 below). Therefore, we do not consider them here.

In Section 3.4, we studied ElGamal encryption and signatures and the digital signature algorithm (DSA). The security of these schemes is based on the discrete logarithm problem in the multiplicative subgroup $\mathbb{Z}_p^*$ of a finite field $\mathbb{Z}_p$ (p a large random prime): discrete exponentiation

$$\mathbb{Z}_{p-1} \longrightarrow \mathbb{Z}_p^*, \ k \longmapsto g^k = \underbrace{g \cdot g \cdot \ldots \cdot g}_{k}$$

is a one-way function if g is a base element with a sufficiently large order $\operatorname{ord}(g)$ (for example, a primitive root g or an element g with $\operatorname{ord}(g) = q$, q a 160-bit prime, as in the DSA).

---

[21] We write $E(\mathbb{F}_q)$ to indicate that we consider only the $\mathbb{F}_q$-rational points, i.e., points (x, y) with coordinates in $\mathbb{F}_q$. We do not consider solutions of the equation in extension fields of $\mathbb{F}_q$.

Elliptic curve cryptography is analogous. Schemes that are based on the discrete logarithm problem, such as the ElGamal and DSA schemes, can be implemented similarly on an elliptic curve E that is defined over a finite field $\mathbb{F}_q$. The set $E(\mathbb{F}_q)$ is an Abelian group, just as $\mathbb{Z}_p^*$ is. The group operation is the addition of points. Any two points $P, Q \in E(\mathbb{F}_q)$ can be added:

$$(P, Q) \mapsto P + Q \in E(\mathbb{F}_q).$$

The description of $E(\mathbb{F}_q)$ given above is not yet complete. There is one more point on the curve $E(\mathbb{F}_q)$, the *point at infinity* $\mathcal{O}$. The point at infinity $\mathcal{O}$ is the zero element of the addition in the Abelian group $E(\mathbb{F}_q)$. See Section A.11.3 for details and a geometric description of this addition.

Elliptic curves are not cyclic in general (see Theorem A.118). The cryptographic algorithms are implemented in a cyclic subgroup. For this purpose, a point $Q \in E(\mathbb{F}_q)$ is selected whose order $\mathrm{ord}(Q)$ is a large prime n. Q generates the subgroup

$$\langle Q \rangle = \{kQ \mid k \in \mathbb{Z}_n\} = \{Q, 2Q, \ldots, (n-1)Q, nQ = 0Q = \mathcal{O}\} \subset E(\mathbb{F}_q).$$

The order of $\langle Q \rangle$ is a prime, and therefore every element $P \in \langle Q \rangle, P \neq \mathcal{O}$, is a generator: $\langle Q \rangle = \langle P \rangle$.

The one-way function that is used in elliptic curve cryptography is the analogue of exponentiation (the operation now is $+$ instead of $\cdot$):

$$\mathbb{Z}_n \longrightarrow \langle Q \rangle, \ k \longmapsto kQ = \underbrace{Q + Q + \ldots + Q}_{k}.$$

This map is bijective, and there is a very efficient algorithm that computes kQ from k. The algorithm works like its analogue over $\mathbb{Z}_p^*$, the fast exponentiation algorithm (see Algorithms A.27 and A.116). The analogue of the discrete logarithm problem also exists. It is called the *elliptic curve discrete logarithm problem*: Given $P \in \langle Q \rangle$, it is practically infeasible to compute $k \in \mathbb{Z}_n$ such that $P = kQ$. The assumption that no efficient algorithm exists for this problem is called the *elliptic curve discrete logarithm assumption*.

In many settings, it is very attractive to use elliptic curve implementations of cryptographic algorithms such as ElGamal, the DSA or Diffie–Hellman key exchange. The reason is that compared with a classical implementation in a field $\mathbb{Z}_p$, one can do all of the computations with numbers of a far shorter binary length, without reducing the level of security. Therefore, elliptic curve implementations are often preferred when the computing resources are very limited, such as in embedded systems. We shall discuss this aspect briefly. Our discussion follows [BlaSerSma99][22].

The best known algorithm for solving the discrete logarithm problem in a prime field $\mathbb{Z}_p$ is the *number field sieve*. The number field sieve can compute discrete logarithms by executing approximately

---

[22] In [BlaSerSma99], algorithms for solving the elliptic curve discrete logarithm problem are studied in detail.

$$T_F(l) = e^{c_0 \cdot l^{1/3} (\ln(l \cdot \ln(2))^{2/3}}$$

basic operations, such as multiplications and squarings. Here, $l = \lfloor \log_2(p) \rfloor + 1$ is the binary length of the field elements, and $c_0 = \frac{4}{3}\sqrt[3]{3}$ is a small constant. The running time is subexponential.

   In contrast, all algorithms for computing discrete logarithms on a general elliptic curve have exponential running time. In a cryptographic application, the elliptic curve E and the base point $Q \in E(\mathbb{F}_q)$ of order n are chosen in such a way that n and $|E(\mathbb{F}_q)|$ differ by only a small factor (see Section 3.7.1 below). The order $|E(\mathbb{F}_q)|$ of the elliptic curve is close to q (see Theorem A.117). Hence, the order n of the base point Q is close to q. This implies that the binary length $m = \lfloor \log_2(q) \rfloor + 1$ of q, which is called the key length,[23] is the same as the binary length of n. Note that the key length gives the binary length of the numbers that must be handled in the computations.

   The best known algorithms for computing the discrete logarithm of elements of $\langle Q \rangle$ have running time proportional to $\sqrt{n}$, which is close to $\sqrt{q} \approx 2^{m/2}$. Let $T_E(m) := 2^{m/2}$.

   To compare the key lengths that are necessary for a comparable level of security, we solve the equation $T_E = T_F$ and express m as a function of l:

$$m(l) = \frac{2}{\ln(2)} \cdot c_0 \cdot l^{1/3} (\ln(l \cdot \ln(2))^{2/3}.$$

The graph of this function (Figure 3.1) shows that elliptic curve systems provide the same security level with substantially shorter keys, and the advantage grows with increasing level of security.

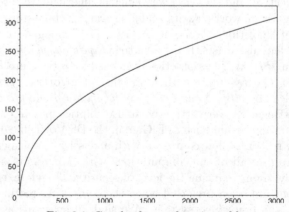

Fig. 3.1: Graph of m as function of l.

   The recommended key lengths for elliptic curve applications are much smaller than for the classical schemes. For the period from 2014 to 2030,

---

[23] The key length of a classical ElGamal or DSA key is the binary length l of the prime p.

NIST recommends a key size of between 224 and 255 for elliptic curves and a key size of 2048 for ElGamal, DSA and RSA ([NIST12, Tables 2 and 4]).

3.7.1 Selecting the Curve and the Base Point

To set up an elliptic curve cryptographic scheme, the first step is to select a base field, a curve and a base point, i.e., we establish the so-called *domain parameters*:

$\mathbb{F}_q$ the finite base field,
$a, b \in \mathbb{F}_q$ the parameters of the curve E,
Q, n the base point $Q \in E(\mathbb{F}_q)$, whose order is a large prime number n,
h the cofactor, defined by $hn = |E(\mathbb{F}_q)|$.

In this section, we describe how to choose suitable domain parameters.

Curves with Nearly Prime Order. The elliptic curve $E(\mathbb{F}_q)$ we are looking for should contain a point Q, whose order is a large prime number n, and n should be close to the group order $|E(\mathbb{F}_q)|$. Such curves have the following important property.

Lemma 3.9. *Let $E(\mathbb{F}_q)$ be an elliptic curve with $|E(\mathbb{F}_q)| = hn$, where n is a prime and $h < n$. Then $E(\mathbb{F}_q)$ has a unique subgroup E_n of order n. E_n is cyclic, and each point in E_n except for the zero element $\mathcal{O}$ is a generator of E_n.*

Proof. There is an isomorphism $E(\mathbb{F}_q) \cong \mathbb{Z}_{d_1} \times \mathbb{Z}_{d_2}$, where d_1 divides d_2 (Theorem A.118). If n were a divisor of d_1, then n would also divide d_2 and n^2 would divide $d_1 d_2 = |E(\mathbb{F}_q)| = hn$. But this is impossible, because $n^2 > hn$. Therefore, the prime n divides d_2, but does not divide d_1. We conclude that the only subgroup E_n of order n is $\{0\} \times G'$, where G' is the subgroup of $\mathbb{Z}_{d_2}$ that is generated by $[d_2/n]$. $\qquad\square$

Definition 3.10. An elliptic curve $E(\mathbb{F}_q)$ has *nearly prime order* if $|E(\mathbb{F}_q)| = hn$, where n is a prime and the cofactor h is small. The unique subgroup E_n of order n is called the *prime-order subgroup* of $E(\mathbb{F}_q)$.

Remark. "Small" means that $h \ll n$, and all prime factors of h can be found efficiently by trial division. In practice, we typically have $h = 1, 2, 3$ or 4. For an elliptic curve $E(\mathbb{F}_q)$ with nearly prime order, the binary length $\lfloor \log_2(n) \rfloor + 1$ of n is close to $\lfloor \log_2(q) \rfloor + 1$, the size of a field element.

Insecure Curves. For special classes of curves, there are algorithms that reduce the computation of a discrete logarithm in E_n to the computation of a discrete logarithm in groups where more efficient algorithms exist. For a given key length, the security level of these curves is significantly lower, and it is reasonable not to select one of these curves. The mathematics behind these algorithms is advanced and sophisticated. But, as we will see, there are simple checks of whether a curve is susceptible to such attacks, and these checks can easily be implemented.

We now give a short overview of the less secure curves and the related algorithms.

1. In [MenOkaVan93], Menezes, Okamoto and Vanstone introduced an attack which later was called the MOV attack. By using the *Weil pairing* on an elliptic curve, they embedded the prime-order subgroup E_n of $E(\mathbb{F}_q)$ into the multiplicative group $\mathbb{F}_{q^l}^*$ of a field extension $\mathbb{F}_{q^l}$, $l \in \mathbb{N}$, of $\mathbb{F}_q$. The smallest l with this property is called the *embedding degree* of E_n. The embedding turns E_n into a subgroup of $\mathbb{F}_{q^l}^*$. Since the order of a subgroup divides the group order, a necessary condition for such an embedding is that $n = |E_n|$ divides $q^l - 1$. The Weil pairing can be efficiently computed. Therefore, the attack amounts to an efficient algorithm for computing discrete logarithms in E_n if discrete logarithms in $\mathbb{F}_{q^l}$ can be efficiently computed. For small values of l, this might be possible (recall that the key length $\lfloor \log_2(q) \rfloor + 1$ is rather small). To ensure that an elliptic curve $E(\mathbb{F}_q)$ is immune to the MOV attack, it is sufficient to check that the order n of E_n does not divide $q^l - 1$ for small values of l. In practice, it suffices to test values of l up to 20. Randomly selected curves pass the test with high probability (see [BalKob98]). Some special classes of elliptic curves, including the supersingular curves, do not pass the test for some $l \leq 6$. Curves defined by $Y^2 + aY = X^3 + bX + c$ are always supersingular. For this reason, we do not consider these equations. The test also excludes curves that are susceptible to a similar attack which is based on the *Tate pairing* (see [FreRue94]).

2. An elliptic curve E over a prime field $\mathbb{F}_p$ is called *prime-field-anomalous* if $|E(\mathbb{F}_p)| = p$. Semaev ([Semaev98]), Smart ([Smart99]), and Satoh and Araki ([SatAra98]) independently developed a polynomial-time algorithm that implements an isomorphism between a prime-field-anomalous curve $E(\mathbb{F}_p)$ and the additive group of $\mathbb{F}_p$. The discrete logarithm problem in the additive group $\mathbb{F}_p$ is easily solved by using the extended Euclidean algorithm. Hence, the elliptic curve discrete logarithm problem can be solved efficiently for a prime-field-anomalous curve.

3. The Weil descent method, first proposed by Frey [Frey98], reduces the elliptic curve discrete logarithm problem to the discrete logarithm problem in the Jacobian of a hyperelliptic curve over a subfield of $\mathbb{F}_q$. In this case, subexponential algorithms for the discrete logarithm problem are known. Gaudry, Hess and Smart ([GauHesSma02]) developed an efficient algorithm to compute the reduction. The method is applicable for curves defined over a binary field $\mathbb{F}_{2^l}$ if l is composite. Menezes and Qu [MenQu01] showed that this attack is not applicable if l is a prime.

Selecting the Field. We make our choice between a prime field $\mathbb{F}_p$ and a binary field $\mathbb{F}_{2^l}$. The binary length $|p|$ of p or the exponent l must meet the security requirements, i.e., typically $|p|$ or l is chosen to be ≥ 224 (see the discussion about the key length above). The prime p is chosen randomly. In the binary case, to avoid known attacks (see item 3 above), a prime number is chosen as the exponent l.

Selecting the Curve. A preferred method for selecting an elliptic curve E is to choose the curve randomly, which means choosing the coefficients a, b of the defining equation at random. If the curve happens to be an insecure one, we reject it and repeat the random choice. Selecting the curve randomly promises the best protection against attacks on special classes of curves that might be discovered in the future.

The following algorithm randomly selects a suitable curve.

Algorithm 3.11.

$RandomCurve(\text{finite field } \mathbb{F}_q)$

1   Select $a, b \in \mathbb{F}_q$ at random
2   if $Y^2 = X^3 + aX + b$ (case $\mathbb{F}_p$) or $Y^2 + XY = X^3 + aX^2 + b$
3   (case $\mathbb{F}_{2^l}$) defines an elliptic curve $E(\mathbb{F}_q)$
4   then compute the order $|E(\mathbb{F}_q)|$
5   else go to 1
6   if $E(\mathbb{F}_q)$ meets one of the conditions (1) or (2) for insecure curves
7   then go to 1
8   if $E(\mathbb{F}_q)$ has nearly prime order n
9   then return (a, b, n, h)
10   else go to 1

Remarks:

1. An elliptic curve is required to be regular. The equation $Y^2 = X^3 + aX + b$ defines an elliptic curve if and only if $4a^3 + 27b^2 \neq 0$, and $Y^2 + XY = X^3 + aX^2 + b$ defines an elliptic curve if and only if $b \neq 0$ (see Proposition A.110). These conditions are easy to verify.
2. The difficult part in the algorithm is the computation of the group order $|E(\mathbb{F}_q)|$. There are sophisticated and efficient algorithms, to do this, however. If the base field is a prime field, the Schoof–Elkies–Atkin (SEA) algorithm can be used. If the base field is binary, extensions of Satoh's algorithm for binary fields can be used (see Section A.11.4).
3. It is easy to check whether the order $|E(\mathbb{F}_q)|$ of the curve is nearly prime. We perform trial divisions with small numbers h, and test if $n = |E(\mathbb{F}_q)|/h$ is a prime number by applying a probabilistic primality test, such as the Miller–Rabin test (Algorithm A.93). The check yields the decomposition $|E(\mathbb{F}_q)| = hn$.
4. There are a sufficient number of elliptic curves with nearly prime order such that a random search for such a curve is efficient in practice.
   If the base field is a prime field, there are many elliptic curves with prime order. If the base field is binary and the curve is defined by $Y^2 + XY =$

$X^3 + aX^2 + b$, then $(0, \sqrt{b}) \in E(\mathbb{F}_q)$ has order 2 (the inverse of a point (x, y) is $(x, x + y)$, see Proposition A.115). Hence, the order of $E(\mathbb{F}_q)$ is always even and never prime.

In [GalMck00], formulae were conjectured for the probability that a randomly chosen elliptic curve has prime or nearly prime order. Heuristic arguments and experimental evidence were given to support these formulae. For example, the formulae say that for the base field $\mathbb{F}_p$ where p is the prime number $2^{200} + 235$, the expected number of random trials until one finds an elliptic curve with prime order is 291. The expected number of trials until one finds a curve with nearly prime order and a cofactor $h \leq 5$ is 127.

Computing Points on the Curve. The projection map

$$\pi : E(\mathbb{F}_q) \setminus \{\mathcal{O}\} \longrightarrow \mathbb{F}_q, \ (x, y) \longmapsto x$$

is not surjective. Not every $x \in \mathbb{F}_q$ occurs as the X-coordinate of a point $(x, y) \in E(\mathbb{F}_q)$ on the curve. The Y-coordinates of the points $(x, y) \in \pi^{-1}(x)$ above x are the solutions of the defining equation

$$Y^2 = x^3 + ax + b$$

if $\mathbb{F}_q = \mathbb{Z}_p$ is a prime field, and

$$Y^2 + xY = x^3 + ax^2 + b$$

if $\mathbb{F}_q = \mathbb{F}_{2^l}$ is a binary field.

A solution exists if and only if the right-hand side of the equation is a square or has trace 0. The solutions can be efficiently computed by applying Algorithm A.67 and the method described in Section A.6, respectively. If there is a solution, then, almost surely, there are exactly two solutions, $\pm y$ and $y, y + x$, respectively. Only if $x^3 + ax + b = 0$, which happens for at most three values of x and for $x = 0$, respectively, is there exactly one solution. Hasse's Theorem (Theorem A.117) says that the number $|E(\mathbb{F}_q)|$ of points on $E(\mathbb{F}_q)$ is close to q – for a typical binary length $|q| = 224$, i.e., $q \approx 2^{224}$, the difference is $\leq 2^{113}$.

This implies that about half of the elements of $\mathbb{F}_q$ occur as the X-coordinate of a point on $E(\mathbb{F}_q)$. Hence, the probability that a randomly chosen $x \in \mathbb{F}_q$ is in $\pi(E(\mathbb{F}_q) \setminus \{\mathcal{O}\})$ is about $1/2$.

To generate a random point P on $E(\mathbb{F}_q)$, we randomly choose the coordinate x from $\mathbb{F}_q$. If $\pi^{-1}(x) \neq \emptyset$, then we compute the points above x by using Algorithm A.67 or the method from Section A.6 and randomly select one of them. Otherwise, we randomly select a new x. We expect to get a random point after two attempts.

Selecting the Base Point. Let $E(\mathbb{F}_q)$ be an elliptic curve having nearly prime order, $|E(\mathbb{F}_q)| = hn$, n a large prime, h small. Knowing how to generate random points on the curve, we are now able to compute a random point Q of order n. The algorithm is similar to the generation of an element of order q in the key generation procedure of the DSA (see Section 3.4.3).

Algorithm 3.12.

$LargeOrderRandomPoint()$
1   Generate a random point $P = (x, y)$ and set $Q \leftarrow hP$
2   if $Q \neq \mathcal{O}$
3   then return (Q)
4   else go to 1

Recall that $hnP = \mathcal{O}$ for every $P \in E(\mathbb{F}_q)$ (Proposition A.24). Hence, $nQ = nhP = \mathcal{O}$ and, since n is a prime, either $\mathrm{ord}(Q) = n$ or $\mathrm{ord}(Q) = 1$, which means $Q = \mathcal{O}$. We see that Q is the desired point of order n as soon as $Q \neq \mathcal{O}$. The probability of selecting a point P with $hP = \mathcal{O}$ is small. The points P with $hP = \mathcal{O}$ form a subgroup, and the order of that subgroup must divide h. Hence, the number of elements with $hP = \mathcal{O}$ is small.

Remark. By using Algorithm 3.12, we can easily verify that the domain parameters are consistent. First, we verify that n is prime. Then, we check if the order of the curve $E(\mathbb{F}_q)$ is indeed hn. If the algorithm does not return a point Q of order n, then the given order is not correct. Otherwise, it returns a point of order n, and then the group order $|E(\mathbb{F}_q)|$ is a multiple of n. By Hasse's Theorem (Theorem A.117), we have $q + 1 - 2\sqrt{q} \leq |E(\mathbb{F}_q)| \leq q + 1 + 2\sqrt{q}$. If $n > 4\sqrt{q}$, which can be immediately checked, then this interval contains only one multiple of n. Hence hn is the correct order if hn lies in this interval. If $n \leq 4\sqrt{q}$, then the order of $E(\mathbb{F}_q)$ is not nearly prime, and we have to choose another curve.

Examples of Random Elliptic Curves. In the FIPS 186-4 standard ([FIPS 186-4]), the National Institute of Standards and Technology (NIST) recommends 15 elliptic curves. Five curves are defined over prime fields and five over binary fields. These curves were generated pseudorandomly. The remaining five curves are so-called Koblitz elliptic curves, which are defined in $\mathbb{F}_{2^l}^2$ by equations with coefficients in $\mathbb{F}_2$.

The curves have varying security levels and should allow an efficient implementation of the arithmetic operations. In the defining equations of the prime-field curves, a is always -3 and only b is chosen pseudorandomly. It is stated that this choice of a for all curves was made for reasons of efficiency. All binary curves have $a = 1$.

The coefficient b was generated using a given seed value and the hash function SHA-1 as a pseudorandom generator (see Section 2.2). Thus, actually the curves are pseudorandom curves. The seed value is published in the standard. The one-way property of SHA-1 makes it possible to verify that b is really pseudorandom, i.e., that b was really generated by that method.

The curves defined over prime fields all have prime order. The random curves over binary fields have nearly prime order, with cofactor 2.

We now give two examples of the NIST curves:

- *Curve P-192* is defined over the prime field $\mathbb{F}_p$ by the equation

$$Y^2 = X^3 - 3X + b,$$

where

$p = 2^{192} - 2^{64} - 1$

    $= 6277101735386680763835789423207666416083908700390324961279,$

$seed = 0x\,3045ae6f\ c8422f64\ ed579528\ d38120ea\ e12196d5,$

    $b = 0x\,64210519\ e59c80e7\ 0fa7e9ab\ 72243049\ feb8deec\ c146b9b1.$

The order $|E(\mathbb{F}_p)|$ of curve P-192 is

    $6277101735386680763835789423176059013767194773182842284081.$

- *Curve B-163* is defined over the binary field $\mathbb{F}_{2^{163}} = \mathbb{F}_2[X]/f(X)$ by the equation

$$Y^2 + XY = X^3 + X^2 + b,$$

where

$f(X) = X^{163} + X^7 + X^6 + X^3 + 1,$

    $seed = 0x\,85e25bfe\ 5c86226c\ db12016f\ 7553f9d0\ e693a268,$

       $b = 0x\,00000002\ 0a601907\ b8c953ca\ 1481eb10\ 512f7874\ 4a3205fd.$

The order $|E(\mathbb{F}_{2^{163}})|$ of curve B-163 is

    $2 \cdot r, \ r = 5846006549323611672814742442876390689256843201587.$

In both examples, the groups are cyclic groups. The curve B-163 consists of the point $\mathcal{O}$ at infinity, $r - 1$ points of order r, $r - 1$ points of order $2r$ and one point $P \ (= (0, \sqrt{b}))$ of order 2.

3.7.2 Diffie–Hellman Key Exchange

The classical Diffie–Hellman key agreement protocol (Section 4.1.2) is readily adapted to elliptic curves. Beforehand, an elliptic curve and a base point are selected, and the domain parameters $(\mathbb{F}_q, a, b, Q, n, h)$ are known publicly.

Alice and Bob can establish a shared secret key by executing the following protocol.

Protocol 3.13.
Diffie–Hellman key agreement:

1. Alice chooses a, $1 \leq a \leq n - 1$, at random, sets $P_A := aQ$ and sends P_A to Bob.
2. Bob chooses b, $1 \leq b \leq n - 1$, at random, sets $P_B := bQ$ and sends P_B to Alice.
3. Alice computes the shared key $k = abQ = aP_B$.
4. Bob computes the shared key $k = abQ = bP_A$.

Remarks:

1. As in the classical setting of $\mathbb{Z}_p^*$, the security of the Diffie–Hellman key agreement depends on the hardness of the Diffie–Hellman assumption – it is practically infeasible to compute $C = abQ$ from $A = aQ$ and $B = bQ$ – and the Diffie–Hellman assumption presupposes the discrete logarithm assumption. Here, too, it is not known whether the two assumptions are equivalent.
2. The Diffie–Hellman key agreement protocol does not provide authentication of the communicating parties and is susceptible to a man-in-the-middle attack. Authentication must be supplemented, as in the classical Diffie–Hellman key agreement (see Section 4.1.4).
3. The elements generated are often referred to as Diffie–Hellman keys. Bob's Diffie–Hellman public key is P_B, and the private key is b.
4. The Diffie–Hellman key agreement can also be used as a one-pass protocol. One of the two Diffie–Hellman keys, say Bob's key (b, P_B), is used as a long-term key and distributed, for example, in a certificate. If Alice wants to encrypt a message m for Bob, she chooses a, $1 \leq a \leq n - 1$, at random, and sets $P_A := aQ$ and $k := aP_B$. Then, she uses k as a key to encrypt m by using a symmetric encryption algorithm (in practice, k is often taken as input to a pseudorandom generator, which then outputs the actual key). Alice sends the ciphertext c and P_A to Bob. By using his secret key b, Bob computes $k = bP_A$ and decrypts c. The Elliptic Curve Integrated Encryption Scheme (ECIES), proposed by Bellare and Rogaway, and the Provably Secure Encryption Curve scheme (PSEC), based on the work of Fujisaki and Okamoto, apply the Diffie–Hellman key agreement in this way (see [HanMenVan04]). Such schemes that include symmetric-key and public-key elements are called *hybrid*.
5. [RFC 4492] specifies the use of elliptic curve cryptography in the Transport Layer Security (TLS) protocol. Both variants of the Diffie–Hellman key agreement are included: either Bob's key is a long-term key and Alice's key is ephemeral, i.e., it is a one-time key used only in the current key exchange, or both keys are ephemeral.

3.7.3 ElGamal Encryption

ElGamal's encryption scheme can be transposed to an elliptic curve $E(\mathbb{F}_q)$ in a straightforward way. The domain parameters $(\mathbb{F}_q, a, b, Q, n, h)$ are assumed to be publicly known. As usual, let $E_n = \langle Q \rangle$ be the subgroup of order n.

Let P_B and s_B be Bob's public and private keys, with $s_B \in \{1, \ldots, n-1\}$ chosen randomly, and $P_B = s_B Q$.

To encrypt a message $m \in E_n$ for Bob, Alice chooses an integer $k, 1 \leq k \leq n - 1$, at random. She computes kQ and kP_B. The ciphertext is

$$(C_1, C_2) := (kQ, m + kP_B) \in E_n^2.$$

To decrypt, Bob computes $s_B C_1 = k s_B Q = k P_B$ from C_1 and obtains the plaintext

$$m = C_2 - s_B C_1.$$

Remarks:

1. The plaintext space is E_n. There are applications where the plaintexts can be considered as elements of E_n in a natural way. A common example is the exchange of a secret shared key. Alice randomly generates a secret key K and sends it ElGamal-encrypted to Bob. Here, Alice can take as the secret a multiple $K := lQ \in E_n$, with $l \in \mathbb{N}$ chosen randomly. But often we want to encrypt some data given as bit strings of length l, or, equivalently, as numbers $0, 1, 2, \ldots, M - 1$ (we interpret the bit strings as numbers). Then, we can use elliptic curve ElGamal encryption only if we can consider the plaintext messages in a unique way as elements of E_n, i.e., we need an encoding $\{0, 1, 2, \ldots, M - 1\} \longrightarrow E_n$ of the message space $\{0, 1, 2, \ldots, M - 1\}$. We discuss such encodings below. The Menezes–Vanstone variant of elliptic curve ElGamal encryption (see below) does not need such an encoding.

2. As in the classical ElGamal encryption in a prime-order subgroup, the plaintexts are restricted to elements of the subgroup. Otherwise, the security of the scheme would be reduced. It would lose ciphertext-indistinguishability (see the second remark in Section 3.4.4).

3. The security of the scheme depends on the hardness of the elliptic curve Diffie–Hellman problem (see Section 3.7.2). It must be infeasible to compute $k s_B Q$ from kQ and $P_B = s_B Q$.

4. As in the classical case, ElGamal's scheme is a probabilistic encryption algorithm. The random number k is chosen independently for each message. The probability that two plaintexts lead to the same ciphertext is negligibly small.

5. A ciphertext consists of two points C_1, C_2 on $E(\mathbb{F}_q)$. To represent these two points, four field elements are needed. Encryption significantly expands the message. An obvious way to reduce message expansion is to

send only the X-coordinates of C_1 and C_2, and then compute the Y-coordinates by using the curve equation. Typically, there are two solutions for y. One extra bit can resolve the ambiguity. This technique is referred to as *point compression*.

Encoding of Plaintexts. We now discuss the encoding of plaintext messages as elements of the prime-order subgroup E_n, $\{0, 1, 2, \ldots, M-1\} \longrightarrow E_n$.

First, we describe the encoding method proposed by Koblitz (see [Koblitz94]). We identify $\mathbb{F}_q$ with the set $\{0, 1, \ldots, q-1\}$ of numbers. We do this in the obvious way if $\mathbb{F}_q = \mathbb{Z}_p$ is a prime field. But this also works for a binary field $\mathbb{F}_{2^l}$. In an implementation, the elements of $\mathbb{F}_{2^l}$ are represented as binary vectors of length l, see Section A.5.3 for a detailed description. Hence, as a set, $\mathbb{F}_{2^l}$ can be identified with $\{0, 1\}^l$ and then, by considering the bit strings as numbers, with $\{0, 1, \ldots, 2^l - 1 = q - 1\}$.

The basic idea of Koblitz's encoding is to cover the interval $[0, q-1]$ of integers with disjoint intervals I_m of length κ, $I_m = [m\kappa, (m+1)\kappa - 1]$. The length κ is chosen large enough such that each interval I_m contains at least one $x \in \mathbb{F}_q$ that is the X-coordinate of an affine point in E_n. Recall that for about half of the elements $x \in \mathbb{F}_q$ there exists a point $(x, y) \in E(\mathbb{F}_q)$ on the curve above x (see "Computing Points on the Curve" on page 96). The cofactor of E_n is h. Therefore, for a fraction of about $1/2h$ of the elements $x \in \mathbb{F}_q$, there is a point $(x, y) \in E_n$ above x. If they are distributed over the interval $[0, q-1]$ in a reasonably uniform way, a suitable, small κ can be chosen. Then, we can encode a plaintext message $m \in \{0, 1, 2, \ldots, M-1\}$, $M := \lfloor q/\kappa \rfloor$, as follows: we choose the first element $x \in I_m$ in the interval I_m that is the X-coordinate of an affine point $(x, y) \in E_n$ and encode m as $P(m) := (x, y)$.

The algorithms for an implementation are described in "Computing Points on the Curve" (Section 3.7.1, page 96). The decoding works in the obvious way.

There are other ways of doing the encoding. For example, we could use the following probabilistic encoding. In order to encode $m \in \{0, 1, 2, \ldots, q-1\}$, we do the following:

1. Choose a random number $r \in \mathbb{Z}_q^*$ and set $x := rm \bmod q$.
2. If possible (the probability is $1/2h$), compute y with $(x, y) \in E_n$, and encode m as $P(m) := (x, y)$; otherwise, repeat 1.

This encoding will work after an expected $2h$ attempts. Then, $P(m)$ is encrypted by applying ElGamal encryption. The random number r is sent to Bob together with the ciphertext. After decrypting, he can decode the message because r is a unit.

A variant of this encoding could apply a random modification of the plaintext in a similar way to OAEP (Section 3.3.3). Consider the elements of $\mathbb{F}_q$ as bit strings in $\{0, 1\}^l$, $l = \lfloor \log_2(q) \rfloor$, and split l according to $l = r + s$. To encode a plaintext $m \in \{0, 1\}^r$, randomly choose $w \in \{0, 1\}^s$ and set

$x := (m \oplus G(w)) \| w$, where $G : \{0,1\}^s \longrightarrow \{0,1\}^r$ is a pseudorandom bit generator. In this variant, the transmission of an additional random value is not necessary.

Menezes–Vanstone Variant. Here, the plaintext m need not be encoded as a point on the elliptic curve. As before, let P_B be Bob's public key and s_B his private key, so that $P_B = s_B Q$. We suppose that the plaintexts are elements of $\mathbb{F}_q^2$.

In order to encrypt a message $m = (m_1, m_2)$ for Bob, Alice randomly chooses an integer $k, 1 \leq k \leq n-1$. She computes kQ and kP_B. Let $(x, y) := kP_B$. If $x = 0$ or $y = 0$, Alice selects another k. The encrypted message is the following element of $E(\mathbb{F}_q) \times \mathbb{F}_q^2$:

$$(C_1, c_2, c_3) := (kQ, xm_1, ym_2).$$

To decrypt, Bob can compute $(x, y) = kP_B = s_B C_1$ from C_1, because he knows s_B. Then he derives

$$m = (m_1, m_2) = (x^{-1}c_2, y^{-1}c_3).$$

The random elements x and y that hide m_1 and m_2 are not independent; (x, y) satisfies the equation of the elliptic curve. If an adversary Eve manages to figure out one component m_i of the plaintext, then she immediately gets the hiding element x or y. By using the curve equation, she can compute the other hiding element and recover the other component of the plaintext. Thus, it is advisable to use only x and encrypt only messages $m \in \mathbb{F}_q$.

3.7.4 Elliptic Curve Digital Signature Algorithm

The Elliptic Curve Digital Signature Algorithm (ECDSA) is the elliptic curve analogue of the Digital Signature Algorithm (Section 3.4.3). The ECDSA is included in several standards, for example IEEE 1363-2000 and ISO/IEC 15946-2. It has great practical importance. For example, it is included in the cipher suites of the Transport Layer Security (TLS) protocol (see [RFC 4492]).

The domain parameters $(\mathbb{F}_q, a, b, Q, n, h)$ are assumed to be known publicly.

Key Generation. Alice randomly chooses an integer $d_A \in \{1, \ldots, n-1\}$ as her secret key. The public key is $P_A = d_A Q$.

Signing. Messages m to be signed by ECDSA must be elements of $\mathbb{Z}_n$. In practice, a hash function H is used to map real messages to $\mathbb{Z}_n$. Alice generates the signature σ for a message $m \in \mathbb{Z}_n$ by using her secret key d_A as follows:

1. She randomly chooses an integer $k, 1 \leq k \leq n-1$.

2. She computes $kQ =: (x_{kQ}, y_{kQ})$, converts the X-coordinate x_{kQ} to an integer and sets $r := x_{kQ} \bmod n$. If $r = 0$, she returns to step 1.
3. She sets $s := k^{-1}(m + rd_A) \bmod n$. If $s = 0$, she returns to step 1.
4. $\sigma := (r, s)$ is the signature for m.

Converting a field element $x \in \mathbb{F}_q$ to an integer in $\{0, 1, \ldots, q-1\}$ is done as before. If $\mathbb{F}_q = \mathbb{Z}_p$ is a prime field, then we identify $\mathbb{Z}_q$ with $\{0, 1, \ldots, q-1\}$, as usual. For a binary field, the elements of $\mathbb{F}_{2^l}$ are represented as binary vectors of length l (see Section A.5.3) and then considered as numbers.

Verification. Bob verifies a signed message (m, r, s) by using Alice's public key P_A as follows:

1. He verifies that $1 \leq r \leq n - 1$ and $1 \leq s \leq n - 1$; if not, then he rejects the signature.
2. He computes the inverse $v := s^{-1}$ of s modulo n, $w_1 := mv \bmod n$, $w_2 := rv \bmod n$ and $R := w_1 Q + w_2 P_A$. If $R = \mathcal{O}$, the signature is rejected.
3. He sets $R = (x_R, y_R)$ and converts x_R to an integer. The signature is accepted if $r = x_R \bmod n$; otherwise, it is rejected.

Proposition 3.14. *If Alice signed the message (m, r, s), then $r = x_R \bmod n$.*

Proof. We have

$$k \equiv s^{-1}(m + rd_A) \equiv s^{-1}m + s^{-1}rd_A \equiv w_1 + w_2 d_A \bmod n.$$

Thus $kQ = w_1 Q + w_2 d_A Q = w_1 Q + w_2 P_A = R$, and $r = x_R \bmod n$ follows.
□

Remarks:

1. The only difference in the signing algorithm between ECDSA and the DSA is in the generation of r. The DSA computes the random element $r' = g^k$ in $\mathbb{Z}_p$ and takes $r = r' \bmod n$.[24] ECDSA generates a random point kQ and takes $r := x_{kQ} \bmod n$.
2. If $r = 0$ were allowed in the signature algorithm, then in this case the signature would not depend on the secret key d_A of Alice.
3. The security of the system depends on the elliptic curve discrete logarithm assumption. Someone who can compute discrete logarithms can get Alice's secret key and thereby break the system totally. Assume that m and r are given. Finding an s such that $R = s^{-1}(mQ + rP_A)$ is equivalent to the computation of discrete logarithms. To forge a signature for a message m, one has to find elements r and s such that r is the X-coordinate of $R = s^{-1}(mQ + rP_A)$. It is not known whether this problem is equivalent to the computation of discrete logarithms. However, it is believed that no efficient algorithm for this problem exists.

---

[24] In the description of the DSA, the letter q (and not n) denotes the order of the generator of the subgroup.

4. The checks $1 \leq r \leq n - 1$ and $1 \leq s \leq n - 1$ are necessary. Assume that the elliptic curve is defined by $Y^2 = X^3 + aX + b$ and that b is a quadratic residue modulo q. If $Q = (0, \sqrt{b})$ is used as the base point (and not a random point), Eve can forge Alice's signature on any message m, $1 \leq m \leq n - 1$, of her choice by setting $r = 0$ and $s = m$. Then $R = Q$ and the verification condition $x_R \bmod n = r$ is fulfilled.

5. The remarks on the security of the DSA and repeated use of the random number k (see Section 3.4.3) also apply to ECDSA.

6. As with the DSA, ECDSA is existentially forgeable when used without a hash function. An adversary Eve can construct a message m and a valid signature (r, s) for m. She selects an integer l, computes $(x, y) := lQ + P_A$, and sets $r := x \bmod n$, $s := r$ and $m := rl \bmod n$. Then

$$
\begin{aligned}
R &= w_1 Q + w_2 P_A \\
  &= rls^{-1}Q + rs^{-1}P_A \\
  &= lQ + P_A.
\end{aligned}
$$

Thus, $x_R = x$ and (m, r, s) is a valid signed message. If used with a cryptographic hash function H, the adversary Eve has to find some meaningful message $\tilde{m}$ with $H(\tilde{m}) = m$. Since H has the one-way property, this is practically infeasible.

Exercises

1. Set up an RSA encryption scheme by generating a pair of public and secret keys. Choose a suitable plaintext and a ciphertext. Encrypt and decrypt them.

2. Let n denote the product of two distinct primes p and q, and let $e \in \mathbb{N}$. Show that e is prime to $\varphi(n)$ if

$$
\mu : \mathbb{Z}_n^* \longrightarrow \mathbb{Z}_n^*, \ x \longmapsto x^e
$$

is bijective.

3. Let $\mathrm{RSA}_{n,e} : \mathbb{Z}_n^* \longrightarrow \mathbb{Z}_n^*, \ x \longmapsto x^e$. Show that

$$
|\{x \in \mathbb{Z}_n^* \mid \mathrm{RSA}_{n,e}(x) = x\}| = \gcd(e - 1, p - 1) \cdot \gcd(e - 1, q - 1).
$$

Hint: Show that $|\{x \in \mathbb{Z}_p^* \mid x^k = 1\}| = \gcd(k, p - 1)$, where p is a prime, and use the Chinese Remainder Theorem (Theorem A.30).

4. Let (n, e) be the public key of an RSA cryptosystem and d be the associated decryption exponent. Construct an efficient probabilistic algorithm A which on input n, e and d computes the prime factors p and q of n with very high probability.

Hint: Use the idea of the Miller–Rabin primality test, especially case 2 in the proof of Proposition A.92.

5. Consider RSA encryption. Discuss, in detail, the advantage you get using, for encryption, a public exponent which has only two 1 digits in its binary encoding and using, for decryption, the Chinese Remainder Theorem.

6. Let p, p', q and q' be prime numbers, with $p' \neq q'$, $p = ap' + 1$, $q = bq' + 1$ and $n := pq$:

   a. Show $|\{x \in \mathbb{Z}_p^* \mid p' \text{ does not divide ord}(x)\}| = a$.

   b. Assume that p' and q' are large (compared to a and b). Let $x \in \mathbb{Z}_n^*$ be a randomly chosen element. Show that the probability that x has large order is $\geq 1 - \left(\frac{1}{p'} + \frac{1}{q'} - \frac{1}{p'q'}\right)$. More precisely, show $|\{x \in \mathbb{Z}_n^* \mid p'q' \text{ does not divide ord}(x)\}| = a(q-1) + b(p-1) - ab$.

7. Consider RSA encryption $\text{RSA}_{n,e} : \mathbb{Z}_n^* \longrightarrow \mathbb{Z}_n^*$, $x \longmapsto x^e$:

   a. Show that $\text{RSA}_e^l = \text{id}_{\mathbb{Z}_n^*}$ for some $l \in \mathbb{N}$.

   b. Consider the decryption-by-iterated-encryption attack (see Section 3.3.1). Let p, p', p'', q, q' and q'' be prime numbers, with $p' \neq q'$, $p = ap' + 1, q = bq' + 1, p' = a'p'' + 1, q' = b'q'' + 1, n := pq$ and $n' := p'q'$. Assume that p' and q' are large (compared to a and b) and that p'' and q'' are large (compared to a' and b'). (This means that the factors of n satisfy the conditions 1 and 3 required for strong primes.)

   Show that the number of iterations necessary to decrypt a ciphertext c is $\geq p''q''$ (and thus very large) for all but an exponentially small fraction of ciphertexts. By exponentially small, we mean $\leq 2^{-|n|/k}$ (k a constant).

8. Set up an ElGamal encryption scheme by generating a pair of public and secret keys.

   a. Choose a suitable plaintext and a ciphertext. Encrypt and decrypt them.

   b. Generate ElGamal signatures for suitable messages. Verify the signatures.

   c. Forge a signature without using the secret key.

   d. Play the role of an adversary Eve, who learns the random number k used to generate a signature, and break the system.

   e. Demonstrate that checking the condition $1 \leq r \leq p - 1$ is necessary in the verification of a signature (r, s).

9. Weak generators (see [Bleichenbacher96]; [MenOorVan96]).

   Let p be a prime, $p \equiv 1 \bmod 4$, and $g \in \mathbb{Z}$ such that $g \bmod p$ is a primitive root in $\mathbb{Z}_p^*$. Let (p, g, x) be Bob's secret key and $(p, g, y = g^x)$ be Bob's public key in an ElGamal signature scheme. We assume that: (1) $p - 1 = gt$ and (2) discrete logarithms can be efficiently computed in the subgroup H of order g in $\mathbb{Z}_p^*$ (e.g. using the Pohlig–Hellman algorithm). To sign a message m, adversary Eve does the following: (1) she sets $r = t$; (2) she computes z such that $g^{tz} = y^t$; and (3) she sets $s = \frac{1}{2}(p-3)(m - tz) \bmod (p-1)$. Show:

a. That it is possible to compute z in step 2.

b. That (r, s) is accepted as Bob's signature for m.

c. How the attack can be prevented.

10. Set up a DSA signature scheme by generating a pair of public and secret keys. Generate and verify signatures for suitable messages. Take small primes p and q.

11. Set up a Rabin encryption scheme by generating a pair of public and secret keys. Choose a suitable plaintext and a ciphertext. Encrypt and decrypt them. Then play the role of an adversary Eve, who succeeds in a chosen-ciphertext attack and recovers the secret key.

12. Adapt the Nyberg–Rueppel signatures of Exercise 12 in Chapter 4, which are defined in a prime-order subgroup of $\mathbb{Z}_p^*$, to the prime-order subgroup E_n of an elliptic curve E. Then solve (a)–(c) of the exercise for the adapted scheme.

4. Cryptographic Protocols

One of the major contributions of modern cryptography is the development of advanced protocols providing high-level cryptographic services, such as secure user identification, voting schemes and digital cash. Cryptographic protocols use cryptographic mechanisms – such as encryption algorithms and digital signature schemes – as basic components.

A protocol is a multi-party algorithm which requires at least two participating parties. Therefore, the algorithm is distributed and invoked in at least two different places. An algorithm that is not distributed, is not called a protocol. The parties of the algorithm must communicate with each other to complete the task. The communication is described by the messages to be exchanged between the parties. These are referred to as the communication interface. The protocol requires precise definitions of the interface and the actions to be taken by each party.

A party participating in a protocol must fulfill the syntax of the communication interface, since not following the syntax would be immediately detected by the other parties. The party can behave honestly and follow the behavior specified in the protocol. Or she can behave dishonestly, only fulfill the syntax of the communication interface and do completely different things otherwise. These points must be taken into account when designing cryptographic protocols.

4.1 Key Exchange and Entity Authentication

Public- and secret-key cryptosystems assume that the participating parties have access to keys. In practice, one can only apply these systems if the problem of distributing the keys is solved.

The security concept for keys, which we describe below, has two levels. The first level embraces long-lived, secret keys, called *master keys*. The keys of the second level are associated with a session, and are called *session keys*. A session key is only valid for the short time of the duration of a session. The master keys are usually keys of a public-key cryptosystem.

There are two main reasons for the two-level concept. The first is that symmetric-key encryption is more efficient than public-key encryption. Thus, session keys are usually keys of a symmetric cryptosystem, and these keys must be exchanged in a secure way, by using other keys. The second, probably more important reason is that the two-level concept provides more security. If a session key is compromised, it affects only that session; other sessions in the past or in the future are not concerned. Given one session key, the number of ciphertexts available for cryptanalysis is limited. Session keys are generated when actually required and discarded after use; they need not be stored. Thus, there is no need to protect a large amount of stored keys.

A master key is used for the generation of session keys. Special care is taken to prevent attacks on the master key. The access to the master key is severely limited. It is possible to store the master key on protected hardware, accessible only via a secure interface. The main focus of this section is to describe how to establish a session key between two parties.

Besides key exchange, we introduce *entity authentication*. Entity authentication prevents impersonation. By entity authentication, Alice can convince her communication partner Bob that, in the current communication, her identity is as declared. This might be achieved, for example, if Alice signs a specific message. Alice proves her identity by her signature on the message. If an adversary Eve intercepts the message signed by Alice, she can use it later to authenticate herself as Alice. Such attacks are called *replay attacks*. A replay attack can be prevented if the message to be signed by Alice varies. For this purpose we introduce two methods. In the first method, Alice puts Bob's name and a time stamp into the message she signs. Bob accepts a message only if it appears the first time. The second method uses random numbers. A random number is chosen by Bob and transmitted to Alice. Then Alice puts the random number into the message, signs the message and returns it to Bob. Bob can check that the returned random number matches the random number he sent and the validity of Alice's signature. The random number is viewed as a challenge. Bob sends a challenge to Alice and Alice returns a response to Bob's challenge. We speak of *challenge-response* identification.

Some of the protocols we discuss provide both entity authentication and key exchange.

4.1.1 Kerberos

Kerberos denotes the distributed authentication service originating from MIT's project Athena. Here we use the term Kerberos in a restricted way: we define it as the underlying protocol that provides both entity authentication and key establishment, by use of symmetric cryptography and a trusted third party.

We continue with our description in [NeuTs'o94]. In that overview article, a simplified version of the basic protocol is introduced to make the basic principles clear. Kerberos is designed to authenticate clients who try to get access to servers in a network. A central role is played by a trusted server called the *Kerberos authentication server*.

The Kerberos authentication server T shares a secret key of a symmetric key encryption scheme E with each client A and each server B in the network, denoted by k_A and k_B respectively. Now, assume that the client A wants to access the server B. At the outset, A and B do not share any secrets. The execution of the Kerberos protocol involves A, B and T, and proceeds as follows:

The client A sends a request to the authentication server T, requesting *credentials* for the server B. T responds with these credentials. The credentials consist of:

1. A *ticket* t for the server B, encrypted with B's secret key k_B.
2. A session key k, encrypted with A's key k_A.

The ticket t contains A's identity and a copy of the session key. It is intended for B. A will forward it to B. The ticket is encrypted with k_B, which is known only to B and T. Thus, it is not possible for A to modify the ticket without being detected. A creates an *authenticator* which also contains A's identity, encrypts it with the session key k (by this encryption A proves to B that she knows the session key embedded in the ticket) and transmits the ticket and the authenticator to B. B trusts the ticket (it is encrypted with k_B, hence it originates from T), decrypts it and gets k. Now B uses the session key to decrypt the authenticator. If B succeeds, he is convinced that A encrypted the authenticator, because only A and the trusted T can know k. Thus A is authenticated to B. Optionally, the session key k can also be used to authenticate B to A. Finally, k may be used to encrypt further communication between the two parties in the current session.

Kerberos protects the ticket and the authenticator against modification by encrypting it. Thus, the encryption algorithm E is assumed to have built-in data integrity mechanisms.

Protocol 4.1.
Basic Kerberos authentication protocol (simplified):

1. A chooses r_A at random[1] and sends (A, B, r_A) to T.
2. T generates a new session key k, and creates a ticket
   $t := (A, k, l)$. Here l defines a validity period (consisting of a starting and an ending time) for the ticket. T sends
   $(E_{k_A}(k, r_A, l, B), E_{k_B}(t))$ to A.
3. A recovers k, r_A, l and B, and verifies that r_A and B match those sent in step 1. Then she creates an authenticator
   $a := (A, t_A)$, where t_A is a time stamp from A's local clock, and sends $(E_k(a), E_{k_B}(t))$ to B.
4. B recovers $t = (A, k, l)$ and $a = (A, t_A)$, and checks that:

---

[1] In this chapter all random choices are with respect to the uniform distribution. All elements have the same probability (see Appendix B.1).

    a. The identifier A in the ticket matches the one in the authenticator.

    b. The time stamp t_A is fresh, i.e., within a small time interval around B's local time.

    c. The time stamp t_A is in the validity period l.

If all checks pass, B considers the authentication of A as successful.

If, in addition, B is to be authenticated to A, steps 5 and 6 are executed:

    5. B takes t_A and sends $E_k(t_A)$ to A.

    6. A recovers t_A from $E_k(t_A)$ and checks whether it matches with the t_A sent in step 3. If yes, she considers B as authenticated.

Remarks:

1. In step 1, a random number is included in the request. It is used to match the response in step 2 with the request. This ensures that the Kerberos authentication server is alive and created the response. In step 3, a time stamp is included in the request to the server. This prevents replay attacks of such requests. To avoid perfect time synchronization, a small time interval around B's local time (called a window) is used. The server accepts the request if its time stamp is in the current window and appears the first time. The use of time stamps means that the network must provide secure and synchronized clocks. Modifications of local time clocks by adversaries must be prevented to guarantee the security of the protocol.

2. The validity period of a ticket allows the reuse of the ticket in that period. Then steps 1 and 2 in the protocol can be omitted. The client can use the ticket t to repeatedly get access to the server for which the ticket was issued. Each time, she creates a new authenticator and executes steps 3 and 4 (or steps 3–6) of the protocol.

3. Kerberos is a popular authentication service. Version 5 of Kerberos (the current version) was specified in [RFC 1510]. Kerberos is based in part on Needham and Schroeder's trusted third-party authentication protocol [NeeSch78].

4. In the non-basic version of Kerberos, the authentication server is only used to get tickets for the ticket-granting server. These tickets are called ticket-granting tickets. The ticket-granting server is a specialized server, granting tickets (server tickets) for the other servers (the ticket-granting server must have access to the servers' secret keys, so usually the authentication server and the ticket granting server run on the same host).

   Client A executes steps 1 and 2 of Protocol 4.1 with the authentication server in order to obtain a ticket-granting ticket. Then A uses the ticket-granting ticket – more precisely, the session key included in the ticket granting ticket – to authenticate herself to the ticket-granting server and to get server tickets. The ticket-granting ticket can be reused during its validity period for the intended ticket-granting server. As long as the same ticket-granting ticket is used, the client's secret key k_A is not used again. This reduces the risk of exposing k_A. We get a three-level key scheme. The first embraces the long-lived secret keys of the participating clients and servers. The keys of the second level are the session keys of the ticket-granting tickets, and the keys of the third level are the session keys of the server tickets.

A ticket-granting ticket is verified by the ticket-granting server in the same way as any other ticket (see above). The ticket-granting server decrypts the ticket, extracts the session key and decrypts the authenticator with the session key. The client uses a ticket from the ticket-granting server as in the basic model to authenticate to a service in the network.

4.1.2 Diffie–Hellman Key Agreement

Diffie–Hellman key agreement (also called exponential key exchange) provided the first practical solution to the key distribution problem. It is based on public-key cryptography. Diffie and Hellman published their fundamental technique of key exchange together with the idea of public-key cryptography in the famous paper, "New Directions in Cryptography", in 1976 in [DifHel76]. Exponential key exchange enables two parties that have never communicated before to establish a mutual secret key by exchanging messages over a public channel. However, the scheme only resists passive adversaries.

Let p be a sufficiently large prime such that it is intractable to compute discrete logarithms in $\mathbb{Z}_p^*$. Let g be a primitive root in $\mathbb{Z}_p^*$. p and g are publicly known. Alice and Bob can establish a secret shared key by executing the following protocol:

Protocol 4.2.
Diffie–Hellman key agreement:

1. Alice chooses a, $1 \leq a \leq p-2$, at random, sets $c := g^a$ and sends c to Bob.
2. Bob chooses b, $1 \leq b \leq p-2$, at random, sets $d := g^b$ and sends d to Alice.
3. Alice computes the shared key $k = d^a = (g^b)^a$.
4. Bob computes the shared key $k = c^b = (g^a)^b$.

Remarks:

1. If an attacker can compute discrete logarithms in $\mathbb{Z}_p^*$, he can compute a from c and then $k = d^a$. However, to get the secret key k, it would suffice to compute g^{ab} from g^a and g^b. This problem is called the *Diffie–Hellman problem*. The security of the protocol relies on the assumption that no efficient algorithms exist to solve this problem. This assumption is called

the *Diffie–Hellman assumption*. It implies the discrete logarithm assumption. For certain primes, the Diffie–Hellman and the discrete logarithm assumption have been proven to be equivalent ([Boer88]; [Maurer94]; [MauWol96]; [MauWol98]; [MauWol00]).

2. Alice and Bob can use the randomly chosen element $k = g^{ab} \in \mathbb{Z}_p^*$ as their session key. The Diffie–Hellman assumption does not guarantee that individual bits or groups of bits of k cannot be efficiently derived from g^a and g^b – this would be a stronger assumption.

   It is recommended to make prime p 1024 bits long. Usually, the length of a session key in a symmetric-key encryption scheme is much smaller, say 128 bits. If we take, for example, the 128 most-significant bits of k as a session key $\tilde{k}$, then $\tilde{k}$ is hard to compute from g^a and g^b. However, we do not know that all the individual bits of $\tilde{k}$ are secure (on the other hand, none of the more significant bits is known to be easy to compute). In [BonVen96] it is shown that computing the $\sqrt{|p|}$ most-significant bits of g^{ab} from g^a and g^b is as difficult as computing g^{ab} from g^a and g^b. For a 1024-bit prime p, this result implies that the 32 most-significant bits of g^{ab} are hard to compute. The problem of finding a more secure random session key can be solved by applying an appropriate hash function h to g^{ab}, and taking $\tilde{k} = h(g^{ab})$.

3. This protocol provides protection against passive adversaries. An active attacker Eve can intercept the message sent to Bob by Alice and then play Bob's role. The protocol does not provide authentication of the opposite party. Combined with authentication techniques, the Diffie–Hellman key agreement can be used in practice (see Section 4.1.4).

4.1.3 Key Exchange and Mutual Authentication

The problem of spontaneous key exchange, like Diffie–Hellman's key agreement, is the authenticity of the communication partners in an open network. Entity authentication (also called entity identification) guarantees the identity of the communicating parties in the current communication session, thereby preventing impersonation. Mutual entity authentication requires some mutual secret, which has been exchanged previously, or access to predistributed authentic material, like the public keys of a digital signature scheme.

The protocol we describe is similar to the X.509 strong three-way authentication protocol (see [ISO/IEC 9594-8]). It provides entity authentication and key distribution, two different cryptographic mechanisms. The term "strong" distinguishes the protocol from simpler password-based schemes. To set up the scheme, a public-key encryption scheme (E, D) and a digital signature scheme $(Sign, Verify)$ are needed. Each user Alice has a key pair (e_A, d_A) for encryption and another key pair (s_A, v_A) for digital signatures. It is assumed that everyone has access to Alice's authentic public keys, e_A and v_A, for encryption and the verification of signatures.

Executing the following protocol, Alice and Bob establish a secret session key. Furthermore, the protocol guarantees the mutual authenticity of the communication parties.

Protocol 4.3.
Strong three-way authentication:

1. Alice chooses r_A at random, sets $t_1 := (B, r_A)$ (where B represents Bob's identity), $s_1 := Sign_{s_A}(t_1)$ and sends (t_1, s_1) to Bob.
2. Bob verifies Alice's signature, checks that he is the intended recipient, chooses r_B and a session key k at random, encrypts the session key with Alice's public key, $c := E_{e_A}(k)$, sets $t_2 := (A, r_A, r_B, c)$, signs t_2 to get $s_2 := Sign_{s_B}(t_2)$ and sends (t_2, s_2) to Alice.
3. Alice verifies Bobs signature, checks that she is the intended recipient and that the r_A she received matches the r_A from step 1 (this prevents replay attacks). If both verifications pass, she is convinced that her communication partner is Bob. Now Alice decrypts the session key k, sets $t_3 := (B, r_B)$, $s_3 := Sign_{s_A}(t_3)$ and sends (t_3, s_3) to Bob.
4. Bob verifies Alice's signature and checks that the r_B he received matches the r_B from step 2 (this again prevents replay attacks). If both verifications pass, Bob and Alice use k as the session key.

Remarks:

1. The protocol identifies the communication partner by checking that she possesses the secret key of the signature scheme. The check is done by the *challenge-response* principle. First the challenge, a random number, used only once, is submitted. If the partner can return a signature of this random number, he necessarily possesses the secret key, thus proving his identity. The messages exchanged (the communication tokens) are signed by the sender and contain the recipient. This guarantees that the token was constructed for the intended recipient by the sender. Three messages are exchanged in the above protocol. Therefore, it is called the *three-way* or *three-pass authentication protocol*.
   There are also *two-way authentication protocols*. To prevent replay attacks, the communication tokens must be stored. Since these communication tokens have to be deleted from time to time, they are given a time stamp and an expiration time. This requires a network with secure and synchronized clocks. A three-way protocol requires more messages to be exchanged, but avoids storing tokens and maintaining secure and synchronized clocks.
2. The session key is encrypted with a public-key cryptosystem. Suppose adversary Eve records all the data that Alice and Bob have exchanged,

hoping that Alice's secret encryption key will be compromised in the future. If Eve really gets Alice's secret key, she can decrypt the data of all sessions she recorded. A key-exchange scheme which resists this attack is said to have *forward secrecy*. The Diffie–Hellman key agreement does not encrypt a session key. Thus, the session key cannot be revealed by a compromised secret key. If we combine the Diffie–Hellman key agreement with the authentication technique of the previous protocol, as in Section 4.1.4, we achieve a key-exchange protocol with authentication and forward secrecy.

4.1.4 Station-to-Station Protocol

The station-to-station protocol, combines Diffie–Hellman key agreement with authentication. It goes back to earlier work on ISDN telephone security, as outlined by Diffie in [Diffie88], in which the protocol is executed between two ISDN telephones (stations). The station-to-station protocol enables two parties to establish a shared secret key k to be used in a symmetric encryption algorithm E. Additionally, it provides mutual authentication.

Let p be a prime such that it is intractable to compute discrete logarithms in $\mathbb{Z}_p^*$. Let g be a primitive root in $\mathbb{Z}_p^*$. p and g are publicly known. Further, we assume that a digital signature scheme (*Sign*, *Verify*) is available. Each user Alice has a key pair (s_A, v_A) for this signature scheme. s_A is the secret key for signing and v_A is the public key for verifying Alice's signatures. We assume that each party has access to authentic copies of the other's public key.

Alice and Bob can establish a secret shared key and authenticate each other if they execute the following protocol:

Protocol 4.4.
Station-to-station protocol:

1. Alice chooses a, $1 \leq a \leq p-2$, at random, sets $c := g^a$ and sends c to Bob.
2. Bob chooses b, $1 \leq b \leq p - 2$, at random, computes the shared secret key $k = g^{ab}$, takes his secret key s_B and signs the concatenation of g^a and g^b to get $s := Sign_{s_B}(g^a \| g^b)$. Then he sends $(g^b, E_k(s))$ to Alice.
3. Alice computes the shared key $k = g^{ab}$, decrypts $E_k(s)$ and verifies Bob's signature. If this verification succeeds, Alice is convinced that the opposite party is Bob. She takes her secret key s_A, generates the signature $s := Sign_{s_A}(g^b \| g^a)$ and sends $E_k(s)$ to Bob.
4. Bob decrypts $E_k(s)$ and verifies Alice's signature. If the verification succeeds, Bob accepts that he actually shares k with Alice.

Remarks:

1. Both Alice and Bob contribute to the random strings $g^a \| g^b$ and $g^b \| g^a$. Thus each string can serve as a challenge.
2. Encrypting the signatures with the key k guarantees that the party who signed also knows the secret key k.

4.1.5 Public-Key Management Techniques

In the public-key-based key-exchange protocols discussed in the previous sections, we assumed that each party has access to the other parties' (authentic) public keys. This requirement can be met by public-key management techniques. A trusted third party C is needed, similar to the Kerberos authentication server in the Kerberos protocol. However, in contrast to Kerberos, the authentication transactions do not include an online communication with C. C prepares information in advance, which is then available to Alice and Bob during the execution of the authentication protocol. We say that C is offline. Offline third parties reduce network traffic, which is advantageous in large networks.

Certification Authority. The offline trusted party is referred to as a *certification authority*. Her tasks are:

1. To verify the authenticity of the entity to be associated with a public key.
2. To bind a public key to a distinguished name and to register it.
3. (Optionally) to generate a party's private and public key.

Certificates play a fundamental role. They enable the storage and forwarding of public keys over insecure media, without danger of undetected manipulation. Certificates are signed by a certification authority, using a public-key signature scheme. Everyone knows the certification authority's public key. The authenticity of this key may be provided by non-cryptographic means, such as couriers or personal acquisition. Another method would be to publish the key in all newspapers. The public key of the certification authority can be used to verify certificates signed by the certification authority. Certificates prove the binding of a public key to a distinguished name. The signature of the certification authority protects the certificate against undetected manipulation. We list some of the most important data stored on a certificate:

1. A distinguished name (the real name or a pseudonym of the owner of the certificate).
2. The owner's public key.
3. The name of the certification authority.
4. A serial number identifying the certificate.
5. A validity period of the certificate.

Creation of Certificates. If Alice wants to get a certificate, she goes to a certification authority C. To prove her identity, Alice shows her passport. Now, Alice needs public- and private-key pairs for encryption and digital signatures. Alice can generate the key pair and hand over a copy of the public key to C. This alternative might be taken if Alice uses a smart card to store her secret key. Smart cards often involve key generation functionality. If the keys are generated inside the smart card, the private key never leaves it. This reduces the risk of theft. Another model is that C generates the key pair and transmits the secret key to Alice. The transmission requires a secret channel. C must of course be trustworthy, because she has the opportunity to steal the secret key. After the key generation, C puts the public key on the certificate, together with all the other necessary information, and signs the certificate with her secret key.

Storing Certificates. Alice can take her certificate and store it at home. She provides the certificate to others when needed, for example for signature verification. A better solution in an open system is to provide a certificate directory, and to store the certificates there. The certificate directory is a (distributed) database, usually maintained by the certification authority. It enables the search and retrieval of certificates.

Usage of Certificates. If Bob wants to encrypt a message for Alice or to verify a signature allegedly produced by Alice, he retrieves Alice's certificate from the certificate directory (or from Alice) and verifies the certification authority's signature. If the verification is successful, he can be sure that he really receives Alice's public key from the certificate and can use it.

For reasons of operational efficiency, multiple certification authorities are needed in large networks. If Alice and Bob belong to different certification authorities, Bob must access an authentic copy of the public key of Alice's certification authority. This is possible if Bob's certification authority C_B has issued a certificate for Alice's certification authority C_A. Then Bob retrieves the certificate for C_A, verifies it and can then trust the public key of C_A. Now he can retrieve and verify Alice's certificate.

It is not necessary that each certification authority issues a certificate for each other certification authority in the network. We may use a directed graph as a model. The vertices correspond to the certification authorities, and an edge from C_A to C_B corresponds to a certificate of C_A for C_B. Then, a directed path should connect any two certification authorities. This is the minimal requirement which guarantees that each user in the network can verify each other user's certificate.

However, the chaining of authentications may reduce the trust in the final result: the more people you have to trust, the greater the risk that you have a cheater in the group.

Revocation of Certificates. If Alice's secret key is compromised, the corresponding public key can no longer be used for encrypting messages. If the key is used in a signature scheme, Alice can no longer sign messages with this key. Moreover, it should be possible for Alice to deny all signatures produced with this key from then on. Therefore, the fact that Alice's secret key is compromised must be publicized. Of course, the certification authority will remove Alice's certificate from the certificate directory. However, certificates may have been retrieved before and may not yet have expired. It is not possible to notify all users possessing copies of Alice's certificate: they are not known to the certification authority. A solution to this problem is to maintain certificate revocation lists. A certificate revocation list is a list of entries corresponding to revoked certificates. To guarantee authenticity, the list is signed by the certification authority.

4.2 Identification Schemes

There are many situations where it is necessary to "prove" one's identity. Typical scenarios are to login to a computer, to get access to an account for electronic banking or to withdraw money from an automatic teller machine. Older methods use passwords or PINs to implement user identification. Though successfully used in certain environments, these methods also have weaknesses. For example, anyone to whom you must give your password to be verified has the ability to use that password and impersonate you. Zero-knowledge (and other) identification schemes provide a new type of user identification. It is possible for you to authenticate yourself without giving to the authenticator the ability to impersonate you. We will see that very efficient implementations of such schemes exist.

4.2.1 Interactive Proof Systems

There are two participants in an interactive proof system, the *prover* and the *verifier*. It is common to call the prover Peggy and the verifier Vic. Peggy knows some fact (e.g., a secret key sk of a public-key cryptosystem or a square root of a quadratic residue x), which we call the *prover's secret*. In an *interactive proof of knowledge*, Peggy wishes to convince Vic that she knows the prover's secret. Peggy and Vic communicate with each other through a communication channel. Peggy and Vic alternately perform *moves* consisting of:

1. Receive a message from the opposite party.
2. Perform some computation.
3. Send a message to the opposite party.

Usually, Peggy starts and Vic finishes the protocol. In the first move, Peggy does not receive a message. The interactive proof may consist of several *rounds*. This means that the protocol specifies a sequence of moves, and this sequence is repeated a specified number of times. Typically, a move consists of a challenge by Vic and a response by Peggy. Vic accepts or rejects

Peggy's proof, depending on whether Peggy successfully answers all of Vic's challenges.

Proofs in interactive proof systems are quite different from proofs in mathematics. In mathematics, the prover of some theorem can sit down and prove the statement by himself. In interactive proof systems, there are two computational tasks, namely producing a proof (Peggy's task) and verifying its validity (Vic's task). Additionally, communication between the prover and the verifier is necessary.

We have the following requirements for interactive proof systems.

1. *(Knowledge) completeness.* If Peggy knows the prover's secret, then Vic will always accept Peggy's proof.
2. *(Knowledge) soundness.* If Peggy can convince Vic with reasonable probability, then she knows the prover's secret.

If the prover and the verifier of an interactive proof system follow the behavior specified in the protocol, they are called an *honest verifier* and an *honest prover*. A prover who does not know the prover's secret and tries to convince the verifier is called a *cheating or dishonest prover*. A verifier who does not follow the behavior specified in the protocol is called a *cheating or dishonest verifier*. Sometimes, the verifier can get additional information from the prover if he does not follow the protocol. Note that each prover (or verifier), whether she is honest or not, fulfills the syntax of the communication interface, because not following the syntax is immediately detected. She may only be dishonest in her private computations and the resulting data that she transmits.

Password Scheme. In a simple password scheme, Peggy uses a secret password to prove her identity. The password is the only message, and it is sent from the prover Peggy to the verifier Vic. Vic accepts Peggy's identity if the transmitted password and the stored password are equal. Here, only one message is transmitted, and obviously the scheme meets the requirements. If Peggy knows the password, Vic accepts. If a cheating prover Eve does not know the password, Vic does not accept. The problem is that everyone who observed the password during communication can use the password.

Identification Based on Public-Key Encryption. In Section 4.1, we considered an identification scheme based on a public-key cryptosystem. We recall the basic scenario. Each user Peggy has a secret key sk only known to her and a public key pk known to everyone. Suppose that everyone who can decrypt a randomly chosen encrypted message must know the secret key. This assumption should be true if the cryptosystem is secure. Hence, the secret key sk can be used to identify Peggy.

Peggy proves her identity to Vic using the following steps:

1. Vic chooses a random message m, encrypts it with the public key pk and sends the cryptogram c to Peggy.

2. Peggy decrypts c with her secret key sk and sends the result m' back to Vic.
3. Vic accepts the identity of Peggy if and only if $m = m'$.

Two messages are exchanged: it is a *two-move* protocol. The completeness of the scheme is obvious. On the other hand, a cheating prover who only knows the public key and a ciphertext should not be able to find the plaintext better than guessing at random. The probability that Vic accepts if the prover does not know the prover's secret is very small. Thus, the scheme is also sound. This reflects Vic's security requirements. Suppose that an adversary Eve observed the exchanged messages and later wants to impersonate Peggy. Vic chooses m at random and computes c. The probability of obtaining the previously observed c is very small. Thus, Eve cannot take advantage of observing the exchanged messages. At first glance, everything seems to be all right. However, there is a security problem if Vic is not honest and does not follow the protocol in step 1. If, instead of a randomly chosen encrypted message, he sends a cryptogram intended for Peggy, then he lets Peggy decrypt the cryptogram. He thereby manages to get the plaintext of a cryptogram which he could not compute by himself. This violates Peggy's security requirements.

4.2.2 Simplified Fiat–Shamir Identification Scheme

Let $n := pq$, where p and q are distinct primes. As usual, QR_n denotes the subgroup of squares in $\mathbb{Z}_n^*$ (see Definition A.54). Let $x \in \mathrm{QR}_n$, and let y be a square root of x. The modulus n and the square x are made public, while the prime factors p, q and y are kept secret. The square root y of x is the secret of prover Peggy. Here we assume that it is intractable to compute a square root of x, without knowing the prime factors p and q. This is guaranteed by the factoring assumption (see Definition 6.9) if p and q are sufficiently large randomly chosen primes, and x is also randomly chosen. Note that the ability to compute square roots is equivalent to the ability to factorize n (Proposition A.70). We assume that Peggy chooses n and y, computes $x = y^2$ and publishes the public key (n, x) to all participants. y is Peggy's secret. Then Peggy may prove her identity by an interactive proof of knowledge by proving that she knows a square root y of x. The computations are done in $\mathbb{Z}_n^*$.

Protocol 4.5.
Fiat–Shamir identification (simplified):

1. Peggy chooses $r \in \mathbb{Z}_n^*$ at random and sets $a := r^2$. Peggy sends a to Vic.
2. Vic chooses $e \in \{0, 1\}$ at random. Vic sends e to Peggy.
3. Peggy computes $b := ry^e$ and sends b to Vic, i.e., Peggy sends r if $e = 0$, and ry if $e = 1$.
4. Vic accepts if and only if $b^2 = ax^e$.

In the protocol, three messages are exchanged – it is a *three-move* protocol:

1. The first message is a commitment by Peggy that she knows a square root of a.
2. The second message is a challenge by Vic. If Vic sends $e = 0$, then Peggy has to open the commitment and reveal r. If $e = 1$, she has to show her secret in encrypted form (by revealing ry).
3. The third message is Peggy's response to the challenge of Vic.

Completeness. If Peggy knows y, and both Peggy and Vic follow the protocol, then the response $b = ry^e$ is a square root of ax^e, and Vic will accept.

Soundness. A cheating prover Eve can convince Vic with a probability of $1/2$ if she behaves as follows:

1. Eve chooses $r \in \mathbb{Z}_n^*$ and $\tilde{e} \in \{0,1\}$ at random, and sets $a := r^2 x^{-\tilde{e}}$. Eve sends a to Vic.
2. Vic chooses $e \in \{0,1\}$ at random. Vic sends e to Eve.
3. Eve sends r to Vic.

If $e = \tilde{e}$, Vic accepts. The event $e = \tilde{e}$ occurs with a probability of $1/2$. Thus, Eve succeeds in cheating with a probability of $1/2$.

On the other hand, $1/2$ is the best probability of success that a cheating prover can reach. Namely, assume that a cheating prover Eve can convince Vic with a probability $> 1/2$. Then Eve knows an a for which she can correctly answer both challenges. This means that Eve can compute b_1 and b_2 such that

$$b_1^2 = a \text{ and } b_2^2 = ax.$$

Hence, she can compute the square root $y = b_2 b_1^{-1}$ of x. Recall that x is a random quadratic residue. Thus Eve has an algorithm A that on input $x \in \mathrm{QR}_n$ outputs a square root y of x. Then Eve can use A to factor n (see Proposition A.70). This contradicts our assumption that the factorization of n is intractable.

Security. We have to discuss the security of the scheme from the prover's and from the verifier's points of view.

The verifier accepts the proof of a cheating prover with a probability of $1/2$. The large probability of success of a cheating prover is too high in practice. It might be decreased by performing t rounds, i.e., by iterating the basic protocol t times sequentially and independently. In this way, the probability of cheating is reduced to 2^{-t}. In Section 4.2.4 we will give a generalized version of the protocol, which decreases the probability of success of a cheating prover.

Now we look at the basic protocol from an honest prover's point of view, and study Peggy's security requirements. Vic chooses his challenges from the small set $\{0,1\}$. He has no chance of producing side effects, as in the identification scheme based on public-key cryptography given above. The only information Peggy "communicates" to Vic is the fact that she knows a square root of x. The protocol has the zero-knowledge property studied in Section 4.2.3.

Remark. In a proof of knowledge, the secret knowledge that enables Peggy to prove her claim is called a *witness*. In the simplified Fiat–Shamir protocol, any square root of x is a witness for the claim to know a square root of x. We have just seen that, in one round, the maximum probability of success of a cheating prover Eve is $1/2$. A better chance for cheating would enable Eve to compute a witness, and this contradicts the factorization assumption.

4.2.3 Zero-Knowledge

In the interactive proof system based on a public-key cryptosystem, which we discussed above, a dishonest verifier Vic can decrypt Peggy's cryptograms by interacting with Peggy. Since Vic is not able to decrypt them without interaction, he learns something new by interacting with Peggy. He obtains *knowledge* from Peggy. This is not desirable, because it might violate Peggy's security requirements as our example shows. It is desirable that interactive proof systems are designed so that no knowledge is transferred from the prover to the verifier. Such proof systems are called zero-knowledge. Informally, an interactive proof system is called zero-knowledge if whatever the verifier can efficiently compute after interacting with the prover, can be efficiently simulated without interaction. Below we define the zero-knowledge property more formally.

We denote the algorithm that the honest prover Peggy executes by P, the algorithm of an honest verifier by V and the algorithm of a general (possibly dishonest) verifier by V^*. The interactive proof system (including the interaction between P and V) is denoted by (P, V). Peggy knows a secret about some object x (e.g., as in the Fiat–Shamir example in Protocol 4.5, the root of a square x). This object x is the common input to P and V.

Each algorithm is assumed to have polynomial running time. It may be partly controlled by random events, i.e., it has access to a source of random bits and thus can make random choices. Such algorithms are called *probabilistic algorithms*. We study this notion in detail in Chapter 5.

Let x be the common input of (P, V). Suppose, the interactive proof takes n moves. A message is sent in each move. For simplicity, we assume that the prover starts with the first move. We denote by m_i the message sent in the i-th move. The messages $m_1, m_3, \ldots$ are sent from the prover to the verifier and the messages $m_2, m_4, \ldots$ are sent from the verifier to the prover. The *transcript* of the joint computation of P and V^* on input x is defined by

$$tr_{P,V^*}(x) := (m_1, \ldots, m_n),$$

where $tr_{P,V^*}(x)$ is called an accepting transcript if V^* accepts after the last move. Note that the transcript $tr_{P,V^*}(x)$ depends on the random bits that the algorithms P and V^* choose. Thus, it is not determined by the input x.

Definition 4.6. An interactive proof system (P, V) is *(perfect) zero-knowledge* if there is a probabilistic *simulator* $S(V^*, x)$, running in expected polynomial time, which for every verifier V^* (honest or not) outputs on input x an accepting transcript t of P and V^* such that these simulated transcripts are distributed in the same way as if they were generated by the honest prover P and V^*.

Remark. The definition of zero-knowledge includes all verifiers (also the dishonest ones). Hence, zero-knowledge is a property of the prover P. It captures the prover's security requirements against attempts to gain "knowledge" by interacting with him.

To understand the definition, we have to clarify what a simulator is. A simulator S is an algorithm which, given some verifier V^*, honest or not, generates valid accepting transcripts for (P, V^*), without communicating with the real prover P. In particular, S does not have any access to computations that rely on the prover's secret. Trying to produce an accepting transcript, S plays the role of P in the protocol and communicates with V^*. Thus, he obtains outgoing messages of V^* which are compliant with the protocol. His task is to fill into the transcript the messages going out from P. Since P computes these messages by use of her secret and S does not know this secret, S applies his own strategy to generate the messages. Necessarily, his probability of obtaining a valid transcript in this way is significantly less than 1. Otherwise, with high probability, S could falsely convince V^* that he knows the secret, and the proof system is not sound. Thus, not every attempt of S to produce an accepting transcript is successful; he fails in many cases. Nevertheless, by repeating his attempts sufficiently often, the simulator is able to generate a valid accepting transcript. It is required that the expected value of the running time which S needs to get an accepting transcript is bounded by a polynomial in the binary length $|x|$ of the common input x.[2]

To be zero-knowledge, the ability to produce accepting transcripts by a simulation is not sufficient. The generation of transcripts, real or simulated, includes random choices. Thus, we have a probability distribution on the set of accepting transcripts. The last condition in the definition means that the probability distribution of the transcripts that are generated by the simulator S and V^* is the same as if they were generated by the honest prover P and V^*. Otherwise, the distribution of transcripts might contain information about the secret and thus reveal some of P's knowledge.

In the following, we will illustrate the notion of zero-knowledge and the simulation of a prover by the simplified version of the Fiat–Shamir identification (Protocol 4.5).

Proposition 4.7. *The simplified version of the Fiat–Shamir identification scheme is zero-knowledge.*

---

[2] In other words, S is a Las Vegas algorithm (see Section 5.2).

Proof. The set of accepting transcripts is

$$\mathcal{T}(x) := \{(a, e, b) \in \mathrm{QR}_n \times \{0, 1\} \times \mathbb{Z}_n^* \mid b^2 = ax^e\}.$$

Let V^* be a general (honest or cheating) verifier. Then, a simulator S with the desired properties is given by the following algorithm.

Algorithm 4.8.

    transcript S(algorithm V^*, int x)
    1   repeat
    2   select $\tilde{e} \in \{0, 1\}$ and $\tilde{b} \in \mathbb{Z}_n^*$ uniformly at random
    3       $\tilde{a} \leftarrow \tilde{b}^2 x^{-\tilde{e}}$
    4       $e \leftarrow V^*(\tilde{a})$
    5   until $e = \tilde{e}$
    6   return $(\tilde{a}, \tilde{e}, \tilde{b})$

The simulator S uses the verifier V^* as a subroutine to get the challenge e. S tries to guess e in advance. If S succeeded in guessing e, he can produce a valid transcript $(\tilde{a}, \tilde{e}, \tilde{b})$. S cannot produce e by himself, because V^* is an arbitrary verifier. Therefore, V^* possibly does not generate the challenges e randomly, as it is specified in (P, V), and S must call V^* to get e. Independent of the strategy that V^* uses to output e, the guess $\tilde{e}$ coincides with e with a probability of $1/2$. Namely, if V^* outputs 0 with a probability of p and 1 with a probability of $1 - p$, the probability that $e = 0$ and $\tilde{e} = 0$ is $p/2$, and the probability that $e = 1$ and $\tilde{e} = 1$ is $(1 - p)/2$. Hence, the probability that one of both events occurs is $1/2$.

The expectation is that S will produce a result after two iterations of the while loop (see Lemma B.12). An element $(\tilde{a}, \tilde{e}, \tilde{b}) \in \mathcal{T}$ returned by S cannot be distinguished from an element (a, e, b) produced by (P, V^*):

1. a and $\tilde{a}$ are random quadratic residues in QR_n.
2. e and $\tilde{e}$ are distributed according to V^*.
3. b and $\tilde{b}$ are random square roots.

This concludes the proof of the proposition. $\square$

4.2.4 Fiat–Shamir Identification Scheme

As in the simplified version of the Fiat–Shamir identification scheme, let $n := pq$, where p and q are distinct primes. Again, computations are performed in $\mathbb{Z}_n$, and we assume that it is intractable to compute square roots of randomly chosen elements in QR_n, unless the factorization of n is known (see Section 4.2.2). In the simplified version of the Fiat–Shamir identification scheme, the verifier accepts the proof of a cheating prover with a probability of $1/2$. To reduce this probability of success, now the prover's secret is a vector $y := (y_1, \ldots, y_t)$ of randomly chosen square roots. The modulus n

and the vector $x := (y_1^2, \ldots, y_t^2)$ are made public. As above, we assume that Peggy chooses n and y, computes x and publishes the public key (n, x) to all participants. Peggy's secret is y.

Protocol 4.9.

Fiat–Shamir Identification:

Repeat the following k times:
1. Peggy chooses $r \in \mathbb{Z}_n^*$ at random and sets $a := r^2$. Peggy sends a to Vic.
2. Vic chooses $e := (e_1, \ldots, e_t) \in \{0, 1\}^t$ at random. Vic sends e to Peggy.
3. Peggy computes $b := r \prod_{i=1}^t y_i^{e_i}$. Peggy sends b to Vic.
4. Vic rejects if $b^2 \neq a \prod_{i=1}^t x_i^{e_i}$, and stops.

Security. The Fiat–Shamir identification scheme extends the simplified scheme in two aspects. First, a challenge $e \in \{0, 1\}$ in the basic scheme is replaced by a challenge $e \in \{0, 1\}^t$. Then the basic scheme is iterated k times. A cheating prover Eve can convince Vic if she guesses Vic's challenge e correctly for each iteration; i.e., if she manages to select the right element from $\{0, 1\}^{kt}$. Her probability of accomplishing this is 2^{-kt}. It can be shown that for $t = O(\log_2(|n|))$ and $k = O(|n|^l)$, the interactive proof system is still zero-knowledge. Observe here that the expected running time of a simulator that is constructed in a similar way as in the proof of Proposition 4.7 is no longer polynomial if t or k are too large.

Completeness. If the legitimate prover Peggy and Vic follow the protocol, then Vic will accept.

Soundness. Suppose a cheating prover Eve can convince (the honest verifier) Vic with a probability $> 2^{-kt}$, where the probability is taken over all the challenges e. Then Eve knows a vector $A = (a^1, \ldots, a^k)$ of commitments a^j (one for each iteration j, $1 \leq j \leq k$) for which she can correctly answer two distinct challenges $E = (e^1, \ldots, e^k)$ and $F = (f^1, \ldots, f^k)$, $E \neq F$, of Vic. There is an iteration j such that $e^j \neq f^j$, and Eve can answer both challenges $e := e^j$ and $f := f^j$ for the commitment $a = a^j$. This means that Eve can compute b_1 and b_2 such that

$$b_1^2 = a \prod_{i=1}^t x_i^{e_i} \text{ and } b_2^2 = a \prod_{i=1}^t x_i^{f_i}.$$

As in Section 4.2.2, this implies that Eve can compute the square root $y = b_2 b_1^{-1}$ of the random square $x = \prod_{i=1}^t x_i^{f_i - e_i}$. This contradicts our assumption that computing square roots is intractable without knowing the prime factors p and q of n.

Remark. The number ν of exchanged bits is $k(2|n|+t)$, the average number μ of multiplications is $k(t+2)$ and the key size s (equal to the size of the prover's secret) is $t|n|$. Choosing k and t appropriately, different values for the three numbers can be achieved. All choices of k and t, with kt constant, lead to the same level of security 2^{-kt}. Keeping the product kt constant and increasing t, while decreasing k, yields smaller values for ν and μ. However, note that the scheme is proven to be zero-knowledge only for small values of t.

4.2.5 Fiat–Shamir Signature Scheme

[FiaSha86] gives a standard method for converting an interactive identification scheme into a digital signature scheme. Digital signatures are produced by the signer without interaction. Thus, the communication between the prover and the verifier has to be eliminated. The basic idea is to take the challenge bits, which in the identification scheme are generated by the verifier, from the message to be signed. It must be guaranteed that the signer makes his commitment before he extracts the challenge bits from the message. This is achieved by the clever use of a publicly known, collision-resistant hash function (see Section 2.2):

$$h : \{0,1\}^* \longrightarrow \{0,1\}^{kt}.$$

As an example, we convert the Fiat–Shamir identification scheme (Section 4.2.4) into a signature scheme.

Signing. To sign a message $m \in \{0,1\}^*$, Peggy proceeds in three steps:

1. Peggy chooses $(r_1,\ldots,r_k) \in \mathbb{Z}_n^{*\,k}$ at random and sets $a_j := r_j^2, 1 \le j \le k$.
2. She computes $h(m\|a_1\|\ldots\|a_k)$ and writes these bits into a matrix, column by column:

$$e := (e_{i,j})_{\substack{1\le i \le t \\ 1 \le j \le k}}.$$

3. She computes

$$b_j := r_j \prod_{i=1}^{t} y_i^{e_{ij}}, \quad 1 \le j \le k,$$

and sets $b = (b_1,\ldots,b_k)$. The signature of m is $s = (b,e)$.

Verification. To verify the signature $s = (b,e)$ of the signed message (m,s), we compute

$$c_j := b_j^2 \prod_{i=1}^{t} x_i^{-e_{ij}}, \quad 1 \le j \le k,$$

and accept if $e = h(m\|c_1\|\ldots\|c_k)$.

Here, the collision resistance of the hash function is needed. The verifier does not get the original values a_j – from step 1 of the protocol – to test $a_j = b_j^2 \prod_{i=1}^{t} x_i^{-e_{ij}}, 1 \le j \le k$.

Remark. The key size is $t|n|$ and the signature size is $k(t + |n|)$. The scheme is proven to be secure in the random oracle model, i.e., under the assumption that the hash function h is a truly random function (see Section 11.1 for a detailed discussion of what the security of a signature scheme means).

4.3 Commitment Schemes

Commitment schemes are of great importance in the construction of cryptographic protocols for practical applications (see Section 4.5.5), as well as for protocols in theoretical computer science. They are used, for example, to construct zero-knowledge proofs for all languages in $\mathcal{NP}$ (see [GolMicWid86]). This result is extended to the larger class of all languages in $\mathcal{IP}$, which is the class of languages that have interactive proofs (see [BeGrGwHåKiMiRo88]).

Commitment schemes enable a party to commit to a value while keeping it secret. Later, the committer provides additional information to open the commitment. It is guaranteed that after committing to a value, this value cannot be changed. No other value can be revealed in the opening step: if you have committed to 0, you cannot open 1 instead of 0, and vice versa. For simplicity (and without loss of generality), we only consider the values 0 and 1 in our discussion.

The following example demonstrates how to use commitment schemes. Suppose Alice and Bob are getting divorced. They have decided how to split their common possessions. Only one problem remains: who should get the car? They want to decide the question by a coin toss. This is difficult, because Alice and Bob are in different places and can only talk to each other by telephone. They do not trust each other to report the correct outcome of a coin toss. This example is attributable to M. Blum. He introduced the problem of tossing a fair coin by telephone and solved it using a bit-commitment protocol (see [Blum82]).

Protocol 4.10.
Coin tossing by telephone:

1. Alice tosses a coin, commits to the outcome b_A (heads = 0, tails = 1) and sends the commitment to Bob.
2. Bob also tosses a coin and sends the result b_B to Alice.
3. Alice opens her commitment by sending the additional information to Bob.

Now, both parties can compute the outcome $b_A \oplus b_B$ of the joint coin toss by telephone. If at least one of the two parties follows the protocol, i.e., sends the result of a true coin toss, the outcome is indeed a truly random bit.

In a commitment scheme, there are two participants, the committer (also called the sender) and the receiver. The commitment scheme defines two steps:

1. *Commit.* The sender sends the bit b he wants to commit to, in encrypted form, to the receiver.
2. *Reveal* or *open.* The sender sends additional information to the receiver, enabling the receiver to recover b.

There are three requirements:

1. *Hiding property.* In the commit step, the receiver does not learn anything about the committed value. He cannot distinguish a commitment to 0 from a commitment to 1.
2. *Binding property.* The sender cannot change the committed value after the commit step. This requirement has to be satisfied, even if the sender tries to cheat.
3. *Viability.* If both the sender and the receiver follow the protocol, the receiver will always recover the committed value.

4.3.1 A Commitment Scheme Based on Quadratic Residues

The commitment scheme based on quadratic residues enables Alice to commit to a single bit. Let QR_n be the subgroup of squares in $\mathbb{Z}_n^*$ (see Definition A.54), and let $J_n^{+1} := \{x \in \mathbb{Z}_n^* \mid \left(\frac{x}{n}\right) = 1\}$ be the units modulo n with Jacobi symbol 1 (see Definition A.61). Let $QNR_n^{+1} := J_n^{+1} \setminus QR_n$ be the non-squares in J_n^{+1}.

Protocol 4.11.
QRCommitment:

1. *System setup.* Alice chooses distinct large prime numbers p and q, and $v \in QNR_n^{+1}$, $n := pq$.
2. *Commit.* To commit to a bit b, Alice chooses $r \in \mathbb{Z}_n^*$ at random, sets $c := r^2 v^b$ and sends n, c and v to Bob.
3. *Reveal.* Alice sends p, q, r and b to Bob. Bob can verify that p and q are primes, $n = pq$, $r \in \mathbb{Z}_n^*$, $v \notin QR_n$ and $c = r^2 v^b$.

Remarks:

1. There is an efficient deterministic algorithm which computes the Jacobi symbol $\left(\frac{x}{n}\right)$ of x modulo n, without knowing the prime factors p and q of n (Algorithm A.65). Thus, it is easy to determine whether a given $x \in \mathbb{Z}_n^*$ is in J_n^{+1}. However, if the factors of n are kept secret, no efficient algorithm is known that can decide whether a randomly chosen element in J_n^{+1} is a square, and it is assumed that no efficient algorithm exists for this question of quadratic residuosity (a precise definition of this quadratic residuosity assumption is given in Definition 6.11). On the other hand, if p and q are known it is easy to check whether $v \in J_n^{+1}$ is a square. Namely, v is a square if and only if $v \bmod p$ and $v \bmod q$ are squares, and this in

turn is true if and only if the Legendre symbols $\left(\frac{v}{p}\right) = v^{(p-1)/2} \bmod p$ and $\left(\frac{v}{q}\right) = v^{(q-1)/2} \bmod q$ are equal to 1 (see Proposition A.58).

2. If Bob could distinguish a commitment to 0 from a commitment to 1, he could decide whether a randomly chosen element in J_n^{+1} is a square. This contradicts the quadratic residuosity assumption stated in the preceding remark.

3. The value c is a square if and only if v^b is a square, i.e., if and only if $b = 0$. Since c is either a square or a non-square, Alice cannot change her commitment after the commit step.

4. Bob needs p and q to check that v is not a square. By not revealing the primes p and q, Alice could use them for several commitments. Then, however, she has to prove that v is not a square. She could do this by an interactive zero-knowledge proof (see Exercise 3).

5. Bob can use Alice's commitment c and commit to the same bit b as Alice, without knowing b. He chooses $\tilde{r} \in \mathbb{Z}_n^*$ at random and sets $\tilde{c} = c\tilde{r}^2$. Bob can open his commitment after Alice has opened her commitment. If commitments are used as subprotocols, it might cause a problem if Bob blindly commits to the same bit as Alice. We can prevent this by asking Bob to open his commitment before Alice does.

4.3.2 A Commitment Scheme Based on Discrete Logarithms

The commitment scheme based on discrete logarithms enables Alice to commit to a message $m \in \{0, \ldots, q-1\}$. This scheme was introduced in [Pedersen91] and is called the *Pedersen commitment scheme*.

Protocol 4.12.

LogCommitment:

1. *System setup.* Bob randomly chooses large prime numbers p and q such that q divides $p - 1$. Then he randomly chooses g and v from the subgroup G_q of order q in $\mathbb{Z}_p^*$, $g, v \neq [1]$.[3] Bob sends p, q, g and v to Alice.

2. *Commit.* Alice verifies that p and q are primes, that q divides $p - 1$ and that g and v are elements of order q. To commit to $m \in \{0, \ldots, q-1\}$, she chooses $r \in \{0, \ldots, q-1\}$ at random, sets $c := g^r v^m$ and sends c to Bob.

3. *Reveal.* Alice sends r and m to Bob. Bob verifies that $c = g^r v^m$.

Remarks:

1. Bob can generate p, q, g and v as in the DSA (see Section 3.4.3).

---

[3] There is a unique subgroup of order q in $\mathbb{Z}_p^*$. It is cyclic and each element $x \in \mathbb{Z}_p^*$ of order q is a generator (see Lemma A.41).

2. If Alice committed to m and could open her commitment as $\tilde{m}$, $\tilde{m} \neq m$, then $g^r v^m = g^{\tilde{r}} v^{\tilde{m}}$ and $\log_g(v) = (\tilde{m} - m)^{-1}(r - \tilde{r})$.[4] Thus, Alice could compute $\log_g(v)$ of a randomly chosen element $v \in G_q$, contradicting the assumption that discrete logarithms of elements in G_q are infeasible to compute (see the remarks on the security of the DSA at the end of Section 3.4.3, and Proposition 4.43).

3. g and v are generators of G_q. g^r is a uniformly chosen random element in G_q, perfectly hiding v^m and m in $g^r v^m$, as in the encryption with a one-time pad (see Section 2.1.1).

4. Bob has no advantage if he knows $\log_g(v)$, for example by choosing $v = g^s$ with a random s.

In the commitment scheme based on quadratic residues, the hiding property depends on the infeasibility of computing the quadratic residues property. The binding property is unconditional. The commitment scheme based on discrete logarithms has an unconditional hiding property, whereas the binding property depends on the difficulty of computing discrete logarithms.

If the binding or the hiding property depends on the complexity of a computational problem, we speak of *computational hiding* or *computational binding*. If the binding or the hiding property does not depend on the complexity of a computational problem, we speak of *unconditional hiding* or *unconditional binding*. The definitions for the hiding and binding properties, given above, are somewhat informal. To define precisely what "cannot distinguish" and "cannot change" means, probabilistic algorithms have to be used, which we introduce in Chapter 5.

It would be desirable to have a commitment scheme which features unconditional hiding and binding. However, as the following considerations show this is impossible. Suppose a deterministic algorithm

$$C : \{0,1\}^n \times \{0,1\} \longrightarrow \{0,1\}^s$$

defines a scheme with both unconditionally hiding and binding. Then when Alice sends a commitment $c = C(r,b)$ to Bob, there must exist an $\tilde{r}$ such that $c = C(\tilde{r}, 1 - b)$. Otherwise, Bob could compute (r,b) if he has unrestricted computational power, violating the unconditional hiding property. However, if Alice also has unrestricted computing power, then she can also find $(\tilde{r}, 1-b)$ and open the commitment as $1 - b$, thus violating the unconditional binding property.

4.3.3 Homomorphic Commitments

Let $\mathrm{Com}(r, m) := g^r v^m$ denote the commitment to m in the commitment scheme based on discrete logarithms. Let $r_1, r_2, m_1, m_2 \in \{0, \ldots, q-1\}$. Then

---

[4] Note that we compute in $G_q \subset \mathbb{Z}_p^*$. Hence, computations like $g^r v^m$ are done modulo p, and since the elements in G_q have order q, exponents and logarithms are computed modulo q (see "Computing Modulo a Prime" on page 412).

$$\mathrm{Com}(r_1, m_1) \cdot \mathrm{Com}(r_2, m_2) = \mathrm{Com}(r_1 + r_2, m_1 + m_2).$$

Commitment schemes satisfying such a property are called *homomorphic commitment schemes*.

Homomorphic commitment schemes have an interesting application in distributed computation: a sum of numbers can be computed without revealing the single numbers. This feature can be used in electronic voting schemes (see Section 4.5). We give an example using the Com function just introduced. Assume there are n voters $V_1, \ldots, V_n$. For simplicity, we assume that only "yes-no" votes are possible. A trusted center T is needed to compute the outcome of the election. The center T is assumed to be honest. If T was dishonest, it could determine each voter's vote. Let E_T and D_T be ElGamal encryption and decryption functions for the trusted center T. To vote on a subject, each voter V_i chooses $m_i \in \{0, 1\}$ as his vote, a random $r_i \in \{0, \ldots, q-1\}$ and computes $c_i := \mathrm{Com}(r_i, m_i)$. Then he broadcasts c_i to the public and sends $E_T(g^{r_i})$ to the trusted center T. T computes

$$D_T(\prod_{i=1}^{n} E_T(g^{r_i})) = \prod_{i=1}^{n} g^{r_i} = g^r,$$

where $r = \sum_{i=1}^{n} r_i$, and broadcasts g^r.

Now, everyone can compute the result s of the vote from the publicly known c_i, $i = 1, \ldots, n$, and g^r:

$$v^s = g^{-r} \prod_{i=1}^{n} c_i,$$

with $s := \sum_{i=1}^{n} m_i$. s can be derived from v^s by computing $v, v^2, \ldots$ and comparing with v^s in each step, because the number of voters is not too large. If the trusted center is honest, the factor g^{r_i} – which hides V_i's vote – is never computed. Although an unconditional hiding commitment is used, the hiding property is only computational because g^{r_i} is encrypted with a cryptosystem that provides at most computational security.

4.4 Secret Sharing

The idea of secret sharing is to start with a secret s and divide it into n pieces called shares. These shares are distributed among n users in a secure way. A coalition of some of the users is able to reconstruct the secret. The secret could, for example, be the password to open a safe shared by five people, with any three of them being able to reconstruct the password and open the safe.

In a (t, n)-*threshold scheme* $(t \leq n)$, a trusted center computes the shares s_j of a secret s and distributes them among n users. Any t of the n users are able to recover s from their shares. It is impossible to recover the secret from $t - 1$ or fewer shares.

Shamir's Threshold Scheme. Shamir's threshold scheme is based on the following properties of polynomials over a finite field k. For simplicity, we take $k = \mathbb{Z}_p$, where p is a prime.

Proposition 4.13. *Let $f(X) = \sum_{i=0}^{t-1} a_i X^i \in \mathbb{Z}_p[X]$ be a polynomial of degree $t-1$, and let $P := \{(x_i, f(x_i)) \mid x_i \in \mathbb{Z}_p, i = 1, \ldots, t, x_i \neq x_j, i \neq j\}$. For $Q \subseteq P$, let $\mathcal{P}_Q := \{g \in \mathbb{Z}_p[X] \mid \deg(g) = t-1, g(x) = y \text{ for all } (x, y) \in Q\}$.*

1. *$\mathcal{P}_P = \{f(X)\}$, i.e., f is the only polynomial of degree $t-1$ whose graph contains all t points in P.*
2. *If $Q \subset P$ is a proper subset and $x \neq 0$ for all $(x, y) \in Q$, then each $a \in \mathbb{Z}_p$ appears with the same frequency as the constant coefficient of a polynomial in $\mathcal{P}_Q$.*

Proof. To find all polynomials $g(X) = \sum_{i=0}^{t-1} b_i X^i \in \mathbb{Z}_p[X]$ of degree $t-1$ through m given points $(x_i, y_i), 1 \leq i \leq m$, we have to solve the following linear equations:

$$\begin{pmatrix} 1 & x_1 & \ldots & x_1^{t-1} \\ & \vdots & & \\ 1 & x_m & \ldots & x_m^{t-1} \end{pmatrix} \begin{pmatrix} b_0 \\ \vdots \\ b_{t-1} \end{pmatrix} = \begin{pmatrix} y_1 \\ \vdots \\ y_m \end{pmatrix}.$$

If $m = t$, then the above matrix (denoted by A) is a Vandermonde matrix, and its determinant obeys

$$\det A = \prod_{1 \leq i < j \leq t} (x_j - x_i) \neq 0$$

if $x_i \neq x_j$ for $i \neq j$. Hence, the system of linear equations has exactly one solution and part 1 of Proposition 4.13 follows.

Now let $Q \subset P$ be a proper subset. Without loss of generality, let Q consist of the points $(x_1, y_1), \ldots, (x_m, y_m)$, $1 \leq m \leq t-1$. We consider the following system of linear equations:

$$\begin{pmatrix} 1 & 0 & \ldots & 0 \\ 1 & x_1 & \ldots & x_1^{t-1} \\ & \vdots & & \\ 1 & x_m & \ldots & x_m^{t-1} \end{pmatrix} \begin{pmatrix} b_0 \\ b_1 \\ \vdots \\ b_{t-1} \end{pmatrix} = \begin{pmatrix} a \\ y_1 \\ \vdots \\ y_m \end{pmatrix}. \tag{4.1}$$

The matrix of the system consists of rows of a Vandermonde matrix (note that all $x_i \neq 0$ by assumption). Thus, the rows are linearly independent and the system (4.1) has solutions for all $a \in \mathbb{Z}_p$. The matrix has rank $m + 1$, independent of a. Hence, the number of solutions is independent of a, and we see that each $a \in \mathbb{Z}_p$ appears as the constant coefficient of a polynomial in $\mathcal{P}_Q$ with the same frequency. $\qquad \square$

Corollary 4.14. *Let $f(X) = \sum_{i=0}^{t-1} a_i X^i \in \mathbb{Z}_p[X]$ be a polynomial of degree $t - 1$, and let $P = \{(x_i, f(x_i)) \mid i = 1, \ldots, t, x_i \neq x_j, i \neq j\}$. Then*

$$f(X) = \sum_{i=1}^{t} f(x_i) \prod_{1 \leq j \leq t, j \neq i} (X - x_j)(x_i - x_j)^{-1}. \tag{4.2}$$

This formula is called the Lagrange interpolation formula.

Proof. The right-hand side of (4.2) is a polynomial g of degree $t - 1$. If we substitute X by x_i in g, we get $g(x_i) = f(x_i)$. Since the polynomial $f(X)$ is uniquely defined by P, the equality holds. $\square$

Now we describe Shamir's (t, n)-threshold scheme. A trusted center T distributes n shares of a secret $s \in \mathbb{Z}$ among n users $P_1, \ldots, P_n$. To set up the scheme, the trusted center T proceeds as follows:

1. T chooses a prime $p > \max(s, n)$ and sets $a_0 := s$.
2. T selects $a_1, \ldots, a_{t-1} \in \{0, \ldots, p-1\}$ independently and at random, and gets the polynomial $f(X) = \sum_{i=0}^{t-1} a_i X^i$.
3. T computes $s_i := f(i)$, $i = 1, \ldots, n$ (we use the values $i = 1, \ldots, n$ for simplicity; any n pairwise distinct values $x_i \in \{1, \ldots, p-1\}$ could also be used) and transfers (i, s_i) to the user P_i in a secure way.

Any group of t or more users can compute the secret. Let $J \subset \{1, \ldots, n\}$, $|J| = t$. From Corollary 4.14 we get

$$s = a_0 = f(0) = \sum_{i \in J} f(i) \prod_{j \in J, j \neq i} j(j - i)^{-1} = \sum_{i \in J} s_i \prod_{j \in J, j \neq i} j(j - i)^{-1}.$$

If only $t - 1$ or fewer shares are available, then each $a \in \mathbb{Z}_p$ is equally likely to be the secret (by Proposition 4.13). Thus, knowing only $t - 1$ or fewer shares provides no advantage over knowing none of them.

Remarks:

1. Suppose that each $a \in \mathbb{Z}_p$ is equally likely to be the secret for someone knowing only $t-1$ or fewer shares, as in Shamir's scheme. Then the (t, n)-threshold scheme is called *perfect*: the scheme provides perfect secrecy in the information-theoretic sense (see Section 9.1). The security does not rely on the assumed hardness of a computational problem.
2. Shamir's threshold scheme is easily extendable to new users. New shares may be computed and distributed without affecting existing shares.
3. It is possible to implement varying levels of control. One user can hold one or more shares.

4.5 Verifiable Electronic Elections

The use of electronic voting technology is of increasing importance. Direct-recording electronic voting machines (called DREs) and optical scanners for paper ballots are widely used for recording votes at supervised polling sites (for example, in the United States presidential elections and in the Russian Federation State Duma elections). Even remote e-voting via the Internet is being applied in legally binding elections; for example, it has been used in the parliamentary elections in Estonia since 2007. An increasing number of states are experimenting with Internet voting for citizens living abroad (examples are the U.S., Mexico and several cantons in Switzerland).

Elections must be general, direct, free, equal and confidential. The application of electronic voting technology must not compromise these basic democratic principles. Therefore, an e-voting system is expected to guarantee the following:

1. Eligible voters can cast their votes.
2. Only eligible voters are allowed to cast votes.
3. *Privacy*: the secrecy of an individual vote must be maintained.
4. *Correctness*: the votes must be counted correctly.
5. *Robustness*: the scheme must work correctly even in the presence of attackers and dishonestly involved parties.

E-voting is highly controversial. There are doubts about the correctness and robustness of the e-voting solutions that are currently used in practice. One main reason is the lack of transparency of most systems. This causes a lack of trust in their security. For example, a voter typically has to believe that his vote will be accurately recorded and counted. He cannot verify that the vote that he entered and saw displayed on the screen of the DRE or his PC is the same as the vote that is stored in the electronic ballot box and then included in the final tally. Moreover, the computation of the tally typically cannot be checked by the public, in contrast to a classical paper-based election, where the ballot-counting is public and is done jointly by the competing parties.

To create trust in e-voting, a crucial requirement is *verifiability* or, more precisely, *end-to-end verifiability* of an electronic election. End-to-end verifiability means that voters and the public (or at least independent observers) are able to verify the casting and counting of votes. There are two steps, the *voter's individual verification* and *result verification*. In every e-voting system, the votes are encoded and encrypted and then stored in an electronic ballot box. In a verifiable election scheme, the ballot box and its content are necessarily visible to the public. The voter's individual verification means that every voter can verify that his ballot is contained in the public ballot box and accurately captures his intended vote ("cast-as-intended"). Only the individual voter knows his intent and can perform this verification. Result verification guarantees that the election result is correctly computed from the ballots in the ballot box ("counted-as-cast"). Here, *universal verifiability* is required: everybody (candidates, voters, political parties, ...) can check that only the votes of legitimate voters have been counted and none of the cast votes has been modified, and everybody is able to reproduce the final tally.

The e-voting systems that are used in practice today typically do not enable true end-to-end verification. For example, DRE machines are black boxes to the voters and the public. They do not allow verification, and serious vulnerabilities have been found in widely used DRE systems. Many U.S. states have adopted requirements to include a voter-verified paper audit trail (VVPAT) in DRE-based voting. This means that a redundant paper ballot is printed by the DRE, checked by the voter and then collected in a ballot box. VVPAT does not mean that the electronic voting process is verifiable. It only enables a traditional recount of votes if doubts about the correct functioning of the electronic voting equipment arise.

Modern cryptographic voting protocols provide convincing solutions for result verification. The universal verifiability of the tally is based either on a homomorphic encryption of the votes and a verifiable threshold decryption of the result, or on a verifiable mixing and decryption of the votes followed by a public counting of the decrypted votes.

In this section, we describe a prominent example of a verifiable voting protocol that relies on the homomorphic encryption of votes. It was proposed by Cramer, Gennaro and Schoenmakers ([CraGenSch97]) and is the basis for many further developments in this field. The protocol could be used for remote elections via the Internet and for polling-site elections with DREs. An example of a verifiable voting protocol that is based on the mixing of votes is given in Section 4.7 below.

The protocol of [CraGenSch97] also enables the voter's individual verification but, nonetheless, it is not perfect. It is not receipt-free. *Receipt-freeness* is an essential property to prevent the selling of votes. It means that a voter is not able to prove to a third party that he has cast a particular vote. The voter must neither obtain nor be able to construct a receipt proving the content of their his. In the [CraGenSch97] protocol, the voter can obtain such a receipt.

In Section 4.7 below, we will do the following:

1. Discuss receipt-freeness in more detail and the stronger property of *coercion resistance*, which was introduced in [JueCatJak05].
2. Describe enhancements of the [CraGenSch97] protocol in order to obtain receipt-freeness.
3. Describe the coercion-resistant protocol of Juels, Catalano and Jakobsson ([JueCatJak05]).

The latter protocol is a typical representative of the class of election protocols where the anonymity of votes is achieved by using mix nets, as already proposed by Chaum in 1981 ([Chaum81]). We will study mix nets in detail in Section 4.6.

In an *electronic voting scheme*, there are two distinct types of participants: the voters casting the votes, and the voting authority (for short, the "authority") that collects the votes and computes the final tally. A single central authority is problematic. The voters' privacy can only be maintained if this authority is honest and we can trust it. The scheme that we describe therefore establishes a group of authorities – you may consider them as election officers. By applying a secret-sharing scheme, they share the core secret of the system which is necessary for decrypting a vote. In this way, privacy is guaranteed, even if some of the authorities are dishonest and collaborate in their efforts to break privacy.

4.5.1 A Multi-authority Election Scheme

For simplicity, we consider only election schemes for yes–no votes. The voters want a majority decision on some subject. In the scheme that we describe, each voter selects his choice (yes or no), encrypts it with a homomorphic encryption algorithm and signs the cryptogram. The signature shows that the vote is from an authorized voter. The votes are collected in a single place, the bulletin board. After all voters have posted their votes, an authority can compute the tally without decrypting the individual votes. This feature depends on the fact that the encryption algorithm used is a homomorphism. It guarantees the secrecy of the votes. But if the authority was dishonest, it could decrypt the individual votes. To reduce this risk, the authority consists of several parties, and the decryption key is shared among these parties by use of a Shamir (t, n)-threshold scheme. Then at least t of the n parties must be dishonest to reveal a vote. First, we assume in our discussion that a trusted center T sets up the scheme. However, the trusted center is not really needed. In Section 4.5.5, we show that it is possible to set up the scheme by a communication protocol which is executed by the parties constituting the authority.

The election scheme that we describe was introduced in [CraGenSch97]. The time and communication complexity of the scheme is remarkably low. A voter simply posts a single encrypted message, together with a compact proof that it contains a valid vote.

The Communication Model. The members of the voting scheme communicate through a *bulletin board*. The bulletin board is best viewed as publicly accessible memory. Each member has a designated section of the memory to post messages. No member can erase any information from the bulletin board. The complete bulletin board can be read by all members (including passive observers). We assume that a public-key infrastructure for digital signatures is used to guarantee the origin of posted messages.

Setting up the Scheme. The votes are encrypted by using ElGamal encryption in a prime-order subgroup, as described in Section 3.4.4. To set up the scheme, we assume for now that there is a trusted center T.[5] The trusted center generates the key pair for this ElGamal encryption, i.e., the trusted center T chooses primes p and q such that q is a large divisor of $p - 1$, and a base element $g \in \mathbb{Z}_p^*$ of order q.

In addition, T chooses the secret key $s \in \mathbb{Z}_q$ at random and publishes the public key (p, q, g, h), where $h := g^s$. A message $m \in G_q$ is encrypted as $(c_1, c_2) = (g^\alpha, h^\alpha m)$, where α is a randomly chosen element of $\mathbb{Z}_q$. Using the secret key s, the plaintext m can be recovered as $m = c_2 c_1^{-s}$ (see Section 3.4.4). The encryption is homomorphic: if (c_1, c_2) and (c_1', c_2') are encryptions of m and m', then

$$(c_1, c_2) \cdot (c_1', c_2') = (c_1 c_1', c_2 c_2') = (g^\alpha g^{\alpha'}, h^\alpha m h^{\alpha'} m') = (g^{\alpha + \alpha'}, h^{\alpha + \alpha'} m m')$$

is an encryption of mm' (see Section 3.6).

Let $A_1, \ldots, A_n$ be the authorities and $V_1, \ldots, V_m$ be the voters in the election scheme. The trusted center T chooses a Shamir (t, n)-threshold scheme. The secret encryption key s is shared among the n authorities. A_j keeps its share (j, s_j) secret. The values $h_j := g^{s_j}$ are published on the bulletin board by the trusted center.

Threshold Decryption. Anyone can decrypt a cryptogram $c := (c_1, c_2) := (g^\alpha, h^\alpha m)$ with some help from the authorities, without reconstructing the secret key s. To do this, the following steps are executed:

1. Each authority A_j posts $w_j := c_1^{s_j}$ to the bulletin board. Here, we assume that the authority A_j is honest and publishes the correct value of w_j. In Section 4.5.2, we will see how to check that it really does so (by a proof of knowledge).

2. Let J be the index set of a subset of t honest authorities. Then, everyone can recover $m = c_2 c_1^{-s}$ as soon as all A_j, $j \in J$, have finished step 1:

$$c_1^s = c_1^{\sum_{j \in J} s_j \lambda_{j,J}} = \prod_{j \in J} (c_1^{s_j})^{\lambda_{j,J}} = \prod_{j \in J} w_j^{\lambda_{j,J}},$$

where

$$\lambda_{j,J} = \prod_{l \in J \setminus \{j\}} l(l - j)^{-1}.$$

Remark. The decryption method which we have just described is independent of its application to elections. The secret ElGamal key is shared by n parties, and a ciphertext can be jointly decrypted by any t of these n parties without reconstructing the secret key. Such a decryption scheme is called a *distributed (t, n)-threshold decryption scheme*. The (t, n)-threshold

---

[5] Assuming the existence of a trusted center simplifies the exposition. But it is not necessary. Later, we will see how the trusted center can be eliminated (Section 4.5.5 below).

decryption algorithm for ElGamal encryption is based on Shamir's (t, n)-threshold secret-sharing scheme, which we studied in detail in Section 4.4. An efficient distributed (t, n)-threshold decryption algorithm is also known for Paillier encryption ([DamJur01]). The correctness of the decryption can be universally verified by means of non-interactive zero-knowledge proofs (see the "authority's proof" in Section 4.5.2 for ElGamal encryption, and [DamJur01] for Paillier encryption). The public key and the shares of the secret key can be jointly generated by the authorities, without assuming a trusted center (see Section 4.5.5 for ElGamal encryption, and [DamKop01], [FouSte01], [FraMaKYun98] for Paillier encryption).

Vote Casting. Each voter V_i makes his choice $v_i \in \{-1, 1\}$, encodes v_i as g^{v_i} and encrypts g^{v_i} by the ElGamal encryption

$$c_i := (c_{i,1}, c_{i,2}) := (g^{\alpha_i}, h^{\alpha_i} g^{v_i}).$$

The voter then signs it to guarantee the origin of the message and posts it to the bulletin board. Here we assume that V_i follows the protocol and correctly forms c_i. The voter has to perform a proof of knowledge which shows that he really has done so (see Section 4.5.2); otherwise the vote is invalid.

Tally Computing. Assume that m votes have been cast. The following is done:

1. Everyone can compute

$$c = (c_1, c_2) = \left(\prod_{i=1}^{m} c_{i,1}, \prod_{i=1}^{m} c_{i,2} \right).$$

   Note that $c = (c_1, c_2)$ is the encryption of g^d, where d is the difference between the numbers of yes votes and no votes, since the encryption is homomorphic.
2. The decryption protocol above is executed to get g^d. After sufficiently many authorities A_j have posted $w_j = c_1^{s_j}$ to the bulletin board, everyone can compute g^d.
3. Now d can be found by computing the sequence $g^{-m}, g^{-m+1}, \ldots$ and comparing it with g^d at each step.

Remarks:

1. Everyone can check whether a voter or an authority is honest (see Section 4.5.2), and discard invalid votes. If someone finds a subset of t honest authorities, he can compute the tally. This implies universal verifiability.
2. No coalition of $t - 1$ or fewer authorities can recover the secret key. This guarantees the robustness of the scheme.
3. Privacy depends on the security of the underlying ElGamal encryption scheme and, hence, on the assumed difficulty of the Diffie–Hellman problem. The scheme provides only computational privacy. A similar scheme

was introduced in [CraFraSchYun96] which provides perfect privacy (in the information-theoretic sense). This is achieved by using a commitment scheme with information-theoretic secure hiding to encrypt the votes.

4. The following remarks concern the communication complexity of the scheme:

   a. Each voter only has to send one message together with a compact proof that the message contains a valid vote (see below). The voter's activities are independent of the number n of authorities.

   b. Each authority has to read m messages from the bulletin board, verify m interactive proofs of knowledge and post one message to the bulletin board.

   c. To compute the tally, one has to read t messages from the bulletin board and verify t interactive proofs of knowledge.

5. It is possible to prepare for an election beforehand. The voter V_i chooses $v_i \in \{-1, 1\}$ at random. The voting protocol is executed with these random v_i values. Later, the voter decides which alternative to choose. The voter selects $\tilde{v}_i \in \{-1, 1\}$ such that $\tilde{v}_i v_i$ is his vote, and posts $\tilde{v}_i$ to the bulletin board. The tally is computed with $\tilde{c}_i = (c_{i,1}^{\tilde{v}_i}, c_{i,2}^{\tilde{v}_i}), i = 1, \ldots, m$.

4.5.2 Proofs of Knowledge

The Authority's Proof. In the decryption protocol above, each authority A_j has to prove that it has really posted $w_j = c_1^{s_j}$, where s_j is the authority's share of the secret key s. Recall that $h_j = g^{s_j}$ is published on the bulletin board. The authority has to prove that w_j and h_j have the same logarithm with respect to the bases c_1 and g and that it knows this logarithm. For this purpose, the classical *Chaum–Pedersen proof* ([ChaPed92]) is applied. In the following, we unify the notation and describe the Chaum–Pedersen proof. The prover Peggy convinces the verifier Vic that she knows the common logarithm x of elements $y_1 = g_1^x$ and $y_2 = g_2^x$. The number $x \in \mathbb{Z}_q$, which is the witness for the claim that y_1 and y_2 have the same logarithm, remains Peggy's secret.

Of course, in our voting scheme it is desirable for practical reasons that the authority can be proven to be honest in a non-interactive way. However, it is easy to convert an interactive proof into a non-interactive one (see Section 4.5.3). Therefore, we first give the interactive version of the proof.

Protocol 4.15.

ProofLogEq$(p, q, g_1, y_1, g_2, y_2)$:

1. Peggy chooses $r \in \mathbb{Z}_q$ at random and sets $a := (a_1, a_2) = (g_1^r, g_2^r)$. Peggy sends a to Vic.
2. Vic chooses $c \in \mathbb{Z}_q$ uniformly at random and sends c to Peggy.
3. Peggy computes $b := r - cx$ and sends b to Vic.
4. Vic accepts if and only if $a_1 = g_1^b y_1^c$ and $a_2 = g_2^b y_2^c$.

The protocol is a three-move protocol. Its structure is very similar to the protocol used in the simplified Fiat–Shamir identification scheme (see Section 4.2.2):

1. The first message is a commitment by Peggy. She commits to the statement that two numbers have the same logarithm with respect to the different bases g_1 and g_2.
2. The second message, c, is a challenge by Vic.
3. The third message is Peggy's response. If $c = 0$, Peggy has to open the commitment (reveal r). If $c \neq 0$, Peggy has to show her secret in encrypted form (reveal $r - cx$).

The Chaum–Pedersen proof is a *public-coin* protocol. This means that

1. all the messages which the verifier Vic sends to the prover Peggy are challenges which Vic has chosen uniformly at random and independently of the messages sent by Peggy, and
2. all the random bits which Vic uses in his computations are included in these challenges.

All the proofs of knowledge which we consider in this book are public-coin protocols.

Next, we study the security properties of the Chaum–Pedersen protocol.

Completeness. If Peggy knows a common logarithm for y_1 and y_2, and both Peggy and Vic follow the protocol, then $a_1 = g_1^b y_1^c$ and $a_2 = g_2^b y_2^c$, and Vic will accept.

Soundness. A cheating prover Eve, who does not know the secret, can convince Vic with a probability of $1/q$ in the following way:

1. Eve chooses $r, \tilde{c} \in \mathbb{Z}_q$ at random, sets $a := (g_1^r y_1^{\tilde{c}}, g_2^r y_2^{\tilde{c}})$ and sends a to Vic.
2. Vic chooses $c \in \mathbb{Z}_q$ at random and sends c to Eve.
3. Eve sends r to Vic.

Vic accepts if and only if $c = \tilde{c}$. The event $c = \tilde{c}$ occurs with a probability of $1/q$. Thus, Eve succeeds in cheating with a probability of $1/q$.

On the other hand, the probability of success of a cheating prover never exceeds $1/q$, which is a negligibly small probability. If a prover is able to convince Vic with a probability greater than $1/q$, then she knows the witness x and therefore, she is a legitimate prover. Namely, in this case, the prover must know at least one commitment $a = (a_1, a_2)$ for which she is able to answer at least two distinct challenges c and $\tilde{c}$ correctly. This means that she can compute b and $\tilde{b}$ such that

$$a_1 = g_1^b y_1^c = g_1^{\tilde{b}} y_1^{\tilde{c}} \text{ and } a_2 = g_2^b y_2^c = g_2^{\tilde{b}} y_2^{\tilde{c}}.$$

This implies that y_1 and y_2 have a common logarithm x,

$$x := \log_{g_1}(y_1) = \log_{g_2}(y_2) = (\tilde{b} - b)(c - \tilde{c})^{-1},$$

and the prover can compute it easily.

Remark. We described an algorithm which, for a given commitment a, computes a witness x from correct responses b and $\tilde{b}$ to two distinct challenges c and $\tilde{c}$. The existence of such an algorithm for extracting a witness in a public-coin protocol is called the property of *special soundness*. Special soundness implies soundness.

Honest-Verifier Zero-Knowledge Property. The above protocol is not known to be zero-knowledge. However, it is zero-knowledge if the verifier is an honest one.

Definition 4.16. An interactive proof system (P, V) is said to be *honest-verifier zero-knowledge* if Definition 4.6 holds for the honest verifier V, but not necessarily for an arbitrary verifier V^*; i.e., there is a simulator S that produces correctly distributed accepting transcripts for executions of the protocol with (P, V).

In our example, it is quite simple to simulate an interaction with an honest verifier. First, we observe that we can simulate correctly distributed transcripts that include a given challenge c by the following algorithm S_1.

Algorithm 4.17.
```
int S₁(int g₁, g₂, y₁, y₂, c)
  1   select b ∈ {0, ..., q − 1} uniformly at random
  2   a₁ ← g₁ᵇy₁ᶜ, a₂ ← g₂ᵇy₂ᶜ
  3   return (a₁, a₂, c, b)
```

The transcripts (a_1, a_2, c, b) returned by S_1 are accepting transcripts and they are not distinguishable from the original transcripts with challenge c:

1. b is uniformly distributed like the original responses b.
2. a_1 and a_2 depend deterministically on (c, b) in the same way as in the original transcripts.

By using S_1, we easily construct a simulator S. The key point is that the honest verifier V chooses the challenge $c \in \mathbb{Z}_q$ independently from (a_1, a_2), uniformly and at random, and this can also be done by S.

Algorithm 4.18.
```
int S(int g₁, g₂, y₁, y₂)
  1   select c ∈ {0, ..., q − 1} uniformly at random
  2   return S₁(g₁, g₂, y₁, y₂, c)
```

The transcripts (a_1, a_2, c, b) returned by S are accepting transcripts and they are not distinguishable from the original transcripts.

Definition 4.19. A public-coin proof of knowledge (P, V) is said to be *special honest-verifier zero-knowledge*, if there exists a simulator S_1, such as Algorithm 4.17, which on input of a challenge c generates accepting transcripts which include c, with the same probability distribution as the transcripts generated by the honest P and V.

Special honest-verifier zero-knowledge proofs are honest-verifier zero-knowledge – the simulator S can always be constructed from S_1 as in our example.

Σ-Proofs. The Chaum-Pedersen proof is a typical example of a Σ-proof. A Σ-*proof* is a three-move public-coin proof (P, V) with the following properties:

1. First, the prover P sends a commitment a. Next, the verifier V generates a challenge c uniformly at random, independently of the prover's commitment a, and sends c to the prover. Then, the prover has to give a consistent response b. The verifier's decision, if he accepts or rejects, is based solely on a, b, c and the common input of (P, V).
2. The proof has the completeness, special soundness and special honest-verifier zero-knowledge properties.

The Voter's Proof. In the vote-casting protocol, each voter has to prove that his encrypted vote $c = (c_1, c_2)$ really contains one of the admissible encoded votes $m_1 := g^{+1}$ or $m_2 := g^{-1}$, i.e., the voter has to prove that $(c_1, c_2) = (g^\alpha, h^\alpha m)$ and $m \in \{m_1, m_2\}$. For this purpose, a proof of knowledge is performed. The voter proves that he knows a witness α, which is either the common logarithm of c_1 and $c_2 m_1^{-1}$ with respect to the bases g and h or the common logarithm of c_1 and $c_2 m_2^{-1}$, i.e., either $c_1 = g^\alpha$ and $c_2 m_1^{-1} = h^\alpha$, or $c_1 = g^\alpha$ and $c_2 m_2^{-1} = h^\alpha$. Each of the two alternatives could be proven as in the authority's proof by applying the Chaum–Pedersen proof (Protocol 4.15). Here, however, the prover's task is more difficult. The proof must not reveal which of the two alternatives is proven.

To solve this problem, we must learn another basic proof technique, the *OR-combination* of Σ-proofs. Let p, q, g, h be as before. Given two pairs (y, z) and $(\tilde{y}, \tilde{z})$, Peggy proves to Vic that she knows an x such that either $y = g^x$ and $z = h^x$ (case 1) or $\tilde{y} = g^x$ and $\tilde{z} = h^x$ (case 2), without revealing which of those cases occurs.

Protocol 4.20.
OneOfTwoPairs$(p, q, g, h, (y, z), (\tilde{y}, \tilde{z}))$:

Case: Peggy knows $\log_g y = \log_h z = x$ (the other case, $\log_g \tilde{y} = \log_h \tilde{z} = x$, follows analogously):
1. Peggy chooses $r, \tilde{r}$ and $\tilde{d} \in \mathbb{Z}_q$ at random and sets $a = (a_1, a_2, \tilde{a}_1, \tilde{a}_2)$ $= (g^r, h^r, g^{\tilde{r}} \tilde{y}^{\tilde{d}}, h^{\tilde{r}} \tilde{z}^{\tilde{d}})$. Peggy sends a to Vic.
2. Vic chooses a challenge $c \in \mathbb{Z}_q$ uniformly at random. Vic sends c to Peggy.

3. Peggy computes

$$d = c - \tilde{d},$$
$$b = r - dx, \ \tilde{b} = \tilde{r},$$

and sends $(b, \tilde{b}, d, \tilde{d})$ to Vic.

4. Vic accepts if and only if

$$c = d + \tilde{d},$$
$$a_1 = g^b y^d, \ a_2 = h^b z^d,$$
$$\tilde{a}_1 = g^{\tilde{b}} \tilde{y}^{\tilde{d}}, \ \tilde{a}_2 = h^{\tilde{b}} \tilde{z}^{\tilde{d}}.$$

Remark. The proof demonstrates the OR-combination of Σ-proofs. Peggy knows the witness x and, therefore, she knows which of the cases is true, say case 1. Now, Peggy combines the proof of case 1, for which she knows a witness, with a simulated proof of case 2. She can do this because Vic's challenge c is split into the components d and $\tilde{d}$, which then serve as challenges in the two cases. Peggy can fix one of the components in advance. Then, she can simulate the proof for this case. In our example, she chooses $\tilde{d}$ by herself and simulates a proof that $\log_g \tilde{y} = \log_h \tilde{z}$. Recall that it is always easy to simulate a proof if the challenge is known in advance (see Algorithms 4.17 and 4.18 above).

4.5.3 Non-interactive Proofs of Knowledge

Public-coin proofs can be converted into non-interactive proofs by applying the *Fiat–Shamir heuristics*. We have already demonstrated this method in Section 4.2.5, where we derived a signature scheme from the Fiat–Shamir identification protocol. The basic idea is the following. In the interactive protocol, Vic's task is to randomly generate a challenge and send it to the prover Peggy. Now, Peggy takes on this task by herself. Peggy computes the challenge by applying a collision-resistant hash function $h : \{0,1\}^* \longrightarrow \mathbb{Z}_q$ to her first message and the public input parameters of the protocol.

As an example, we shall apply the heuristics to the Chaum–Pedersen proof (Protocol 4.15). We get the non-interactive proof

$$(c, b) = \text{ProofLogEq}_h(p, q, g_1, y_1, g_2, y_2)$$

for $y_1 = g_1^x$ and $y_2 = g_2^x$ in the following way. The prover Peggy chooses $r \in \mathbb{Z}_q$ at random and sets $a := (a_1, a_2) = (g_1^r, g_2^r)$. Then she computes the challenge $c := h(p\|q\|g_1\|y_1\|g_2\|y_2\|a_1\|a_2)$ and sets $b := r - cx$. The verification condition is

$$c = h(p\|q\|g_1\|y_1\|g_2\|y_2\|g_1^b y_1^c\|g_2^b y_2^c).$$

The verifier does not need to know a to compute the verification condition. If we trust the collision resistance of h, we can conclude that $u = v$ from $h(u) = h(v)$.

Recall that cryptographic hash functions are approximations of random oracles (Section 2.2.4). This means that in the non-interactive protocol, the verifier represented by the hash function can be assumed to be honest. Therefore, it is sufficient for the zero-knowledge property that the interactive proof is honest-verifier zero-knowledge. The resulting non-interactive proofs are *non-interactive zero-knowledge proofs*. Their soundness can be proven in the random oracle model (Section 3.3.5, [BelRog93]).

If we include a message m in the hash value, then the non-interactive proof becomes dependent on the message and we obtain a digital signature for m. Some examples of such signature schemes are Fiat–Shamir signatures (Section 4.2.5 above) and Schnorr signatures (Section 4.8.1 and Protocol 4.39 below).

The Fiat–Shamir heuristics can also be applied to public-coin proofs with more than three moves. The challenges are generated by applying the hash function to the transcript of the protocol so far. An example is the shuffle argument in Section 4.6.4 below.

In our election protocol, each authority and each voter complete their message with a non-interactive proof which convinces everyone that they have followed the protocol.

4.5.4 Extension to Multi-way Elections

We now describe how to extend the scheme if a choice between several, say $l > 2$, options is to be possible.

To encode the votes $v_1, \ldots, v_l$, we independently choose l generators g_j, $j = 1, \ldots, l$, of G_q, and encode v_j with g_j. The voter V_i encrypts his vote v_j as

$$c_i = (c_{i,1}, c_{i,2}) = (g^\alpha, h^\alpha g_j).$$

Each voter shows – by an interactive proof of knowledge – that he knows α with

$$c_{i,1} = g^\alpha \text{ and } g_j^{-1} c_{i,2} = h^\alpha,$$

for exactly one j. We refer the interested reader to [CraGenSch97] for some hints about the proof technique used.

The problem of computing the final tally turns out to be more complicated. The result of the tally computation is

$$c_2 c_1^{-s} = g_1^{d_1} g_2^{d_2} \cdots g_l^{d_l},$$

where d_j is the number of votes for v_j. The exponents d_j are uniquely determined by $c_2 c_1^{-s}$ in the following sense: computing a different solution $(\tilde{d}_1, \ldots, \tilde{d}_l)$ would contradict the discrete logarithm assumption, because the generators were chosen independently (see Proposition 4.43 in Section 4.8.3). Again as above, $d = (d_1, \ldots, d_l)$ can be found for small values of l and d_i by searching.

4.5.5 Eliminating the Trusted Center

The trusted center sets up an ElGamal cryptosystem, generates a secret key, publishes the corresponding public key and shares the secret key among n users using a (t, n)-threshold scheme. To eliminate the trusted center, all these activities must be performed by the group of users (in our case, the authorities). For the communication between the participants, we assume below that a bulletin board, mutually secret communication channels and a commitment scheme exist.

Setting up an ElGamal Cryptosystem. We need large primes p, q such that q is a divisor of $p - 1$, and a generator g of the subgroup G_q of order q of $\mathbb{Z}_p^*$. These are generated jointly by the group of users (in our case, the group of authorities). This can be achieved if each party runs the same generation algorithm (see Section 3.4.1). The random input to the generation algorithm, must be generated jointly. To do this, the users $P_1, \ldots, P_n$ execute the following protocol:

1. Each user P_i chooses a string r_i of random bits of sufficient length, computes a commitment $c_i = C(r_i)$ and posts the result to the bulletin board.
2. After all users have posted their commitments, each user opens his commitment.
3. The users take $r = \oplus_{i=1}^n r_i$ as the random input.

Publishing the Public Key. To generate and distribute the public key, the users $P_1, \ldots, P_n$ execute the following protocol:

1. Each user P_i chooses $x_i \in \{0, \ldots, q-1\}$ at random, and computes $h_i := g^{x_i}$ and a commitment $c_i := C(h_i)$ for h_i. Then the user posts c_i to the bulletin board.
2. After all users have posted their commitments, each user opens his commitment.
3. Everyone can compute the public key $h := \prod_{i=1}^n h_i$.

Sharing the Secret Key. The corresponding secret key,

$$x := \sum_{i=1}^n x_i,$$

must be shared. The basic idea is that each user constructs a Shamir (t, n)-threshold scheme to share his part x_i of the secret key. These schemes are combined to get a scheme for sharing x.

Let $f_i(X) \in \mathbb{Z}_{\tilde{p}}[X]$ be the polynomial of degree $t - 1$ used for sharing x_i, and let

$$f(X) = \sum_{i=1}^n f_i(X).$$

Then $f(0) = x$ and f can be used for a (t, n)-threshold scheme, provided $\deg(f) = t - 1$. The shares $f(j)$ can be computed from the shares $f_i(j)$:

$$f(j) = \sum_{i=1}^{n} f_i(j).$$

The group of users executes the following protocol to set up the threshold scheme:

1. The group chooses a prime $\tilde{p} > \max(x, n)$.
2. Each user P_i randomly chooses $f_{i,j} \in \{0, \ldots, \tilde{p} - 1\}$, $j = 1, \ldots, t - 1$, and sets $f_{i,0} = x_i$ and $f_i(X) = \sum_{j=0}^{t-1} f_{i,j} X^j$. The user posts $F_{i,j} := g^{f_{i,j}}$, $j = 1, \ldots, t - 1$, to the bulletin board. Note that $F_{i,0} = h_i$ is P_i's piece of the public key and hence is already known.
3. After all users have posted their encrypted coefficients, each user tests whether $\sum_{i=1}^{n} f_i(X)$ has degree $t - 1$, by checking

$$\prod_{i=1}^{n} F_{i,t-1} \neq [1].$$

In the rare case in which the test fails, they return to step 2. If $f(X)$ passes the test, the degree of $f(X)$ is $t - 1$.

4. Each user P_i distributes the shares $s_{i,l} = f_i(l)$, $l = 1, \ldots, n$, of his piece x_i of the secret key to the other users over secure communication channels.
5. Each user P_i verifies for $l = 1, \ldots, n$ that the share $s_{l,i}$ received from P_l is consistent with the previously posted committed coefficients of P_l's polynomial:

$$g^{s_{l,i}} = \prod_{j=0}^{t-1} (F_{l,j})^{i^j}.$$

If the test fails, the user stops the protocol by broadcasting the message

$$\text{"failure}, (l, i), s_{l,i}\text{"}.$$

6. Finally, P_i computes his share s_i of x,

$$s_i = \sum_{l=1}^{n} s_{l,i},$$

signs the public key h and posts his signature to the bulletin board.

After all members have signed h, the group use h as their public key. If all participants have followed the protocol, the corresponding secret key is shared among the group. The protocol given here is described in [Pedersen91]. It only works if all parties are honest. [GenJarKraRab99] introduced an improved protocol, which works if a majority of the participants are honest.

4.6 Mix Nets and Shuffles

Mix nets, first introduced by Chaum ([Chaum81]), are used in applications that require anonymity, such as electronic voting. They guarantee, for example, that a vote cannot be linked with the voter who cast it. Mix nets are the preferred tool for implementing *anonymous communication channels*. Chaum originally used mix nets to obtain untraceable electronic mail. Web anonymizers and anonymous remailers are often based on mix nets (see the remarks at the end of Section 4.6.1 below).

A *mix server* (or *mix* for short) M collects encrypted input messages into a batch and then executes a *shuffle*, in which it re-encrypts or partially decrypts the input ciphertexts and outputs them in a permuted ("mixed") order. M chooses the permutation that it uses to re-order the list of messages randomly, and keeps it secret. The random re-ordering and the re-encryption (or decryption) ensure that the output messages of a mix server cannot be linked with the input messages: an observer is not able to find out how the outputs correspond to the inputs.

Mix servers and, better, mix cascades are used to send messages anonymously. A *mix cascade* is a series of mix servers $M_1, M_2, \ldots, M_r$, arranged in a chain. A user who wants to send a message anonymously encrypts his message by using a public key and sends it to the first mix, M_1. M_1 collects the inputs into a batch. In this way, M_1 receives as input a list L_0 of ciphertexts. Each mix M_i takes as input the output L_{i-1} of M_{i-1} (M_1 takes L_0), shuffles the input and sends its output list L_i to the next mix, M_{i+1}. The shuffles ensure that an observer is not able to trace a message – the observer cannot find out which sender an output message originates from.

Using a series of mix servers and not only one mix enhances the probability that sender anonymity is achieved. If at least one of the mix servers behaves as required, in particular, if it really chooses the permutation at random and keeps it secret, then an observer is not able to link a message at the output of the mix cascade with an input message or a sender. The sender remains anonymous.

A *mix net* is a network of mix servers, and mix cascades can be selected from this network. There are, for example, applications where the mix net consists of a single, fixed mix cascade, and there are applications where *source routing* is applied. Here sender S – the message source – selects a mix chain from a larger mix net and then sends messages through this chain. The sender may, for example, use that cascade for some time, say 10 minutes (as in the Web anonymizer Tor), or may choose mix chains on a per-message basis.

It is essential to use a secure probabilistic public-key encryption algorithm,[6] such as prime-order-subgroup ElGamal or Paillier encryption. An

---

[6] It should be semantically secure, i.e., ciphertext-indistinguishable (see Definition 9.14).

observer must not be able to find any link between plaintexts and cipher-texts. Otherwise, information about the secret permutation would leak.

In the following, let E be a secure probabilistic public-key encryption algorithm. We can take, for example, ElGamal encryption in a prime-order subgroup, Paillier encryption or probabilistic RSA encryption with OAEP (see Sections 3.4.4, 3.6.2 and 3.3.3). As usual, we use E_{pk} to denote encryption with the public key pk and D_{sk} to denote decryption with the secret key sk.

We distinguish two types of mix nets, decryption and re-encryption mix nets.

4.6.1 Decryption Mix Nets

Decryption mix nets are usually built by using a hybrid encryption scheme $\tilde{E}$ that combines the probabilistic public-key encryption algorithm E and a symmetric-key encryption algorithm F (e.g., a block cipher together with a suitable mode of operation). A typical structure for $\tilde{E}$ is the following. Assume that Bob has a public–secret key pair (pk, sk) for E. To compute the ciphertext $\tilde{E}_{pk}(m)$ of a message $m \in \{0,1\}^*$ for Bob, Alice randomly generates an ephemeral key k for F and sets

$$\tilde{E}_{pk}(m) := E_{pk}(k) \| F_k(m),$$

i.e., m is encrypted symmetrically and the symmetric key is encrypted asymmetrically. Bob decrypts a ciphertext c in the obvious way: He splits c into the two parts $c_1 = E_{pk}(k)$ and $c_2 = F_k(m)$.[7] By decrypting c_1 with his secret key sk, he obtains k. Then he decrypts c_2 and gets m.

The Decryption Shuffle. In a decryption mix net, each mix server M has its own key pair (pk_M, sk_M) for E. M receives ciphertexts as input, which are encrypted by using $\tilde{E}$ and M's public key pk_M. It collects the incoming ciphertexts into a batch and then shuffles the list $(c_1, c_2, \ldots, c_n)$ of input ciphertexts: it decrypts the list of input ciphertexts by using its secret key sk_M, randomly permutes the list and outputs it. More precisely, M randomly generates a permutation $\pi : \{1, 2, \ldots, n\} \longrightarrow \{1, 2, \ldots, n\}$, sets

$$c'_k := \tilde{D}_{sk_M}(c_{\pi(k)}), \quad 1 \leq k \leq n,$$

and outputs $(c'_1, c'_2, \ldots, c'_n)$. The permutation π is kept secret.

Now, we consider a decryption mix cascade $M_1, M_2, \ldots, M_r$. Let pk_i be the public key of $M_i, i = 1, 2, \ldots, r$. Anonymously sending messages through the mix cascade works as follows:

1. Each sender S iteratively encrypts his message m, which includes the address of the recipient, with the public keys $pk_r, pk_{r-1}, \ldots, pk_1$ and sends the ciphertext

---

[7] Note that the public-key ciphertexts c_1 have a fixed binary length.

$$c = \tilde{E}_{pk_1} \left(\tilde{E}_{pk_2} \left(\ldots \left(\tilde{E}_{pk_{r-2}} \left(\tilde{E}_{pk_{r-1}} \left(\tilde{E}_{pk_r}(m) \right) \right) \right) \ldots \right) \right)$$

to the first mix M_1.

2. The mix M_i collects the incoming ciphertexts, then shuffles its input list $c_1, c_2, \ldots, c_n$ of ciphertexts and sends the output list $c_1', c_2', \ldots, c_n'$ to the next mix, M_{i+1} ($1 \le i \le r - 1$), or, for $i = r$, to the intended recipients of the messages.

The ciphertexts have the structure of an onion. The sender builds the initial onion, and each mix removes one layer of the onion. The last mix in the chain M_r recovers the original messages m including the addresses of the intended recipients.

It is common to wrap the control information and the payload separately. The sender S randomly generates secret keys k_i for the symmetric cipher F, one for each mix server M_i. The message that S submits to the mix net consists of the two onions

$$\omega_1 := \tilde{E}_{pk_1} \left(k_1 \| s_1 \| \left(\tilde{E}_{pk_2} \left(k_2 \| s_2 \| \left(\tilde{E}_{pk_3} \ldots \right. \right. \right. \right.$$
$$\left. \left. \left. \ldots \| \left(\tilde{E}_{pk_{r-1}} \left(k_{r-1} \| s_{r-1} \| \left(\tilde{E}_{pk_r}(k_r \| s_r) \right) \right) \right) \ldots \right) \right) \right),$$
$$\omega_2 := F_{k_1} \left(F_{k_2} \left(\ldots \left(F_{k_{r-2}} \left(F_{k_{r-1}} \left(F_{k_r}(m) \right) \right) \right) \ldots \right) \right).$$

The message m is encrypted symmetrically by using the keys $k_r, k_{r-1}, \ldots, k_1$. The first onion, ω_1, is the header onion of the message. Besides the key k_i, there may be more control information s_i. For example, the address of the intended recipient may be part of s_r, and in a source-routed mix net the address of the next mix, M_{i+1}, in the chain is included in s_i. Each mix server decrypts and removes its layer in the obvious way. The length of the header onion increases with the number of mixes. Since hybrid encryption is used, messages of any length can be wrapped and sent through the mix net, and there is no upper bound on the length of the mix cascade. At the output of the mix cascade, the plaintext of m appears. Of course, m itself can be a message that is encrypted, for example with the public key of the recipient.

Let Alice send a message m to Bob anonymously through a mix net. Assume that Alice expects an answer from Bob, and she still wants to keep her identity secret as the recipient of Bob's answer. Therefore, she needs anonymity as the recipient of a message. We assume in our discussion that message and control information are transmitted in separate onions, as described before. Then Alice's recipient anonymity can be accomplished by including an *untraceable return address* in her message to Bob ([Chaum81]). Alice can build such a return address (also called a reply block) as follows, for example.[8] Alice selects a cascade $M_1', M_2', \ldots, M_t'$ of mix servers through which Bob's answer will be routed. Then, she generates secret keys

---

[8] We sketch a construction here that is similar to that in [Chaum81].

k_i', $1 \le i \le t$, for the symmetric cipher F. Let pk_i' be the public key of M_i'. Now, Alice generates the header onion ω for Bob's answer:

$$\omega := \tilde{E}_{pk_1'} \left(k_1' \| s_1' \| \left(\tilde{E}_{pk_2'} \ldots \| \left(\tilde{E}_{pk_{t-1}'} \left(k_{t-1}' \| s_{t-1}' \| \left(\tilde{E}_{pk_t'} (k_t' \| s_t') \right) \right) \right) \ldots \right) \right).$$

The address of M_{i+1}' is part of the control information s_i', and s_t' contains the address of Alice. The reply block consists of the address of M_1' and ω. Alice includes it in her message to Bob. Bob submits (ω, ω') to M_1', where $\omega' = m'$ is Bob's answer. This input is then processed by the mix chain $M_1', \ldots, M_t'$ in the usual way – each mix removes its layer from the onions. This transforms ω' to

$$F_{k_t'}^{-1}(F_{k_{t-1}'}^{-1}(\ldots (F_{k_1'}^{-1}(m')))),$$

which is then sent to Alice by M_t'. Since Alice knows the keys $k_1', \ldots, k_t'$, she can recover Bob's answer m'. In this solution, however, the answer m' becomes known to observers and M_1'. Often this is undesirable, for example because it gives a hint to potential recipients. In this case Alice generates a one-time public–secret key pair (pk_m, sk_m) – it is used only once, for the answer to m – and includes pk_m in the return address. Bob encrypts his answer m' by using pk_m and submits $\omega' := \tilde{E}_{pk_m}(m')$.

Remarks:

1. Anonymity can only be ensured if the input lists to the mix servers are sufficiently long. Otherwise, an observer might simply guess links between inputs and outputs with a probability that is unacceptably high.
2. Encryption algorithms do not usually hide the length of the plaintexts. Therefore, to prevent an observer from linking inputs and outputs, it must be ensured that all the inputs or all the outputs of a mix server have the same length, by applying appropriate paddings.
3. Replay attacks must be prevented in decryption networks. If an attacker manages to submit an input ciphertext c for a second time to a mix, he will be able to identify the corresponding output messages, because these output messages will be identical. Therefore, decryption mixes must store the inputs that they have already processed and refuse to decrypt the same ciphertext a second time. It is sufficient to store an identifier instead of the complete input. One can take, for example, the hash value of the message or the control information as an identifier (as in [DanGol09]). Various strategies can be applied to limit the storage capacity required, for example timestamps. Alternatively, the key pair of a mix may be changed at suitable intervals. Since replay is only dangerous as long as the same key is used, the mix only has to remember messages that have been processed with the current key pair.
4. Many public-key encryption schemes and symmetric ciphers are *malleable*. This means that an adversary may be able to modify the plaintext purposefully by modifying the ciphertext without knowing the plaintext.

If malleable encryption algorithms are used in a mix net, an adversary might be able to identify messages at the output of the mix net. Additional provision then has to be made to avoid such attacks. See, for example, the discussion of "tagging attacks" in [DanDinMat03]. For re-encryption mix nets, we discuss this aspect in Section 4.6.3 below.

5. Examples of sophisticated message formats for decryption mixes with proven security features have been presented by [Moeller03] and [DanGol09].

6. Since the outputs of a mix server do not contain any hint about the inputs, a mix server might remove, add or duplicate messages. This is critical in applications such as electronic voting, where encrypted votes are mixed. *Robustness* ([JakJueRiv02]) of the mix net is therefore desirable. This means that the mix net provides a proof or at least strong evidence that it has operated correctly. We will describe below how a re-encryption mix net can give a zero-knowledge proof that it has worked correctly. Any interested party is able to check the validity of the proof. For certain types of decryption mix nets, robustness can also be verified by zero-knowledge proofs (see, e.g., [Wikström05]).

7. The *anonymous remailers* Mixmaster and Mixminion ([DanDinMat03]) and the *Web anonymizer* JonDonym use decryption mix nets. The Web anonymizer Tor ("The Onion Router") is based on a source-routed decryption network. The nodes do not permute the messages. Tor's security relies on the assumption that an attacker will not be able to watch all the nodes in a chain.

4.6.2 Re-encryption Mix Nets

As before, we assume that the public-key encryption algorithm E is semantically secure and, as a consequence, probabilistic (see Definition 9.14 and the subsequent remarks). In addition, we assume that E is a homomorphic encryption algorithm (Section 3.6). Examples are ElGamal encryption in a prime-order subgroup (Section 3.4.4) and Paillier encryption (Section 3.6.2). In our notation, we explicitly state the randomness ω that is used for the encryption of a plaintext: $E_{pk}(m, \omega)$. Recall that $E_{pk}(m, \omega) = (g^\omega, y^\omega m)$ for ElGamal encryption ($\omega \in \mathbb{Z}_q$) and $E_{pk}(m, \omega) = g^m \omega^n$ ($\omega \in \mathbb{Z}_n^*$) for Paillier encryption.

We consider a re-encryption mix net and a mix cascade $M_1, M_2, \ldots, M_r$ in this network. A key pair (pk, sk) for E is randomly generated for the mix net. All messages through the mix net are encrypted by using E and pk. The last mix in the mix cascade outputs E_{pk}-encrypted ciphertexts. Therefore, a final decryption step is necessary to obtain the messages. It depends on the application when and where this decryption is done. For example, the secret key sk may be shared by the mix servers, and the mix servers jointly decrypt the ciphertexts (see Section 4.5.1 for such a threshold decryption of ElGamal ciphertexts). Alternatively, suppose that the mix cascade is used for anonymizing votes in an electronic voting system. Then the final output of the mix cascade is the encrypted, anonymized votes. Typically, they are written to a public bulletin board, and, later, when the election is closed and the tally is computed, they are jointly decrypted by the election authorities (see Section 4.7.2 for such a protocol). Alternatively, all messages may be intended for a single recipient, who is the owner of the secret key sk and decrypts the ciphertexts.

The Re-encryption Shuffle. A re-encryption mix server M receives ciphertexts as input, which are encrypted by using E and the public key pk of the mix net. It collects the incoming ciphertexts for a while and then executes a *re-encryption shuffle* on the list $(c_1, c_2, \ldots, c_n)$ of input ciphertexts: it re-encrypts the ciphertexts and randomly permutes them. More precisely, M randomly generates elements $\rho_1, \rho_2, \ldots, \rho_n$ ("randomness for re-encryption"), randomly generates a permutation $\pi : \{1, 2, \ldots, n\} \longrightarrow \{1, 2, \ldots, n\}$, sets

$$c'_k := E_{pk}(e, \rho_k) \cdot c_{\pi(k)}, \quad 1 \le k \le n,$$

and outputs the permuted list $(c'_1, c'_2, \ldots, c'_n)$. Here, e denotes the unit element in the plaintext group of the encryption algorithm ($e = [1]$ for ElGamal and $e = [0]$ for Paillier encryption, see Section 3.6.3 for a discussion of re-encryption). The random values $\rho_1, \rho_2, \ldots, \rho_n$ and the permutation π are kept secret.

The anonymous sending of messages through the mix cascade $M_1, M_2, \ldots, M_r$ works as follows:

1. Each sender S randomly generates ω, encrypts his message m by using the public key pk of the mix net and ω and sends the ciphertext

$$c = E_{pk}(m, \omega)$$

   to the first mix, M_1.
2. The mix M_i collects the incoming ciphertexts, then executes a re-encryption shuffle on its input list $c_1, c_2, \ldots, c_n$ of ciphertexts and sends the output list $c'_1, c'_2, \ldots, c'_n$ to the next mix, M_{i+1} ($1 \le i \le r - 1$), or, for $i = r$, to the final decryption step.

Remarks:

1. All encryptions and re-encryptions are done with E. Therefore, only messages that can be encoded as elements in the domain of E can be sent through the mix net. In typical applications of re-encryption mix nets such as voting systems, this is not a problem because the messages (for example, votes or credentials) are rather short. Recall that the domain of ElGamal encryption in a prime-order subgroup is the unique subgroup G_q of order q of $\mathbb{Z}_p^*$ (p, q large primes such that q divides $p - 1$), and the domain of Paillier encryption is $\mathbb{Z}_n$ ($n = pq$, p, q distinct large primes).

2. Assume that the secret key sk is known to a single individual Bob. Then Bob is able to decrypt and trace all messages in the mix net. This is only acceptable if the identity of the senders need not be kept secret from Bob. Therefore, in most applications the secret key is shared by using a (t, n)-threshold secret-sharing scheme (Section 4.4). Then t shareholders must collaborate to restore the secret, and the anonymity of the senders is preserved as long as no coalition of t dishonest shareholders exists.

3. Since re-encryption with a probabilistic homomorphic encryption algorithm is used, the re-encryption shuffle does not increase the length of the ciphertexts. The messages are of constant length throughout the whole mix net.

4. As observed before, anonymity can only be ensured if the input lists to the mix servers are sufficiently long. Otherwise, an observer might simply guess links between inputs and outputs with a probability that is unacceptably high.

Robustness and Public Verifiability. The outputs of a mix server do not contain any hint about the inputs. Therefore, a malicious mix server could substitute messages without being detected. This is critical in the typical applications of re-encryption mix nets, for example electronic voting, where encrypted votes are mixed. We require, therefore, that a re-encryption mix cascade is *robust* ([JakJueRiv02]). This means that the mix cascade provides a proof or at least strong evidence that it operates correctly, i.e., that its output corresponds to a permutation of its input. *Randomized partial checking* ([JakJueRiv02]) is a cut-and-choose technique to obtain such evidence.

Rigorous proofs are based on zero-knowledge protocols. In Section 4.6.4 below, we will describe how a re-encryption mix server can generate a non-interactive zero-knowledge proof that its shuffle was executed correctly. The robustness of a mix cascade $M_1, M_2, \ldots, M_r$ is then guaranteed if each mix server $M_i, 1 \leq i \leq r$, in the cascade publishes such a proof. Since the proofs (including the inputs and outputs) of all mix servers are published, any interested party is able to verify that the mix cascade operates correctly by checking the validity of the proofs. This property is called *public verifiability* ([JakJueRiv02]).

Malleability of the Input Ciphertexts. The homomorphic public-key encryption algorithms that we use in re-encryption mix nets are *malleable*. An adversary can modify the plaintext purposefully by modifying the ciphertext without knowing the plaintext. By exploiting the malleability of ciphertexts, an active attacker can identify messages at the mix cascade's output and in this way reveal the identity of senders.

We illustrate this problem with two examples. Let S send a message m, encrypted as c, to a re-encryption mix cascade. An adversary Eve intercepts c and submits a modification c' of c to the cascade, which is then mixed together with c.

1. Assume that ElGamal encryption in a prime-order subgroup with public key $pk = (p, q, g, y)$ is used in the mix net. Let $c = (g^\omega, y^\omega m)$ be S's ciphertext. Eve chooses some exponent $e \in \mathbb{Z}_q$ and some randomness $\rho \in \mathbb{Z}_q$. She raises c to the e-th power and re-encrypts it:

$$c' := E_{pk}([1], \rho)c^e = (g^\rho, y^\rho) \cdot (g^\omega, y^\omega m)^e = (g^{\rho + e\omega}, y^{\rho + e\omega} m^e).$$

c' is an encryption of m^e. Eve analyzes the output of the mix cascade after the final decryption and searches for elements z, z' with $z' = z^e$.

2. Assume that Paillier encryption with public key $pk = (n, g)$ is used in the mix net. Eve chooses some multiplier $k \in \mathbb{Z}_n$ (e.g., $k = 2$) and some randomness $\rho \in \mathbb{Z}_n^*$. She raises c to the k-th power and re-encrypts it:

$$c' := E_{pk}([0], \rho)c^k.$$

c' is an encryption of $k \cdot m \in \mathbb{Z}_n$. Eve analyzes the output of the mix cascade after the final decryption and searches for elements z, z' with $z' = k \cdot z$.

In this way, Eve can find out, with high probability, which of the output messages results from c and hence from S.

We can prevent such attacks by using a robust mix cascade and requiring that each sender has to prove that he knows the plaintext of the ciphertext which he is submitting to the mix cascade. This prevents an adversary from intercepting an input ciphertext c and submitting a modification of c to the cascade.

In Section 4.6.3, we explain how a sender can generate a non-interactive zero-knowledge proof that he knows the plaintext.

4.6.3 Proving Knowledge of the Plaintext

We describe here how the sender of a ciphertext c can prove *plaintext awareness*, i.e., that the sender knows the plaintext contained in c. We use ElGamal encryption. The proof for Paillier encryption is analogous (see Exercise 10).

Given a public key $pk = (p, q, g, y)$ and a ciphertext c, Peggy can interactively prove to Vic that she knows a plaintext m and a randomness value ρ such that $c = E_{pk}(m, \rho) = (g^\rho, y^\rho m)$. (m, ρ) is Peggy's witness.

Protocol 4.21.
KnowledgeOfThePlaintext(p, q, g, y, c):

1. Peggy randomly chooses a plaintext m' and a randomness ρ', computes $c' = E_{pk}(m', \rho') = (g^{\rho'}, y^{\rho'} m')$ and sends c' to Vic.
2. Vic randomly selects a challenge $e \in \mathbb{Z}_q$ and sends e to Peggy.
3. Peggy computes $\tilde{m} := m^e m'$ and $\tilde{\rho} := e\rho + \rho'$ and sends $\tilde{m}, \tilde{\rho}$ to Vic.
4. Vic accepts if $c^e c' = E_{pk}(\tilde{m}, \tilde{\rho}) = (g^{\tilde{\rho}}, y^{\tilde{\rho}} \tilde{m})$.

Proposition 4.22. *The above protocol is a special honest-verifier zero-knowledge proof of knowledge.*

Proof. Obviously, the interactive proof is complete. To prove soundness, assume that a prover P^*, honest or not, is able to convince Vic with probability $> 1/q$. Then, P^* is able to send an element c' in the first move such that she is able to compute correct answers $(\tilde{m}_1, \tilde{\rho}_1)$ and $(\tilde{m}_2, \tilde{\rho}_2)$ for at least two distinct challenges $e_1, e_2 \in \mathbb{Z}_q$, $e_1 \neq e_2$. Since ElGamal encryption is homomorphic, $\tilde{m}_1 \tilde{m}_2^{-1}$ is the plaintext $m^{e_1-e_2}$ of $c^{e_1-e_2}$. Therefore, P^* immediately derives

$$m = \left(\tilde{m}_1 \tilde{m}_2^{-1} \right)^{(e_1-e_2)^{-1}},$$

which means that P^* is an honest prover who knows the plaintext.

Given a challenge $e \in \mathbb{Z}_q$, the following simulator S generates correctly distributed accepting transcripts $(c', e, \tilde{m}, \tilde{\rho})$: S randomly generates elements $\tilde{m} \in G_q, \tilde{\rho} \in \mathbb{Z}_q$ and sets $c' := E_{pk}(\tilde{m}, \tilde{\rho})c^{-e}$. Therefore, the protocol is a special honest-verifier zero-knowledge protocol. $\square$

By applying the Fiat–Shamir heuristics, the interactive proof can be converted into a non-interactive zero-knowledge proof (Section 4.5.3).

4.6.4 Zero-Knowledge Proofs of Shuffles

In mix nets, semantically secure encryption algorithms are applied. Therefore, the shuffle output does not reveal any information about the permutation and the messages. This means that a malicious mix server could substitute some of the ciphertexts without being detected. In an election protocol, for instance, it could replace a large number of ciphertexts with encrypted votes for its favorite candidate. In critical applications, we therefore require robust and verifiable mix nets: each mix server must prove the correctness of its shuffle, without revealing any information about the permutation or the randomizers in the re-encryption.

We describe in this section how an ElGamal-based re-encryption mix server can prove interactively that its shuffle was executed correctly. The proof is an honest-verifier zero-knowledge proof. Therefore, by applying the

Fiat–Shamir heuristics (Section 4.5.3), the mix server can deliver a non-interactive zero-knowledge proof.

There is a lot of research on zero-knowledge proofs of the correctness of shuffles (e.g., [Neff01]; [FurSak01]; [Furukawa04]; [Groth03]; [GroLu07]; [Wikström05]; [Wikström09]; [TerWik10]).

Here, we explain a simplified version of the recent protocol of Bayer and Groth ([BayGro12]). A challenge in the construction of correctness proofs of shuffles is efficiency, because the number of shuffled messages can be huge in voting applications, for example. By applying sophisticated methods from Linear Algebra – messages, ciphertexts and commitments are arranged in matrices – Bayer and Groth achieved a remarkable efficiency in the original protocol. The communication complexity is sublinear, and the computation costs are low.

The Generalized Pedersen Commitment Scheme. We need a generalization of the homomorphic Pedersen commitment scheme described in Section 4.3.2. In this generalization ([BayGro12]), Alice commits not just to one value $a \in \mathbb{Z}_q$, but simultaneously to a whole vector $\boldsymbol{a} = (a_1, a_2, \ldots, a_n)$ of values. The receiver of the commitment is Bob.

Protocol 4.23.
Generalized Pedersen Commitment:

1. *Key Generation.* Bob randomly chooses large prime numbers p and q such that q divides $p - 1$. Then he randomly chooses elements $h, g_1, g_2, g_3, \ldots, g_n$ from the subgroup G_q of order q in $\mathbb{Z}_p^*$, with h and all of the $g_i \neq [1]$. Bob publishes the commitment key $ck = (p, q, h, g_1, g_2, \ldots, g_n).$[9]

2. *Commit.* To commit to $\boldsymbol{a} = (a_1, a_2, \ldots, a_n) \in \mathbb{Z}_q^n$, Alice chooses $r \in \mathbb{Z}_q$ at random, sets $c := \mathrm{com}_{ck}(r, \boldsymbol{a}) := h^r g_1^{a_1} g_2^{a_2} \cdots g_n^{a_n} = h^r \prod_{k=1}^n g_k^{a_k}$ and sends c to Bob.

3. *Reveal.* Alice sends r and $\boldsymbol{a} = (a_1, a_2, \ldots, a_n)$ to Bob. Bob verifies that $c = h^r \prod_{k=1}^n g_k^{a_k}$.

Remarks:

1. The commitment key can be reused for many commitments.
2. The scheme can be used to commit to fewer than n values. To commit to a vector $(a_1, \ldots, a_m), m < n$, we simply set $a_{m+1} = \ldots = a_n = 0$.
3. The generalized Pedersen commitment scheme is homomorphic:

$$\mathrm{com}_{ck}(r, \boldsymbol{a}) \cdot \mathrm{com}_{ck}(s, \boldsymbol{b})$$
$$= \left(h^r \prod_{k=1}^n g_k^{a_k} \right) \cdot \left(h^s \prod_{k=1}^n g_k^{b_k} \right) = h^{r+s} \prod_{k=1}^n g_k^{a_k + b_k}.$$

---

[9] He can generate the key as in prime-order-subgroup ElGamal encryption or as in the DSA (see Sections 3.4.4 and 3.4.3). Since the elements are chosen randomly, the probability that two of them are equal is negligibly small.

$$= \mathrm{com}_{ck}(r + s, \boldsymbol{a} + \boldsymbol{b}).$$

4. The commitment scheme is computationally binding. If Alice had committed to $\boldsymbol{a}$ and could open her commitment as $\tilde{\boldsymbol{a}}$, $\boldsymbol{a} \neq \tilde{\boldsymbol{a}}$, then $h^r \prod_{k=1}^{n} g_k^{a_k} = h^{\tilde{r}} \prod_{k=1}^{n} g_k^{\tilde{a}_k}$ and Alice could compute a non-trivial representation $[1] = h^{r-\tilde{r}} \prod_{k=1}^{n} g_k^{a_k - \tilde{a}_k}$ of $[1]$. This contradicts the assumption that discrete logarithms are infeasible to compute (see Section 4.8.3 and the proof of Proposition 4.43).

5. The commitment scheme hides perfectly. The committed values are perfectly hidden by the random element h^r.

A Simple Version of the Schwartz–Zippel Lemma. We need a very simple version of the *Schwartz–Zippel Lemma*.[10]

Lemma 4.24. *Let* $f(X) := a_0 + a_1 X + \ldots + a_d X^d \in \mathbb{F}_q[X]$ *be a polynomial of degree* d *over a finite field* $\mathbb{F}_q$ *with* q *elements (for example,* q *may be a prime number and* $\mathbb{F}_q = \mathbb{Z}_q$*). If* $f \neq 0$ *and* $x \in \mathbb{F}_q$ *is chosen uniformly at random, then the probability that* $f(x) = 0$ *is* $\leq d/q$.

Proof. If $f \neq 0$, then f has at most d zeros, i.e., for at most d out of q elements x we have $f(x) = 0$. $\square$

Corollary 4.25 *(Simple Schwartz–Zippel Lemma). Let* $d, q \in \mathbb{Z}$, q *a prime. Assume that* d/q *is negligibly small.*[11] *Let* $f, g \in \mathbb{Z}_q[X]$ *be polynomials of degree* $\leq d$. *Assume that* $f(x) = g(x)$ *for a randomly chosen* $x \in \mathbb{Z}_q$. *Then the polynomials* f *and* g *are equal,* $f = g$, *with overwhelming probability, i.e., the probability that* $f \neq g$ *is negligibly small.*

In our setting, q is sufficiently large such that d/q is negligibly small for all of the polynomials that occur.

The Shuffle Argument. The *shuffle argument* is an interactive proof which enables an ElGamal-based re-encryption mix server to prove the correctness of its shuffle. In the proof, two subprotocols are executed, the *exponentiation argument* and the *product argument*. The subprotocols are described later (Protocols 4.30 and 4.28) .

We consider ciphertexts that are encrypted by using ElGamal encryption in a prime-order subgroup and a fixed public ElGamal key pk. Let p, q be the prime numbers in pk. Recall that q divides $p - 1$ and G_q denotes the unique subgroup of order q of $\mathbb{Z}_p^*$.

Given ElGamal ciphertexts $c_1, c_2, \ldots, c_n \in G_q^2$ and $\tilde{c}_1, \tilde{c}_2, \ldots, \tilde{c}_n \in G_q^2$, Peggy can prove interactively to Vic that she knows $\rho_1, \ldots, \rho_n \in \mathbb{Z}_q$ and a permutation $\pi : \{1, \ldots, n\} \longrightarrow \{1, \ldots, n\}$ such that

$$\tilde{c}_k = E_{pk}([1], \rho_k) c_{\pi(k)}, \ 1 \leq k \leq n.$$

---

[10] The original Schwartz–Zippel Lemma covers the case of multivariate polynomials.

[11] More precisely, we assume that $|d| = O(\log(|q|))$.

For her commitments, Peggy uses a generalized Pedersen commitment scheme com_{ck}. The commitment key $ck = (h, g_1, g_2, \ldots, g_n)$ is supplied by Vic. The parameters p, q of the commitment scheme are the same as in the ElGamal encryption key.

Protocol 4.26.
ProofOfCorrectShuffle$((c_1, c_2, \ldots, c_n), (\tilde{c}_1, \tilde{c}_2, \ldots, \tilde{c}_n))$:

1. Peggy randomly chooses $r \in \mathbb{Z}_q$, computes the commitment
$$c_\pi := \mathrm{com}_{ck}(r, (\pi(1), \pi(2), \ldots, \pi(n)))$$
   and sends c_π to Vic.
2. Vic randomly selects a challenge $x \in \mathbb{Z}_q$ and sends it to Peggy.
3. Peggy randomly selects $s \in \mathbb{Z}_q$, computes the commitment
$$c_{x,\pi} := \mathrm{com}_{ck}(s, (x^{\pi(1)}, x^{\pi(2)}, \ldots, x^{\pi(n)}))$$
   and sends $c_{x,\pi}$ to Vic.
4. Vic randomly selects challenges $y, z \in \mathbb{Z}_q$ and sends y, z to Peggy.
5. The numbers
$$c_{-z} := \mathrm{com}_{ck}(0, (-z, -z, \ldots, -z)), \quad c_d := c_\pi^y c_{x,\pi},$$
$$b := \prod_{k=1}^{n}(yk + x^k - z), \quad c := \prod_{k=1}^{n} c_k^{x^k},$$
   which appear in the following steps, are known to both Peggy and Vic. Both can compute them easily.
   a. Peggy proves to Vic by a product argument (Protocol 4.30 below) that she knows $t, e_1, \ldots, e_n$ such that
$$b = \prod_{k=1}^{n} e_k \quad \text{and} \quad c_d c_{-z} = \mathrm{com}_{ck}(t, (e_1, \ldots, e_n)).$$

   b. Peggy proves to Vic by an exponentiation argument (Protocol 4.28 below) that she knows $s, \rho, b_1, \ldots, b_n$ such that
$$c = E([1], \rho) \prod_{k=1}^{n} \tilde{c}_k^{b_k} \quad \text{and} \quad c_{x,\pi} = \mathrm{com}_{ck}(s, (b_1, \ldots, b_n)).$$

6. Vic checks that $c_\pi, c_{x,\pi} \in G_q$ and accepts if the verification conditions in the product and exponentiation arguments are satisfied.

The Non-interactive Proof. By using a cryptographic hash function h, the proof can be turned into a non-interactive zero-knowledge proof. Peggy does the following:

1. She computes c_π.
2. She computes the challenge $x := h(c_\pi \| c_1 \| c_2 \| \ldots \| c_n \| \tilde{c}_1 \| \tilde{c}_2 \| \ldots \| \tilde{c}_n)$.
3. She computes $c_{x,\pi}$ as before.
4. She computes the challenges y, z:

$$y \| z := h(c_\pi \| c_{x,\pi} \| c_1 \| c_2 \| \ldots \| c_n \| \tilde{c}_1 \| \tilde{c}_2 \| \ldots \| \tilde{c}_n).$$

5. She computes c_{-z}, c_d, b, c as before.
6. She generates a non-interactive product argument $(x', c_{\boldsymbol{\Delta}}, \tilde{r}, (\tilde{a}_1, \ldots, \tilde{a}_n), \tilde{s}, (\tilde{b}_1, \ldots, \tilde{b}_n))$ for $b, c_d c_{-z}$ (see below for details).
7. She generates a non-interactive exponentiation argument $(x'', r', (a'_1, \ldots, a'_n), \rho')$ for $c, c_{x,\pi}$ (see below for details).

The proof consists of $(x, y, z, c_\pi, c_{x,\pi})$ and the non-interactive product and exponentiation arguments.

Vic verifies the proof by validating the product and exponentiation arguments and by checking that $y \| z = h(c_\pi \| c_{x,\pi} \| c_1 \| c_2 \| \ldots \| c_n \| \tilde{c}_1 \| \tilde{c}_2 \| \ldots \| \tilde{c}_n)$.

Remarks:

1. The product argument (Step 5a of Protocol 4.26) and the ability of the prover to compute an opening of $c_{x,\pi}$ in the exponentiation argument (Step 5b) together guarantee that

   - the committed values in c_π are indeed a permutation $(\pi(1), \ldots, \pi(n))$ of $(1, \ldots, n)$,
   - $c_{x,\pi}$ contains the values $(x^1, x^2, \ldots, x^n)$ permuted in the same way, i.e., $b_k = x^{\pi(k)}, 1 \le k \le n$,
   - the prover knows this permutation.

   The randomly chosen challenges y, z and the Schwartz–Zippel Lemma are needed for these conclusions (for details, see the soundness proof of Proposition 4.27). Since the commitment c_π is sent in the first move, before Vic selects the challenge x, the permutation is fixed from the beginning and is independent of the challenge x.

2. The exponentiation argument (Step 5b) then shows that $\prod_{k=1}^n c_k^{x^k} = E([1], \rho) \prod_{k=1}^n \tilde{c}_k^{x^{\pi(k)}}$. Since x is a random element, the Schwartz–Zippel Lemma then implies that $\tilde{c}_k$ is a re-encryption of $c_{\pi(k)}, 1 \le k \le n$, which means that the shuffle is correct (for details, see again the soundness proof of Proposition 4.27).

3. The proofs a and b in step 5 are three-move protocols and can be executed in parallel. Thus, the shuffle proof is a seven-move protocol.

Proposition 4.27. *Protocol 4.26 is a special honest-verifier zero-knowledge proof of knowledge.*

Proof. Completeness. The honest prover Peggy can successfully complete the product and exponentiation arguments (Protocols 4.30 and 4.28). In the product argument, she proves to Bob that she knows the opening $(t, e_1, \ldots, e_n)$ of $c_d c_{-z}$, where

$$t := yr + s, \ e_k := d_k - z, \ d_k := y\pi(k) + x^{\pi(k)}, \ 1 \leq k \leq n,$$

without revealing any of these values. Note that $c_d = c_\pi^y c_{x,\pi} = \operatorname{com}_{ck}(t, (d_1, \ldots, d_n))$ and $c_d c_{-z} = \operatorname{com}_{ck}(t, (d_1 - z, \ldots, d_n - z))$, because the commitment scheme is homomorphic.

In the exponentiation argument, she proves to Bob that she knows the opening $(s, b_1, \ldots, b_n)$ of $c_{x,\pi}$ and the randomness value ρ, where

$$\rho := -\sum_{k=1}^{n} \rho_k x^{\pi(k)} \quad \text{and} \quad b_k := x^{\pi(k)}, 1 \leq k \leq n,$$

without revealing any of these values.

Soundness. Assume that a prover P^*, honest or not, is able to convince Vic with a non-negligible probability w. "Non-negligible" means that there is some $\nu \in \mathbb{N}$ such that $w \geq 1/|p|^\nu$. We will show that P^* can, with overwhelming probability, efficiently compute a witness, consisting of a permutation π and $\rho_1, \ldots, \rho_n \in \mathbb{Z}_q$ such that

$$\tilde{c}_k = E_{pk}([1], \rho_k) c_{\pi(k)}, \ 1 \leq k \leq n.$$

Thus, with overwhelming probability, P^* is an honest prover.

P^* is able to execute the product and exponentiation arguments convincingly. This includes the fact that P^* is able to compute openings $(e_1, \ldots, e_n)$ and $(b_1, \ldots, b_n)$ of the commitments $c_d c_{-z}$ and $c_{x,\pi}$ (see the proofs of Propositions 4.31 and 4.29 below for details), and hence also the opening $(d_1, \ldots, d_n)$ of c_d, where $d_k := e_k + z, 1 \leq k \leq n$. Since $c_d = c_\pi^y c_{x,\pi}$, P^* can then easily derive an opening $(a_1, \ldots, a_n)$ of c_π:

$$a_k := y^{-1}(d_k - b_k) \quad \text{resp.} \quad d_k = ya_k + b_k, 1 \leq k \leq n.$$

The product argument shows that $\prod_{k=1}^{n}(d_k - z) = \prod_{k=1}^{n}(yk + x^k - z)$. Since z was chosen randomly by Vic, we conclude from Corollary 4.25 that, with overwhelming probability, there is a permutation π with $d_k = ya_k + b_k = y\pi(k) + x^{\pi(k)}$. By applying Corollary 4.25 again, now utilizing the fact that y is chosen randomly, we see that, with overwhelming probability, $a_k = \pi(k)$ and $b_k = x^{\pi(k)}$.

The exponentiation argument shows that

$$\prod_{k=1}^{n} c_k^{x^k} = E([1], \rho) \prod_{k=1}^{n} \tilde{c}_k^{x^{\pi(k)}}.$$

Since the encryption algorithm is homomorphic, decryption preserves this equation, i.e.,

$$\prod_{k=1}^{n} m_k^{x^k} = \prod_{k=1}^{n} \tilde{m}_k^{x^{\pi(k)}},$$

where m_k and $\tilde{m}_k$ are the plaintexts of c_k and $\tilde{c}_k$, respectively. Taking the logarithm on both sides, we get

$$\sum_{k=1}^{n} \log_g(m_k) x^k = \sum_{k=1}^{n} \log_g(\tilde{m}_k) x^{\pi(k)}.$$

Since x is chosen randomly, Corollary 4.25 says that

$$\log_g(m_{\pi(k)}) = \log_g(\tilde{m}_k), \quad \text{and hence} \quad m_{\pi(k)} = \tilde{m}_k, \ 1 \le k \le n.$$

Then, $\tilde{c}_k$ is a re-encryption of $c_{\pi(k)}$, i.e., there exists $\rho_k \in \mathbb{Z}_q$ such that

$$\tilde{c}_k = E_{pk}([1], \rho_k) c_{\pi(k)}, \ 1 \le k \le n.$$

P^* is able to compute the randomness elements $\rho_k, 1 \le k \le n$. Namely, P^* provides the element $c_\pi \in G_q$ in the first move, and is able to answer randomly chosen challenges x correctly with probability $\ge {}^1/_{|p|^\nu}$. Therefore, we expect that P^* will answer convincingly for at least one out of $|p|^\nu$ random challenges x. This implies that P^* has computed correct answers to n pairwise distinct challenges $x_1, x_2, \ldots, x_n$ after polynomially many steps. The permutation π is committed by c_π and hence is the same for all x_i. P^*'s convincing answer in the exponentiation argument for x_i implies that, with overwhelming probability, she is able to compute elements $b_{i,k}, 1 \le k \le n$, and a randomness element λ_i such that

$$\prod_{k=1}^{n} c_k^{x_i^k} = E_{pk}([1], \lambda_i) \prod_{k=1}^{n} \tilde{c}_k^{b_{i,k}}.$$

We have seen above that $b_{i,k} = x_i^{\pi(k)}, 1 \le k \le n$. Hence

$$\prod_{k=1}^{n} c_k^{x_i^k} = E_{pk}([1], \lambda_i) \prod_{k=1}^{n} \tilde{c}_k^{x_i^{\pi(k)}} = E_{pk}([1], \lambda_i) \prod_{k=1}^{n} \tilde{c}_{\pi^{-1}(k)}^{x_i^k}.$$

From

$$\prod_{k=1}^{n} c_k^{x_i^k} = \prod_{k=1}^{n} c_{\pi(k)}^{x_i^{\pi(k)}} = \prod_{k=1}^{n} \left(E_{pk}([1], -\rho_k) \tilde{c}_k \right)^{x_i^{\pi(k)}}$$

$$= \prod_{k=1}^{n} E_{pk}\left([1], -x_i^{\pi(k)} \rho_k\right) \tilde{c}_k^{x_i^{\pi(k)}} = E_{pk}\left([1], -\sum_{k=1}^{n} x_i^{\pi(k)} \rho_k\right) \prod_{k=1}^{n} \tilde{c}_k^{x_i^{\pi(k)}}$$

$$= E_{pk}\left([1], -\sum_{k=1}^{n} x_i^k \rho_{\pi^{-1}(k)}\right) \prod_{k=1}^{n} \tilde{c}_k^{x_i^{\pi(k)}},$$

we conclude that $\lambda_i = -\sum_{k=1}^{n} x_i^k \rho_{\pi^{-1}(k)}$ or

$$(\lambda_1, \ldots, \lambda_n) = -(\rho_{\pi^{-1}(1)}, \ldots, \rho_{\pi^{-1}(n)})X, \quad \text{where} \quad X := \begin{pmatrix} x_1^1 & \cdots & x_n^1 \\ \vdots & & \vdots \\ x_1^n & \cdots & x_n^n \end{pmatrix}.$$

The matrix X is a submatrix of a Vandermonde matrix and hence invertible. Thus, P^* can compute the randomness values ρ_k by

$$\left(\rho_{\pi^{-1}(1)}, \ldots, \rho_{\pi^{-1}(n)}\right) = -(\lambda_1, \ldots, \lambda_n)X^{-1}.$$

Special Honest-Verifier Zero-Knowledge Property. We rely on the fact that the product and the exponentiation arguments have this property. On input of a challenge $(x, y, z) \in \mathbb{Z}_q^3$ and challenges $x', x'' \in \mathbb{Z}_q$ for the product and exponentiation arguments, the simulator S does the following:

1. It randomly generates elements $c_\pi, c_{x,\pi} \in G_q$.
2. It sets
$$c_d := c_\pi^y c_{x,\pi}, \ c_{-z} := \text{com}_{ck}(0, (-z, -z, \ldots, -z)).$$

3. It calls the simulator of the product argument with parameters $(c_d c_{-z}, \prod_{k=1}^{n}(yk + x^k - z))$ and challenge x'.
4. It calls the simulator of the exponentiation argument with parameters $\left(c_1, c_2, \ldots, c_n, \prod_{k=1}^{n} c_k^{x^k}, c_{x,\pi}\right)$ and challenge x''.

The transcripts generated are accepting transcripts by their very construction. They are identically distributed as in the real proof by Peggy, because this is true for the simulated transcripts of the product and exponentiation arguments and $c_\pi, c_{x,\pi}$ are uniformly distributed elements of G_q.   □

The Exponentiation Argument. As before, we consider ciphertexts that are encrypted by using ElGamal encryption in a prime-order subgroup and a fixed public ElGamal key pk, and p, q are the prime numbers in pk. The commitments com_{ck} are generalized Pedersen commitments. The commitment key $ck = (h, g_1, g_2, \ldots, g_n)$ is supplied by Vic. The parameters p, q of the commitment scheme are the same as in the ElGamal key.

Given ElGamal ciphertexts $c_1, c_2, \ldots, c_n \in G_q^2$ and $c \in G_q^2$ and a commitment value $c_a \in G_q$, Peggy can prove interactively to Vic that she knows an opening $(r, a), a = (a_1, a_2, \ldots, a_n)$, of c_a (i.e., $c_a = \text{com}_{ck}(r, (a_1, a_2, \ldots, a_n)))$ and a randomness value $\rho \in \mathbb{Z}_q$ such that

$$c = E_{pk}([1], \rho) \prod_{k=1}^{n} c_k^{a_k}.$$

Protocol 4.28.
Exponentiation Argument$((c_1, c_2, \ldots, c_n), c, c_a)$:

1. Peggy randomly chooses $\tilde{r}, \tilde{a}_1, \ldots, \tilde{a}_n, \tilde{\rho} \in \mathbb{Z}_q$, computes

$$c_{\tilde{a}} := \mathrm{com}_{ck}(\tilde{r}, (\tilde{a}_1, \ldots, \tilde{a}_n)), \ \tilde{c} := E_{pk}([1], \tilde{\rho}) \prod_{k=1}^{n} c_k^{\tilde{a}_k}$$

and sends $c_{\tilde{a}}, \tilde{c}$ to Vic.
2. Vic randomly chooses a challenge $x \in \mathbb{Z}_q$ and sends x to Peggy.
3. Peggy computes

$$r' := xr + \tilde{r}, \ a_k' = xa_k + \tilde{a}_k, 1 \le k \le n, \ \rho' := x\rho + \tilde{\rho}$$

and sends $r', (a_1', \ldots, a_n'), \rho'$ to Vic.
4. Vic accepts if $c_{\tilde{a}} \in G_q, \tilde{c} \in G_q^2, r', a_1', \ldots, a_n', \rho' \in \mathbb{Z}_q$ and

$$c_a^x c_{\tilde{a}} = \mathrm{com}_{ck}(r', (a_1', \ldots, a_n')), \ c^x \tilde{c} = E([1], \rho') \prod_{k=1}^{n} c_k^{a_k'}.$$

Proposition 4.29. *Protocol 4.28 is a special honest-verifier zero-knowledge proof of knowledge.*

Proof. The completeness property is obvious.
Soundness. Assume that a prover P^*, honest or not, is able to convince Vic with a non-negligible probability w. "Non-negligible" means that there is some $\nu \in \mathbb{N}$ such that $w \ge 1/|p|^\nu$. We will show that P^* can, with overwhelming probability, efficiently compute a witness $(r, \boldsymbol{a}, \rho), \boldsymbol{a} = (a_1, \ldots, a_n)$, with the desired properties

$$c = E([1], \rho) \prod_{k=1}^{n} c_k^{a_k}, \ c_{\boldsymbol{a}} = \mathrm{com}_{ck}(r, (a_1, \ldots, a_n)).$$

Thus, with overwhelming probability, P^* is an honest prover.

P^* is able to present elements $c_{\tilde{a}} \in G_q, \tilde{c} \in G_q^2$ in the first move such that she can answer randomly chosen challenges x correctly with probability $\ge 1/|p|^\nu$. If P^* is challenged with randomly chosen elements $x \in \mathbb{Z}_q$, we expect that P^* will give a convincing answer for at least one out of $|p|^\nu$ challenges x. Therefore, after polynomially many steps, P^* will have answered two distinct challenges $x_1, x_2, x_1 \ne x_2$, correctly. She has been able to compute answers

$$r_1', (a_{11}', \ldots, a_{1n}'), \rho_1' \text{ for } x_1 \text{ and}$$

$$r_2', (a_{21}', \ldots, a_{2n}'), \rho_2' \text{ for } x_2$$

in such a way that Vic's verification conditions are satisfied.

From the openings $(r_1', (a_{11}', \ldots, a_{1n}'))$ and $(r_2', (a_{21}', \ldots, a_{2n}'))$ of $c_a^{x_1} c_{\tilde{a}}$ and $c_a^{x_2} c_{\tilde{a}}$, P^* can efficiently derive openings $(r, (a_1, \ldots, a_n))$ and $(\tilde{r}, (\tilde{a}_1, \ldots, \tilde{a}_n))$ of c_a and $c_{\tilde{a}}$ (note that $c_a^{x_1} c_{\tilde{a}} \, (c_a^{x_2} c_{\tilde{a}})^{-1} = c_a^{x_1 - x_2}$):

$$r = (x_1 - x_2)^{-1}(r_1' - r_2'), \quad a_k = (x_1 - x_2)^{-1}(a_{1k}' - a_{2k}'), \ 1 \le k \le n,$$

$$\tilde{r} := r_1' - r x_1, \quad \tilde{a}_k := a_{1k}' - a_k x_1, \ 1 \le k \le n.$$

Moreover, P^* sets $\rho := (x_1 - x_2)^{-1}(\rho_1' - \rho_2')$. By dividing the valid verification conditions

$$c^{x_1} \tilde{c} = E([1], \rho_1') \prod_{k=1}^{n} c_k^{a_{1k}'} \quad \text{and} \quad c^{x_2} \tilde{c} = E([1], \rho_2') \prod_{k=1}^{n} c_k^{a_{2k}'},$$

we get

$$c^{x_1 - x_2} = E([1], \rho_1' - \rho_2') \prod_{k=1}^{n} c_k^{a_{1k}' - a_{2k}'},$$

and hence

$$c = E([1], (x_1 - x_2)^{-1}(\rho_1' - \rho_2')) \prod_{k=1}^{n} c_k^{(x_1 - x_2)^{-1}(a_{1k}' - a_{2k}')},$$

i.e.,

$$c = E([1], \rho) \prod_{k=1}^{n} c_k^{a_k},$$

and we see that P^* has indeed computed a witness.

Special Honest-Verifier Zero-Knowledge Property. On input of a challenge $x \in \mathbb{Z}_q$, the simulator S does the following:

1. It randomly selects $r', a_1', \ldots, a_n', \rho' \in \mathbb{Z}_q$.
2. It sets

$$c_{\tilde{a}} := c_a^{-x} \operatorname{com}_{ck}(r', (a_1', \ldots, a_n')),$$

$$\tilde{c} := c^{-x} \cdot E([1], \rho') \prod_{k=1}^{n} c_k^{a_k'}.$$

The transcripts generated are accepting transcripts by their very construction, and they are identically distributed as in the real proof by Peggy: $r', a_1', \ldots, a_n', \rho'$ are uniformly distributed independent elements of $\mathbb{Z}_q$, and $c_{\tilde{a}}, \tilde{c}$ are then uniquely determined by the verification conditions. $\square$

The Non-interactive Proof. Peggy computes $c_{\tilde{a}}, \tilde{c}$ as in the interactive proof. Then she derives the challenge

$$x := h(c_1 \| c_2 \| \ldots \| c_n \| c \| c_a \| c_{\tilde{a}} \| \tilde{c})$$

and the response $r', (a'_1, \ldots, a'_n), \rho'$. The non-interactive proof is $(x, r', (a'_1, \ldots, a'_n), \rho')$. Vic verifies it by checking that

$$x = h\left(c_1 \| c_2 \| \ldots \| c_n \| c \| c_a \| c_a^{-x} \operatorname{com}_{ck}(r', (a'_1, \ldots, a'_n)) \| c^{-x} E([1], \rho') \prod_{k=1}^{n} c_k^{a'_k}\right).$$

The Product Argument. As before, the commitments com_{ck} are generalized Pedersen commitments. The underlying prime numbers are p, q, and the commitment key ck is supplied by Vic.

Given a commitment value $c_a \in G_q$ and a product value $b \in \mathbb{Z}_q$, Peggy can prove interactively to Vic that she knows $r \in \mathbb{Z}_q$ and $a = (a_1, a_2, \ldots, a_n) \in \mathbb{Z}_q^n$ such that

$$c_a = \operatorname{com}_{ck}(r, (a_1, a_2, \ldots, a_n)) \text{ and } b = \prod_{k=1}^{n} a_k,$$

i.e., Peggy proves that she knows an opening (r, a) of c_a such that b is the product of the committed values. The tuple $(r, (a_1, a_2, \ldots, a_n))$ is Peggy's witness.

Protocol 4.30.
ProductArgument(c_a, b):

1. Peggy computes $b_1 := a_1, b_2 := a_1 a_2, b_3 := a_1 a_2 a_3, \ldots, b_n := \prod_{k=1}^{n} a_k$. She randomly selects elements $d_1, d_2, \ldots, d_n$ and r_d from $\mathbb{Z}_q$, sets $\delta_1 := d_1, \delta_n := 0$, randomly selects elements $\delta_2, \ldots, \delta_{n-1}, s_1, s_2 \in \mathbb{Z}_q$ and computes

$$c_d := \operatorname{com}_{ck}(r_d, (d_1, \ldots, d_n)),$$

$$c_\delta := \operatorname{com}_{ck}(s_1, (-\delta_1 d_2, -\delta_2 d_3, \ldots, -\delta_{n-1} d_n))$$

$$c_\Delta := \operatorname{com}_{ck}(s_2, (\delta_2 - a_2\delta_1 - b_1 d_2, \ldots, \ldots, \delta_n - a_n\delta_{n-1} - b_{n-1}d_n)).$$

   She sends the commitments c_d, c_δ, c_Δ to Vic.
2. Vic randomly selects a challenge $x \in \mathbb{Z}_q$ and sends it to Peggy.
3. Peggy computes her answer

$$\tilde{r} := xr + r_d, \quad \tilde{a}_k := xa_k + d_k, 1 \le k \le n,$$

$$\tilde{s} := xs_2 + s_1, \quad \tilde{b}_k := xb_k + \delta_k, 1 \le k \le n,$$

and sends $\tilde{r}, (\tilde{a}_1, \ldots, \tilde{a}_n), \tilde{s}, (\tilde{b}_1, \ldots, \tilde{b}_n)$ to Vic.

4. Vic accepts if $c_{\boldsymbol{d}}, c_{\boldsymbol{\delta}}, c_{\boldsymbol{\Delta}} \in G_q$, which means that the commitments are elements of order q, and $\tilde{r}, \tilde{a}_1, \ldots, \tilde{a}_n, \tilde{s}, \tilde{b}_1, \ldots, \tilde{b}_n \in \mathbb{Z}_q$ and

$$c_{\boldsymbol{a}}^x c_{\boldsymbol{d}} = \mathrm{com}_{ck}(\tilde{r}, (\tilde{a}_1, \tilde{a}_2, \ldots, \tilde{a}_n)),$$

$$c_{\boldsymbol{\Delta}}^x c_{\boldsymbol{\delta}} = \mathrm{com}_{ck}(\tilde{s}, (x\tilde{b}_2 - \tilde{b}_1\tilde{a}_2, x\tilde{b}_3 - \tilde{b}_2\tilde{a}_3, \ldots, x\tilde{b}_n - \tilde{b}_{n-1}\tilde{a}_n)),$$

$$\tilde{b}_1 = \tilde{a}_1, \ \tilde{b}_n = xb.$$

Remark. To understand the proof of knowledge and its completeness, it is useful to observe that for $1 \le k \le n - 1$

$$\begin{aligned}
\tilde{b}_{k+1} &- \tilde{b}_k \tilde{a}_{k+1} \\
&= x(xb_{k+1} + \delta_{k+1}) - (xb_k + \delta_k)(xa_{k+1} + d_{k+1}) \\
&= (b_{k+1} - b_k a_{k+1})x^2 + (\delta_{k+1} - \delta_k a_{k+1} - b_k d_{k+1})x - \delta_k d_{k+1} \quad (*) \\
&= (\delta_{k+1} - \delta_k a_{k+1} - b_k d_{k+1})x - \delta_k d_{k+1},
\end{aligned}$$

assuming that Peggy is the honest prover and computes all values as required by the protocol (which includes that $b_{k+1} - b_k a_{k+1} = 0$).

Proposition 4.31. *Protocol 4.30 is a special honest-verifier zero-knowledge proof of knowledge.*

Proof. Completeness. If Peggy knows the witness $(r, (a_1, \ldots, a_n))$ and computes all values as required, then $\tilde{b}_1 = \tilde{a}_1$ and $\tilde{b}_n = xb$.

Moreover, $c_{\boldsymbol{a}} = \mathrm{com}_{ck}(r, (a_1, \ldots, a_n))$. From the homomorphic property of the commitment scheme, we conclude that

- $c_{\boldsymbol{a}}^x c_{\boldsymbol{d}}$ is a commitment for $(\tilde{a}_k = xa_k + d_k \mid 1 \le k \le n)$, and
- $c_{\boldsymbol{\Delta}}^x c_{\boldsymbol{\delta}}$ is a commitment for $((\delta_{k+1} - \delta_k a_{k+1} - b_k d_{k+1})x - \delta_k d_{k+1} \mid 1 \le k \le n - 1)$, which is equal to $(x\tilde{b}_{k+1} - \tilde{b}_k \tilde{a}_{k+1} \mid 1 \le k \le n - 1)$, according to $(*)$.

Soundness. Assume that a prover P^*, honest or not, is able to convince Vic with a non-negligible probability w. "Non-negligible" means that there is some $\nu \in \mathbb{N}$ such that $w \ge 1/|p|^\nu$. We will show that P^* can, with overwhelming probability, efficiently compute a witness $(r, \boldsymbol{a}), \boldsymbol{a} = (a_1, \ldots, a_n)$, with the desired properties

$$\prod_{k=1}^n a_k = b, \ c_{\boldsymbol{a}} = \mathrm{com}_{ck}(r, \boldsymbol{a}).$$

Thus, with overwhelming probability, P^* is an honest prover.

P^* is able to provide elements $c_{\boldsymbol{d}}, c_{\boldsymbol{\delta}}, c_{\boldsymbol{\Delta}} \in G_q$ in the first move such that she can answer randomly chosen challenges x correctly with probability $\ge 1/|p|^\nu$. If P^* is challenged with randomly chosen elements $x \in \mathbb{Z}_q$, we expect that P^* will give a convincing answer for at least one out of $|p|^\nu$ challenges x. Therefore, after polynomially many steps, P^* will have answered two distinct challenges $x_1, x_2, x_1 \neq x_2$, correctly. She has been able to compute answers

$$\tilde{r}_1, \tilde{a}_1 = (\tilde{a}_{11}, \tilde{a}_{12}, \ldots, \tilde{a}_{1n}), \tilde{s}_1, \tilde{b}_1 = (\tilde{b}_{11}, \tilde{b}_{12}, \ldots, \tilde{b}_{1n})$$

for x_1 and

$$\tilde{r}_2, \tilde{a}_2 = (\tilde{a}_{21}, \tilde{a}_{22}, \ldots, \tilde{a}_{2n}), \tilde{s}_2, \tilde{b}_2 = (\tilde{b}_{21}, \tilde{b}_{22}, \ldots, \tilde{b}_{2n})$$

for x_2 in such a way that Vic's verification conditions are satisfied. We will see now that P^* is then also able to compute openings of all of the commitments.

1. We have
$$c_a^{x_1} c_d = \mathrm{com}_{ck}(\tilde{r}_1, \tilde{a}_1), \quad c_a^{x_2} c_d = \mathrm{com}_{ck}(\tilde{r}_2, \tilde{a}_2).$$
Hence, since the commitment is homomorphic,
$$c_a^{x_1 - x_2} = \mathrm{com}_{ck}(\tilde{r}_1 - \tilde{r}_2, \tilde{a}_1 - \tilde{a}_2)$$
and then
$$c_a = \mathrm{com}_{ck}((x_1 - x_2)^{-1}(\tilde{r}_1 - \tilde{r}_2), (x_1 - x_2)^{-1}(\tilde{a}_1 - \tilde{a}_2)).$$
So, P^* opens c_a with (r, a), where
$$r := (x_1 - x_2)^{-1}(\tilde{r}_1 - \tilde{r}_2), \quad a := (x_1 - x_2)^{-1}(\tilde{a}_1 - \tilde{a}_2).$$

2. Then she opens c_d with $(r_d, d = (d_1, \ldots, d_n))$, where
$$r_d = \tilde{r}_1 - x_1 r \ (= \tilde{r}_2 - x_2 r), \quad d := \tilde{a}_1 - x_1 a \ (= \tilde{a}_2 - x_2 a).$$

3. From
$$c_\Delta^{x_1} c_\delta = \mathrm{com}_{ck}(\tilde{s}_1, (x_1 \tilde{b}_{12} - \tilde{b}_{11} \tilde{a}_{12}, \ldots, x_1 \tilde{b}_{1n} - \tilde{b}_{1(n-1)} \tilde{a}_{1n})),$$
$$c_\Delta^{x_2} c_\delta = \mathrm{com}_{ck}(\tilde{s}_2, (x_2 \tilde{b}_{22} - \tilde{b}_{21} \tilde{a}_{22}, \ldots, x_2 \tilde{b}_{2n} - \tilde{b}_{2(n-1)} \tilde{a}_{2n})),$$
she opens, analogously to step 1, the commitment c_Δ with (s_2, Δ), where
$$s_2 := (x_1 - x_2)^{-1}(\tilde{s}_1 - \tilde{s}_2),$$
$$\Delta := (\Delta_1, \ldots, \Delta_{n-1}), \quad \Delta_k := (x_1 - x_2)^{-1}(y_{1k} - y_{2k}),$$
$$y_{ik} := x_i \tilde{b}_{i(k+1)} - \tilde{b}_{ik} \tilde{a}_{i(k+1)}, \ 1 \leq i \leq 2, \ 1 \leq k \leq n - 1.$$

4. Finally, she opens c_δ with $(s_1, \overline{\delta})$, where
$$s_1 := \tilde{s}_1 - x_1 s_2 \ (= \tilde{s}_2 - x_2 s_2),$$
$$\overline{\delta} := (\overline{\delta}_1, \ldots \overline{\delta}_{n-1}), \overline{\delta}_k := y_{1k} - x_1 \Delta_k \ (= y_{2k} - x_2 \Delta_k), 1 \leq k \leq n - 1,$$
and sets
$$\delta_k := -\overline{\delta}_k d_{k+1}^{-1}, \ 1 \leq k \leq n - 1, \ \delta_n := 0,$$
$$b_k := x_1^{-1}(\tilde{b}_{1k} - \delta_k) \ (= x_2^{-1}(\tilde{b}_{2k} - \delta_k)), \ 1 \leq k \leq n.$$

It remains to be proven that

$$\prod_{k=1}^{n} a_i = b.$$

Whenever P^*'s answer $\tilde{r}, \tilde{\boldsymbol{a}} = (\tilde{a}_1, \ldots, \tilde{a}_n), \tilde{s}, \tilde{\boldsymbol{b}} = (\tilde{b}_1, \ldots, \tilde{b}_n)$ to a randomly chosen challenge x is correct (for example, $x = x_1$), it satisfies, with overwhelming probability, the equations

$$\tilde{a}_k = xa_k + d_k,$$

$$x\tilde{b}_{k+1} - \tilde{b}_k\tilde{a}_{k+1} = \Delta_k x + \bar{\delta}_k \text{ or, equivalently,}$$

$$x\tilde{b}_{k+1} = \tilde{b}_k\tilde{a}_{k+1} + p_k(x), \ 1 \le k \le n-1,$$

where $p_k(x) := \Delta_k x + \bar{\delta}_k$ is a polynomial of degree 1 in x. This follows from the verification conditions

$$c_{\boldsymbol{a}}^x c_d = \text{com}_{ck}(\tilde{r}, (\tilde{a}_1, \tilde{a}_2, \ldots, \tilde{a}_n)) \text{ and}$$

$$c_{\boldsymbol{\Delta}}^x c_\delta = \text{com}_{ck}(\tilde{s}, (x\tilde{b}_2 - \tilde{b}_1\tilde{a}_2, x\tilde{b}_3 - \tilde{b}_2\tilde{a}_3, \ldots, x\tilde{b}_n - \tilde{b}_{n-1}\tilde{a}_n))$$

and the binding property of the commitment scheme. Moreover, we have $\tilde{b}_1 = \tilde{a}_1$ and $\tilde{b}_n = xb$, and hence

$$\begin{aligned}
x^n b &= x^{n-1}\tilde{b}_n = x^{n-2}(\tilde{b}_{n-1}\tilde{a}_n + p_{n-1}(x)) = x^{n-3}x\tilde{b}_{n-1}\tilde{a}_n + x^{n-2}p_{n-1}(x) \\
&= x^{n-3}(\tilde{b}_{n-2}\tilde{a}_{n-1} + p_{n-2}(x))\tilde{a}_n + x^{n-2}p_{n-1}(x) \\
&= x^{n-3}\tilde{b}_{n-2}\tilde{a}_{n-1}\tilde{a}_n + x^{n-3}p_{n-2}(x)(xa_n + d_n) + x^{n-2}p_{n-1}(x) \\
&= \ldots \\
&= \tilde{b}_1\tilde{a}_2 \ldots \tilde{a}_{n-1}\tilde{a}_n + q(x) \\
&= \prod_{k=1}^{n} \tilde{a}_k + q(x) = \prod_{k=1}^{n}(xa_k + d_k) + q(x) \\
&= x^n \prod_{k=1}^{n} a_k + r(x),
\end{aligned}$$

where $q(x)$ and $r(x)$ are polynomials in x of degree $\le n-1$, whose coefficients are expressions in the committed values $\Delta_k, \bar{\delta}_k, a_k, d_k$.

A randomly chosen x satisfies the polynomial equation $x^n b = x^n \prod_{k=1}^{n} a_k + r(x)$ whenever P^* answers x correctly, which happens with probability $w \gg 1/q$. Thus, by Corollary 4.25, $\prod_{k=1}^{n} a_k = b$, and P^* has indeed computed a witness.

Special Honest-Verifier Zero-Knowledge Property. On input of a challenge $x \in \mathbb{Z}_q$, the simulator S does the following:

1. It randomly selects $\tilde{r}, \tilde{a}_1, \ldots, \tilde{a}_n, s_2, \tilde{s}, \tilde{b}_2, \ldots, \tilde{b}_{n-1} \in \mathbb{Z}_q$.

2. It sets

$$c_d := c_a^{-x} \operatorname{com}_{ck}(\tilde{r}, (\tilde{a}_1, \ldots, \tilde{a}_n)),$$
$$c_\Delta := \operatorname{com}_{ck}(s_2, (0, \ldots, 0)),$$
$$\tilde{b}_1 := \tilde{a}_1, \ \tilde{b}_n := xb,$$
$$c_\delta := c_\Delta^{-x} \operatorname{com}_{ck}(\tilde{s}, (x\tilde{b}_2 - \tilde{b}_1\tilde{a}_2, \ldots, x\tilde{b}_n - \tilde{b}_{n-1}\tilde{a}_n)).$$

The transcripts generated are accepting transcripts by their very construction, and they are identically distributed as in the real proof by Peggy: c_Δ, $\tilde{r}, \tilde{a}_1, \ldots, \tilde{a}_n, \tilde{s}, \tilde{b}_2, \ldots, \tilde{b}_{n-1}$ are uniformly distributed elements from G_q and $\mathbb{Z}_q$, and $\tilde{b}_1, \tilde{b}_n, c_d, c_\delta$ are then uniquely determined by the verification conditions. □

The Non-interactive Proof. Peggy computes c_d, c_δ, c_Δ as in the interactive proof. Then she derives the challenge

$$x := h(c_a \| b \| c_d \| c_\delta \| c_\Delta)$$

and the responses $\tilde{r}, (\tilde{a}_1, \ldots, \tilde{a}_n), \tilde{s}, (\tilde{b}_1, \ldots, \tilde{b}_n)$. The non-interactive proof is $(x, c_\Delta, \tilde{r}, (\tilde{a}_1, \ldots, \tilde{a}_n), \tilde{s}, (\tilde{b}_1, \ldots, \tilde{b}_n))$. Vic verifies it by checking that $c_\Delta \in G_q$ and $\tilde{r}, \tilde{a}_1, \ldots, \tilde{a}_n, \tilde{s}, \tilde{b}_1, \ldots, \tilde{b}_n \in \mathbb{Z}_q$, and $\tilde{b}_n b^{-1} = x$, $\tilde{b}_1 = \tilde{a}_1$ and

$$x = h(c_a \| b \| c_a^{-x} \operatorname{com}_{ck}(\tilde{r}, (\tilde{a}_1, \tilde{a}_2, \ldots, \tilde{a}_n))$$
$$\| c_\Delta^{-x} \operatorname{com}_{ck}(\tilde{s}, (x\tilde{b}_2 - \tilde{b}_1\tilde{a}_2, x\tilde{b}_3 - \tilde{b}_2\tilde{a}_3, \ldots, x\tilde{b}_n - \tilde{b}_{n-1}\tilde{a}_n)) \| c_\Delta).$$

4.7 Receipt-Free and Coercion-Resistant Elections

Receipt-freeness is required to prevent voters from selling their votes. Vote buying and voter coercion are not new threats. They were not introduced by online elections. But electronic voting and, in particular, Internet voting are exposed to these dangers to a greater degree, because an attacker might have comprehensive abilities to collect information about voters and their votes. If the election protocol allows the implementation of tools for verifying voters' behavior, vote buying and voter coercion could be practiced on a large scale.

In Section 4.5, we studied the sophisticated voting protocol of Cramer, Gennaro and Schoenmakers ([CraGenSch97]). Based on the homomorphic encryption of votes, it guarantees voters' privacy, and enables end-to-end verifiability. But it is not receipt-free. In Section 4.7.1, we explain how the missing receipt-freeness can be obtained by randomly re-encrypting the votes.

In Section 4.7.2, we then study the even stronger property of *coercion-resistance*, which was introduced in [JueCatJak05], and give as an example the coercion-resistant protocol of Juels, Catalano and Jakobsson ([JueCatJak05]; [JueCatJak10][12]).

---

[12] [JueCatJak10] is an updated and enhanced version of [JueCatJak05].

We shall learn various basic concepts and techniques that are very useful in the design of cryptographic protocols: designated-verifier proofs, reencryption proofs, diverted proofs, untappable channels and plaintext equivalence tests.

4.7.1 Receipt-Freeness by Randomized Re-encryption

Receipt-freeness is an essential property to prevent the selling of votes. It means that a voter is not able to prove to a third party that he has cast a particular vote. The voter must neither obtain nor be able to construct a receipt proving the content of his vote. In the [CraGenSch97] election protocol, a voter V can obtain such a receipt.

We will see in this section how receipt-freeness can be achieved in the [CraGenSch97] protocol (or similar protocols) by introducing an independent third party R, called the *"randomizer"*. The randomizer re-randomizes the encrypted votes by re-encrypting them. Our exposition follows [Hirt10] and [Hirt01].[13] Other examples of randomizer-based protocols have been presented in [BaFoPoiPouSt01], [MagBurChr01], [LeeKim03] and [Schweisgut07, Chapter 5].

We consider the [CraGenSch97] protocol, as introduced in Section 4.5.1, and use the same notation as before: p, q are large primes with $q \mid p - 1$, $g, h \in \mathbb{Z}_p^*$ are elements of order q, and $h = g^s, s \in \mathbb{Z}_q$. $pk := (p, q, g, h)$ is the joint public key of the election authorities and s is their secret key.

Obtaining a Receipt. Each voter V encrypts his vote v, digitally signs the encrypted vote and posts his ballot to the public bulletin board. The ballot (c, σ) consists of V's ElGamal-encrypted vote

$$c := (c_1, c_2) := (g^\alpha, h^\alpha g^v)$$

and V's signature σ. Here, $\alpha \in \mathbb{Z}_q$ is the random value that V has generated for encrypting his vote v (see Section 4.5.1).

The scheme is not receipt-free. The voter can deliver the randomness α to a vote buyer as a receipt. The vote buyer reads V's ballot (c, σ) from the bulletin board. Using α, the vote buyer can decrypt c and obtains V's vote: $g^v = c_2 \cdot h^{-\alpha}$.

By comparing g^α and c_1, the vote buyer can check that V has delivered the correct value of α. V's ballot is easily identifiable on the bulletin board, because each ballot on the bulletin board contains the voter's signature. Therefore, it can be immediately linked to the voter.

Re-encryption of Votes. In order to enable receipt-freeness, the election protocol is extended in the following way.

1. The voter V encrypts his vote. Then the voter sends the encrypted vote $c = (c_1, c_2) = (g^\alpha, h^\alpha g^v)$ to the randomizer R.

---

[13] [Hirt01] contains a previous version of [Hirt10].

2. The randomizer re-encrypts c, i.e., R randomly generates $\omega \in \mathbb{Z}_q$ and sets

$$c' = (c'_1, c'_2) = E_{pk}([1], \omega) \cdot c = (g^\omega, y^\omega) \cdot (g^\alpha, y^\alpha g^v) = (g^\alpha g^\omega, y^\alpha y^\omega g^v)$$

$$= (g^{\alpha+\omega}, y^{\alpha+\omega} g^v) = E_{pk}(g^v, \alpha + \omega),$$

(homomorphic re-encryption was discussed in Section 3.6.3).

3. R generates a non-interactive zero-knowledge proof prf_1 that c' is indeed a re-encryption of c. The proof must be a designated-verifier proof, with the intended verifier being V (see below).

4. R sends c' and prf_1 to the voter V.

5. V checks the proof. If V accepts it, V digitally signs c' and posts the encrypted vote c' together with his signature to the bulletin board.

6. V also has to post a non-interactive zero-knowledge proof prf_2 to the bulletin board that c' is the encryption of g^v, where v is one of the valid votes, i.e., $v \in \{+1, -1\}$ in our setting (the "voter's proof" in Section 4.5.1). V can generate this validity proof jointly with the randomizer R (see below).

The other elements of the voting protocol remain unchanged. We have to assume that the communication channel between the voter and the randomizer is untappable (we will discuss this property below).

Remarks:

1. The zero-knowledge re-encryption proof is necessary to preserve the individual verifiability of the vote. The voter must be sure that the ciphertext c' really contains his intended vote v.

2. It is obvious that the voter V is no longer able to obtain a receipt as before. The randomness of the encryption is now $\alpha + \omega$, and V has access only to α, whereas ω is the randomizer's secret. To avoid the possibility that α, c and prf_1 are a receipt that c' is an encryption of g^v, the proof prf_1 is a *designated-verifier proof*. This means that the proof is convincing for the voter V, who is the intended verifier, but it is not convincing when presented to other people (such as a vote buyer).

3. Since individual voters do not know the randomness $\alpha + \omega$ of the encryption, they are not able to prove the validity of the encrypted vote c' by themselves, as described in the voter's proof in Section 4.5.1. They need the assistance of the randomizer (see the "diverted validity proof" below).

We will now discuss the proofs prf_1 and prf_2. The designated-verifier proof requires that the designated verifier, here the voter, owns a public–private key pair. This requirement is fulfilled in our setting: the voter has such a key pair, which he uses to digitally sign his encrypted vote. We assume that the voter's key is an ElGamal key pair, with public key $pk = (p, q, g, y)$ and secret key $sk = x$, where $y = g^x$. To simplify the discussion, we assume that the parameters p, q, g are the same as in the ElGamal key of the election authorities.

The designated-verifier proof (introduced in [JakSakImp96]) is the OR-combination of two proofs, a proof of the re-encryption and a proof that the designated verifier, here the voter, knows his secret key. We first study these components.

Proof of Knowledge of the Secret Key. We assume that the prover Peggy has an ElGamal key pair with public key $pk = (p, q, g, y)$ and secret key $sk = x$, where $y = g^x$. She proves to the verifier Vic that she knows the secret key x. She can do this by executing Schnorr's identification protocol (see Protocol 4.39 below).

Protocol 4.32.

KnowSecretKey(p, q, g, y):

1. Peggy randomly chooses $r \in \mathbb{Z}_q$, sets $a := g^r$ and sends the commitment a to the verifier Vic.
2. Vic chooses his challenge $e \in \mathbb{Z}_q$ at random and sends it to Peggy.
3. Peggy computes $b := r - ex$ and sends the response b to Vic.
4. Vic accepts if the verification condition $g^b y^e = a$ holds.

The above protocol is a special honest-verifier zero-knowledge proof. Given a challenge e, everyone can easily generate correctly distributed accepting transcripts $S(e) := (a, e, b)$ by choosing a random b and setting $a := g^b y^e$. In particular, the protocol is honest-verifier zero-knowledge. We can perfectly simulate accepting transcripts by first randomly choosing e and then computing $S(e)$.

For Paillier encryption, there is an analogous proof, in which Peggy proves that she is able to compute n-th roots modulo n, where n is the modulus in the Paillier key.

The Standard Re-encryption Proof. Let $pk = (p, q, g, y)$ be a public ElGamal key and c be a ciphertext encrypted by using this key. Assume that Peggy has randomly re-encrypted $c = (c_1, c_2)$ by using the randomness $\omega \in \mathbb{Z}_q$:

$$c' = (c_1', c_2') = (g^\omega c_1, y^\omega c_2).$$

The prover Peggy convinces the verifier Vic that c' is a re-encryption of c, without divulging any information about the randomness ω. For this purpose, she executes the classical Chaum–Pedersen proof (see Protocol 4.15) and shows that the logarithms of $c_1' c_1^{-1}$ and $c_2' c_2^{-1}$, with respect to the bases g and y, are equal.

Protocol 4.33.
ReEncryptionProof(p, q, g, y, c, c′):

1. Peggy chooses $r \in \mathbb{Z}_q$ at random and sets $a := (a_1, a_2) = (g^r, y^r)$. Peggy sends a to Vic.
2. Vic chooses $e \in \mathbb{Z}_q$ uniformly at random and sends e to Peggy.
3. Peggy computes $b := r - e\omega$ and sends b to Vic.
4. Vic accepts if and only if $a_1 = g^b \left(c_1' c_1^{-1}\right)^e$ and $a_2 = y^b \left(c_2' c_2^{-1}\right)^e$.

This protocol is a special honest-verifier zero-knowledge proof, as we have already observed above (Section 4.5.2).

The Designated-Verifier Re-encryption Proof. We now transform the re-encryption proof into a designated-verifier proof. We consider the same setting as in the standard re-encryption proof. Moreover, we assume that the designated verifier Vic has an ElGamal key, with public key $pk_V = (p, q, \tilde{g}, \tilde{y})$ and secret key $x \in \mathbb{Z}_q$. As before, the prover Peggy convinces the designated verifier Vic that c' is a re-encryption of c. The proof is the OR-combination of the standard re-encryption proof and the proof of knowledge of Vic's secret key. The technique of OR-combining Σ-proofs of knowledge was introduced in Section 4.5.2 (the "voter's proof").

Protocol 4.34.
DesignatedVerifierReEncryptionProof(p, q, g, y, $\tilde{g}$, $\tilde{y}$, c, c′):

1. As in the standard re-encryption proof, Peggy randomly chooses $r \in \mathbb{Z}_q$ and computes the commitment $a := (a_1, a_2) = (g^r, y^r)$. Then, she chooses a challenge $\tilde{d} \in \mathbb{Z}_q$ and, for this challenge $\tilde{d}$, she simulates a proof of knowledge of Vic's secret key sk_V, i.e., she randomly chooses $\tilde{r} \in \mathbb{Z}_q$ and computes the "fake" commitment

$$\tilde{a} := g^{\tilde{r}} y^{\tilde{d}}.$$

   She sends the commitment $(a, \tilde{a})$ to the verifier.
2. Vic chooses a challenge $e \in \mathbb{Z}_q$ at random and sends it to Peggy.
3. Peggy computes $d := e - \tilde{d}$. Then, she computes the response b to the challenge d that is required by the re-encryption proof, i.e., $b := r - d\omega$. The correct response to the challenge $\tilde{d}$ in the simulated proof is $\tilde{b} := \tilde{r}$. Peggy responds $d, \tilde{d}$ and $b, \tilde{b}$ to the verifier.
4. The verifier accepts if
   a. $d + \tilde{d} = e$; and
   b. the verification condition of the re-encryption proof holds for (a, d, b), i.e.,

$$a_1 = g^b \left(c_1' c_1^{-1}\right)^d, \quad a_2 = y^b \left(c_2' c_2^{-1}\right)^d; \text{ and}$$

c. the verification condition of the proof of knowledge of the secret key holds for $(\tilde{a}, \tilde{d}, \tilde{b})$, i.e.,

$$\tilde{a} = \tilde{g}^{\tilde{b}} \tilde{y}^{\tilde{d}}.$$

As usual, by applying the Fiat–Shamir heuristics, the designated-verifier re-encryption proof can be converted into a non-interactive zero-knowledge proof, which is generated solely by Peggy. Peggy performs the same computations as in the interactive proof. She obtains the challenge e by applying a collision-resistant hash function H to the commitment $(a, \tilde{a})$ and the input parameters of the proof:[14]

$$e := H(p\|q\|g\|y\|\tilde{g}\|\tilde{y}\|c\|c'\|a_1\|a_2\|\tilde{a}).$$

The non-interactive proof is the tuple $(d, \tilde{d}, b, \tilde{b})$. It is verified by the condition

$$d + \tilde{d} = H(p\|q\|g\|y\|\tilde{g}\|\tilde{y}\|c\|c'\|g^b \left(c_1' c_1^{-1}\right)^d \|y^b \left(c_2' c_2^{-1}\right)^d \|\tilde{g}^{\tilde{b}} \tilde{y}^{\tilde{d}}).$$

The proof convinces the designated verifier Vic that c' is a re-encryption of c. He knows that the prover cannot have simulated the re-encryption proof, because she does not know his secret key x. But, if Vic transfers the proof to another verifier, then it is no longer convincing. In fact, the logical OR of the statements "c' is a re-encryption of c" and "I know the secret key x" is proven. By simulating the re-encryption proof and proving knowledge of his secret key, the designated verifier Vic can generate valid proofs for elements c' of his choice, whether c' is a re-encryption of c or not. Vic's proof runs as follows:

1. As in a proof of knowing his secret key, Vic, who now acts as the prover, randomly chooses $\tilde{r} \in \mathbb{Z}_q$ and sets $\tilde{a} := \tilde{g}^{\tilde{r}}$. Then he chooses a challenge $d \in \mathbb{Z}_q$. For this challenge d, he simulates a proof that c' is a re-encryption of c, i.e., he randomly chooses $r \in \mathbb{Z}_q$ and sets

$$a := (a_1, a_2) := \left(g^r \left(c_1' c_1^{-1}\right)^d, y^r \left(c_2' c_2^{-1}\right)^d\right).$$

He sends the commitment $(a, \tilde{a})$ to the verifier.
2. The verifier chooses a challenge $e \in \mathbb{Z}_q$ at random and sends it to Vic.
3. Vic computes $\tilde{d} := e - d$ and computes the correct response $\tilde{b}$ to the challenge $\tilde{d}$ that is required in the proof of knowing the secret key, i.e.,

$$\tilde{b} := \tilde{r} - \tilde{d}x.$$

Moreover, he sets $b := r$. Then he responds $d, \tilde{d}$ and $b, \tilde{b}$ to the verifier.

---

[14] We use the capital letter H to denote the hash function here, because h is already used in the public ElGamal key of the election authorities.

Untappable Channels. In step 4 of the election protocol, the voter gets the non-interactive designated-verifier re-encryption proof $prf_1 = (e, d, \tilde{d}, b, \tilde{b})$ from the randomizer. If the voter were able to prove to a vote buyer that the proof prf_1 was generated by the randomizer and not by the voter, then the voter could prove the content of his vote c'. Namely, the voter could then deliver c, c', α, prf_1 as a receipt to the vote buyer.

We see that the receipt-freeness of the election protocol requires that the voter must not be able to prove the origin of prf_1. We have to require that the communication channel between the voter and the randomizer is untappable. We now give an informal definition of an untappable channel (for a more precise definition, see [Okamoto97]).

Definition 4.35. A (bidirectional) *untappable channel* is a secret communication channel with the additional property that not even the communication partners are able to prove to a third party which messages have been sent through the channel.

Of course, each communication partner can record all received data. But he must not be able to deliver a proof to a third party that he received a particular message. Untappability is a requirement that is hard to satisfy. If at all possible, one should attempt to achieve it by physical means. In an insecure environment, there are no perfect implementations of untappable channels (think of malware on the voter's PC). Some election protocols ([MagBurChr01]; [LeeKim03]; [Schweisgut07]) try to achieve untappability by implementing the randomizer by means of a tamper-resistant hardware device (called an "observer"), such as a smart card.[15]

The Diverted Validity Proof. The voter Peggy has to prove that her encrypted vote $c = (c_1, c_2)$ really contains one of the admissible encoded votes $m_1 := g^{+1}$ or $m_2 := g^{-1}$. In the basic voting protocol (see the voter's proof in Section 4.5.2, Protocol 4.20), she proves that she knows a witness α such that either $y := c_1 = g^\alpha$ and $z := c_2 m_1^{-1} = h^\alpha$, or $\tilde{y} := c_1 = g^\alpha$ and $\tilde{z} := c_2 m_2^{-1} = h^\alpha$, without revealing which of those cases occurs. α is the randomness that Peggy used when encrypting her vote. Now, the vote c is re-encrypted by the randomizer R, and the re-encrypted vote $c' = (c_1', c_2')$ is posted to the bulletin board. The randomness of the ciphertext c' is $\alpha + \omega$. α is known only to the voter, and ω only to the randomizer. Therefore, the voter Peggy and the randomizer must interact to generate a validity proof. They must jointly prove that either $y' := c_1'$ and $z' := c_2' m_1^{-1}$ or $\tilde{y}' := c_1'$ and $\tilde{z}' := c_2' m_2^{-1}$ have a common logarithm with respect to the bases g and h.

During this interaction, the randomizer must not learn anything about the vote or α, and the voter must not learn anything about the randomization witness ω. The voter's proof in Protocol 4.20 has to be diverted from

---

[15] In a classical poll-site election with paper ballot sheets, an untappable channel from the voter to the ballot box is implemented by the voting booths and the supervision of the poll workers.

the prover to the randomizer; we speak of a *diverted proof*. The voter and the randomizer construct a non-interactive diverted version of the proof in Protocol 4.20. In non-interactive proofs, the verifier's challenge is computed by applying a cryptographic hash function (using the Fiat–Shamir heuristics; see Section 4.5.3). Let H be a collision-resistant hash function. The proof is derived from the interactive proof with Peggy as prover and the randomizer as verifier.

Protocol 4.36.
DivertedOneOfTwoPairsProof $(p, q, g, h, (y', z'), (\tilde{y}', \tilde{z}'))$:

   Case: Peggy knows $\log_g y = \log_h z = \alpha$ (the other case follows analogously):

    1. The prover Peggy chooses $r, \tilde{r}$ and $\tilde{d} \in \mathbb{Z}_q$ at random and sets

$$a_1 := g^r,\ a_2 := h^r,\ \tilde{a}_1 := g^{\tilde{r}} \tilde{y}^{\tilde{d}},\ \tilde{a}_2 := h^{\tilde{r}} \tilde{z}^{\tilde{d}}.$$

     Then, Peggy sends $(a_1, a_2, \tilde{a}_1, \tilde{a}_2)$ to the randomizer.

    2. The randomizer randomly chooses elements $\beta, \tilde{\beta}, \gamma \in \mathbb{Z}_q$, sets

$$a_1' := a_1 g^\beta y'^\gamma,\ a_2' := a_2 h^\beta z'^\gamma,$$

$$\tilde{a}_1' := \tilde{a}_1 g^{\tilde{\beta}} \tilde{y}'^{-\gamma},\ \tilde{a}_2' := \tilde{a}_2 h^{\tilde{\beta}} \tilde{z}'^{-\gamma}$$

     and sends $(a_1', a_2', \tilde{a}_1', \tilde{a}_2')$ to Peggy.

    3. Peggy computes the challenge e by applying the hash function:

$$e := H(g\|h\|y'\|z'\|\tilde{y}'\|\tilde{z}'\|a_1'\|a_2'\|\tilde{a}_1'\|\tilde{a}_2').$$

     She sets

$$d := e - \tilde{d},\ b := r - d\alpha,\ \tilde{b} := \tilde{r}$$

     and sends $(d, \tilde{d}, b, \tilde{b})$ to the randomizer.

    4. The randomizer computes

$$d' := d + \gamma,\ \tilde{d}' := \tilde{d} - \gamma,\ b' := b + \beta - d\omega,\ \tilde{b}' := \tilde{b} + \tilde{\beta} - \tilde{d}\omega$$

     and sends $(d', \tilde{d}', b', \tilde{b}')$ to the verifier.
   The desired validity proof is $(d', \tilde{d}', b', \tilde{b}')$. It is verified by checking the condition

$$d' + \tilde{d}' = H\left(g\|h\|y'\|z'\|\tilde{y}'\|\tilde{z}'\|g^{b'} y'^{d'}\|h^{b'} z'^{d'}\|g^{\tilde{b}'} \tilde{y}'^{\tilde{d}'}\|h^{\tilde{b}'} \tilde{z}'^{\tilde{d}'}\right).$$

Remarks:

1. As before, the channel between the voter and the randomizer is assumed to be untappable. The messages exchanged are known only to the voter and to the randomizer, and neither the voter nor the randomizer can prove to a third party which messages were exchanged.

2. The basic structure of the protocol is that Peggy proves the validity of the vote to the randomizer. In this protocol, the randomizer would accept if

$$e = d + \tilde{d}, \ a_1 = g^b y^d, \ a_2 = h^b z^d, \ \tilde{a}_1 = g^{\tilde{b}} \tilde{y}^{\tilde{d}}, \tilde{a}_2 = h^{\tilde{b}} \tilde{z}^{\tilde{d}},$$

which becomes

$$e = d' + \tilde{d}', a_1' = g^{b'} y'^{d'}, \ a_2' = h^{b'} z'^{d'}, \ \tilde{a}_1' = g^{\tilde{b}'} \tilde{y}'^{\tilde{d}'}, \tilde{a}_2' = h^{\tilde{b}'} \tilde{z}'^{\tilde{d}'}.$$

3. The randomizer learns nothing about the voter's vote. The underlying protocol – the non-diverted voter's validity proof – is an interactive honest-verifier zero-knowledge proof of knowledge. Therefore, the verifier, here the randomizer, does not get any information about the secret if the challenge is chosen honestly at random. In the diverted proof, the challenge is generated by the hash function H, and the hash function is applied by the voter. Hence, the random oracle assumption (Section 3.3.5) implies that the challenge is indeed chosen uniformly at random.

4. The interaction does not give the voter any information that might enable him to construct a receipt. The elements $\beta, \tilde{\beta}, \gamma \in \mathbb{Z}_q$ are chosen randomly and independently by the randomizer, and hence $d', b', \tilde{b}'$ are independent random elements. Therefore, the proof $(d', \tilde{d}', b', \tilde{b}')$ is a uniformly selected random element from among all validity proofs for c', i.e., among all tuples from $\mathbb{Z}_q^4$ that satisfy the verification condition $d' + \tilde{d}' = H \left(p \| q \| g \| h \| y' \| z' \| \tilde{y}' \| \tilde{z}' \| g^{b'} y'^{d'} \| h^{b'} z'^{d'} \| g^{\tilde{b}'} \tilde{y}'^{\tilde{d}'} \| h^{\tilde{b}'} \tilde{z}'^{\tilde{d}'} \right)$. This means that the proof is uniformly random and independent of all of the voter's information. Beyond the diverted proof, the randomizer sends only random numbers to the verifier.

4.7.2 A Coercion-Resistant Protocol

When discussing vote-buying attacks, we usually assume that it is the voter's free decision to sell his vote. There is another threat, however: a voter might be intimidated or coerced to vote in a particular manner. He might be forced to abstain from voting or to disclose his secret voting credentials. Coercion is a critical threat in Internet-based elections. Using automated surveillance tools, a coercer might be able to influence the behavior of voters on a large scale. Consider, for example, the election scheme of Cramer, Gennaro and Schoenmakers ([CraGenSch97]), which we described in Section 4.5. A coercer can furnish a voter with an encrypted vote for a particular candidate, and then verify that the voter has posted a ballot containing that ciphertext. Alternatively, the coercer might demand the private signing key of the voter, verify its correctness by using the corresponding public key and then cast the vote instead of the legitimate voter.

These attacks are also possible in many *mix-net-based election protocols*. Mixing is a preferred method for ensuring the privacy of votes. Typically, in such an election scheme, the voter encrypts his vote, digitally signs the ciphertext and posts the ballot, which consists of 'the ciphertext and the signature, to a bulletin board. When all ballots have been collected, the signatures are checked and, for example, the eligibility of the voters is verified by comparing them with the voter roll. To anonymize the votes, the ciphertexts are separated from the signatures and sent through a mix cascade. After the mixing, nobody can tell which vote was cast by which voter. The votes are then decrypted and tallied. This approach was outlined in Chaum's classic paper [Chaum81]. Other protocols of this kind have been presented in, for example, [FurSak01], [BonGol02] and [JakJueRiv02].

In the preceding section, we introduced an enhanced receipt-free version of the [CraGenSch97] protocol. The encrypted votes are re-randomized by an independent party. This prevents one of the attacks described previously: if a coercer provides a voter with an encrypted vote, the coercer can no longer verify that the voter really has submitted that vote. However, the other attack – demanding the private key and casting the vote instead of the legitimate voter – still works.

We see that receipt-freeness does not prevent all kinds of coercion. Randomization, forced-abstention and simulation attacks might still be possible, as Juels, Catalano and Jakobsson observed in [JueCatJak05].

The Forced-Abstention Attack. The attacker coerces a voter to refrain from voting. The election schemes described above are vulnerable to this attack. The votes are digitally signed and posted to the bulletin board. The voter roll, with the public keys (or certificates) of the eligible voters, is also published. This enables verification that the ballots were cast by authorized voters. But it also enables an attacker to see who has voted. If necessary, for example if pseudonyms are used, the attacker can first coerce the voter to divulge his public key or pseudonym.

The Simulation Attack. The attacker coerces a voter to divulge his secret key material that authenticates him as an eligible voter. In the election schemes described previously, the attacker can verify the correctness of the secret key by checking that it fits with the corresponding public key, which is published on the voter roll. The attacker can then simulate the legitimate voter and vote on his behalf. Note that, in general, receipt-freeness does not prevent an attacker from buying secret keys from voters.

The Randomization Attack. In this attack, the attacker is unable to learn which candidate a voter has cast a ballot for. But the attacker is able to coerce voters to submit randomly composed ballots, which results in votes for randomly selected candidates (or invalid votes). If applied on a large scale, the attack can nullify the choice of the voters. Think of a district where a large majority of voters favors one party. The randomization attack was initially observed by Schoenmakers as a possible attack against an early version of a voting protocol presented by Hirt and Sako (see [Hirt01]). Typical examples that are vulnerable to this attack are voting schemes where the candidates are encoded, individually for each voter, in a random order. In this case, the attacker can demand, for example, "Select the candidate in the first position." An easily understandable example of this type is Punchscan ([FisCarShe06]). Note, however, that Punchscan is a paper-based and not an Internet voting scheme.

All remote elections, whether Internet voting or classical mail voting, are faced with the problem that voters may be coerced by an adversary who can observe them at the moment of vote casting.

Coercion Resistance. An election scheme which not only is receipt-free but also resists all these other kinds of coercion is called *coercion-resistant*. We quote an informal definition from [JueCatJak05], [JueCatJak10]: "We allow the adversary to demand of coerced voters that they vote in a particular manner, abstain from voting, or even disclose their secret keys. We define a scheme to be coercion-resistant if it is infeasible for the adversary to determine if a coerced voter complies with the demands." A precise, formal definition of coercion resistance is given in [JueCatJak05, Section 3] and [JueCatJak10, Section 3].

The Distributed Plaintext Equivalence Test (PET). We consider an ElGamal encryption scheme in a prime-order subgroup with public key $pk = (p, q, g, y)$ and secret key $sk = s \in \mathbb{Z}_q^*$. Recall that g is an element of order q and generates the subgroup G_q of order q in $\mathbb{Z}_p^*$, and $y = g^s \bmod p$ (Section 3.4.4). Let $A_1, \ldots, A_n$ be parties who share the secret key s in a Shamir (t, n)-threshold secret-sharing scheme (Section 4.4). We shall explain how these parties are able to find out, by a joint computation, whether two ElGamal ciphertexts that are encrypted with pk contain the same plaintext. No information about the plaintexts leaks. Such a *plaintext equivalence test* is applied in the coercion-resistant protocol by Juels, Catalano and Jakobsson.

Let $c = (c_1, c_2)$ and $c' = (c_1', c_2')$ be ElGamal ciphertexts, computed with the public key pk.

Definition 4.37. c and c' are *plaintext equivalent* if and only if c and c' are encryptions of the same plaintext, i.e., if and only if there are a plaintext $m \in G_q$ and randomness values $\alpha, \alpha' \in \mathbb{Z}_q$ such that

$$c = (g^\alpha, y^\alpha m), \quad c' = (g^{\alpha'}, y^{\alpha'} m).$$

The following test ([JakJue00]), performed as a distributed computation of $A_1, \ldots, A_n$, finds out whether two ciphertexts c and c' are plaintext equivalent. No information about the plaintexts leaks, except for a single bit that specifies if they are equal or not. The probability that the test gives a wrong result is negligibly small. The basic idea is the following: The parties jointly decrypt d^z, where $d := cc'^{-1}$ and $z \in \mathbb{Z}_q$ is a random exponent, jointly generated by the parties. The ciphertexts are encryptions of the same plaintext if the resulting plaintext is 1. The only-if direction is also true except in the case $z = 0$, which happens only with the negligible probability $1/q$.

Protocol 4.38.

PlainTextEquiv(p, q, g, y, c, c'):

Let $d := (d_1, d_2) := (c_1 c_1'^{-1}, c_2 c_2'^{-1})$ be the quotient of the ciphertexts.
1. Each party A_i randomly selects $z_i \in \mathbb{Z}_q$ and commits to z_i by publishing $a_i := g^{z_i}$.
2. Each party A_i computes $\tilde{d}_i = d^{z_i} = (d_1^{z_i}, d_2^{z_i})$ and broadcasts it.
3. A_i delivers a non-interactive zero-knowledge proof that she has correctly computed $\tilde{d}_i$, i.e., she proves that $\log_g(a_i) = \log_{d_i}(\tilde{d}_i)$ by generating a Chaum–Pedersen proof (see the "authority's proof" in Section 4.5.2).
4. The parties jointly decrypt $\tilde{d} := \prod_{i=1}^{n} \tilde{d}_i$ (see "Threshold Decryption" in Section 4.5.1).
5. If the resulting plaintext is 1, then the parties conclude that c and c' are plaintext equivalent. Otherwise, they conclude that the plaintexts are different.

Remarks:

1. The test may give a wrong result only if $z = \sum_{i=1}^{n} z_i = [0]$. This happens only with the negligible probability $1/q$.
2. If at least one of the parties is honest and chooses her exponent z_i randomly, then z is a random element, and hence d^z is a random element of G_q and the decryption of d^z does not contain any information about the plaintexts of d, c and c'.
3. The privacy of pk-encrypted ciphertexts and the threshold decryption are guaranteed if at least t of the n parties are honest and no coalition of t parties behaves dishonestly.

The Protocol. We now describe the foundational protocol of Juels, Catalano and Jakobsson ([JueCatJak05]; [JueCatJak10]). This is the basis of many further research efforts to develop a coercion-resistant, practical scheme (see, e.g., [Smith05]; [ClaChoMye07]; [WebAraBuc07]; [AraFouTra10]). In [JueCatJak10], a modified ElGamal encryption is applied, which may be viewed as a simplified version of the Cramer–Shoup public-key encryption scheme (Section 9.5.2). The modification of ElGamal encryption is necessary for the formal security proof in [JueCatJak10]. In our exposition, we simplify the situation by using classical ElGamal encryption.

As before (Section 4.5.1), the participants are the following:

1. A group of election authorities.
2. The voters $V_1, \ldots, V_m$.
3. The public, as observer and verifier. The scheme is universally verifiable. Everybody can check the correctness of the election process and the election result.

The group of authorities is subdivided into the following:

1. A registrar R, who registers the eligible voters and issues credentials to them.
2. Talliers $A_1, \ldots, A_n$, who process the ballots and compute the final tally.
3. Mixers, who anonymize the ballots by operating a publicly verifiable re-encryption mix net. They provide non-interactive zero-knowledge proofs that they do their work correctly (see Sections 4.6.2 and 4.6.4 above).

We assume that each voter owns an ElGamal key pair. As in the protocols described previously, the participants in the voting scheme communicate through a bulletin board, and the integrity of the tallying process is ensured by a (t, n)-threshold scheme: fraud is possible only if there is a coalition of at least t dishonest talliers.

To simplify our description, we assume that the registrar is a single entity, and we assume that R is honest. These strict assumptions are not really necessary. The registrar's task can be distributed among a group of registration authorities, and the security of the registration process can be based on a threshold scheme (see the remark below).

Setup. To set up an ElGamal cryptosystem in a prime-order subgroup, the authorities jointly generate random large primes p, q such that q divides $p - 1$, and a random element $g \in \mathbb{Z}_p^*$ of order q (see Section 4.5.5). Then, the registrar R generates an ElGamal key pair $(pk_R, sk_R) = ((p, q, g, h_R), s_R)$. The talliers $A_1, \ldots, A_n$ jointly generate one ElGamal key pair $(pk_T, sk_T) = ((p, q, g, h_T), s_T)$ and share the secret key using a (t, n)-threshold scheme (again, see Section 4.5.5). Recall that $s_R, s_T \in \mathbb{Z}_q$, $h_R = g^{s_R} \in \mathbb{Z}_p^*$ and $h_T = g^{s_T} \in \mathbb{Z}_p^*$. The public keys pk_R and pk_T are published on the bulletin board.

Registration. The registrar R authenticates each voter V and verifies his eligibility. This authentication could be based on V's ElGamal key and a challenge-and-response protocol. Then, R generates a random $\sigma_V \in G_q$ that serves as the credential of V. R encrypts the credential with the public key pk_T, i.e.,

$$\tau_V := (g^{\alpha_V}, h_T^{\alpha_V} \sigma_V) \quad (\alpha_V \text{ randomly selected from } \mathbb{Z}_q),$$

signs τ_V by using his secret key sk_R and publishes τ_V together with his signature on the voter roll, which is part of the bulletin board. The credential σ_V is sent to V, together with a designated-verifier proof that τ_V is indeed an encryption of σ_V (Protocol 4.34).[16] The communication channel from the registrar to the voter is assumed to be untappable.

Publication of the Candidate Slate. The authorities publish the names of the candidates and a unique identifier $v_k \in G_q$ for each candidate C_k. In a yes–no election, the identifiers can be g^{+1} and g^{-1}, as in Section 4.5.

---

[16] This proof can be omitted if the registrar R is a single, honest authority. But recall that R can be replaced by a group of authorities, and some of these authorities might be dishonest.

Vote Casting. Each voter V makes his choice $v \in \{v_1, v_2, \ldots, v_l\}$. The voter does the following:

1. He encrypts his vote v and his credential σ_V by using the public key pk_T and randomly generated elements $\alpha, \beta \in \mathbb{Z}_q$:

$$c_1 := (g^\alpha, h_T^\alpha v), \quad c_2 := (g^\beta, h_T^\beta \sigma_V).$$

2. He generates a non-interactive zero-knowledge proof prf_1 that c_1 is the encryption of a valid candidate identifier (analogous to the voter's proof in Section 4.5.2).
3. He generates a non-interactive zero-knowledge proof prf_2 that he knows the plaintexts of c_1 and c_2 (see Section 4.6.3).

Then, the voter adds a timestamp ts and posts the ballot $(c_1, c_2, prf_1, prf_2, ts)$ through an anonymous channel to the bulletin board. The voter can submit ballots repeatedly during the voting phase.

The anonymous channel can be implemented by a mix net (Section 4.6). The timestamps are used to identify the last ballot cast if a voter posts more than one ballot with the same credential. In this case, only the last ballot cast counts (see below). If the anonymous channel preserves the chronology of postings, the timestamps can be omitted and the order of the postings can be used, as proposed in the original protocol [JueCatJak10]. However, mix nets destroy the order of messages.[17]

Tallying. After the vote-casting phase, the tallying authorities perform the following steps:

1. *Checking proofs:* The proofs prf_1 and prf_2 are verified. Ballots with invalid proofs are discarded.
2. *Eliminating duplicates:* By executing pairwise plaintext equivalence tests (see Protocol 4.38 above) on the encrypted credentials, duplicates of credentials are detected. If there are ballots with duplicates of the same credential, all but one of them are removed. Only the most recent ballot is kept. The timestamp is used for this purpose. This policy means that a voter can cast a vote repeatedly, but only the last ballot counts. Before performing the next processing step, the timestamps and validity proofs are separated from the encrypted votes and credentials. Let $A = ((a_1, b_1), (a_2, b_2), \ldots)$ be the list of the remaining encrypted (vote, credential) pairs (c_1, c_2).
3. *Mixing:* The list A is randomly and secretly permuted by sending it through a publicly verifiable ElGamal re-encryption mix cascade (Sections 4.6.2 and 4.6.4). We denote the resulting list by A'. Let B' be the list consisting only of the encrypted credentials, in the same order as A'.

---

[17] It is noted in [JueCatJak05] that here it might be sufficient to realize the anonymous channel in some other practical way.

4. *Checking credentials:* The list L of encrypted credentials from the voter roll is randomly and secretly permuted by sending it through a publicly verifiable ElGamal re-encryption mix cascade (Sections 4.6.2 and 4.6.4). Let L' be the permuted list. Each ciphertext in B' is compared with the ciphertexts of valid credentials in L' by applying the plaintext equivalence test. Invalid voter credentials are detected in this way. The ballots with invalid credentials are removed from A'.

5. *Computing the tally:* The vote component a_i of the remaining entries (a_i, b_i) in A' is jointly decrypted by applying the (t, n)-threshold decryption method (Section 4.5.1, page 136). Then the tally is computed by counting. The talliers give non-interactive zero-knowledge correctness proofs of their decryption (Section 4.5.2).

A complete transcript of all actions during the tallying (proof verification results, plaintext equivalence tests, removed ballots, validity proofs, decryption results, correctness proofs of the mixes and talliers, ...) is posted to the bulletin board.

Remark. By applying techniques from [GenJarKraRab99] (for a simplified version, see Section 4.5.5), the single honest registrar R can be replaced by a group $R_1, R_2, \ldots, R_l$ of registrars. Security is guaranteed if the majority of the registrars behave honestly.

We now discuss the essential security aspects of the election protocol.

1. The ballots are sent to the bulletin board through an anonymous channel. This makes it impossible to find out whether a voter V has submitted a ballot, at least if a coercer cannot observe V continuously, which seems to be impossible in an online election, where voters can use any Internet-connected computer to cast their votes. The anonymous channel to the bulletin board defeats the forced-abstention attack.

2. Verifying the validity proof prf_1 ensures that only valid candidate choices appear when the votes are decrypted. An adversary cannot deprive a voter of his right to vote freely by forcing him to submit an identifiable invalid choice.

3. By exploiting the malleability of ElGamal encryption, an adversary could take a voter V's ballot, substitute c_1 with a modification and submit the modified ballot to the bulletin board. This might enable the adversary to identify V's vote in the final tally. The validity proof prf_2 prevents this attack (see Sections 4.6.2 and 4.6.3).

4. Since a complete transcript of all actions is published on the bulletin board, everyone can check the correctness of the tallying process. The election result is universally verifiable.

5. By inspecting the bulletin board, each voter can check that his ballot(s) have arrived unmodified at the ballot box and are included in the tally. Hence, the scheme enables the voter's individual verification.

6. Although the ballots are submitted through an anonymous channel, a coercer might track the ballot of a voter V on the bulletin board up to the mixing step. If the coercer knows the time when V cast his vote, then the timestamp might enable the coercer to spot V's ballot. Alternatively, the coercer might mount a "1009 attack" ([Smith05]), where the coercer forces the voter to submit a large number, say 1009, identical copies of his ballot. Then, the ballot can be identified because 1009 duplicates are detected. Both methods – the latter also works if no timestamps are used – can also be used by a voter to convincingly show his ballot on the bulletin board to a vote buyer.

7. The mixing step guarantees that an element in A' cannot be assigned to the corresponding element in A – no information about this relation leaks. The encrypted credentials on the voter roll can be linked to the voters. But they are also mixed before the credentials of the ballots are checked. Hence, this path for linking a credential or vote in A' to a voter is also cut. In particular, the privacy of a vote is guaranteed.

8. Credentials are never decrypted during the tallying process. Voter V obtains his credential σ_V through an untappable channel from the registrar. Therefore, V cannot prove to a third party, for example a vote buyer, that σ_V is valid. Even with the voter's help, an observer is not able to distinguish valid from fake credentials. The voter can easily generate fake credentials – the voter simply generates a random element in G_q. In particular, a voter cannot prove his vote to a third party. Hence, the protocol is receipt-free.

9. If coerced to vote in a specific manner, the voter can intentionally cast an invalid ballot by using a fake credential. If the voter uses a valid credential, he can later recast a vote with his true choice. If an attacker coerces a voter to divulge his credential, the voter can deliver a fake credential. A simulation attack is prevented in this way. The coercion resistance of the protocol is based mainly on the coercer's inability to distinguish between fake and valid credentials.

10. A critical point in the protocol, which threatens the receipt-freeness, is the secure transfer of the credential from the authorities to the voter. The credential is assumed to be communicated through an untappable channel. In practice, such a channel can only be implemented approximately – postal mail or the registration of voters in person has been proposed.

The Efficiency Problem. The blind comparison of credentials is a core feature of the protocol. The plaintext equivalent test which is applied for this purpose is inefficient. The elimination of duplicates requires $O(k^2)$ tests, where k is the number of ballots submitted, and the checking of the credentials involves $O(km)$ tests, where m is the number of eligible voters. For example, for 10^8 voters, a multiple of 10^{16} exponentiations in $\mathbb{Z}_p^*$ are necessary. A huge number of proofs of failed plaintext equivalence tests have to be stored on the bulletin board.

There are many enhancements of the protocol aimed at improving the efficiency of the blind comparison of credentials. The voters are assigned to blocks of a bounded size in the proposal of [ClaChoMye07], and the ballots are processed independently in each block. A block is the analogue of a voting precinct in a present-day election. Smith introduced a blind comparison mechanism that is based on deterministic fingerprints and works in linear time ([Smith05], [WebAraBuc07]). But there are weaknesses in the proposed comparison mechanisms ([AraFouTra07]; [ClaChoMye07]). Araújo et al. ([AraFouTra07]; [AraFouTra10]) therefore propose the use of composite credentials with an internal algebraic relation. This relation can easily be compared blindly, and the validity can be checked blindly and efficiently by a joint computation done by the authorities. A drawback of this approach is that credentials cannot be invalidated. However, in typical approaches to online voting, voters can overwrite their online votes by recasting their vote on election day at an election office or by traditional mail voting. In that case, invalidating their credentials must be possible.

4.8 Digital Cash

The growth of electronic commerce on the Internet requires *digital cash*. Today, credit cards are used to pay on the Internet. Transactions are online; i.e., all participants – the customer, the shop and the bank – are involved at the same time (for simplicity, we assume only one bank). This requires that the bank is available even during peak traffic times, which makes the scheme very costly. Exposing the credit card number to the vendor provides the vendor with the ability to impersonate the customer in future purchases. The bank can easily observe who pays which amount to whom and when, so the customer cannot pay anonymously, as she can with ordinary money.

A payment with ordinary money requires three different steps. First, the customer fetches some money from the bank, and his account is debited. Then the customer can pay anonymously in a shop. Later, the vendor can bring the money to the bank, and his account is credited.

Ordinary money has an acceptable level of security and functions well for its intended task. Its security is based on a complicated and secret manufacturing process. However, it is not as secure in the same mathematical sense as some of the proposed digital cash schemes.

Digital cash schemes are modeled on ordinary money. They involve three interacting parties: the bank, the customer and the shop. The customer and the shop have accounts with the bank. A digital cash system transfers money in a secure way from the customer's account to the shop's account. In the following, the money is called an *electronic coin*, or *coin* for short.

As with ordinary money, paying with digital cash requires three steps:

1. The customer fetches the coin from the bank: the customer and the bank execute the withdrawal protocol.

2. The customer pays the vendor: the customer and the vendor execute the payment protocol.
3. The vendor deposits the coin on his account: the vendor and the bank execute the deposit protocol.

In an offline system, each step occurs in a separate transaction, whereas in an online system, steps 2 and 3 take place in a single transaction involving all three parties.

The bank, the shop and the customer have different security requirements:

1. The bank is assured that only a previously withdrawn coin can be deposited. It must be impossible to deposit a coin twice without being detected.
2. The customer is assured that the shop will accept previously withdrawn coins and that he can pay anonymously.
3. In an offline system, the vendor is assured that the bank will accept a payment he has received from the customer.

It is not easy to enable anonymous payments, for which it must be impossible for the bank to trace a coin, i.e., to link a coin from a withdrawal with the corresponding coin in the deposit step. This requirement protects an honest customer's privacy, but it also enables the misuse by criminals.

Thus, to make anonymous payment systems practical, they must implement a mechanism for tracing a coin. It must be possible to revoke the customer's anonymity under certain well-defined conditions. Such systems are sometimes called *fair payment systems*.

In the scheme we describe, anonymity may be revoked by a trusted third party called the trusted center. During the withdrawal protocol, the customer has to provide data which enables the trusted center to trace the coin. The trusted center is only needed if someone asks to revoke the anonymity of a customer. The trusted center is not involved if an account is opened or a coin is withdrawn, paid or deposited. Using the secret sharing techniques from Section 4.4, it is easy to distribute the ability to revoke a customer's anonymity among a group of trusted parties.

A customer's anonymity is achieved by using blind signatures (see Section 4.8.1). The customer has to construct a message of a special form, and then he hides the content. The bank signs the hidden message, without seeing its content. Older protocols use the "cut and choose method" to ensure that the customer formed the message correctly: the customer constructs, for example, 1000 messages and sends them to the bank. The bank selects one of the 1000 messages to sign it. The customer has to open the remaining 999 messages. If all these messages are formed correctly, then, with high probability, the message selected and signed by the bank is also correct. In the system that we describe, the customer proves that he constructed the message m correctly. Therefore, only one message is constructed and sent to the bank. This makes digital cash efficient.

Ordinary money is transferable: the shop does not have to return a coin to the bank after receiving it. He can transfer the money to a third person. The digital cash system we introduce does not have this feature, but there are electronic cash systems which do implement transferable electronic coins (e.g., [OkaOht92]; [ChaPed92]).

4.8.1 Blindly Issued Proofs

The payment system which we describe provides customer anonymity by using a blind digital signature.[18] Such a signature enables the signer (the bank) to sign a message without seeing its content (the content is hidden). Later, when the message and the signature are revealed, the signer is not able to link the signature with the corresponding signing transaction. The bank's signature can be verified by everyone, just like an ordinary signature.

The Basic Signature Scheme. Let p and q be large primes such that q divides $p - 1$. Let G_q be the subgroup of order q in $\mathbb{Z}_p^*$, g a generator of G_q[19] and $h : \{0,1\}^* \longrightarrow \mathbb{Z}_q$ be a collision-resistant hash function. The signer's secret key is a randomly chosen $x \in \mathbb{Z}_q$, and the public key is (p, q, g, y), where $y = g^x$.

The basic protocol in this section is *Schnorr's identification protocol*. It is an interactive proof of knowledge. The prover Peggy proves to the verifier Vic that she knows x, the discrete logarithm of y.

Protocol 4.39.
ProofLog(g, y):

1. Peggy randomly chooses $r \in \mathbb{Z}_q$, computes $a := g^r$ and sends it to Vic.
2. Vic chooses $c \in \mathbb{Z}_q$ at random and sends it to Peggy.
3. Peggy computes $b := r - cx$ and sends it to Vic.
4. Vic accepts the proof if $a = g^b y^c$; otherwise he rejects it.

To achieve a signature scheme, the protocol is converted into a non-interactive proof of knowledge ProofLog$_h$ using the same method as shown in Sections 4.5.3 and 4.2.5: the challenge c is computed by means of the collision-resistant hash function h.

A signature $\sigma(m)$ of a message m consists of a non-interactive proof that the signer (prover) knows the secret key x. The proof depends on the message m, because when computing the challenge c, the message m serves as an additional input to h:

---

[18] Blind signatures were introduced by D. Chaum in [Chaum82] to enable untraceable electronic cash.

[19] As stated before, there is a unique subgroup of order q of $\mathbb{Z}_p^*$. It is cyclic and each element $x \in \mathbb{Z}_p^*$ of order q is a generator (see Lemma A.41 and "Computing Modulo a Prime" on page 412).

$$\sigma(m) = (c, b) = \text{ProofLog}_h(m, g, y),$$

where $c := h(m\|a), a := g^r, r \in \mathbb{Z}_q$ chosen at random, and $b := r - cx$. The signature is verified by checking the condition

$$c = h(m\|g^b y^c).$$

Here, the collision resistance of the hash function is needed. The verifier does not get a to test $a = g^b y^c$. If m is the empty string, we omit m and write

$$(c, b) = \text{ProofLog}_h(g, y).$$

In this way, we attain a non-interactive proof that the prover knows $x = \log_g(y)$.

Remarks:

1. As in the commitment scheme in Section 4.3.2 and the election protocol in Section 4.5, the security relies on the assumption that discrete logarithms of elements in G_q are infeasible to compute (also see the remarks on the security of the DSA at the end of Section 3.4.3).
2. If the signer uses the same r (i.e., he uses the same commitment $a = g^r$) to sign two different messages m_1 and m_2, then the secret key x can easily be computed:[20]
   Let $\sigma(m_i) := (c_i, b_i), i = 1, 2$. We have $g^r = g^{b_1} y^{c_1} = g^{b_2} y^{c_2}$ and derive $x = (b_1 - b_2)(c_2 - c_1)^{-1}$. Note that $c_1 \neq c_2$ for $m_1 \neq m_2$, since h is collision resistant.

The Blind Signature Scheme. The basic signature scheme can be transformed into a blind signature scheme. To understand the ideas, we first recall our application scenario. The customer (Vic) would like to submit a coin to the shop (Alice). The coin is signed by the bank (Peggy). Alice must be able to verify Peggy's signature. Later, when Alice brings the coin to the bank, Peggy should not be able to recognize that she signed the coin for Vic. Therefore, Peggy has to sign the coin blindly. Vic obtains the blind signature for a message m from Peggy by executing the interactive protocol ProofLog with Peggy. In step 2 of the protocol, he deviates a little from the original protocol: as in the non-interactive version ProofLog$_h$, the challenge is not chosen randomly, but computed by means of the hash function h (with m as part of the input). We denote the transcript of this interaction by $(\bar{a}, \bar{c}, \bar{b})$; i.e., $\bar{a}$ is Peggy's commitment in step 1, $\bar{c}$ is Vic's challenge in step 2 and $\bar{b}$ is sent from Peggy to Vic in step 3. $(\bar{c}, \bar{b})$ is a valid signature with $\bar{a} = g^{\bar{b}} y^{\bar{c}}$.

Now Peggy may store Vic's identity and the transcript $\bar{\tau} = (\bar{a}, \bar{c}, \bar{b})$, but later she should not be able to recognize the signature $\sigma(m)$ of m. Therefore, Vic must transform Peggy's signature $(\bar{c}, \bar{b})$ into another valid signature (c, b)

---

[20] The elements in G_q have order q, hence all exponents and logarithms are computed modulo q. See "Computing Modulo a Prime" on page 412.

of Peggy that Peggy is not able to link with the original transcript $(\overline{a}, \overline{c}, \overline{b})$. The idea is that Vic transforms the given transcript $\overline{\tau}$ into another accepting transcript $\tau = (a, c, b)$ of ProofLog by the following transformation:

$$\beta_{(u,v,w)} : G_q \times \mathbb{Z}_q^2 \longrightarrow G_q \times \mathbb{Z}_q^2, \ (\overline{a}, \overline{c}, \overline{b}) \longmapsto (a, c, b), \ \text{where}$$

$$a := \overline{a}^u g^v y^w,$$
$$c := u\overline{c} + w,$$
$$b := u\overline{b} + v.$$

We have
$$a = \overline{a}^u g^v y^w = (g^{\overline{b}} y^{\overline{c}})^u g^v y^w = g^{u\overline{b}+v} y^{u\overline{c}+w} = g^b y^c.$$

Thus, τ is indeed an accepting transcript of ProofLog. If Vic chooses $u, v, w \in \{0, \ldots, q\}$ at random, then, $\overline{\tau}$ and τ are independent, and Peggy cannot get any information about $\overline{\tau}$ by observing the transcript τ. Namely, given $\overline{\tau} = (\overline{a}, \overline{c}, \overline{b})$, each (a, c, b) occurs exactly q times among the $\beta_{(u,v,w)}(\overline{a}, \overline{c}, \overline{b})$, where $(u, v, w) \in \mathbb{Z}_q^3$. Thus, the probability that $\tau = (a, b, c)$ is Vic's transformation of $(\overline{a}, \overline{c}, \overline{b})$ is

$$\frac{|\{(u, v, w) \in \mathbb{Z}_q^3 \mid \beta_{(u,v,w)}(\overline{a}, \overline{c}, \overline{b}) = (a, c, b)\}|}{|\mathbb{Z}_q^3|} = q^{-2},$$

and this is the same as the probability of $\tau = (a, b, c)$ if we randomly (and uniformly) select a transcript from the set $\mathcal{T}$ of all accepting transcripts of ProofLog. We see that Peggy's signature is really blind – she has no better chance than guessing at random to link her signature (c, b) with the original transcript $(\overline{a}, \overline{c}, \overline{b})$. Peggy receives no information about the transcript τ from knowing $\overline{\tau}$ (in the information-theoretic sense, see Appendix B.5).

On the other hand, the following arguments give evidence that for Vic, the only way to transform $(\overline{a}, \overline{c}, \overline{b})$ into another accepting, randomly looking transcript (a, b, c) is to randomly choose (u, v, w) and to apply $\beta_{(u,v,w)}$. First, we observe that Vic has to set $a = \overline{a}^u g^v y^w$ (for some u, v, w). Namely, assume that Vic sets $a = \overline{a}^u g^v y^w g'$ with some randomly chosen g'. Then he gets

$$a = \overline{a}^u g^v y^w g' = (g^{\overline{b}} y^{\overline{c}})^u g^v y^w g' = g^{u\overline{b}+v} y^{u\overline{c}+w} g',$$

and, since (a, c, b) is an accepting transcript, it follows that

$$g^{u\overline{b}+v} y^{u\overline{c}+w} g' = g^b y^c$$

or

$$g^{b-(u\overline{b}+v)+x(c-(u\overline{c}+w))} = g'.$$

This equation shows that Peggy and Vic together could compute $\log_g(g')$. This contradicts the discrete logarithm assumption, because g' was chosen at random.

If (a, b, c) is an accepting transcript, we get from $a = \overline{a}^u g^v y^w$ that

$$g^b y^c = a = \overline{a}^u g^v y^w = \left(g^{\overline{b}} y^{\overline{c}}\right)^u g^v y^w = g^{u\overline{b}+v} y^{u\overline{c}+w},$$

and conclude that

$$g^{(u\overline{b}+v)-b} = y^{c-(u\overline{c}+w)}.$$

This implies that

$$b = u\overline{b} + v \text{ and } c = u\overline{c} + w,$$

because otherwise Vic could compute Peggy's secret key x as $(u\overline{b} + v - b)(c - u\overline{c} - w)^{-1}$.

Our considerations lead to *Schnorr's blind signature* scheme. In this scheme, the verifier Vic gets a blind signature for m from the prover Peggy by executing the following protocol.

Protocol 4.40.

BlindLogSig$_h$(m):

1. Peggy randomly chooses $\overline{r} \in \mathbb{Z}_q$, computes $\overline{a} := g^{\overline{r}}$ and sends it to Vic.
2. Vic chooses $u, v, w \in \mathbb{Z}_q$, $u \neq 0$, at random and computes $a = \overline{a}^u g^v y^w$, $c := h(m\|a)$ and $\overline{c} := (c - w)u^{-1}$. Vic sends $\overline{c}$ to Peggy.
3. Peggy computes $\overline{b} := \overline{r} - \overline{c}x$ and sends it to Vic.
4. Vic verifies whether $\overline{a} = g^{\overline{b}} y^{\overline{c}}$, computes $b := u\overline{b} + v$ and gets the signature $\sigma(m) := (c, b)$ of m.

The verification condition for a signature (c, b) is $c = h(m\|g^b y^c)$.

A dishonest Vic may try to use a blind signature (c, b) for more than one message m. This is prevented by the collision resistance of the hash function h. In Section 4.8.2 we will use the blind signatures to form digital cash. A coin is simply the blind signature (c, b) issued by the bank for a specific message m. Thus, there is another basic security requirement for blind signatures: Vic should have no chance of deriving more than one (c, b) from one transcript $(\overline{a}, \overline{c}, \overline{b})$, i.e., from one execution of the protocol. Otherwise, in our digital cash example, Vic could derive more than one coin from the one coin issued by the bank. The Schnorr blind signature scheme seems to fulfill this security requirement. Namely, since h is collision resistant, Vic can work with at most one a. Moreover, Vic can know at most one triplet (u, v, w), with $a = \overline{a}^u g^v y^w$. This follows from Proposition 4.43 below, because $\overline{a}, g$ and y are chosen independently. Finally, the transformations $\beta_{(u,v,w)}$ are the only ones Vic can apply, as we saw above. However, we only gave convincing evidence for the latter statement, not a rigorous mathematical proof.

There is no mathematical security proof for either Schnorr's blind signature scheme or the underlying Schnorr identification scheme. A modification of the Schnorr identification scheme, the Okamoto–Schnorr identification scheme, is proven to be secure under the discrete logarithm assumption

([Okamoto92]). [PoiSte00] gives a security proof for the Okamoto–Schnorr blind signature scheme, derived from the Okamoto–Schnorr identification scheme. It is shown that no one can derive more than l signed messages after receiving l blind signatures from the signer. The proof is in the so-called random oracle model: the hash function is assumed to behave like a truly random function. Below we use Schnorr's blind signature scheme, because it is a bit easier.

If we use BlindLogSig_h as a subprotocol, we simply write

$$(c, b) = \text{BlindLogSig}_h(m).$$

The Blindly Issued Proof That Two Logarithms Are Equal. As before, let p and q be large primes such that q divides $p - 1$, and let G_q be the subgroup of order q in $\mathbb{Z}_p^*$, g a generator of G_q and $h : \{0,1\}^* \longrightarrow \mathbb{Z}_q$ be a collision-resistant hash function. Peggy's secret is a randomly chosen $x \in \mathbb{Z}_q$, whereas $y = g^x$ is publicly known.

In Section 4.5.2 we introduced an interactive proof $\text{ProofLogEq}(g, y, \tilde{g}, \tilde{y})$ that two logarithms are equal. Given a $\tilde{y}$ with $\tilde{y} = \tilde{g}^x$, Peggy can prove to the verifier Vic that she knows the common logarithm x of y and $\tilde{y}$ (with respect to the bases g and $\tilde{g}$). This proof can be transformed into a blindly issued proof. Here it is the goal of Vic to obtain, by interacting with Peggy, a z with $z = m^x$ for a given message $m \in G_q$, together with a proof (c, b) of this fact which he may then present to somebody else (recall that x is Peggy's secret and is not revealed). Now, Peggy should issue this proof blindly, i.e., she should not see m and z during the interaction, and later she should not be able to link (m, z, c, b) with this issuing transaction. We proceed in a similar way as in the blind signature scheme. Vic must not give m to Peggy, so he sends a cleverly transformed $\overline{m}$. Peggy computes $\overline{z} = \overline{m}^x$, and then both execute the ProofLogEq protocol (Section 4.5.2), with the analogous modification as above: Vic does not choose his challenge randomly, but computes it by means of the hash function h (with m as part of the input). This interaction results in a transcript $\overline{\tau} = (\overline{m}, \overline{z}, \overline{a}_1, \overline{a}_2, \overline{c}, \overline{b})$. Finally, Vic transforms $\overline{\tau}$ into an accepting transcript $\tau = (m, z, a_1, a_2, c, b)$ of the non-interactive version of ProofLogEq (see Section 4.5.3). Since τ appears completely random to Peggy, she cannot link τ to the original $\overline{\tau}$. Vic uses the following transformation:

$$
\begin{aligned}
a_1 &:= \overline{a}_1^u g^v y^w, \\
a_2 &:= (\overline{a}_2^u \overline{m}^v \overline{z}^w)^s (\overline{a}_1^u g^v y^w)^t, \\
c &:= u\overline{c} + w, \\
b &:= u\overline{b} + v, \\
m &:= \overline{m}^s g^t, \\
z &:= \overline{z}^s y^t.
\end{aligned}
$$

As above, a straightforward computation shows that $a_1 = g^b y^c$ and $a_2 = m^b z^c$. Thus, the transcript τ is indeed an accepting one. Analogous arguments as those given for the blind signature scheme show that the given transformation is the only way for Vic to obtain an accepting transcript, and that $\overline{\tau}$ and τ are independent if Vic chooses $u, v, w, s, t \in \mathbb{Z}_q$ at random. Hence, the proof is really blind – Peggy gets no information about the transcript τ from knowing $\overline{\tau}$.

Our considerations lead to the following blind signature scheme. The signature is issued blindly by Peggy. If Vic wants to get a signature for m, he transforms m to $\overline{m}$ and sends it to Peggy. Peggy computes $\overline{z} = \overline{m}^x$ and executes the ProofLogEq protocol with Vic to prove that $\log_{\overline{m}}(\overline{z}) = \log_g(y)$. Vic derives z and a proof that $\log_m(z) = \log_g(y)$ from $\overline{z}$ and the proof that $\log_{\overline{m}}(\overline{z}) = \log_g(y)$. The signature of m consists of $z = m^x$ and the proof that $\log_m(z) = \log_g(y)$.

Protocol 4.41.

BlindLogEqSig$_h$ (g, y, m):

1. Vic chooses $s, t \in \mathbb{Z}_q$, $s \neq 0$, at random, computes $\overline{m} := m^{1/s} g^{-t/s}$ and sends $\overline{m}$ to Peggy.[21]
2. Peggy randomly chooses $\overline{r} \in \mathbb{Z}_q$ and computes $\overline{z} := \overline{m}^x$ and $\overline{a} := (\overline{a}_1, \overline{a}_2) = (g^{\overline{r}}, \overline{m}^{\overline{r}})$. Peggy sends $(\overline{z}, \overline{a})$ to Vic.
3. Vic chooses $u, v, w \in \mathbb{Z}_q$, $u \neq 0$, at random, and computes $a_1 := \overline{a}_1^u g^v y^w$, $a_2 := (\overline{a}_2^u \overline{m}^v \overline{z}^w)^s (\overline{a}_1^u g^v y^w)^t$ and $z := \overline{z}^s y^t$. Then Vic computes $c := h(m\|z\|a_1\|a_2)$ and $\overline{c} := (c - w)u^{-1}$, and sends $\overline{c}$ to Peggy.
4. Peggy computes $\overline{b} := \overline{r} - \overline{c}x$ and sends it to Vic.
5. Vic verifies whether $\overline{a}_1 = g^{\overline{b}} y^{\overline{c}}$ and $\overline{a}_2 = \overline{m}^{\overline{b}} \overline{z}^{\overline{c}}$, computes $b := u\overline{b} + v$ and receives (z, c, b) as the final result.

If Vic presents the proof to Alice, then Alice may verify the proof by checking the verification condition $c = h(m\|z\|g^b y^c\|m^b z^c)$.

Below we will use BlindLogEqSig$_h$ as a subprotocol. Then we simply write

$$(z, c, b) = \text{BlindLogEqSig}_h(g, y, m).$$

Remarks:

1. Again the collision resistance of h implies that Vic can use the proof (z, c, b) for only one message m. As before, in the blind signature protocol *BlindLogSig*, we see that Vic cannot derive two different signatures (z, c, b) and $(\tilde{z}, \tilde{c}, \tilde{b})$ from one execution of the protocol, i.e., from one transcript $(\overline{m}, \overline{z}, \overline{a}_1, \overline{a}_2, \overline{c}, \overline{b})$.

---

[21] As is common practice, we denote the s-th root $x^{s^{-1}}$ of an element x of order q by $x^{1/s}$. Here, s^{-1} is the inverse of s modulo q. See "Computing Modulo a Prime" on page 412.

2. The protocols BlindLogEqSig$_h$ and BlindLogSig$_h$ may be merged to yield not only a signature of m but also a signature of an additionally given message M. Namely, if Vic computes $c = h(M\|m\|z\|a_1\|a_2)$ in step 3, then (c, b) is also a blind signature of M, formed in the same way as the signatures of BlindLogSig$_h$. We denote this merged protocol by

$$(z, c, b) = \text{BlindLogEqSig}_h(M, g, y, m),$$

and call it the proof BlindLogEqSig$_h$ dependent on the message M. It simultaneously gives signatures of M and m (consisting of z and a proof that $\log_g(y) = \log_m(z)$).

4.8.2 A Fair Electronic Cash System

The payment scheme we describe is published in [CamMauSta96]. A coin is a bit string that is (blindly) signed by the bank. For simplicity, we restrict the scheme to a single denomination of coins: the extension to multiple denominations is straightforward, with the bank using a separate key pair for each denomination. As discussed in the introduction of Section 4.8, in a fair electronic cash system the tracing of a coin and the revoking of the customer's anonymity must be possible under certain well-defined conditions; for example to track kidnappers who obtain a ransom as electronic cash. Here, anonymity may be revoked by a trusted third party, the trusted center.

System Setup. As before, let p and q be large primes such that q divides $p - 1$. Let G_q be the subgroup of order q in $\mathbb{Z}_p^*$. Let g, g_1 and g_2 be randomly and independently chosen generators of G_q. The security of the scheme requires that the discrete logarithm of none of these elements with respect to another of these elements is known. Since g, g_1 and g_2 are chosen randomly and independently, this is true with a very high probability. Let $h : \{0, 1\}^* \longrightarrow \mathbb{Z}_q$ be a collision-resistant hash function.

1. The bank chooses a secret key $x \in \mathbb{Z}_q$ at random and publishes $y = g^x$.
2. The trusted center T chooses a secret key $x_T \in \mathbb{Z}_q$ at random and publishes $y_T = g_2^{x_T}$.
3. The customer – we call her Alice – has the secret key $x_C \in \mathbb{Z}_q$ (randomly selected) and the public key $y_C = g_1^{x_C}$.
4. The shop's secret key x_S is randomly selected in $\mathbb{Z}_q$ and $y_S = g_1^{x_S}$ is the public key.

By exploiting data observed by the bank in the withdrawal protocol, the trusted center can provide information which enables the recognition of a coin withdrawn by Alice in the deposit step. This trace mechanism is called *coin tracing*. Moreover, from data observed by the bank in the deposit protocol, the trusted center can compute information which enables the identification of the customer. This tracing mechanism is called *owner tracing*.

Opening an Account. When the customer Alice opens an account, she proves her identity to the bank. She can do this by executing the protocol ProofLog(g_1, y_C) with the bank. The bank then opens an account and stores y_C in Alice's entry in the account database.

The Online Electronic Cash System. We first discuss the online system. The trusted center is involved in every withdrawal transaction, and the bank is involved in every payment.

The Withdrawal Protocol. As before, Alice has to authenticate herself to the bank. She can do this by proving that she knows $x_C = \log_{g_1}(y_C)$. To get a coin, Alice executes the *withdrawal* protocol with the bank. It is a modification of the BlindLogSig$_h$ protocol given in Section 4.8.1. Essentially, a coin is a signature of the empty string blindly issued by the bank.

Protocol 4.42.
Withdrawal:

1. The bank randomly chooses $\overline{r} \in \mathbb{Z}_q$ and computes $\overline{a} = g^{\overline{r}}$. The bank sends $\overline{a}$ to Alice.
2. Alice chooses $u, v, w \in \mathbb{Z}_q$, $u \neq 0$, at random, and computes $a := \overline{a}^u g^v y^w$ and $c := h(a), \overline{c} := (c - w)u^{-1}$. Alice sends (u, v, w) encrypted with the trusted center's public key and $\overline{c}$ to the bank.
3. The bank sends $\overline{a}$ and $\overline{c}$ and the encrypted (u, v, w) to the trusted center T.
4. T checks whether $u\overline{c} + w = h(\overline{a}^u g^v y^w)$, and sends the result to the bank.
5. If the result is correct, the bank computes $\overline{b} := \overline{r} - \overline{c}x$ and sends it to Alice. Alice's account is debited.
6. Alice verifies whether $\overline{a} = g^{\overline{b}} y^{\overline{c}}$, computes $b := u\overline{b} + v$ and gets as the coin the signature $\sigma := (c, b)$ of the empty message.

Payment and Deposit. In online payment, the shop must be connected to the bank when the customer spends the coin. Payment and deposit form one transaction. Alice spends a coin by sending it to a shop. The shop verifies the coin, i.e., it verifies the signature by checking $c = h(g^b y^c)$. If the coin is valid, it passes the coin to the bank. The bank also verifies the coin and then compares it with all previously spent coins (which are stored in the database). If the coin is new, the bank accepts it and inserts it into the database. The shop's account is credited.

Coin and Owner Tracing. The trusted center can link $(\overline{a}, \overline{c}, \overline{b})$ and (c, b), which enables coin and owner tracing.

Customer Anonymity. The anonymity of the customer relies on the fact that the used signature scheme is blind and on the security of the encryption scheme used to encrypt the blinding factors u, v and w.

The Offline Electronic Cash System. In the offline system, the trusted center is not involved in the withdrawal transaction, and the bank is not involved in the payment protocol. To achieve an offline trusted center, the immediate check that is performed by the trusted center in the withdrawal protocol above is replaced by a proof that Alice correctly provides the information necessary for tracing the coin. This proof can be checked by the bank. To obtain such a proof, the BlindLogSig$_h$-protocol is replaced by the BlindLogEqSig$_h$ protocol (see Section 4.8.1).

Essentially, a coin consists of a pair (m, z) with $m = g_1 g_2^s$, where s is chosen at random and $z = m^x$, and a proof of this fact which is issued blindly by the bank. For this purpose the BlindLogEqSig$_h$ protocol is executed: the bank sees $(\overline{m} = m^{1/s}, \overline{z} = z^{1/s})$. We saw above that the general blinding transformation in the BlindLogEqSig$_h$ protocol is $\overline{m} = m^{1/s} g^{-t}$. Here Alice chooses $t = 0$, i.e., $m = \overline{m}^s$; otherwise Alice could not perform ProofLog$_h$ $(M, m g_1^{-1}, g_2)$ as required in the payment protocol (see below). The blinding exponent s is encrypted by $d = y_T^s = g_2^{x_T s}$, where y_T is the trusted center's public key. This enables the trusted center to revoke anonymity later. It can get m by decrypting d (note that $m = g_1 g_2^s = g_1 d^{1/x_T}$), and the coin can be traced by linking m and $\overline{m}$.

The Withdrawal Protocol. As before, first Alice has to authenticate herself to the bank. The withdrawal protocol in the offline system consists of two steps. In the first of these steps, Alice generates an encryption of the message m to enable anonymity revocation by the trusted center. In the withdrawal step, Alice executes a message-dependent proof BlindLogEqSig$_h$ with the bank to obtain a (blind) signature on her coin:

1. *Enable coin and owner tracing.* Coin tracing means that, starting from the information gathered during the withdrawal the bank recognizes a specific coin in the deposit protocol. Owner tracing identifies the withdrawer of a coin, starting from the deposited coin. Tracings require the cooperation of the trusted center. To enable coin and owner tracing, Alice encrypts the message m with the trusted center's public key y_T and proves that she correctly performs this encryption:

   a. Alice chooses $s \in \mathbb{Z}_q$ at random, then computes

   $$m = g_1 g_2^s, d = y_T^s, \overline{m} = m^{1/s} = g_1^{1/s} g_2 \text{ and}$$
   $$(c, b) = \text{ProofLogEq}_h \left(\overline{m} g_2^{-1}, g_1, y_T, d \right).$$

   By the proof, Alice shows that $\log_{\overline{m} g_2^{-1}}(g_1) = \log_{y_T}(d)$ and that she knows this logarithm. Alice sends $(c, b, \overline{m} g_2^{-1}, g_1, y_T, d)$ to the bank.

   b. The bank verifies the proof. If the verification condition holds, then the bank stores d in Alice's entry in the withdrawal database for a possible later anonymity revocation.

2. *The withdrawal of a coin.* Alice chooses $r \in \mathbb{Z}_q$ at random, computes the so-called coin number $c\# = g^r$ and executes

$$(z, c_1, b_1) = \text{BlindLogEqSig}_h(c\#, g, y, m)$$

with the bank. Here, in the first step of BlindLogEqSig_h, Alice takes the s from the coin and owner tracing step 1 and $t = 0$. Thus, she sends the $\overline{m} = m^{1/s} = g_1^{1/s} g_2$ from step 1 to the bank. The variant of BlindLogEqSig_h is used that, in addition to the signature of m, gives a signature of $c\#$ by the bank (see the remarks after Protocol 4.41). The coin number $c\#$ is needed in the payment below. The bank debits Alice's account. The coin consists of $(c_1, b_1, c\#, g, y, m, z)$ and some additional information Alice has to supply when spending it (see the payment protocol below).

Payment. In an offline system, double spending cannot be prevented. It can only be detected by the bank in the deposit protocol. An additional mechanism in the protocols is necessary to decide whether the customer or the shop doubled the coin. If the coin was doubled by the shop, the bank refuses to credit the shop's account. If the customer doubled the coin, her anonymity should be revoked without the help of the trusted center. This can be achieved by the payment protocol. For this purpose, Alice, when paying with the coin, has to sign the message $M = y_S \| time \| (c_1, b_1)$, where y_S is the public key of the shop, $time$ is the time of the payment and (c_1, b_1) is the bank's blind signature on the coin from above. Alice has to sign by means of the basic signature scheme. There she has to use $s = \log_{g_2}(mg_1^{-1})$ as her secret and the coin number $c\# = g^r$, which she computed in the withdrawal protocol, as her commitment a.

$$\sigma(M) = (c_2, b_2) = \text{ProofLog}_h\left(M, mg_1^{-1}, g_2\right)$$

Now, if Alice spent the same coin $(c_1, b_1, c\#, g, y, m, z)$ twice, she would provide two signatures $\sigma(M)$ and $\sigma(M')$ of different messages M and M' (at least the times differ!). Both signatures are computed with the same commitment $a = c\#$. Then, it is easy to identify Alice (see below). The coin submitted to the shop is defined by:

$$coin = \left((c_1, b_1, c\#, g, y, m, z), \left(c_2, b_2, M, g_2, mg_1^{-1}\right)\right).$$

The shop verifies the coin, i.e., it verifies:

1. The correct form of M.
2. Whether $c_2 = h(M \| c\#)$.
3. The proof $(z, c_1, b_1) = \text{BlindLogEqSig}_h(c\#, g, y, m)$.
4. The proof $(c_2, b_2) = \text{ProofLog}_h\left(M, mg_1^{-1}, g_2\right)$, by testing

$$c_2 = h\left(M \| (mg_1^{-1})^{b_2} g_2^{c_2}\right).$$

Since h is collision resistant and the shop checks $c_2 = h(M\|c\#)$, Alice necessarily has to use the coin number $c\#$ in the second proof. The shop accepts if the coin passes the verification.

Deposit. The shop sends the coin to the bank. The bank verifies the coin and searches in the database for an identical coin. If she finds an identical coin, she refuses to credit the shop's account. If she finds a coin with identical first and different second component, the bank revokes the customer's anonymity (see below).

Coin and Owner Tracing.

1. *Coin tracing.* If the bank provides the trusted center T with $d = y_T^s$ produced by Alice in the withdrawal protocol, T computes m:
$$g_1 d^{1/x_T} = g_1 g_2^s - m.$$
   This value can be used to recognize the coin in the deposit protocol.

2. *Owner tracing.* If the bank provides the trusted center T with the m of a spent coin, T computes d:
$$\left(mg_1^{-1}\right)^{x_T} = (g_2^s)^{x_T} = y_T^s = d.$$
   This value can be used for searching in the withdrawal database.

Security.

1. *Double spending.* If Alice spends a coin twice (at different shops, or at the same shop but at different times), she produces signatures of two different messages. Both signatures are computed with the same commitment $c\#$ (the coin number). This reveals the signer's secret, which is the blinding exponent s in our case (see the remarks after Protocol 4.39). Knowing the blinding exponent s, the bank can derive $\overline{m} = m^{1/s}$. Searching in the withdrawal database yields Alice's identity. If the shop doubled the coin, the bank detects an identical coin in the database. Then she refuses to credit the shop's account.

2. *Customer's anonymity.* The anonymity of the customer is ensured. Namely, BlindLogEqSig$_h$ is a perfectly blind signature scheme, as we observed above. Moreover, since ProofLogEq is honest-verifier zero-knowledge (see Section 4.5.2), the bank also cannot get any information about the blinding exponent s from the ProofLogEq$_h$ in the withdrawal protocol. Note that any information about s could enable the bank to link $\overline{m}$ with m, and hence the withdrawal of the coin with the deposit of the coin. The bank could also establish this link, if she were able to determine (without computing the logarithms) that
$$\log_{g_2}\left(mg_1^{-1}\right) = \log_{\overline{m}g_2^{-1}}(g_1). \qquad (4.3)$$

Then she could link the proofs ProofLogEq$_h$ $\left(\overline{m}g_2^{-1}, g_1, y_T, d\right)$ (from the withdrawal transaction) and ProofLog$_h$ $\left(M, g_2, mg_1^{-1}\right)$ (from the deposit transaction). However, to find out (4.3) contradicts the decision Diffie–Hellman assumption (see Section 4.8.3).

3. *Security for the trusted center.* The trusted center makes coin and owner tracing possible. The tracings require that Alice correctly forms $m = g_1 g_2^s$ and $\overline{m} = m^{1/s} = g_1^{1/s} g_2$, and that it is really y_T^s which she sends as d to the bank (in the withdrawal transaction).

Now in the withdrawal protocol, Alice proves that $\overline{m} = g_1^{1/s} g_2$ and that $d = y_T^s$. In the payment protocol, Alice proves that $m = g_1 g_2^{\tilde{s}}$ with $\tilde{s}$ known to her. It is not clear a priori that $\tilde{s} = s$ or, equivalently, that $\overline{m} = m^{1/s}$. However, as we observed before in the blind signature scheme BlindLogEqSig$_h$, the only way to transform $\overline{m}$ into m is to choose σ and τ at random and to compute $m = \overline{m}^\sigma g^\tau$. From this we conclude that indeed $\tilde{s} = s$ and $m = \overline{m}^s$:

$$g_1 g_2^{\tilde{s}} = m = \overline{m}^\sigma g^\tau = (g_1^{1/s} g_2)^\sigma g^\tau = g_1^{\sigma/s} g_2^\sigma g^\tau.$$

Hence, $\tau = 0$ and $\sigma = \tilde{s} = s$ (by Proposition 4.43, below).

4.8.3 Underlying Problems

The Representation Problem. Let p and q be large primes such that q divides $p - 1$. Let G_q be the subgroup of order q in $\mathbb{Z}_p^*$.

Let $r \geq 2$ and let $g_1, \ldots, g_r$ be pairwise distinct generators of G_q.[22] Then $g = (g_1, \ldots, g_r) \in G_q^r$ is called a *generator of length* r. Let $y \in G_q$. $a = (a_1, \ldots, a_r) \in \mathbb{Z}_q^r$ is a *representation* of y (with respect to g) if

$$y = \prod_{i=1}^{r} g_i^{a_i}.$$

To represent y, the elements $a_1, \cdots, a_{r-1}$ can be chosen arbitrarily; a_r is then uniquely determined. Therefore, each $y \in G_q$ has q^{r-1} different representations. Given y, the probability that a randomly chosen $a \in \{0, \ldots, q\}^r$ is a representation of y is only $1/q$.

Proposition 4.43. *Assume that it is infeasible to compute discrete logarithms in G_q. Then no polynomial algorithm can exist which, on input of a randomly chosen generator of length $r \geq 2$, outputs $y \in G_q$ and two different representations of y.*

Remark. A stronger probabilistic version of Proposition 4.43 can be proven. There cannot even exist a polynomial algorithm that outputs an element $y \in G_q$ and two different representations of y with some non-negligible probability. Otherwise, there would be a polynomial algorithm that computes discrete logarithms with an overwhelmingly high probability (see Exercise 4 in Chapter 6).

---

[22] Note that every element of G_q except $[1]$ is a generator of G_q (see Lemma A.41).

Proof (Proposition 4.43). Assume that such an algorithm exists. On input of a randomly chosen generator, it outputs $y \in G_q$ and two different representations $a = (a_1, \ldots, a_r)$ and $b = (b_1, \ldots, b_r)$ of y. Then, $a - b$ is a non-trivial representation of $[1]$. Thus, we have a polynomial algorithm A which on input of a randomly chosen generator outputs a non-trivial representation of $[1]$. We may use A to define an algorithm B that on input of $g \in G_q, g \neq [1]$, and $z \in G_q$, computes the discrete logarithm of z with respect to g.

Algorithm 4.44.

```
int B(int g, z)
1   repeat
2       select i ∈ {1,...,r} and
3       uⱼ ∈ {1,...,q − 1}, 1 ≤ j ≤ r, uniformly at random
4       gᵢ ← zᵘⁱ, gⱼ ← gᵘ ʲ, 1 ≤ j ≠ i ≤ r
5       (a₁,...,aᵣ) ← A(g₁,...,gᵣ)
6   until aᵢuᵢ ≢ 0 mod q
7   return − (aᵢuᵢ)⁻¹ (∑ⱼ≠ᵢ aⱼuⱼ) mod q
```

We have

$$z^{-a_i u_i} = \prod_{j \neq i} g^{a_j u_j}.$$

Hence, the value returned is indeed the logarithm of z. Since at least one a_j returned by A is $\neq 0$ modulo q, the probability that $a_i \neq 0$ modulo q is $1/r$. Hence, we expect that the repeat until loop will terminate after r iterations. If r is bounded by a polynomial in the binary length $|p|$ of p, the expected running time of B is polynomial in $|p|$.   □

The Decision Diffie–Hellman Problem. Let p and q be large primes such that q divides $p - 1$. Let G_q be the subgroup of order q in $\mathbb{Z}_p^*$. Let $g \in G_q$ and $a, b \in \mathbb{Z}_q$ be randomly chosen. Then, the Diffie–Hellman assumption (Section 4.1.2) says that it is impossible to compute g^{ab} from g^a and g^b.

Let $g_1 = g^a; g_2 = g^b$ and g_3 be given. The *decision Diffie–Hellman problem* is to decide if

$$g_3 = g^{ab}.$$

This is equivalent to deciding whether

$$\log_g(g_3) = \log_g(g_1) \log_g(g_2), \text{ or}$$
$$\log_{g_2}(g_3) = \log_g(g_1).$$

The *decision Diffie–Hellman assumption* says that no efficient algorithm exists to solve the decision Diffie–Hellman problem if a, b and g_3 (g_1, g_2 and g_3, respectively) are chosen at random (and independently). The decision Diffie–Hellman problem is random self-reducible (see the remark on page 222). If you can solve it with an efficient probabilistic algorithm A, then you can also solve it, if $g \in G_q$ is any element of G_q and only g_1, g_2, g_3 are chosen randomly. Namely, let $g \in G_q$, then (g, g_1, g_2, g_3) has the Diffie–Hellman property if and only if $(g^s, g_1^s, g_2^s, g_3^s)$ has the Diffie–Hellman property, with s randomly chosen in $\mathbb{Z}_q^*$.

The representation problem and the decision Diffie–Hellman problem are studied, for example, in [Brands93].

Exercises

1. Let p be a sufficiently large prime such that it is intractable to compute discrete logarithms in $\mathbb{Z}_p^*$. Let g be a primitive root in $\mathbb{Z}_p^*$. p and g are publicly known. Alice has a secret key x_A and a public key $y_A := g^{x_A}$. Bob has a secret key x_B and a public key $y_B := g^{x_B}$. Alice and Bob establish a secret shared key by executing the following protocol (see [MatTakIma86]):

   Protocol 4.45.
   A variant of the Diffie–Hellman key agreement protocol:

   1. Alice randomly chooses a, $1 \le a \le p - 2$, sets $c := g^a$ and sends c to Bob.
   2. Bob randomly chooses b, $1 \le b \le p - 2$, sets $d := g^b$ and sends d to Alice.
   3. Alice computes the shared key $k = d^{x_A} y_B{}^a = g^{bx_A + ax_B}$.
   4. Bob computes the shared key $k = c^{x_B} y_A{}^b = g^{ax_B + bx_A}$.

   Does the protocol provide entity authentication? Discuss the security of the protocol.

2. Let $n := pq$, where p and q are distinct primes and $x_1, x_2 \in \mathbb{Z}_n^*$. Assume that at least one of x_1 and x_2 is in QR_n. Peggy wants to prove to Vic that she knows a square root of x_i for at least one $i \in \{1, 2\}$ without revealing i. Modify Protocol 4.5 to get an interactive zero-knowledge proof of knowledge.

3. Besides interactive proofs of knowledge, there are interactive proofs for proving the membership in a language. The completeness and soundness conditions for such proofs are slightly different. Let (P, V) be an interactive proof system. P and V are probabilistic algorithms, but only V is assumed to have polynomial running time. By P^* we denote a general (possibly dishonest) prover. Let $\mathcal{L} \subseteq \{0, 1\}^*$ ($\mathcal{L}$ is called a *language*). Bit strings $x \in \{0, 1\}^*$ are supplied to (P, V) as common input. (P, V) is called an *interactive proof system for the language* $\mathcal{L}$ if the following conditions are satisfied:

   a. *Completeness.* If $x \in \mathcal{L}$, then the probability that the verifier V accepts, if interacting with the honest prover P, is $\ge 3/4$.

    b. *Soundness.* If $x \notin \mathcal{L}$, then the probability that the verifier V accepts, if interacting with any prover P^*, is $\leq 1/2$.

Such an interactive proof system (P, V) is *(perfect) zero-knowledge* if there is a probabilistic *simulator* $S(V^*, x)$, running in expected polynomial time such that for every verifier V^* (honest or not) and for every $x \in \mathcal{L}$ the distributions of the random variables $S(V^*, x)$ and $(P, V^*)(x)$ are equal.

The class of languages that have interactive proof systems is denoted by $\mathcal{IP}$. It generalizes the complexity class $\mathcal{BPP}$ (see Exercise 3 in Chapter 5).

As in Section 4.3.1, let $n := pq$, with p and q distinct primes, and let $J_n^{+1} := \{x \in \mathbb{Z}_n^* \mid \left(\frac{x}{n}\right) = 1\}$ be the units with Jacobi symbol 1. Let $\mathrm{QNR}_n^{+1} := J_n^{+1} \setminus \mathrm{QR}_n$ be the quadratic non-residues in J_n^{+1}.

The following protocol is an interactive proof system for the language QNR_n^{+1} (see [GolMicRac89]). The common input x is assumed to be in J_n^{+1} (whether or not $x \in \mathbb{Z}_n$ is in J_n^{+1} can be efficiently determined using a deterministic algorithm; see Algorithm A.65).

Protocol 4.46.

Quadratic non-residuosity:

    Let $x \in J_n^{+1}$.

      1. Vic chooses $r \in \mathbb{Z}_n^*$ and $\sigma \in \{0, 1\}$ uniformly at random and sends $a = r^2 x^\sigma$ to Peggy.

      2. Peggy computes $\tau := \begin{cases} 0 \text{ if } a \in \mathrm{QR}_n \\ 1 \text{ if } a \notin \mathrm{QR}_n \end{cases}$ and sends τ to Vic.

      (Note that it is not assumed that Peggy can solve this in polynomial time. Thus, she can find out whether $a \in \mathrm{QR}_n$, for example, by an exhaustive search.)

      3. Vic accepts if and only if $\sigma = \tau$.

  Show:

    a. If $x \in \mathrm{QNR}_n^{+1}$ and both follow the protocol, Vic will always accept.

    b. If $x \notin \mathrm{QNR}_n^{+1}$, then Vic accepts with probability $\leq 1/2$.

    c. Show that the protocol is not zero-knowledge (under the quadratic residuosity assumption; see Remark 1 in Section 4.3.1).

    d. The protocol is honest-verifier zero-knowledge.

    e. Modify the protocol to get a zero-knowledge proof for quadratic non-residuosity.

4. We consider the identification scheme based on public-key encryption introduced in Section 4.2.1. In this scheme a dishonest verifier can obtain knowledge from the prover. Improve the scheme.

5. We modify the commitment scheme based on quadratic residues.

Protocol 4.47.

QRCommitment:

1. *System setup.* Alice chooses distinct large prime numbers $p, q \equiv 3 \bmod 4$ and sets $n := pq$. (Note $-1 \in J_n^{+1} \setminus QR_n$, see Proposition A.59.)
2. *Commit to $b \in \{0, 1\}$.* Alice chooses $r \in \mathbb{Z}_n^*$ at random, sets $c := (-1)^b r^2$ and sends c to Bob.
3. *Reveal.* Alice sends p, q, r and b to Bob. Bob can verify that p and q are primes $\equiv 3 \bmod 4$, $r \in \mathbb{Z}_n^*$, and $c := (-1)^b r^2$.

Show:

   a. If c is a commitment to b, then $-c$ is a commitment to $1 - b$.
   b. If c_i is a commitment to b_i, $i = 1, 2$, then $c_1 c_2$ is a commitment to $b_1 \oplus b_2$.
   c. Show how Alice can prove to Bob that two commitments c_1 and c_2 commit to equal or distinct values, without opening them.

6. Let $P = \{P_i \mid i = 1, \ldots, 6\}$. Set up a secret sharing system such that exactly the groups $\{P_1, P_2\}, \{Q \subset P \mid |Q| \geq 3, P_1 \in Q\}$ and $\{Q \subset P \mid |Q| \geq 4, P_2 \in Q\}$ are able to reconstruct the secret.

7. Let $P = \{P_1, P_2, P_3, P_4\}$. Is it possible to set up a secret sharing system by use of Shamir's threshold scheme such that the members of a group $Q \subset P$ are able to reconstruct the secret if and only if $\{P_1, P_2\} \subset Q$ or $\{P_3, P_4\} \subset Q$?

8. In the voting scheme of Section 4.5, it is necessary that each authority and each voter proves that he really follows the protocol. Explain why.

9. We consider the problem of *vote duplication*. This means that a voter can duplicate the vote of another voter who has previously posted his vote. He can do this without knowing the content of the other voter's ballot. Discuss this problem for the voting scheme of Section 4.5.

10. In Section 4.6.3 we describe a special honest-verifier zero-knowledge proof of knowing the plaintext of an ElGamal ciphertext. Give the analogous proof for Paillier encryption.

11. **Blind RSA signatures.** Construct a blind signature scheme based on the fact that the RSA function is a homomorphism.

12. **Nyberg–Rueppel Signatures.** Let p and q be large primes such that q divides $p - 1$. Let G be the subgroup of order q in $\mathbb{Z}_p^*$, and let g be a generator of G. The secret key of the signer is a randomly chosen $x \in \mathbb{Z}_q$, the public key is $y := g^x$.

   Signing. We assume that the message m to be signed is an element in $\mathbb{Z}_p^*$. The signed message is produced using the following steps:
   1. Select a random integer k, $1 \leq k \leq q - 1$.
   2. Set $r := mg^k$ and $s := xr + k \bmod q$.

3. (m, r, s) is the signed message.

Verification. If $1 \leq r \leq p - 1$, $1 \leq s \leq q - 1$ and $m = ry^r g^{-s}$, accept the signature. If not, reject it.

   a. Show that the verification condition holds for a signed message.
   b. Show that it is easy to produce forged signatures.
   c. How can you prevent this attack?
   d. Show that the condition $1 \leq r \leq p - 1$ has to be checked to detect forged signatures, even if the scheme is modified as in item c.

13. **Blind Nyberg–Rueppel Signatures** (see also [CamPivSta94]). In the situation of Exercise 12, Bob gets a blind signature for a message $m \in \{1, \ldots, q - 1\}$ from Alice by executing the following protocol:

Protocol 4.48.
BlindNybergRueppelSig(m):

   1. Alice chooses $\tilde{k}$ at random, $1 \leq \tilde{k} \leq q - 1$, and sets $\tilde{a} := g^{\tilde{k}}$. Alice sends $\tilde{a}$ to Bob.
   2. Bob chooses α, β at random with $1 \leq \alpha \leq q - 1$ and $0 \leq \beta \leq q - 1$, sets $\tilde{m} := m \tilde{a}^{\alpha-1} g^{\beta} \alpha^{-1}$, and sends $\tilde{m}$ to Alice.
   3. Alice computes $\tilde{r} := \tilde{m} g^{\tilde{k}}$, $\tilde{s} := \tilde{r}x + \tilde{k} \bmod q$, and sends $\tilde{r}$ and $\tilde{s}$ to Bob.
   4. Bob checks whether $(\tilde{m}, \tilde{r}, \tilde{s})$ is a valid signed message. If it is, then he sets $r := \tilde{r}\alpha$ and $s := \tilde{s}\alpha + \beta$.

Show that (m, r, s) is a signed message and that the protocol is really blind.

14. **Proof of Knowledge of a Representation** (see [Okamoto92]). Let p and q be large primes such that q divides $p - 1$. Let G be the subgroup of order q in $\mathbb{Z}_p^*$, and g_1 and g_2 be independently chosen generators. The secret is a randomly chosen $(x_1, x_2) \in \{0, \ldots, q - 1\}^2$, and the public key is (p, q, g_1, g_2, y), where $y := g_1^{x_1} g_2^{x_2}$ of G.
How can Peggy convince Vic by an interactive proof of knowledge that she knows (x_1, x_2), which is a representation of y with respect to (g_1, g_2)?

15. Convert the interactive proof of Exercise 14 into a blind signature scheme.

5. Probabilistic Algorithms

Probabilistic algorithms are important in cryptography. On the one hand, the algorithms used in encryption and digital signature schemes often include random choices (as in Vernam's one-time pad or the DSA) and therefore are probabilistic. On the other hand, when studying the security of cryptographic schemes, adversaries are usually modeled as probabilistic algorithms. The subsequent chapters, which deal with provable security properties, require a thorough understanding of this notion. Therefore, we clarify what is meant precisely by a probabilistic algorithm, and discuss the underlying probabilistic model.

The output y of a *deterministic algorithm* A is completely determined by its input x. In a deterministic way, y is computed from x by a sequence of steps decided in advance by the programmer. A behaves like a mathematical mapping: applying A to the same input x several times always yields the same output y. Therefore, we may use the mathematical notation of a mapping, $A : X \longrightarrow Y$, for a deterministic algorithm A, with inputs from X and outputs in Y. There are various equivalent formal models for such algorithms. A popular one is the description of algorithms by Turing machines (see, for example, [HopUll79]). Turing machines are state machines, and deterministic algorithms are modeled by Turing machines with deterministic behavior: the state transitions are completely determined by the input.

A probabilistic algorithm A is an algorithm whose behavior is partly controlled by random events. The computation of the output y on input x depends on the outcome of a finite number of random experiments. In particular, applying A to the same input x twice may yield two different outputs.

5.1 Coin-Tossing Algorithms

Probabilistic algorithms are able to toss coins. The control flow depends on the outcome of the coin tosses. Therefore, probabilistic algorithms exhibit random behavior.

Definition 5.1. Given an input x, a *probabilistic (or randomized) algorithm* A may toss a coin a finite number of times during its computation of the output y, and the next step may depend on the results of the preceding coin tosses. The number of coin tosses may depend on the outcome of the previous ones, but it is bounded by some constant t_x for a given input x. The coin tosses are independent and the coin is a fair one, i.e., each side appears with probability $1/2$.

Examples. The encryption algorithms in Vernam's one-time pad (Section 2.1.1), OAEP (Section 3.3.3) and ElGamal's scheme (Section 3.4) include random choices, and thus are probabilistic, as well as the signing algorithms in PSS (Section 3.3.5), ElGamal's scheme (Section 3.4.2) and the DSA (Section 3.4.3). Other examples of probabilistic algorithms are the algorithm for computing square roots in $\mathbb{Z}_p^*$ (see Algorithm A.67) and the probabilistic primality tests discussed in Appendix A.10. Many examples of probabilistic algorithms in various areas of application can be found, for example, in [MotRag95].

Remarks and Notations:

1. A formal definition of probabilistic algorithms can be given by the notion of probabilistic Turing machines ([LeeMooShaSha55]; [Rabin63]; [Santos69]; [Gill77]; [BalDiaGab95]).[1] In a probabilistic Turing machine, the state transitions are determined by the input and the outcome of coin tosses. Probabilistic Turing machines should not be confused with non-deterministic machines. A non-deterministic Turing machine is "able to simply guess the solution to the given problem" and thus, in general, is not something that can be implemented in practice. A probabilistic machine (or algorithm) is able to find the solution by use of its coin tosses, with some probability. Thus, it is something that can be implemented in practice.

   Of course, we have to assume (and will assume in the following) that a random source of independent fair coin tosses is available. To implement such a source, the inherent randomness in physical phenomena can be exploited (see [MenOorVan96] and [Schneier96] for examples of sources which might be used in a computer).

   To derive perfectly random bits from a natural source is a non-trivial task. The output bits may be biased (i.e., the probability that 1 is emitted is different from $1/2$) or correlated (the probability of 1 depends on the previously emitted bits). The outcomes of physical processes are often affected by previous outcomes and the circumstances that led to these outcomes. If the bits are independent, the problem of biased bits can be easily solved using the following method proposed by John von Neumann ([von Neumann63]): break the sequence of bits into pairs, discard pairs 00 and 11, and interpret 01 as 0 and 10 as 1 (the pairs 01 and 10 have the same probability). Handling a correlated bit source is more difficult.

---

[1] All algorithms are assumed to have a finite description (as a Turing machine) which is independent of the size of the input. We do not consider non-uniform algorithms in this book.

However, there are effective means of generating truly random sequences of bits from a biased and correlated source. For example, Blum developed a method for a source which produces bits according to a known Markov chain ([Blum84]). Vazirani ([Vazirani85]) shows how almost independent, unbiased bits can be derived from two independent "slightly-random" sources. For a discussion of slightly random sources and their use in randomized algorithms, see [Papadimitriou94], for example.

2. The output y of a probabilistic algorithm A depends on the input x and on the binary string r, which describes the outcome of the coin tosses. Usually, the coin tosses are considered as internal operations of the probabilistic algorithm. A second way to view a probabilistic algorithm A is to consider the outcome of the coin tosses as an additional input, which is supplied by an external coin-tossing device. In this view, the model of a probabilistic algorithm is a deterministic machine. We call the corresponding deterministic algorithm A_D the *deterministic extension* of A. It takes as inputs the original input x and the outcome r of the coin tosses.

3. Given x, the output $A(x)$ of a probabilistic algorithm A is not a single constant value, but a random variable. "A outputs y on input x" is a random event, and by $\mathrm{prob}(A(x) = y)$ we mean the probability of this event. More precisely, we have

$$\mathrm{prob}(A(x) = y) := \mathrm{prob}(\{r \mid A_D(x, r) = y\}).^2$$

Here a question arises: what probability distribution of the coin tosses is meant? The question is easily answered if, as in our definition of probabilistic algorithms, the number of coin tosses is bounded by some constant t_x for a given x. In this case, adding some dummy coin tosses, if necessary, we may assume that the number of coin tosses is exactly t_x. Then the possible outcomes r of the coin tosses are the binary strings of length t_x, and since the coin tosses are independent, we have the uniform distribution of $\{0,1\}^{t_x}$. The probability of an outcome r is $1/2^{t_x}$, and hence

$$\mathrm{prob}(A(x) = y) = \frac{|\{r \mid A_D(x, r) = y\}|}{2^{t_x}}.$$

It is sufficient for all our purposes to consider only probabilistic algorithms with a bounded number of coin tosses, for a given x. In most parts of this book we consider algorithms whose running time is bounded by a function $f(|x|)$, where $|x|$ is the size of the input x. For these algorithms, the assumption is obviously true.

4. Given x, the probabilities $\mathrm{prob}(A(x) = y), y \in Y$, define a probability distribution on the range Y. We denote it by $p_{A(x)}$. The random variable $A(x)$ samples Y according to the distribution $p_{A(x)}$.

---

[2] If the probability distribution is determined by the context, we often do not specify the distribution explicitly and simply write $\mathrm{prob}(e)$ for the probability of an element or event e (see Appendix B.1).

5. The setting where a probabilistic algorithm A is executed may include further random events. Now, tossing a fair coin in A is assumed to be an independent random experiment. Therefore, the outcome of the coin tosses of A on input x is independent of all further random events in the given setting. In the following items, we apply this basic assumption.

6. Suppose that the input $x \in X$ of a probabilistic algorithm A is randomly generated. This means that a probability distribution p_X is given for the domain X (e.g. the uniform distribution). We may consider the random experiment "Randomly choose $x \in X$ according to p_X and compute $y = A(x)$". If the outputs of A are in Y, then the experiment is modeled by a joint probability space (XY, p_{XY}). The coin tosses of $A(x)$ are independent of the random choice of x. Thus, the probability that $x \in X$ is chosen and that $y = A(x)$ is

$$\text{prob}(x, A(x) = y) = p_{XY}(x, y) = p_X(x) \cdot \text{prob}(A(x) = y).$$

The probability $\text{prob}(A(x) = y)$ is the conditional probability $\text{prob}(y\,|\,x)$ of the outcome y, assuming the input x.

7. Each execution of A is a new independent random experiment: the coin tosses during one execution of A are independent of the coin tosses in other executions of A. In particular, when executing A twice, with inputs x and x', we have

$$\text{prob}(A(x) = y, A(x') = y') = \text{prob}(A(x) = y) \cdot \text{prob}(A(x') = y').$$

If probabilistic algorithms A and B are applied to inputs x and x', then the coin tosses of $A(x)$ and $B(x')$ are independent, unless $B(x')$ is called as a subroutine of $A(x)$ (such that the coin tosses of $B(x')$ are contained in the coin tosses of $A(x)$), or vice versa:

$$\text{prob}(A(x) = y, B(x') = y') = \text{prob}(A(x) = y) \cdot \text{prob}(B(x') = y').$$

8. Let A be a probabilistic algorithm with inputs from X and outputs in Y. Let $h : X \longrightarrow Z$ be a map yielding some property $h(x)$ for the elements in X (e.g. the least-significant bit of x). Let B be a probabilistic algorithm which on input $y \in Y$ outputs $B(y) \in Z$. Assume that a probability distribution p_X is given on X. B might be an algorithm trying to invert A or at least trying to determine the property $h(x)$ from $y := A(x)$. We are interested in the random experiment "Randomly choose x, compute $y = A(x)$ and $B(y)$, and check whether $B(y) = h(x)$". The random choice of x, the coin tosses of $A(x)$ and the coin tosses of $B(y)$ are independent random experiments. Thus, the probability that $x \in X$ is chosen, and that $A(x) = y$ and $B(y)$ correctly computes $h(x)$ is

$$\text{prob}(x, A(x) = y, B(y) = h(x))$$
$$= \text{prob}(x) \cdot \text{prob}(A(x) = y) \cdot \text{prob}(B(y) = h(x)).$$

9. Let p_X be a probability distribution on the domain X of a probabilistic algorithm A with outputs in Y. Randomly selecting an $x \in X$ and computing $A(x)$ is described by the joint probability space XY (see above). We can project to Y, $(x, y) \mapsto y$, and calculate the probability distribution p_Y on Y:

$$p_Y(y) := \sum_{x \in X} p_{XY}(x, y) = \sum_{x \in X} p_X(x) \cdot \mathrm{prob}(A(x) = y).$$

We call p_Y the image of p_X under A. p_Y is also the image of the joint distribution of X and the coin tosses r under the deterministic extension A_D of A:

$$p_Y(y) = \mathrm{prob}(\{(x, r) \mid A_D(x, r) = y\})$$
$$= \sum_{x \in X, r \in \{0,1\}^{t_x} : A_D(x,r)=y} p_X(x) \cdot \mathrm{prob}(r).$$

Let A be a probabilistic algorithm with inputs from X and outputs in Y. Then A (as a Turing machine) has a finite binary description. In particular, we can assume that both the domain X and the range Y are subsets of $\{0,1\}^*$. The time and space complexity of an algorithm A (corresponding to the running time and memory requirements) are measured as functions of the binary length $|x|$ of the input x.

Definition 5.2. A *probabilistic polynomial algorithm* is a probabilistic algorithm A such that the running time of $A(x)$ is bounded by $P(|x|)$, where $P \in \mathbb{Z}[X]$ is a polynomial (the same for all inputs x). The running time is measured as the number of steps in our model of algorithms, i.e., the number of steps of the probabilistic Turing machine. Tossing a coin is one step in this model.

Remark. The randomness in probabilistic algorithms is caused by random events in a very specific probability space, namely $\{0,1\}^{t_x}$ with the uniform distribution, and at first glance this might look restrictive. Actually, it is a very general model.

For example, suppose you want to control an algorithm A on input x by r_x random events with probabilities $p_{x,1}, \ldots, p_{x,r_x}$ (the deterministic extension A_D takes as inputs x and one of the events).[3] Assume that always one of the events occurs (i.e., $\sum_{i=1}^{r_x} p_{x,i} = 1$) and that the probabilities $p_{x,i}$ have a finite binary representation $p_{x,i} = \sum_{j=1}^{t_x} a_{x,i,j} \cdot 2^{-j}$ ($a_{x,i,j} \in \{0,1\}$). Further, assume that r_x, t_x and the probabilities $p_{x,i}$ are computable by deterministic (polynomial) algorithms with input x. The last assumption is satisfied, for example, if the events and probabilities are the same for all x.

---

[3] For example, think of an algorithm A that attacks an encryption scheme. If all possible plaintexts and their probability distribution are known, then A might be based on the random choice of correctly distributed plaintexts.

Then the random behavior of A can be implemented by coin tosses, i.e., A can be implemented as a probabilistic (polynomial) algorithm in the sense of Definitions 5.1 and 5.2. Namely, let S be the coin-tossing algorithm which on input x:

(1) t_x times tosses the coin and obtains a binary number $b := b_{t_x-1} \ldots b_1 b_0$, with $0 \leq b < 2^{t_x}$, and

(2) returns $S(x) := i$, if $2^{t_x} \sum_{j=1}^{i-1} p_{x,j} \leq b < 2^{t_x} \sum_{j=1}^{i} p_{x,j}$.

The outputs of $S(x)$ are in $\{1, \ldots, r_x\}$ and $\mathrm{prob}(S(x) = i) = p_{x,i}$, for $1 \leq i \leq r_x$. The probabilistic (polynomial) algorithm S can be used to produce the random inputs for A_D.

5.2 Monte Carlo and Las Vegas Algorithms

The running time of a probabilistic polynomial algorithm A is required to be bounded by a polynomial P, for all inputs x. Assume A tries to compute the solution to a problem. Due to its random behavior, A might not reach its goal with certainty, but only with some probability. Therefore, the output might not be correct in some cases. Such algorithms are also called Monte Carlo algorithms if their probability of success is not too low. They are distinguished from Las Vegas algorithms.

$time_A(x)$ denotes the running time of A on input x, i.e., the number of steps A needs to generate the output $A(x)$ for the input x. As before, $|x|$ denotes the binary length of x.

Definition 5.3. Let $\mathcal{P}$ be a computational problem.

1. A *Monte Carlo algorithm* A for $\mathcal{P}$ is a probabilistic algorithm A, whose running time is bounded by a polynomial Q and which yields a correct answer to $\mathcal{P}$ with a probability of at least $2/3$:

$$time_A(x) \leq Q(|x|) \text{ and } \mathrm{prob}(A(x) \text{ is a correct answer to } \mathcal{P}) > \frac{2}{3},$$

for all instances x of $\mathcal{P}$.

2. A probabilistic algorithm A for $\mathcal{P}$ is called a *Las Vegas algorithm* if its output is always a correct answer to $\mathcal{P}$, and if the expected value for the running time is bounded by a polynomial Q:

$$\mathrm{E}(time_A(x)) = \sum_{t=1}^{\infty} t \cdot \mathrm{prob}(time_A(x) = t) < Q(|x|),$$

for all instances x of $\mathcal{P}$.

Remarks:

1. The probabilities are computed assuming a fixed input x; they are only taken over the coin tosses during the computation of $A(x)$. The distribution of the inputs x is not considered. For example,

$$\text{prob}(A(x) \text{ is a correct answer to } \mathcal{P}) = \sum_{y \in Y_x} \text{prob}(A(x) = y),$$

with the sum taken over the set Y_x of correct answers for input x.

2. The running time of a Monte Carlo algorithm is bounded by one polynomial Q, for all inputs. A Monte Carlo algorithm may sometimes fail to produce a correct answer to the given problem. However, the probability of such a failure is bounded. In contrast, a Las Vegas algorithm always gives a correct answer to the given problem. The running time may vary substantially and is not necessarily bounded by a single polynomial. However, the expected value of the running time is polynomial.

Examples:

1. Typical examples of Monte Carlo algorithms are the probabilistic primality tests discussed in Appendix A.10. They check whether an integer is a prime or not.
2. Algorithm A.67, which computes square roots modulo a prime p, is a Las Vegas algorithm.
3. To prove the zero-knowledge property of an interactive proof system, it is necessary to simulate the prover by a Las Vegas algorithm (see Section 4.2.3).

A Las Vegas algorithm may be turned into a Monte Carlo algorithm simply by stopping it after a suitable polynomial number of steps. To state this fact more precisely, we use the notion of positive polynomials.

Definition 5.4. A polynomial $P(X) = \sum_{i=0}^{n} a_i X^i \in \mathbb{Z}[X]$ in one variable X, with integer coefficients $a_i, 0 \le i \le n$, is called a *positive polynomial* if $P(x) > 0$ for $x > 0$, i.e., P has positive values for positive inputs.

Examples. The polynomials X^n are positive. More generally, each polynomial whose non-zero coefficients are positive is a positive polynomial.

Proposition 5.5. *Let $A(x)$ be a Las Vegas algorithm for a problem $\mathcal{P}$ with an expected running time $\le Q(|x|)$ (Q a polynomial). Let P be a positive polynomial. Let $\tilde{A}$ be the algorithm obtained by stopping $A(x)$ after at most $P(|x|)$ steps. Then $\tilde{A}$ is a Monte Carlo algorithm for $\mathcal{P}$, which gives a correct answer to $\mathcal{P}$ with probability $\ge 1 - Q(|x|)/P(|x|)$.*

Proof. We have

$$P(|x|) \cdot \text{prob}(time_A(x) \geq P(|x|)) \leq \sum_{t=P(|x|)}^{\infty} t \cdot \text{prob}(time_A(x) = t)$$

$$\leq \text{E}(time_A(x)) \leq Q(|x|).$$

Thus, $\tilde{A}$ gives a correct answer to $\mathcal{P}$ with probability $\geq 1 - Q(|x|)/P(|x|)$. $\square$

Remark. $\tilde{A}(x)$ might return "I don't know" if $A(x)$ did not yet terminate after $P(|x|)$ steps and is stopped. Then the answer of $\tilde{A}$ is never false, though it is not always a solution to $\mathcal{P}$.

The choice of the bound $2/3$ in our definition of Monte Carlo algorithms is somewhat arbitrary. Other bounds could be used equally well, as Proposition 5.6 below shows. We can increase the probability of success of an algorithm $A(x)$ by repeatedly applying the original algorithm and by making a majority decision. This works if the probability that A produces a correct result exceeds $1/2$ by a non-negligible amount.

Proposition 5.6. *Let P and Q be positive polynomials. Let A be a probabilistic polynomial algorithm which computes a function $f : X \longrightarrow Y$, with*

$$\text{prob}(A(x) = f(x)) \geq \frac{1}{2} + \frac{1}{P(|x|)} \text{ for all } x \in X.$$

Then, by repeating the computation $A(x)$ and returning the most frequent result, we get a probabilistic polynomial algorithm $\tilde{A}$ such that

$$\text{prob}(\tilde{A}(x) = f(x)) > 1 - \frac{1}{Q(|x|)} \text{ for all } x \in X.$$

Proof. Consider the algorithms A_t ($t \in \mathbb{N}$), defined on input $x \in X$ as follows:

1. Execute $A(x)$ t times, and get the set $Y_x := \{y_1, \ldots, y_t\}$ of outputs.
2. Select an $i \in \{1, \ldots, t\}$, with $|\{y \in Y_x \mid y = y_i\}|$ maximal.
3. Set $A_t(x) := y_i$.

We expect that more than half of the results of A coincide with $f(x)$, and hence $A_t(x) = f(x)$ with high probability. More precisely, define the binary random variables $S_j, 1 \leq j \leq t$:

$$S_j := \begin{cases} 1 \text{ if } y_j = f(x), \\ 0 \text{ otherwise.} \end{cases}$$

The expected values $\text{E}(S_j)$ are equal to

$$\text{E}(S_j) = \text{prob}(A(x) = f(x)) \geq \frac{1}{2} + \frac{1}{P(|x|)},$$

and we conclude, by Corollary B.26, that

$$\operatorname{prob}(A_t(x) = f(x)) \geq \operatorname{prob}\left(\sum_{j=1}^{t} S_j > \frac{t}{2}\right) \geq 1 - \frac{P(|x|)^2}{4t}.$$

For $t > \frac{1}{4} \cdot P(|x|)^2 \cdot Q(|x|)$, the probability of success of A_t is $> 1 - \frac{1}{Q(|x|)}$.
$\square$

Remark. By using the Chernoff bound from probability theory, the preceding proof can be modified to get a polynomial algorithm $\tilde{A}$ whose probability of success is exponentially close to 1 (see Exercise 5). We do not need this stronger result. If a polynomial algorithm which checks the correctness of a solution is available, we can do even better.

Proposition 5.7. *Let P, Q and R be positive polynomials. Let A be a probabilistic polynomial algorithm computing solutions to a given problem $\mathcal{P}$ with*

$$\operatorname{prob}(A(x) \text{ is a correct answer to } \mathcal{P}) \geq \frac{1}{P(|x|)}, \text{ for all inputs } x.$$

Let D be a deterministic polynomial algorithm, which checks whether a given answer to $\mathcal{P}$ is correct. Then, by repeating the computation $A(x)$ and by checking the results with D, we get a probabilistic polynomial algorithm $\tilde{A}$ for $\mathcal{P}$ such that

$$\operatorname{prob}(\tilde{A}(x) \text{ is a correct answer to } \mathcal{P}) > 1 - 2^{-Q(|x|)}, \text{ for all inputs } x.$$

Proof. Consider the algorithms A_t $(t \in \mathbb{N})$, defined on input x as follows:

1. Repeat the following at most t times:
   a. Execute $A(x)$, and get the answer y.
   b. Apply D to check whether y is a correct answer.
   c. If D says "correct", stop the iteration.
2. Return y.

The executions of $A(x)$ are t independent repetitions of the same experiment. Hence, the probability that all t executions of A yield an incorrect answer is $< (1 - \frac{1}{P(|x|)})^t$, and we obtain

$$\left(1 - \frac{1}{P(|x|)}\right)^t = \left(\left(1 - \frac{1}{P(|x|)}\right)^{P(|x|)}\right)^{t/P(|x|)} < \left(e^{-1}\right)^{t/P(|x|)}$$

$$= e^{-t/P(|x|)} \leq e^{-\ln(2)Q(|x|)} = 2^{-Q(|x|)},$$

for $t \geq \ln(2)P(|x|)Q(|x|)$.
$\square$

Remark. The iterations A_t in the preceding proof may be used to construct a Las Vegas algorithm that always gives a correct answer to the given problem (see Exercise 1).

Exercises

1. Let P be a positive polynomial. Let A be a probabilistic polynomial algorithm which on input $x \in X$ computes solutions to a given problem $\mathcal{P}$ with

$$\text{prob}(A(x) \text{ is a correct answer to } \mathcal{P}) \geq \frac{1}{P(|x|)} \text{ for all } x \in X.$$

   Assume that there is a deterministic algorithm D, which checks in polynomial time whether a given solution y to $\mathcal{P}$ on input x is correct.

   Construct a Las Vegas algorithm that always gives the correct answer and whose expected running time is $\leq P(|x|)(R(|x|) + S(|x| + R(|x|)))$, where R and S are polynomial bounds for the running times of A and D (use Lemma B.12).

2. Let $\mathcal{L} \subset \{0,1\}^*$ be a decision problem in the complexity class $\mathcal{BPP}$ of *bounded-error probabilistic polynomial-time problems*. This means that there is a probabilistic polynomial algorithm A with input $x \in \{0,1\}^*$ and output $A(x) \in \{0,1\}$ which solves the membership decision problem for $\mathcal{L} \subset \{0,1\}^*$, with a bounded error probability. More precisely, there are a positive polynomial P and a constant a, $0 < a < 1$, such that

$$\text{prob}(A(x) = 1) \geq a + \frac{1}{P(|x|)}, \text{ for } x \in \mathcal{L}, \text{ and}$$

$$\text{prob}(A(x) = 1) \leq a - \frac{1}{P(|x|)}, \text{ for } x \notin \mathcal{L}.$$

   Let Q be another positive polynomial. Show how to obtain a probabilistic polynomial algorithm $\tilde{A}$, with

$$\text{prob}(\tilde{A}(x) = 1) \geq 1 - \frac{1}{Q(|x|)}, \text{ for } x \in \mathcal{L}, \text{ and}$$

$$\text{prob}(\tilde{A}(x) = 1) \leq \frac{1}{Q(|x|)}, \text{ for } x \notin \mathcal{L}.$$

   (Use a similar technique as that used, for example, in the proof of Proposition 5.6.)

3. A decision problem $\mathcal{L} \subset \{0,1\}^*$ belongs to the complexity class $\mathcal{RP}$ of *randomized probabilistic polynomial-time problems* if there exists a probabilistic polynomial algorithm A which on input $x \in \{0,1\}^*$ outputs $A(x) \in \{0,1\}$, and a positive polynomial Q such that $\text{prob}(A(x) = 1) \geq 1/Q(|x|)$, for $x \in \mathcal{L}$, and $\text{prob}(A(x) = 1) = 0$, for $x \notin \mathcal{L}$.

   A decision problem $\mathcal{L}$ belongs to the complexity class $\mathcal{NP}$ if there is a deterministic polynomial algorithm $M(x,y)$ and a polynomial L such that $M(x,y) = 1$ for some $y \in \{0,1\}^*$, with $|y| \leq L(|x|)$, if and only if $x \in \mathcal{L}$ (y is called a certificate for x). Show that:

    a. $\mathcal{RP} \subseteq \mathcal{BPP}$.

    b. $\mathcal{RP} \subseteq \mathcal{NP}$.

4. A decision problem $\mathcal{L} \subset \{0,1\}^*$ belongs to the complexity class $\mathcal{ZPP}$ of *zero-sided probabilistic polynomial-time problems* if there exists a Las Vegas algorithm $A(x)$ such that $A(x) = 1$ if $x \in \mathcal{L}$, and $A(x) = 0$ if $x \notin \mathcal{L}$.

    Show that $\mathcal{ZPP} \subseteq \mathcal{RP}$.

5. If $S_1, \ldots, S_n$ are independent repetitions of a binary random variable X and $p := \mathrm{prob}(X = 1) = \mathrm{E}(X)$, then the *Chernoff bound* holds for $0 < \varepsilon \leq p(1 - p)$:

$$\mathrm{prob}\left(\left| \frac{1}{n} \sum_{i=1}^{n} S_i - p \right| < \varepsilon \right) \geq 1 - 2e^{-n\varepsilon^2/2},$$

(see, e.g., [Rényi70], Section 7.4). Use the Chernoff bound to derive an improved version of Proposition 5.6.

6. One-Way Functions and the Basic Assumptions

In Chapter 3 we introduced the notion of one-way functions. As the examples of RSA encryption and Rabin signatures show, one-way functions play the key role in asymmetric cryptography.

Speaking informally, a *one-way function* is a map $f : X \longrightarrow Y$ which is easy to compute but hard to invert. There is no efficient algorithm that computes pre-images of $y \in Y$. If we want to use a one-way function f for encryption in the straightforward way (applying f to the plaintext, as, for example, in RSA encryption), then f must belong to a special class of one-way functions. Knowing some information ("the trapdoor information": e.g. the factorization of the modulus n in RSA schemes), it must be easy to invert f, and f is one way only if the trapdoor information is kept secret. These functions are called *trapdoor functions*.

Our notion of one-way functions introduced in Chapter 3 was a rather informal one: we did not specify precisely what we mean by "efficiently computable", "infeasible" or "hard to invert". Now, in this chapter, we will clarify these terms and give a precise definition of one-way functions. For example, "efficiently computable" means that the solution can be computed by a probabilistic polynomial algorithm, as defined in Chapter 5.

We discuss three examples in some detail: the discrete exponential function, modular powers and modular squaring. The first is not a trapdoor function. Nevertheless, it has important applications in cryptography (e.g. pseudorandom bit generators, see Chapter 8; ElGamal's encryption and signature scheme and the DSA, see Chapter 3).

Unfortunately, there is no proof that these functions are really one way. However, it is possible to state the basic assumptions precisely, which guarantee the one-way feature. It is widely believed that these assumptions are true.

In order to define the one-way feature (and in a way that naturally matches the examples), we have to consider not only single functions, but, more generally, families of functions defined over appropriate index sets.

In the preliminary (and very short) Section 6.1, we introduce an intuitive notation for probabilities that will be used subsequently.

6.1 A Notation for Probabilities

The notation

$$\text{prob}(B(x) = 1 : x \leftarrow X) := \text{prob}(\{x \in X \mid B(x) = 1\}),$$

introduced for Boolean predicates[1] $B : X \longrightarrow \{0,1\}$ in Appendix B.1 (p. 466), intuitively suggests that we mean the probability of $B(x) = 1$ if x is randomly chosen from X. We will use an analogous notation for probabilistic algorithms.

Let A be a probabilistic algorithm with inputs from X and outputs in Y, and let $B : X \times Y \longrightarrow \{0,1\}$, $(x,y) \longmapsto B(x,y)$ be a Boolean predicate. Let p_X be a probability distribution on X. As in Section 5.1, let A_D be the deterministic extension of A, and let t_x denote the number of coin tosses of A on input x:

$$\text{prob}(B(x, A(x)) = 1 : x \xleftarrow{p_X} X)$$
$$:= \sum_{x \in X} \text{prob}(x) \cdot \text{prob}(B(x, A(x)) = 1)$$
$$= \sum_{x \in X} \text{prob}(x) \cdot \text{prob}(\{r \in \{0,1\}^{t_x} \mid B(x, A_\text{D}(x, r)) = 1)\}$$
$$= \sum_{x \in X} \text{prob}(x) \cdot \frac{|\{r \in \{0,1\}^{t_x} \mid B(x, A_\text{D}(x, r)) = 1)\}|}{2^{t_x}}.$$

The notation is typically used in the following situation. A Monte Carlo algorithm A tries to compute a function $f : X \longrightarrow Y$, and $B(x,y) := B_f(x,y) := 1$ if $f(x) = y$, and $B(x,y) := B_f(x,y) := 0$ if $f(x) \neq y$. Then

$$\text{prob}(A(x) = f(x) : x \xleftarrow{p_X} X) := \text{prob}(B_f(x, A(x)) = 1 : x \xleftarrow{p_X} X)$$

is the probability that A succeeds if the input x is randomly chosen from X (according to p_X).

We write $x \xleftarrow{u} X$ if the distribution on X is the uniform one, and often we simply write $x \leftarrow X$ instead of $x \xleftarrow{p_X} X$.

In cryptography, we often consider probabilistic algorithms whose domain X is a joint probability space $X_1 X_2 \ldots X_r$ constructed by iteratively joining fibers $X_{j,x_1 \ldots x_{j-1}}$ to $X_1 \ldots X_{j-1}$ (Appendix B.1, p. 465). In this case, the notation is

$$\text{prob}(B(x_1, \ldots, x_r, A(x_1, \ldots, x_r)) = 1 :$$
$$x_1 \leftarrow X_1, x_2 \leftarrow X_{2,x_1}, x_3 \leftarrow X_{3,x_1 x_2}, \quad \ldots \quad , x_r \leftarrow X_{r,x_1 \ldots x_{r-1}}).$$

Now, the notation suggests that we mean the probability of the event $B(x_1, \ldots, x_r, A(x_1, \ldots, x_r)) = 1$ if first x_1 is randomly chosen, then x_2, then

---

[1] Maps with values in $\{0,1\}$ are called *Boolean predicates*.

x_3, then ....

A typical example is the discrete logarithm assumption in Section 6.2 (Definition 6.1).

The distribution $x_j \leftarrow X_{j,x_1...x_{j-1}}$ is the conditional distribution of $x_j \in X_{j,x_1...x_{j-1}}$, assuming $x_1, \ldots, x_{j-1}$. The probability can be computed as follows (we consider $r = 3$ and the case where A computes a function f and B is the predicate $A(x) = f(x)$):

$$\text{prob}(A(x_1, x_2, x_3) = f(x_1, x_2, x_3) : x_1 \leftarrow X_1, x_2 \leftarrow X_{2,x_1}, x_3 \leftarrow X_{3,x_1 x_2})$$

$$= \sum_{x_1, x_2, x_3} \text{prob}(x_1, x_2, x_3) \cdot \text{prob}(A(x_1, x_2, x_3) = f(x_1, x_2, x_3))$$

$$= \sum_{x_1 \in X_1} \text{prob}(x_1) \cdot \sum_{x_2 \in X_{1,x_1}} \text{prob}(x_2 \mid x_1)$$

$$\cdot \sum_{x_3 \in X_{1,x_1,x_2}} \text{prob}(x_3 \mid x_2, x_1) \cdot \text{prob}(A(x_1, x_2, x_3) = f(x_1, x_2, x_3)).$$

Here $\text{prob}(x_2 \mid x_1)$ (resp. $\text{prob}(x_3 \mid x_2, x_1)$) denotes the conditional probability of x_2 (resp. x_3) assuming x_1 (resp. x_1 and x_2); see Appendix B.1 (p. 462). The last probability, $\text{prob}(A(x_1, x_2, x_3) = f(x_1, x_2, x_3))$, is the probability that the coin tosses of A on input (x_1, x_2, x_3) yield the result $f(x_1, x_2, x_3)$.

In Section 5.1, we introduced the image p_Y of the distribution p_X under a probabilistic algorithm A from X to Y. We have

$$p_Y(y) = \text{prob}(A(x) = y : x \overset{p_X}{\leftarrow} X)$$

for each $y \in Y$.

For each $x \in X$, we have the distribution $p_{A(x)}$ on Y:

$$p_{A(x)}(y) = \text{prob}(A(x) = y).$$

We write $y \leftarrow A(x)$ instead of $y \overset{p_{A(x)}}{\leftarrow} Y$. This notation suggests that y is generated by the random variable $A(x)$.[2] With this notation, we have

$$\text{prob}(A(x) = f(x) : x \leftarrow X) = \text{prob}(f(x) = y : x \leftarrow X, y \leftarrow A(x)).$$

6.2 Discrete Exponential Function

The notion of one-way functions can be precisely defined using probabilistic algorithms. As a first example we consider the discrete exponential function. Let $I := \{(p, g) \mid p \text{ a prime number}, g \in \mathbb{Z}_p^* \text{ a primitive root}\}$. We call the family of discrete exponential functions

---

[2] We generally use this notation for random variables, see Appendix B.1, p. 466.

$$\text{Exp} := (\text{Exp}_{p,g} : \mathbb{Z}_{p-1} \longrightarrow \mathbb{Z}_p^*, \ x \longmapsto g^x)_{(p,g) \in I}$$

the *Exp family*. Since g is a primitive root, $\text{Exp}_{p,g}$ is an isomorphism between the additive group $\mathbb{Z}_{p-1}$ and the multiplicative group $\mathbb{Z}_p^*$. The family of inverse functions

$$\text{Log} := (\text{Log}_{p,g} : \mathbb{Z}_p^* \longrightarrow \mathbb{Z}_{p-1})_{(p,g) \in I}$$

is called the *Log family*.

The algorithm of modular exponentiation computes $\text{Exp}_{p,g}(x)$ efficiently (see Algorithm A.27). It is unknown whether an efficient algorithm for the computation of the discrete logarithm function exists. All known algorithms have exponential running time, and it is widely believed that, in general, $\text{Log}_{p,g}$ is not efficiently computable. We state this assumption on the one-way property of the Exp family by means of probabilistic algorithms.

Definition 6.1. Let $I_k := \{(p, g) \in I \mid |p| = k\}$, with $k \in \mathbb{N}$,[3] and let $Q(X) \in \mathbb{Z}[X]$ be a positive polynomial. Let $A(p, g, y)$ be a probabilistic polynomial algorithm. Then there exists a $k_0 \in \mathbb{N}$ such that

$$\text{prob}(A(p, g, y) = \text{Log}_{p,g}(y) : (p, g) \xleftarrow{u} I_k, y \xleftarrow{u} \mathbb{Z}_p^*) \leq \frac{1}{Q(k)}$$

for $k \geq k_0$.
This is called the *discrete logarithm assumption*.

Remarks:

1. The probabilistic algorithm A models an attacker who tries to compute the discrete logarithm or, equivalently, to invert the discrete exponential function. The discrete logarithm assumption essentially states that for a sufficiently large size k of the modulus p, the probability of A successfully computing $\text{Log}_{p,g}(y)$ is smaller than $1/Q(k)$. This means that Exp cannot be inverted by A for all but a negligible fraction of the inputs. Therefore, we call Exp a family of *one-way functions*. The term "negligible" is explained more precisely in a subsequent remark.
2. When we use the discrete exponential function in a cryptographic scheme, such as ElGamal's encryption scheme (see Section 3.4.1), selecting a function $\text{Exp}_{p,g}$ from the family means to choose a public key $i = (p, g)$ (actually, i may be only one part of the key).
3. The index set I is partitioned into disjoint subsets: $I = \bigcup_{k \in \mathbb{N}} I_k$. k may be considered as the security parameter of $i = (p, g) \in I_k$. The one-way property requires a sufficiently large security parameter. The security parameter is closely related to the binary length of i. Here, $k = |p|$ is half the length of i.

---

[3] As usual, $|p|$ denotes the binary length of p.

4. The probability in the discrete logarithm assumption is also taken over the random choice of a key i with a given security parameter k. Hence, the meaning of the probability statement is: choosing both the key $i = (p, g)$ with security parameter k and $y = g^x$ randomly, the probability that A correctly computes the logarithm x from y is small. The statement is not related to a particular key i. In practice, however, a public key is chosen and then fixed for a long time, and it is known to the adversary. Thus, we are interested in the conditional probability of success, assuming a fixed public key i. Even if the security parameter k is very large, there may be keys (p, g) such that A correctly computes $\mathrm{Log}_{p,g}(y)$ with a significant chance. However, as we will see below, the number of such keys (p, g) is negligibly small compared to all keys with security parameter k. Choosing (p, g) at random (and uniformly) from I_k, the probability of obtaining one for which A has a significant chance of success is negligibly small (see Proposition 6.3 for a precise statement). Indeed, if $p - 1$ has only small prime factors, an efficient algorithm developed by Pohlig and Hellman computes the discrete logarithm function (see [PohHel78]).

Remark. In this book, we often consider families $\varepsilon = (\varepsilon_k)_{k \in \mathbb{N}}$ of quantities $\varepsilon_k \in \mathbb{R}$, as the probabilities in the discrete logarithm assumption. We call them *negligible* or *negligibly small* if, for every positive polynomial $Q \in \mathbb{Z}[X]$, there is a $k_0 \in \mathbb{N}$ such that $|\varepsilon_k| \leq {}^1\!/_{Q(k)}$ for $k \geq k_0$. "Negligible" means that the absolute value is asymptotically smaller than any polynomial bound.

Remark. In order to simplify, definitions and results are often stated asymptotically (as the discrete logarithm assumption or the notion of negligible quantities). Polynomial running times or negligible probabilities are not specified more precisely, even if it were possible. A typical situation is as follows. A cryptographic scheme is based on a one-way function f (e.g. the Exp family). Let g be a function that describes a property of the cryptographic scheme (e.g. g predicts the next bit of the discrete exponential pseudorandom bit generator; see Chapter 8). It is desirable that this property g cannot be efficiently computed by an adversary. Sometimes, this can be proven. Typically a proof runs by a contradiction. We assume that a probabilistic polynomial algorithm A_1 which successfully computes g with probability ε_1 is given. Then, a probabilistic polynomial algorithm A_2 is constructed which calls A_1 as a subroutine and inverts the underlying one-way function f, with probability ε_2. Such an algorithm is called a *polynomial-time reduction* of f to g. If ε_2 is non-negligible, we get a contradiction to the one-way assumption (e.g. the discrete logarithm assumption).

In our example, a typical statement would be as follows. If discrete logarithms cannot be computed in polynomial time with non-negligible probability (i.e., if the discrete logarithm assumption is true), then a polynomial-time adversary cannot predict, with non-negligible probability, the next bit of the discrete exponential pseudorandom bit generator.

Actually, in many cases the statement could be made more precise, by performing a detailed analysis of the reduction algorithm A_2. The running time of A_2 can be described as an explicit function of $\varepsilon_1, \varepsilon_2$ and the running time of A_1 (see, e.g., the results in Chapter 7).

As in the Exp example, we often meet families of functions indexed on a set of keys which may be partitioned according to a security parameter. Therefore we propose the notion of indexes, whose binary lengths are measured by a security parameter as specified more precisely by the following definition.

Definition 6.2. Let $I = \bigcup_{k \in \mathbb{N}} I_k$ be an infinite index set which is partitioned into finite disjoint subsets I_k. Assume that the indexes are binarily encoded. As always, we denote by $|i|$ the binary length of i.
I is called a *key set with security parameter k* or an *index set with security parameter k*, if:

1. The security parameter k of $i \in I$ can be derived from i by a deterministic polynomial algorithm.
2. There is a constant $m \in \mathbb{N}$ such that

$$k^{1/m} \leq |i| \leq k^m \text{ for } i \in I_k.$$

We usually write $I = (I_k)_{k \in \mathbb{N}}$ instead of $I = \bigcup_{k \in \mathbb{N}} I_k$.

Remarks:

1. The second condition means that the security parameter k is a measure for the binary length $|i|$ of the elements $i \in I_k$. In particular, statements such as:
   (1) "There is a polynomial P with $\ldots \leq P(|i|)$", or
   (2) "For every positive polynomial Q, there is a $k_0 \in \mathbb{N}$ such that $\ldots \leq 1/Q(|i|)$ for $|i| \geq k_0$",
   are equivalent to the corresponding statements in which $|i|$ is replaced by the security parameter k. In almost all of our examples, we have $k \leq |i| \leq 3k$ for $i \in I_k$.
2. The index set I of the Exp family is a key set with security parameter. As with all indexes occurring in this book, the indexes of the Exp family consist of numbers in $\mathbb{N}$ or residues in some residue class ring $\mathbb{Z}_n$. Unless otherwise stated, we consider them as binarily encoded in the natural way (see Appendix A): the binary encoding of $x \in \mathbb{N}$ is its standard encoding as an unsigned number, and the encoding of a residue class $[a] \in \mathbb{Z}_n$ is the encoding of its representative x with $0 \leq x \leq n - 1$.
3. If $I = \bigcup_{k \in \mathbb{N}} I_k$ satisfies only the second condition, then we can easily modify it and turn it into a key set with a security parameter, which also satisfies the first condition. Namely, let $\tilde{I}_k := \{(i, k) \mid i \in I_k\}$ and replace I by $\tilde{I} := \bigcup_{k \in \mathbb{N}} \tilde{I}_k$.

In the discrete logarithm assumption, we do not consider a single fixed key i: the probability is also taken over the random choice of the key. In the following proposition, we relate this average probability to the conditional probabilities, assuming a fixed key.

Proposition 6.3. *Let $I = (I_k)_{k \in \mathbb{N}}$ be a key set with security parameter k. Let $f = (f_i : X_i \longrightarrow Y_i)_{i \in I}$ be a family of functions and A be a probabilistic polynomial algorithm with inputs $i \in I$ and $x \in X_i$ and output in Y_i. Assume that probability distributions are given on I_k and X_i for all k, i (e.g. the uniform distributions). Then the following statements are equivalent:*

1. *For every positive polynomial P, there is a $k_0 \in \mathbb{N}$ such that for all $k \geq k_0$*

$$\mathrm{prob}(A(i,x) = f_i(x) : i \leftarrow I_k, x \leftarrow X_i) \leq \frac{1}{P(k)}.$$

2. *For all positive polynomials Q and R, there is a $k_0 \in \mathbb{N}$ such that for all $k \geq k_0$*

$$\mathrm{prob}\left(\left\{i \in I_k \,\middle|\, \mathrm{prob}(A(i,x) = f_i(x) : x \leftarrow X_i) > \frac{1}{Q(k)}\right\}\right) \leq \frac{1}{R(k)}.$$

Proof. Let

$$p_i := \mathrm{prob}(A(i,x) = f_i(x) : x \leftarrow X_i)$$

be the conditional probability of success of A assuming a fixed i.

We first prove that statement 2 implies statement 1. Let P be a positive polynomial. By statement 2, there is some $k_0 \in \mathbb{N}$ such that for all $k \geq k_0$

$$\mathrm{prob}\left(\left\{i \in I_k \,\middle|\, p_i > \frac{1}{2P(k)}\right\}\right) \leq \frac{1}{2P(k)}.$$

Hence

$$\mathrm{prob}(A(i,x) = f_i(x) : i \leftarrow I_k, x \leftarrow X_i)$$
$$= \sum_{i \in I_k} \mathrm{prob}(i) \cdot p_i$$
$$= \sum_{p_i \leq 1/(2P(k))} \mathrm{prob}(i) \cdot p_i + \sum_{p_i > 1/(2P(k))} \mathrm{prob}(i) \cdot p_i$$
$$\leq \sum_{p_i \leq 1/(2P(k))} \mathrm{prob}(i) \cdot \frac{1}{2P(k)} + \sum_{p_i > 1/(2P(k))} \mathrm{prob}(i) \cdot 1$$
$$= \mathrm{prob}\left(\left\{i \in I_k \,\middle|\, p_i \leq \frac{1}{2P(k)}\right\}\right) \cdot \frac{1}{2P(k)}$$
$$\qquad + \mathrm{prob}\left(\left\{i \in I_k \,\middle|\, p_i > \frac{1}{2P(k)}\right\}\right)$$
$$\leq \frac{1}{2P(k)} + \frac{1}{2P(k)} = \frac{1}{P(k)},$$

for $k \geq k_0$.

Conversely, assume that statement 1 holds. Let Q and R be positive polynomials. Then there is a $k_0 \in \mathbb{N}$ such that for all $k \geq k_0$

$$\frac{1}{Q(k)R(k)} \geq \mathrm{prob}(A(i,x) = f_i(x) : i \leftarrow I_k, x \leftarrow X_i)$$

$$= \sum_{i \in I_k} \mathrm{prob}(i) \cdot p_i$$

$$\geq \sum_{p_i > 1/Q(k)} \mathrm{prob}(i) \cdot p_i$$

$$> \frac{1}{Q(k)} \cdot \mathrm{prob}\left(\left\{i \in I_k \,\middle|\, p_i > \frac{1}{Q(k)}\right\}\right).$$

This inequality implies statement 2.   □

Remark. A nice feature of the discrete logarithm problem is that it is *random self-reducible*. This means that solving the problem for arbitrary inputs can be reduced to solving the problem for randomly chosen inputs. More precisely, let $(p,g) \in I_k := \{(p,g) \mid p \text{ a prime}, |p| = k, g \in \mathbb{Z}_p^* \text{ a primitive root}\}$. Assume that there is a probabilistic polynomial algorithm A such that

$$\mathrm{prob}(A(p,g,y) = \mathrm{Log}_{p,g}(y) : y \xleftarrow{u} \mathbb{Z}_p^*) > \frac{1}{Q(k)} \qquad (6.1)$$

for some positive polynomial Q; i.e., $A(p,g,y)$ correctly computes the discrete logarithm with a non-negligible probability if the input y is randomly selected. Since y is chosen uniformly, we may rephrase this statement: $A(p,g,y)$ correctly computes the discrete logarithm for a polynomial fraction of inputs $y \in \mathbb{Z}_p^*$.

Then, however, there is also a probabilistic polynomial algorithm $\tilde{A}$ which correctly computes the discrete logarithm for every input $y \in \mathbb{Z}_p^*$, with an overwhelmingly high probability. Namely, given $y \in \mathbb{Z}_p^*$, we apply a slight modification A_1 of A. On input (p,g,y), A_1 randomly selects $r \xleftarrow{u} \mathbb{Z}_{p-1}$ and returns

$$A_1(p,g,y) := (A(p,g,yg^r) - r) \bmod (p-1).$$

Then $\mathrm{prob}(A_1(p,g,y) = \mathrm{Log}_{p,g}(y)) > 1/Q(k)$ for every $y \in \mathbb{Z}_p^*$. Now we can apply Proposition 5.7. For every positive polynomial P, we obtain – by repeating the computation of $A_1(p,g,y)$ a polynomial number of times and by checking each time whether the result is correct – a probabilistic polynomial algorithm $\tilde{A}$, with

$$\mathrm{prob}(\tilde{A}(p,g,y) = \mathrm{Log}_{p,g}(y)) > 1 - 2^{-P(k)},$$

for every $y \in \mathbb{Z}_p^*$. The existence of a random self-reduction enhances the credibility of the discrete logarithm assumption. Namely, assume that the discrete logarithm assumption is not true. Then by Proposition 6.3 there is a probabilistic polynomial algorithm A such that for infinitely many k, the inequality (6.1) holds for a polynomial fraction of keys (p, g); i.e.,

$$\frac{|\{(p, g) \in I_k \mid \text{ inequality (6.1) holds}\}|}{|I_k|} > \frac{1}{R(k)}$$

(with R a positive polynomial). For these keys $\tilde{A}$ computes the discrete logarithm for every $y \in \mathbb{Z}_p^*$ with an overwhelmingly high probability, and the probability of obtaining such a key is $> 1/R(k)$ if the keys are selected uniformly at random.

6.3 Uniform Sampling Algorithms

In the discrete logarithm assumption (Definition 6.1), the probabilities are taken with respect to the uniform distributions on I_k and $\mathbb{Z}_p^*$. Stating the assumption in this way, we tacitly assumed that it is possible to sample uniformly over I_k (during key generation) and $\mathbb{Z}_p^*$, by using efficient algorithms. In practice it might be difficult to construct a probabilistic polynomial sampling algorithm that selects the elements exactly according to the uniform distribution. However, as in the present case of discrete logarithms (see Proposition 6.6), we are often able to find practical sampling algorithms which sample in a "virtually uniform" way. Then the assumptions stated for the uniform distribution, such as the discrete logarithm assumption, apply. This is shown by the following considerations.

Definition 6.4. Let $J = (J_k)_{k \in \mathbb{N}}$ be an index set with security parameter k (see Definition 6.2). Let $X = (X_j)_{j \in J}$ be a family of finite sets:

1. A probabilistic polynomial algorithm S_X with input $j \in J$ is called a *sampling algorithm* for X if $S_X(j)$ outputs an element in X_j with a probability $\geq 1 - \varepsilon_k$ for $j \in I_k$, where $\varepsilon = (\varepsilon_k)_{k \in \mathbb{N}}$ is negligible; i.e., given a positive polynomial Q, there is a k_0 such that $\varepsilon_k \leq 1/Q(k)$ for $k \geq k_0$.
2. A sampling algorithm S_X for X is called *(virtually) uniform* if the distributions of $S_X(j)$ and the uniform distributions on X_j are polynomially close (see Definition B.30). This means that the statistical distance is negligibly small; i.e., given a positive polynomial Q, there is a k_0 such that the statistical distance (see Definition B.27) between the distribution of $S_X(j)$ and the uniform distribution on X_j is $\leq 1/Q(k)$, for $k \geq k_0$ and $j \in J_k$.

Remark. If S_X is a virtually uniform sampling algorithm for $X = (X_j)_{j \in J}$, we usually do not need to distinguish between the virtually uniform distribution of $S_X(j)$ and the truly uniform distribution when we compute a probability involving $x \leftarrow S_X(j)$. Namely, consider probabilities

$$\text{prob}(B_j(x, y) = 1 : x \leftarrow S_X(j), y \leftarrow Y_{j,x}),$$

where $(Y_{j,x})_{x \in X_j}$ is a family of probability spaces and B_j is a Boolean predicate. Then for every positive polynomial P, there is a $k_0 \in \mathbb{N}$ such that

$$| \text{prob}(B_j(x, y) = 1 : x \leftarrow S_X(j), y \leftarrow Y_{j,x})$$
$$- \text{prob}(B_j(x, y) = 1 : x \xleftarrow{u} X_j, y \leftarrow Y_{j,x}) | < \frac{1}{P(k)},$$

for $k \geq k_0$ and $j \in J_k$ (by Lemmas B.29 and B.32), and we see that the difference between the probabilities is negligibly small.

Therefore, we usually do not distinguish between perfectly and virtually uniform sampling algorithms and simply talk of *uniform sampling algorithms*.

We study an example. Suppose we want to construct a uniform sampling algorithm S for $(\mathbb{Z}_n)_{n \in \mathbb{N}}$. We have $\mathbb{Z}_n \subseteq \{0, 1\}^{|n|}$, and could proceed as follows. We toss the coin $|n|$ times and obtain a binary number $x := b_{|n|-1} \ldots b_1 b_0$, with $0 \leq x < 2^{|n|}$. We can easily verify whether $x \in \mathbb{Z}_n$, by checking $x < n$. If the answer is affirmative, we return $S(n) := x$. Otherwise, we repeat the coin tosses. Since S is required to have a polynomial running time, we have to stop after at most $P(|n|)$ iterations (P a polynomial). Thus, $S(n)$ does not always succeed to return an element in $\mathbb{Z}_n$. The probability of a failure is, however, negligibly small.[4]

Our construction, which derives a uniform sampling algorithm for a subset, works, if the membership in this subset can be efficiently tested. It can be applied in many situations. Therefore, we state the following lemma.

Lemma 6.5. *Let $J = (J_k)_{k \in \mathbb{N}}$ be an index set with security parameter k. Let $X = (X_j)_{j \in J}$ and $Y = (Y_j)_{j \in J}$ be families of finite sets with $Y_j \subseteq X_j$ for all $j \in J$. Assume that there is a polynomial Q such that $|Y_j| \cdot Q(k) \geq |X_j|$ for $j \in J_k$.*

Let S_X be a uniform sampling algorithm for $(X_j)_{j \in J}$ which on input $j \in J$ outputs $x \in X_j$[5] and some additional information $aux(x)$ about x. Let $A(j, x, aux(x))$ be a Monte Carlo algorithm which decides the membership in Y_j; i.e., on input $j \in J, x \in X_j$ and $aux(x)$, it yields 1 if $x \in Y_j$, and 0 if $x \notin Y_j$. Assume that the error probability of A is negligible; i.e., for every positive polynomial P, there is a k_0 such that the error probability is $\leq 1/P(k)$ for $k \geq k_0$.
Then there exists a uniform sampling algorithm S_Y for $(Y_j)_{j \in J}$.

---

[4] We could construct a Las Vegas algorithm in this way, which always succeeds. See Section 5.2.

[5] Here and in what follows we use this formulation, though the sampling algorithm may sometimes yield elements outside of X_j. However, as stated in Definition 6.4, this happens only with a negligible probability.

Proof. Let S_Y be the probabilistic polynomial algorithm which on input $j \in J$ repeatedly computes $x := S_X(j)$ until $A(j, x, aux(x)) = 1$. To get a polynomial algorithm, we stop S_Y after at most $\ln(2)kQ(k)$ iterations.

We now show that S_Y has the desired properties. We first assume that $S_X(j) \in X_j$ with certainty and that A has an error probability of 0. By Lemma B.5, we may also assume that S_Y has found an element in Y_j (before being stopped), because this event has a probability $\geq 1 - (1 - 1/Q(k))^{kQ(k)} > 1 - 2^{-k}$, which is exponentially close to 1 (see the proof of Proposition 5.7 for an analogous estimate).

By construction, we have for $V \subset Y_j$ that

$$\mathrm{prob}(S_Y(j) \in V) = \mathrm{prob}(S_X(j) \in V \mid S_X(j) \in Y_j) = \frac{\mathrm{prob}(S_X(j) \in V)}{\mathrm{prob}(S_X(j) \in Y_j)}.$$

Thus, we have for all subsets $V \subset Y_j$ that

$$\mathrm{prob}(S_Y(j) \in V) = \frac{\frac{|V|}{|X_j|} + \varepsilon_j(V)}{\frac{|Y_j|}{|X_j|} + \varepsilon_j(Y_j)},$$

with a negligibly small function ε_j. Then $\mathrm{prob}(S_Y(j) \in V) - \frac{|V|}{|Y_j|}$ is also negligibly small (you can see this immediately by Taylor's formula for the real function $(x, y) \mapsto x/y$). Hence, S_Y is a uniform sampling algorithm for $(Y_j)_{j \in J}$.

The general case, where $S_X(j) \notin X_j$ with a negligible probability and A has a negligible error probability, follows by applying Lemma B.5. $\qquad\square$

Example. Let $Y_n := \mathbb{Z}_n$ or $Y_n := \mathbb{Z}_n^*$. Then Y_n is a subset of $X_n := \{0, 1\}^{|n|}, n \in \mathbb{N}$. Obviously, $\{0, 1\}^{|n|}$ can be sampled uniformly by $|n|$ coin tosses. The membership of x in $\mathbb{Z}_n$ is checked by $x < n$, and the Euclidean algorithm tells us whether x is a unit. Thus, there are (probabilistic polynomial) uniform sampling algorithms for $(\mathbb{Z}_n)_{n \in \mathbb{N}}$ and $(\mathbb{Z}_n^*)_{n \in \mathbb{N}}$, which on input $n \in \mathbb{N}$ output an element $x \in \mathbb{Z}_n$ (or $x \in \mathbb{Z}_n^*$).

To apply Lemma 6.5 in this example, let $J := \mathbb{N}$ and $J_k := \{n \in \mathbb{N} \mid |n| = k\}$.

Example. Let Primes_k be the set of primes p whose binary length $|p|$ is k; i.e., $\mathrm{Primes}_k := \{p \in \mathrm{Primes} \mid |p| = k\} \subseteq \{0, 1\}^k$. The number of primes $< 2^k$ is $\approx 2^k/k\ln(2)$ (Theorem A.82). By iterating a probabilistic primality test (e.g. Miller-Rabin's test, see Appendix A.10), we can, with a probability $> 1 - 2^{-k}$, correctly test the primality of an element x in $\{0, 1\}^k$. Thus, there is a (probabilistic polynomial) uniform sampling algorithm S which on input 1^k yields a prime $p \in \mathrm{Primes}_k$.

To apply Lemma 6.5 in this example, let $J_k := \{1^k\}$ and $J := \mathbb{N} = \bigcup_{k \in \mathbb{N}} J_k$, i.e., the index set is the set of natural numbers. However, an index $k \in \mathbb{N}$ is not encoded in the standard way; it is encoded as the constant bit string 1^k (see the subsequent remark on 1^k).

Remark. 1^k denotes the constant bit string $11 \ldots 1$ of length k. Using it as input for a polynomial algorithm means that the number of steps in the algorithm is bounded by a polynomial in k. If we used k (encoded in the standard way) instead of 1^k as input, the bound would be a polynomial in $\log_2(k)$.

We return to the example of discrete exponentiation.

Proposition 6.6. *Let* $I := \{(p, g) \mid p$ *prime number*, $g \in \mathbb{Z}_p^*$ *primitive root*$\}$ *and* $I_k := \{(p, g) \in I \mid |p| = k\}$. *There is a probabilistic polynomial uniform sampling algorithm for* $I = (I_k)_{k \in \mathbb{N}}$, *which on input* 1^k *yields an element* $(p, g) \in I_k$.

Proof. We want to apply Lemma 6.5 to the index set $J := \mathbb{N} = \bigcup_{k \in \mathbb{N}} J_k, J_k := \{1^k\}$, and the families of sets $X_k := \text{Primes}_k \times \{0, 1\}^k, Y_k := I_k \subseteq X_k (k \in \mathbb{N})$. The number of primitive roots in $\mathbb{Z}_p^*$ is $\varphi(p-1)$, where φ is the Eulerian totient function (see Theorem A.37). For $x \in \mathbb{N}$, we have

$$\varphi(x) = x \prod_{i=1}^{r} \left(1 - \frac{1}{p_i}\right) = x \prod_{i=1}^{r} \frac{p_i - 1}{p_i},$$

where $p_1, \ldots, p_r$ are the primes dividing x (see Corollary A.31). Since $\prod_{i=1}^{r} \frac{p_i-1}{p_i} \geq \prod_{i=2}^{r+1} \frac{i-1}{i} = \frac{1}{r+1}$ and $r + 1 \leq |x|$, we immediately see that $\varphi(x) \cdot |x| \geq x.$[6] In particular, we have $\varphi(p-1) \cdot k \geq p - 1 \geq 2^{k-1}$ for $p \in \text{Primes}_k$, and hence $2k \cdot |Y_k| \geq |X_k|$.

Given a prime $p \in \text{Primes}_k$ and all prime numbers $q_1, \ldots, q_r$ dividing $p - 1$, we can efficiently verify whether $g \in \{0, 1\}^k$ is in $\mathbb{Z}_p^*$ and whether it is a primitive root. Namely, we first test $g < p$ and then apply the criterion for primitive roots (see Algorithm A.40), i.e., we check whether $g^{(p-1)/q} \neq 1$ for all prime divisors q of $p - 1$.

We may apply Lemma 6.5 if there is a probabilistic polynomial uniform sampling algorithm for $(\text{Primes}_k)_{k \in \mathbb{N}}$ which not only outputs a prime p, but also the prime factors of $p - 1$. Bach's algorithm (see [Bach88]) yields such an algorithm: it generates uniformly distributed k-bit integers n, along with their factorization. We may repeatedly generate such numbers n until $n + 1$ is a prime. $\qquad\square$

6.4 Modular Powers

Let $I := \{(n, e) \mid n = pq, p \neq q \text{ primes}, 1 < e < \varphi(n), e \text{ prime to } \varphi(n)\}$. The family

$$\text{RSA} := (\text{RSA}_{n,e} : \mathbb{Z}_n^* \longrightarrow \mathbb{Z}_n^*, x \longmapsto x^e)_{(n,e) \in I}$$

---

[6] Actually, $\varphi(x)$ is much closer to x. It can be shown that $\varphi(x) > \frac{x}{6 \log(|x|)}$ (see Appendix A.2).

is called the *RSA family*.

Consider an $(n, e) \in I$, and let $d \in \mathbb{Z}_{\varphi(n)}^*$ be the inverse of $e \bmod \varphi(n)$. Then we have $x^{ed} = x^{ed \bmod \varphi(n)} = x$ for $x \in \mathbb{Z}_n^*$, since $x^{\varphi(n)} = 1$ (see Proposition A.26). This shows that $\mathrm{RSA}_{n,e}$ is bijective and that the inverse function is also an RSA function, namely $\mathrm{RSA}_{n,d} : \mathbb{Z}_n^* \longrightarrow \mathbb{Z}_n^*, \ x \longmapsto x^d$.

$\mathrm{RSA}_{n,e}$ can be computed by modular exponentiation, an efficient algorithm. d can be easily computed by the extended Euclidean algorithm A.6, if $\varphi(n) = (p-1)(q-1)$ is known. No algorithm to compute $\mathrm{RSA}_{n,e}^{-1}$ in polynomial time is known, if p, q and d are kept secret. We call d (or p, q) the *trapdoor information* for the RSA function.

All known attacks to break RSA, if implemented by an efficient algorithm, would deliver an efficient algorithm for factoring n. All known factoring algorithms have exponential running time. Therefore, it is widely believed that *RSA* cannot be efficiently inverted. The following assumption makes this more precise.

Definition 6.7. Let $I_k := \{(n, e) \in I \mid n = pq, |p| = |q| = k\}$, with $k \in \mathbb{N}$, and let $Q(X) \in \mathbb{Z}[X]$ be a positive polynomial. Let $A(n, e, y)$ be a probabilistic polynomial algorithm. Then there exists a $k_0 \in \mathbb{N}$ such that

$$\mathrm{prob}(A(n, e, y) = \mathrm{RSA}_{n,d}(y) : (n, e) \xleftarrow{u} I_k, y \xleftarrow{u} \mathbb{Z}_n^*) \leq \frac{1}{Q(k)}$$

for $k \geq k_0$.
This is called the *RSA assumption*.

Remarks:

1. The assumption states the one-way property of the RSA family. The algorithm A models an adversary, who tries to compute $x = \mathrm{RSA}_{n,d}(y)$ from $y = \mathrm{RSA}_{n,e}(x) = x^e$ (in $\mathbb{Z}_n^*$) without knowing the trapdoor information d. By using Proposition 6.3, we may interpret the RSA assumption in an analogous way to the discrete logarithm assumption (Definition 6.1). The fraction of keys (n, e) in I_k, for which the adversary A has a significant chance to succeed, is negligibly small if the security parameter k is sufficiently large.

2. $\mathrm{RSA}_{n,e}$ is bijective, and its range and domain coincide. Therefore, we also speak of a family of *one-way permutations* (or a family of *trapdoor permutations*).

3. Here and in what follows, we restrict the key set I of the RSA family and consider only those functions $\mathrm{RSA}_{n,e}$, where $n = pq$ is the product of two primes of the same binary length. Instead, we could also define a stronger version of the assumption, where I_k is the set of pairs (n, e), with $|n| = k$ (the primes may have different length). However, our statement is closer to normal practice. To generate keys with a given security parameter k, usually two primes of length k are chosen and multiplied.

4. The RSA problem – computing x from x^e – is random self-reducible (see the analogous remark on the discrete logarithm problem on p. 222).

Stating the RSA assumption as above, we assume that the set I of keys can be uniformly sampled by an efficient algorithm.

Proposition 6.8. *There is a probabilistic polynomial uniform sampling algorithm for $I = (I_k)_{k \in \mathbb{N}}$, which on input 1^k yields a key $(n, e) \in I_k$ along with the trapdoor information (p, q, d).*

Proof. Above (see the examples after Lemma 6.5), we saw that Primes_k can be uniformly sampled by a probabilistic polynomial algorithm. Thus, there is a probabilistic polynomial uniform sampling algorithm for

$$(X_k := \{n = pq \mid p, q \text{ distinct primes}, |p| = |q| = k\} \times \{0, 1\}^{2k})_{k \in \mathbb{N}}.$$

In the proof of Proposition 6.6, we observed that $|x| \cdot \varphi(x) \geq x$, and we immediately conclude that

$$|\mathbb{Z}^*_{\varphi(n)}| = \varphi(\varphi(n)) \geq \frac{\varphi(n)}{|\varphi(n)|} \geq \frac{n}{|n| \cdot |\varphi(n)|} \geq \frac{n}{4k^2} \geq \frac{2^{2k-2}}{4k^2} = \frac{2^{2k}}{16k^2}.$$

Thus, we can apply Lemma 6.5 to $Y_k := I_k \subseteq X_k$ and obtain the desired sampling algorithm. It yields (p, q, e). The inverse d of e in $\mathbb{Z}^*_{\varphi(n)}$ can be computed using the extended Euclidean algorithm (Algorithm A.6).   □

Remarks:

1. The uniform sampling algorithm for $(I_k)_{k \in \mathbb{N}}$ which we derived in the proof of Proposition 6.8 is constructed by the method given in Lemma 6.5. Thus, it chooses triples (p, q, e) uniformly and then tests whether $e < \varphi(n) = (p - 1)(q - 1)$ and whether e is prime to $\varphi(n)$. If this test fails, a new triple (p, q, e) is selected. It would be more natural and more efficient to first choose a pair (p, q) uniformly, and then, with $n = pq$ fixed, to choose an exponent e uniformly from $\mathbb{Z}^*_{\varphi(n)}$. Then, however, the statistical distance between the distribution of the elements (n, e) and the uniform distribution is not negligible. The sampling algorithm is not uniform. Note that even for fixed k, there is a rather large variance of the cardinalities $|\mathbb{Z}^*_{\varphi(n)}|$. Nevertheless, this more natural sampling algorithm is an admissible key generator for the RSA family; i.e., the one-way condition is preserved if the keys are sampled by it (see Definition 6.13, of admissible key generators, and Exercise 1).
   An analogous remark applies to the sampling algorithm, given in the proof of Proposition 6.6.

2. We can generate the primes p and q by uniformly selecting numbers of length k and testing their primality by using a probabilistic prime number test (see the examples after Lemma 6.5). There are also other very efficient algorithms for the generation of uniformly distributed primes (see, e.g., [Maurer95]).

6.5 Modular Squaring

Let $I := \{n \mid n = pq,\ p, q \text{ distinct prime numbers}, |p| = |q|\}$. The family

$$\mathrm{Sq} := (\mathrm{Sq}_n : \mathbb{Z}_n^* \longrightarrow \mathbb{Z}_n^*,\ x \longmapsto x^2)_{n \in I}$$

is called the *Square family*.[7] Sq_n is neither injective nor surjective. If $\mathrm{Sq}_n^{-1}(x) \neq \emptyset$, then $|\mathrm{Sq}_n^{-1}(x)| = 4$ (see Proposition A.68).

Modular squaring can be done efficiently. Square roots modulo p are computable by a probabilistic polynomial algorithm if p is a prime number (see Algorithm A.67). Applying the Chinese Remainder Theorem (Theorem A.30), it is then easy to derive an efficient algorithm that computes square roots in $\mathbb{Z}_n^*$ if $n = pq$ (p and q are distinct prime numbers) and if the factorization of n is known.

Conversely, given an efficient algorithm for computing square roots in $\mathbb{Z}_n^*$, an efficient algorithm for the factorization of n can be derived (see Proposition A.70).

All known factoring algorithms have exponential running time. Therefore, it is widely believed that the factors of n (or, equivalently, square roots modulo n) cannot be computed efficiently. We make this statement more precise by the following assumption.

Definition 6.9. Let $I_k := \{n \in I \mid n = pq, |p| = |q| = k\}$, with $k \in \mathbb{N}$, and let $Q(X) \in \mathbb{Z}[X]$ be a positive polynomial. Let $A(n)$ be a probabilistic polynomial algorithm. Then there exists a $k_0 \in \mathbb{N}$ such that

$$\mathrm{prob}(A(n) = p : n \xleftarrow{u} I_k) \leq \frac{1}{Q(k)}$$

for $k \geq k_0$.
This is called the *factoring assumption*.

Stating the factoring assumption, we again assume that the set I of keys may be uniformly sampled by an efficient algorithm.

Proposition 6.10. *There is a probabilistic polynomial uniform sampling algorithm for $I = (I_k)_{k \in \mathbb{N}}$, which on input 1^k yields a number $n \in I_k$, along with its factors p and q.*

Proof. The algorithm chooses integers p and q with $|p| = |q| = k$ at random, and applies a probabilistic primality test (see Appendix A.10) to check whether p and q are prime. By repeating the probabilistic primality test sufficiently often, we can, with a probability $> 1 - 2^{-k}$, correctly test the primality

---

[7] As above in the RSA family, we only consider moduli n which are the product of two primes of equal binary length; see the remarks after the RSA assumption (Definition 6.7).

of an element x in $\{0,1\}^k$. This sampling algorithm is uniform (Lemma 6.5).

□

Restricting the range and the domain to the set QR_n of squares modulo n (called the quadratic residues modulo n, see Definition A.54), the modular squaring function can be made bijective in many cases. Of course, each $x \in \mathrm{QR}_n$ has a square root. If p and q are distinct primes with $p, q \equiv 3 \bmod 4$ and $n := pq$, then exactly one of the four square roots of $x \in \mathrm{QR}_n$ is an element in QR_n (see Proposition A.72). Taking as key set

$$I := \{n \mid n = pq,\ p, q \text{ distinct prime numbers}, |p| = |q|, p, q \equiv 3 \bmod 4\},$$

we get a family

$$\mathrm{Square} := (\mathrm{Square}_n : \mathrm{QR}_n \longrightarrow \mathrm{QR}_n,\ x \longmapsto x^2)_{n \in I}$$

of bijective functions, also called the *Square family*. Since the range and domain are of the same set, we speak of a family of permutations. The family of inverse maps is denoted by

$$\mathrm{Sqrt} := (\mathrm{Sqrt}_n : \mathrm{QR}_n \longrightarrow \mathrm{QR}_n)_{n \in I}.$$

Sqrt_n maps x to the square root of x which is an element of QR_n.

The same considerations as those on $\mathrm{Sq}_n : \mathbb{Z}_n^* \longrightarrow \mathbb{Z}_n^*$ above show that Square_n is efficiently computable, and that computing Sqrt_n is equivalent to factoring n. Square is a family of trapdoor permutations with trapdoor information p and q.

6.6 Quadratic Residuosity Property

If p is a prime and $x \in \mathbb{Z}_p^*$, the Legendre symbol $\left(\frac{x}{p}\right)$ tells us whether x is a quadratic residue modulo p: $\left(\frac{x}{p}\right) = 1$ if $x \in \mathrm{QR}_p$, and $\left(\frac{x}{p}\right) = -1$ if $x \notin \mathrm{QR}_p$ (Definition A.57). The Legendre symbol can be easily computed using Euler's criterion (Proposition A.58): $\left(\frac{x}{p}\right) = x^{(p-1)/2} \bmod p$.

Now, let p and q be distinct prime numbers and $n := pq$. The Jacobi symbol $\left(\frac{x}{n}\right)$ is defined as $\left(\frac{x}{n}\right) := \left(\frac{x}{p}\right) \cdot \left(\frac{x}{q}\right)$ (Definition A.61). It is efficiently computable for every element $x \in \mathbb{Z}_n^*$ – without knowing the prime factors p and q of n (see Algorithm A.65). The Jacobi symbol cannot be used to decide whether $x \in \mathrm{QR}_n$. If $\left(\frac{x}{n}\right) = -1$, then x is not in QR_n. However, if $\left(\frac{x}{n}\right) = 1$, both cases $x \in \mathrm{QR}_n$ and $x \notin \mathrm{QR}_n$ are possible. $x \in \mathbb{Z}_n^*$ is a quadratic residue if and only if both $x \bmod p \in \mathbb{Z}_p^*$ and $x \bmod q \in \mathbb{Z}_q^*$ are quadratic residues, which is equivalent to $\left(\frac{x}{p}\right) = \left(\frac{x}{q}\right) = 1$.

Let $I := \{n \mid n = pq,\ p, q \text{ distinct prime numbers}, |p| = |q|\}$ and let

$$J_n^{+1} := \left\{ x \in \mathbb{Z}_n^* \,\middle|\, \left(\frac{x}{n}\right) = +1 \right\}$$

be the elements with Jacobi symbol $+1$. QR_n is a proper subset of J_n^{+1}. Consider the functions

$$PQR_n : J_n^{+1} \longrightarrow \{0,1\}, PQR_n(x) := \begin{cases} 1 \text{ if } x \in QR_n, \\ 0 \text{ otherwise.} \end{cases}$$

The family $PQR := (PQR_n)_{n \in I}$ is called the *quadratic residuosity family*.

It is believed that there is no efficient algorithm which, without knowing the factors of n, is able to decide whether $x \in J_n^{+1}$ is a quadratic residue. We make this precise in the following assumption.

Definition 6.11. Let $I_k := \{n \in I \mid n = pq, |p| = |q| = k\}$, with $k \in \mathbb{N}$, and let $Q(X) \in \mathbb{Z}[X]$ be a positive polynomial. Let $A(n, x)$ be a probabilistic polynomial algorithm. Then there exists a $k_0 \in \mathbb{N}$ such that

$$\text{prob}(A(n,x) = PQR_n(x) : n \xleftarrow{u} I_k, x \xleftarrow{u} J_n^{+1}) \leq \frac{1}{2} + \frac{1}{Q(k)}$$

for $k \geq k_0$.
This is called the *quadratic residuosity assumption*.

Remark. The assumption states that there is not a significant chance of computing the predicate PQR_n if the factors of n are secret. It differs a little from the previous assumptions: the adversary algorithm A now has to compute a predicate. Since exactly half of the elements in J_n^{+1} are quadratic residues (see Proposition A.71), A can always predict the correct value with probability $1/2$, simply by tossing a coin. However, her probability of success is at most negligibly more than $1/2$.

Remark. The factoring assumption follows from the RSA assumption and also from the quadratic residuosity assumption. Hence, each of these two assumptions is stronger than the factoring assumption.

6.7 Formal Definition of One-Way Functions

As our examples show, one-way functions actually are families of functions. We give a formal definition of such families.

Definition 6.12. Let $I = (I_k)_{k \in \mathbb{N}}$ be a key set with security parameter k. Let K be a probabilistic polynomial sampling algorithm for I, which on input 1^k outputs $i \in I_k$.

A family

$$f = (f_i : D_i \longrightarrow R_i)_{i \in I}$$

of functions between finite sets D_i and R_i is a *family of one-way functions* (or, for short, a *one-way function*) with *key generator* K if and only if:

1. f can be computed by a Monte Carlo algorithm $F(i, x)$.
2. There is a uniform sampling algorithm S for $D := (D_i)_{i \in I}$, which on input $i \in I$ outputs $x \in D_i$.
3. f is not invertible by any efficient algorithm if the keys are generated by K. More precisely, for every positive polynomial $Q \in \mathbb{Z}[X]$ and every probabilistic polynomial algorithm $A(i, y)$ ($i \in I, y \in R_i$), there is a $k_0 \in \mathbb{N}$ such that for all $k \geq k_0$

$$\text{prob}(f_i(A(i, f_i(x))) = f_i(x) : i \leftarrow K(1^k), x \xleftarrow{u} D_i) \leq \frac{1}{Q(k)}.$$

If K is a uniform sampling algorithm for I, then we call f a family of one-way functions (or a one-way function), without explicitly referring to a key generator.

If f_i is bijective for all $i \in I$, then f is called a *bijective one-way function*, and if, in addition, the domain D_i coincides with the range R_i for all i, we call f a *family of one-way permutations* (or simply a *one-way permutation*).

Examples: The examples studied earlier in this chapter are families of one-way functions, provided Assumptions 6.1, 6.7 and 6.9 are true:

1. Discrete exponential function.
2. Modular powers.
3. Modular squaring.

Our considerations on the "random generation of the key" above (Propositions 6.6 and 6.8, 6.10) show that there are uniform key generators for these families. There are uniform sampling algorithms S for the domains $\mathbb{Z}_{p-1}$ and $\mathbb{Z}_n^*$, as we have seen in the examples after Lemma 6.5. Squaring a uniformly selected $x \in \mathbb{Z}_n^*$, $x \mapsto x^2$, we get a uniform sampling algorithm for the domains QR_n of the Square family.

Modular powers is a one-way permutation, as well as the modular squaring function Square. The discrete exponential function is a bijective one-way function.

Remarks:

1. Selecting an index i, for example, to use f_i as an encryption function, is equivalent to choosing a public key. Recall from Definition 6.2 that for $i \in I_k$, the security parameter k is a measure of the key length in bits.
2. Condition 3 means that pre-images of $y := f_i(x)$ cannot be computed in polynomial time if x is randomly and uniformly chosen from the domain (all inputs have the same probability), or equivalently, if y is random with respect to the image distribution induced by f. f is only called a one-way function if the random and uniform choice of elements in the domain can be accomplished by a probabilistic algorithm in polynomial time (condition 2).

Definition 6.12 can be immediately generalized to the case of one-way functions, where the inputs $x \in D_i$ are generated by any probabilistic polynomial – not necessarily uniform – sampling algorithm S (see, e.g., [Goldreich01]). The distribution $x \xleftarrow{u} D_i$ is replaced by $x \leftarrow S(i)$.

In this book, we consider only families of one-way functions with uniformly distributed inputs. The keys generated by the key generator K may be distributed in a non-uniform way.

3. The definition can be easily extended to formally define *families of trapdoor functions* (or, for short, *trapdoor functions*). We only sketch this definition. A bijective one-way function $f = (f_i)_{i \in I}$ is a trapdoor function if the inverse family $f^{-1} := (f_i^{-1})_{i \in I}$ can be computed by a Monte Carlo algorithm $F^{-1}(i, t_i, y)$, which takes as inputs the (public) key i, the (secret) trapdoor information t_i for f_i and a function value $y := f_i(x)$. It is required that the key generator K generates the trapdoor information t_i along with i.

   The RSA and the Square families are examples of trapdoor functions (see above).

4. The probability of success of the adversary A in the "one-way condition" (condition 3) is taken over the random choice of a key of a given security parameter k. It says that over all possibly generated keys, A has on average only a small probability of success. An adversary A usually knows the public key i when performing her attack. Thus, in a concrete attack the probability of success is given by the conditional probability assuming a fixed i, and this conditional probability might be high even if the average probability is negligibly small, as stated in condition 3. However, according to Proposition 6.3, the probability of such insecure keys is negligibly small. Thus, when randomly generating a key i by K, the probability of obtaining one for which A has a significant chance of succeeding is negligibly small (see Proposition 6.3 for a precise statement).

5. Condition 2 implies, in particular, that the binary length of the elements in D_i is bounded by the running time of $S(i)$, and hence is $\leq P(|i|)$ if P is a polynomial bound for the running time of S.

6. In all our examples, the one-way function can be computed using a deterministic polynomial algorithm. Computable by a Monte Carlo algorithm (see Definition 5.3) means that there is a probabilistic polynomial algorithm $F(i, x)$ with

$$\text{prob}(F(i, x) = f_i(x)) \geq 1 - 2^{-k} \quad (i \in I_k)$$

   (see Proposition 5.6 and Exercise 5 in Chapter 5).

7. Families of one-way functions, as defined here, are also called collections of strong one-way functions. They may be considered as a single one-way function $\{0, 1\}^* \longrightarrow \{0, 1\}^*$, defined on the infinite domain $\{0, 1\}^*$ (see [GolBel01]; [Goldreich01]). For the notion of weak one-way functions, see Exercise 3.

The key generator of a one-way function f is not uniquely determined: there are more suitable key generation algorithms (see Proposition 6.14 below). We call them "admissible generators".

Definition 6.13. Let $f = (f_i : D_i \longrightarrow R_i)_{i \in I}, I = (I_k)_{k \in \mathbb{N}}$, be a family of one-way functions with key generator K. A probabilistic polynomial algorithm $\tilde{K}$ that on input 1^k outputs a key $i \in I_k$ is called an *admissible key generator* for f if the one-way condition 3 of Definition 6.12 is satisfied for $\tilde{K}$.

Proposition 6.14. *Let* $f = (f_i : D_i \longrightarrow R_i)_{i \in I}, I = (I_k)_{k \in \mathbb{N}}$, *be a family of one-way functions with key generator* K. *Let* $\tilde{K}$ *be a probabilistic polynomial sampling algorithm for* I, *which on input* 1^k *yields* $i \in I_k$. *Assume that the family of distributions* $i \leftarrow \tilde{K}(1^k)$ *is polynomially bounded by the family of distributions* $i \leftarrow K(1^k)$ *(see Definition B.33).*
Then $\tilde{K}$ *is also an admissible key generator for* f.

Proof. This is a consequence of Proposition B.34. Namely, apply Proposition B.34 to $J := \mathbb{N} := \bigcup_{k \in \mathbb{N}} J_k, J_k := \{1^k\}, X_k := \{(i,x) \mid i \in I_k, x \in D_i\}$ and the probability distributions $(i \leftarrow \tilde{K}(1^k), x \xleftarrow{u} D_i)$ and $(i \leftarrow K(1^k), x \xleftarrow{u} D_i)$, $k \in \mathbb{N}$. The first family of distributions is polynomially bounded by the second. Assume as event $\mathcal{E}_k$ that $f_i(A(i, f_i(x))) = f_i(x)$.   □

Example. Let $f = (f_i : D_i \longrightarrow R_i)_{i \in I}$ be a family of one-way functions (with uniform key generation), and let $J \subseteq I$ with $|J_k| \cdot Q(k) \geq |I_k|$, for some polynomial Q. Let $\tilde{K}$ be a uniform sampling algorithm for J. Then $i \leftarrow \tilde{K}(1^k)$ is polynomially bounded by the uniform distribution $i \xleftarrow{u} I_k$. Thus, $\tilde{K}$ is an admissible key generator for f. This fact may be restated as: $f = (f_i : D_i \longrightarrow R_i)_{i \in J}$ is also a one-way function.

Example. As a special case of the previous example, consider the RSA one-way function (Section 6.4). Take as keys only pairs $(n,e) \in I_k$ (notation as above), with e a prime number in $\mathbb{Z}^*_{\varphi(n)}$. Since the number of primes in $\mathbb{Z}_{\varphi(n)}$ is of the same order as $\varphi(n)/k$ (by the Prime Number Theorem, Theorem A.82), we get an admissible key generator in this way. In other words, the classical RSA assumption (Assumption 6.7) implies an RSA assumption, with $(n,e) \xleftarrow{u} I_k$ replaced by $(n,e) \xleftarrow{u} J_k$, where $J_k := \{(n,e) \in I_k \mid e \text{ prime}\}$.

Example. As already mentioned in Section 6.4, the key generator that first uniformly chooses an $n = pq$ and then, in a second step, uniformly chooses an exponent $e \in \mathbb{Z}^*_{\varphi(n)}$ is an admissible key generator for the RSA one-way function. The distribution given by this generator is polynomially bounded by the uniform distribution.

Similarly, we get an admissible key generator for the discrete exponential function (Section 6.2) if we first uniformly generate a prime p (together with a factorization of $p-1$) and then, for this fixed p, repeatedly select $g \xleftarrow{u} \mathbb{Z}^*_p$ until g happens to be a primitive root (see Exercise 1).

6.8 Hard-Core Predicates

Given a one-way function f, it is impossible to compute a pre-image x from $y = f(x)$ using an efficient algorithm. Nevertheless, it is often easy to derive single bits of the pre-image x from $f(x)$. For example, if f is the discrete exponential function, the least-significant bit of x is derived from $f(x)$ in a straightforward way (see Chapter 7). On the other hand, since f is one way there should be other bits of the pre-image x, or more generally, properties of x stated as Boolean predicates, which are very hard to derive from $f(x)$. Examples of such hard-core predicates are studied thoroughly in Chapter 7.

Definition 6.15. Let $I = (I_k)_{k \in \mathbb{N}}$ be a key set with security parameter k. Let $f = (f_i : D_i \longrightarrow R_i)_{i \in I}$ be a family of one-way functions with key generator K, and let $B = (B_i : D_i \longrightarrow \{0, 1\})_{i \in I}$ be a family of Boolean predicates. B is called a *family of hard-core predicates* (or, for short, a *hard-core predicate*) of f if and only if:

1. B can be computed by a Monte Carlo algorithm $A_1(i, x)$.
2. $B(x)$ is not computable from $f(x)$ by an efficient algorithm; i.e., for every positive polynomial $Q \in \mathbb{Z}[X]$ and every probabilistic polynomial algorithm $A_2(i, y)$, there is a $k_0 \in \mathbb{N}$ such that for all $k \geq k_0$

$$\mathrm{prob}(A_2(i, f_i(x)) = B_i(x) : i \leftarrow K(1^k), x \xleftarrow{u} D_i) \leq \frac{1}{2} + \frac{1}{Q(k)}.$$

Remarks:

1. As above in the one-way conditions, we do not consider a single fixed key in the probability statement of the definition. The probability is also taken over the random generation of a key i, with a given security parameter k. Hence, the meaning of statement 2 is: choosing both the key i with security parameter k and $x \in D_i$ randomly, the adversary A_2 does not have a significantly better chance of finding out the bit $B_i(x)$ from $f_i(x)$ than by simply tossing a coin, provided the security parameter is sufficiently large. In practice, the public key is known to the adversary, and we are interested in the conditional probability of success assuming a fixed public key i. Even if the security parameter k is very large, there may be keys i such that A_2 has a significantly better chance than $1/2$ of determining $B_i(x)$. The probability of such insecure keys is, however, negligibly small. This is shown in Proposition 6.17.
2. Considering a family of one-way functions with hard-core predicate B, we call a key generator K admissible only if it guarantees the "hard-core" condition 2 of Definition 6.15, in addition to the one-way condition 3 of Definition 6.12. Proposition 6.14 remains valid for one-way functions with hard-core predicates. This immediately follows from Proposition 6.17.

The *inner product bit* yields a generic hard-core predicate for all one-way functions.

Theorem 6.16. *Let* $f = (f_i : D_i \longrightarrow R_i)_{i \in I}$ *be a family of one-way functions,* $D_i \subset \{0,1\}^*$ *for all* $i \in I$. *Extend the functions* f_i *to functions* $\tilde{f}_i$ *which on input* $x \in D_i$ *and* $y \in \{0,1\}^{|x|}$ *return the concatenation* $f_i(x) \| y$. *Let*

$$B_i(x, y) := \left(\sum_{j=1}^{l} x_j \cdot y_j \right) \bmod 2,$$

where $l = |x|$, $x = x_1 \ldots x_l$, $y = y_1 \ldots y_l$, $x_j, y_j \in \{0,1\}$,

be the inner product modulo 2. Then $B := (B_i)_{i \in I}$ *is a hard-core predicate of* $\tilde{f} := (\tilde{f}_i)_{i \in I}$.

For a proof, see [GolLev89], [Goldreich99] or [Luby96].

Proposition 6.17. *Let* $I = (I_k)_{k \in \mathbb{N}}$ *be a key set with security parameter* k. *Let* $f = (f_i : X_i \longrightarrow Y_i)_{i \in I}$ *be a family of functions between finite sets, and let* $B = (B_i : X_i \longrightarrow \{0,1\})_{i \in I}$ *be a family of Boolean predicates. Assume that both* f *and* B *can be computed by a Monte Carlo algorithm, with inputs* $i \in I$ *and* $x \in X_i$. *Let probability distributions be given on* I_k *and* X_i *for all* k *and* i. *Assume there is a probabilistic polynomial sampling algorithm* S *for* $X := (X_i)_{i \in I}$ *which, on input* $i \in I$, *randomly chooses an* $x \in X_i$ *with respect to the given distribution on* X_i, *i.e.,* $\mathrm{prob}(S(i) = x) = \mathrm{prob}(x)$. *Then the following statements are equivalent:*

1. *For every probabilistic polynomial algorithm* A *with inputs* $i \in I$ *and* $y \in Y_i$ *and output in* $\{0,1\}$, *and every positive polynomial* P, *there is a* $k_0 \in \mathbb{N}$ *such that for all* $k \geq k_0$

$$\mathrm{prob}(A(i, f_i(x)) = B_i(x) : i \leftarrow I_k, x \leftarrow X_i) \leq \frac{1}{2} + \frac{1}{P(k)}.$$

2. *For every probabilistic polynomial algorithm* A *with inputs* $i \in I$ *and* $x \in X_i$ *and output in* $\{0,1\}$, *and all positive polynomials* Q *and* R, *there is a* $k_0 \in \mathbb{N}$ *such that for all* $k \geq k_0$

$$\mathrm{prob}\left(\left\{ i \in I_k \;\middle|\; \mathrm{prob}(A(i, f_i(x)) = B_i(x) : x \leftarrow X_i) > \frac{1}{2} + \frac{1}{Q(k)} \right\} \right)$$

$$\leq \frac{1}{R(k)}.$$

Statement 2 implies statement 1, even if we omit "for every algorithm" in both statements and instead consider a fixed probabilistic polynomial algorithm A.

Proof. An analogous computation as in the proof of Proposition 6.3 shows that statement 2 implies statement 1.

Now, assume that statement 1 holds, and let $A(i, y)$ be a probabilistic polynomial algorithm with output in $\{0, 1\}$. Let Q and R be positive polynomials. We abbreviate the probability of success of A, conditional on a fixed i, by p_i:

$$p_i := \text{prob}(A(i, f_i(x)) = B_i(x) : x \leftarrow X_i).$$

Assume that for k in an infinite subset $\mathcal{K} \subseteq \mathbb{N}$,

$$\text{prob}\left(\left\{i \in I_k \,\middle|\, p_i > \frac{1}{2} + \frac{1}{Q(k)}\right\}\right) > \frac{1}{R(k)}. \tag{6.2}$$

If all $p_i, i \in I_k$, were $\geq 1/2$, we could easily conclude that their average also significantly exceeds $1/2$ and obtain a contradiction to statement 1. Unfortunately, it might happen that many of the probabilities p_i are significantly larger than $1/2$ and many are smaller than $1/2$, whereas their average is close to $1/2$. The basic idea now is to modify A and replace A by an algorithm $\tilde{A}$. We want to replace the output $A(i, y)$ by its complementary value $1 - A(i, y)$ if $p_i < 1/2$. In this way we could force all the probabilities to be $\geq 1/2$. At this point we face another problem. We want to define $\tilde{A}$ as

$$(i, y) \mapsto \begin{cases} A(i, y) & \text{if } p_i \geq 1/2, \\ 1 - A(i, y) & \text{if } p_i < 1/2, \end{cases}$$

and see that we have to determine by a polynomial algorithm whether $p_i \geq 1/2$. At least we can compute the correct answer to this question with a high probability in polynomial time. By Proposition 6.18, there is a probabilistic polynomial algorithm $C(i)$ such that

$$\text{prob}\left(|C(i) - p_i| < \frac{1}{4Q(k)R(k)}\right) \geq 1 - \frac{1}{4Q(k)R(k)}.$$

We define that $Sign(i) := +1$ if $C(i) \geq 1/2$, and $Sign(i) := -1$ if $C(i) < 1/2$. Then $Sign$ computes the sign $\sigma_i \in \{+1, -1\}$ of $(p_i - 1/2)$ with a high probability if the distance of p_i from $1/2$ is not too small:

$$\text{prob}(Sign(i) = \sigma_i) \geq 1 - \frac{1}{4Q(k)R(k)} \text{ for all } i \text{ with } \left|p_i - \frac{1}{2}\right| \geq \frac{1}{4Q(k)R(k)}.$$

Now the modified algorithm $\tilde{A} = \tilde{A}(i, y)$ is defined as follows. Let

$$\tilde{A}(i, y) := \begin{cases} A(i, y) & \text{if } Sign(i) = +1, \\ 1 - A(i, y) & \text{if } Sign(i) = -1, \end{cases}$$

for $i \in I$ and $y \in Y_i$. Similarly as before, we write

$$\tilde{p}_i := \text{prob}(\tilde{A}(i, f_i(x)) = B_i(x) : x \leftarrow X_i)$$

for short. By the definition of $\tilde{A}$, $\tilde{p}_i = p_i$ if $Sign(i) = +1$, and $\tilde{p}_i = 1 - p_i$ if $Sign(i) = -1$. Hence $|p_i - 1/2| = |\tilde{p}_i - 1/2|$. Moreover, we have $\tilde{p}_i \geq 1/2$, if $Sign(i) = \sigma_i$, and $\mathrm{prob}(Sign(i) \neq \sigma_i) \leq 1/(4Q(k)R(k))$ if $|p_i - 1/2| \geq 1/(4Q(k)R(k))$.

Let

$$I_{1,k} := \left\{ i \in I_k \,\middle|\, p_i - \frac{1}{2} > \frac{1}{Q(k)} \right\},$$

$$I_{2,k} := \left\{ i \in I_k \,\middle|\, \left| p_i - \frac{1}{2} \right| > \frac{1}{4Q(k)R(k)} \right\}.$$

We compute

$$\mathrm{prob}(\tilde{A}(i, f_i(x)) = B_i(x) : i \leftarrow I_k, x \leftarrow X_i) - \frac{1}{2}$$

$$= \sum_{i \in I_k} \mathrm{prob}(i) \cdot \left(\tilde{p}_i - \frac{1}{2} \right)$$

$$= \sum_{i \in I_{2,k}} \mathrm{prob}(i) \cdot \left(\tilde{p}_i - \frac{1}{2} \right) + \sum_{i \notin I_{2,k}} \mathrm{prob}(i) \cdot \left(\tilde{p}_i - \frac{1}{2} \right)$$

$$=: (1).$$

For $i \notin I_{2,k}$, we have $|\tilde{p}_i - 1/2| = |p_i - 1/2| \leq 1/(4Q(k)R(k))$ and hence $\tilde{p}_i - 1/2 \geq -1/(4Q(k)R(k))$. Thus

$$(1) \geq \sum_{i \in I_{2,k}} \mathrm{prob}(i) \cdot \left(\tilde{p}_i - \frac{1}{2} \right) - \frac{1}{4Q(k)R(k)}$$

$$= \sum_{i \in I_{2,k}} \mathrm{prob}(Sign(i) = \sigma_i) \cdot \mathrm{prob}(i) \cdot \left(\tilde{p}_i - \frac{1}{2} \right)$$

$$+ \sum_{i \in I_{2,k}} \mathrm{prob}(Sign(i) \neq \sigma_i) \cdot \mathrm{prob}(i) \cdot \left(\tilde{p}_i - \frac{1}{2} \right) - \frac{1}{4Q(k)R(k)}$$

$$=: (2).$$

We observed before that $\tilde{p}_i \geq 1/2$ if $Sign(i) = \sigma_i$, and $\mathrm{prob}(Sign(i) \neq \sigma_i) \leq 1/(4Q(k)R(k))$ for $i \in I_{2,k}$. Moreover, we obviously have $I_{1,k} \subseteq I_{2,k}$, and our assumption (6.2) means that

$$\mathrm{prob}(I_{1,k}) = \sum_{i \in I_{1,k}} \mathrm{prob}(i) > \frac{1}{R(k)},$$

for the infinitely many $k \in \mathcal{K}$. Therefore, we may continue our computation for $k \in \mathcal{K}$ in the following way:

$$(2) \geq \left(1 - \frac{1}{4Q(k)R(k)}\right) \sum_{i \in I_{2,k}} \mathrm{prob}(i) \cdot \left|p_i - \frac{1}{2}\right| - \frac{1}{4Q(k)R(k)} - \frac{1}{4Q(k)R(k)}$$

$$\geq \left(1 - \frac{1}{4Q(k)R(k)}\right) \sum_{i \in I_{1,k}} \mathrm{prob}(i) \cdot \left(p_i - \frac{1}{2}\right) - \frac{1}{4Q(k)R(k)} - \frac{1}{4Q(k)R(k)}$$

$$\geq \left(1 - \frac{1}{4Q(k)R(k)}\right) \cdot \frac{1}{R(k)} \cdot \frac{1}{Q(k)} - \frac{1}{4Q(k)R(k)} - \frac{1}{4Q(k)R(k)}$$

$$= \frac{1}{2Q(k)R(k)} - \frac{1}{4Q^2(k)R^2(k)}$$

$$\geq \frac{1}{4Q(k)R(k)}.$$

Since $\mathcal{K}$ is infinite, we obtain a contradiction to statement 1 applied to $\tilde{A}$ and $P = 4QR$, and the proof that statement 1 implies statement 2 is finished. $\square$

As in the preceding proof, we sometimes want to compute probabilities with a probabilistic polynomial algorithm, at least approximately.

Proposition 6.18. *Let $B(i,x)$ be a probabilistic polynomial algorithm that on input $i \in I$ and $x \in X_i$ outputs a bit $b \in \{0,1\}$. Assume that a probability distribution is given on X_i, for every $i \in I$. Further assume that there is a probabilistic polynomial sampling algorithm S which on input $i \in I$ randomly chooses an element $x \in X_i$ with respect to the distribution given on X_i, i.e., $\mathrm{prob}(S(i) = x) = \mathrm{prob}(x)$. Let P and Q be positive polynomials.*
Let $p_i := \mathrm{prob}(B(i,x) = 1 : x \leftarrow X_i)$ be the probability of $B(i,x) = 1$, assuming that i is fixed. Then there is a probabilistic polynomial algorithm A that approximates the probabilities $p_i, i \in I$, with high probability; i.e.,

$$\mathrm{prob}\left(|A(i) - p_i| < \frac{1}{P(|i|)}\right) \geq 1 - \frac{1}{Q(|i|)}.$$

Proof. We first observe that $\mathrm{prob}(B(i, S(i)) = 1) = p_i$ for every $i \in I$, since $S(i)$ samples (by use of its coin tosses) according to the distribution given on X_i. We define the algorithm A on input i as follows:

1. Let t be the smallest $n \in \mathbb{N}$ with $n \geq \frac{1}{4} \cdot P(|i|)^2 \cdot Q(|i|)$.
2. Compute $B(i, S(i))$ t times, and obtain the results
   $b_1, \ldots, b_t \in \{0,1\}$.
3. Let

$$A(i) := \frac{1}{t} \sum_{i=1}^{t} b_i.$$

Applying Corollary B.25 to the t independent computations of the random variable $B(i, S(i))$, we get

$$\mathrm{prob}\left(|A(i) - p_i| < \frac{1}{P(|i|)}\right) \geq 1 - \frac{P(|i|)^2}{4t} \geq 1 - \frac{1}{Q(|i|)},$$

as desired. $\square$

Exercises

1. Let S be the key generator for

   a. the RSA family, which on input 1^k first generates random prime numbers p and q of binary length k, and then repeatedly generates random exponents $e \in \mathbb{Z}_{\varphi(n)}, n := pq$, until it finds an e prime to $\varphi(n)$.

   b. the Exp family, which on input 1^k first generates a random prime p of binary length k, together with the prime factors of $p-1$, and then randomly chooses elements $g \in \mathbb{Z}_p^*$, until it finds a primitive root g (see the proof of Proposition 6.6).

   Show that S is an admissible key generator for the RSA family and the Exp family (see the remark after Proposition 6.8).

2. Compare the running times of the uniform key generator for RSA keys constructed in Proposition 6.8 and the (more efficient) admissible key generator S in Exercise 1.

3. Consider the Definition 6.12 of "strong" one-way functions. If we replace the probability statement in condition 3 by "There is a positive polynomial Q such that for all probabilistic polynomial algorithms A

$$\mathrm{prob}(f_i(A(i, f_i(x))) = f_i(x) : i \leftarrow K(1^k), x \xleftarrow{u} D_i) \leq 1 - \frac{1}{Q(k)},$$

   for sufficiently large k", then $f = (f_i)_{i \in I}$ is called a family of *weak one-way functions*.[8]
   Let $X_j := \{2, 3, \ldots, 2^j - 1\}$ be the set of numbers > 1 of binary length $\leq j$. Let $D_n := \bigcup_{j=2}^{n-2} X_j \times X_{n-j}$ (disjoint union), and let $R_n := X_n$. Show that

$$f := (f_n : D_n \longrightarrow R_n, \ (x, y) \longmapsto x \cdot y)_{n \in \mathbb{N}}$$

   is a weak – not a strong – one-way function, if the factoring assumption (Definition 6.9) is true.

4. We consider the representation problem (Section 4.8.3).
   Let $A(p, q, g_1, \ldots, g_r)$ be a probabilistic polynomial algorithm ($r \geq 2$). The inputs are primes p and q such that q divides $p - 1$, and elements $g_1, \ldots, g_r \in \mathbb{Z}_p^*$ of order q. A outputs two tuples $(a_1, \ldots, a_r)$ and $(a_1', \ldots, a_r')$ of integer exponents. Assume that r is polynomially bounded by the binary length of p, i.e., $r \leq T(|p|)$ for some polynomial T. We denote by G_q the subgroup of order q of $\mathbb{Z}_p^*$. Recall that G_q is cyclic and each $g \in G_q, g \neq 1$, is a generator (Lemma A.41). Let P be a positive polynomial, and let $\mathcal{K}$ be the set of pairs (p, q) such that

---

[8] A weak one-way function f yields a strong one-way function by the following construction: $(x_1, \ldots, x_{2kQ(k)}) \mapsto (f_i(x_1), \ldots, f_i(x_{2kQ(k)}))$ (where $i \in I_k$). See [Luby96] for a proof.

$$\text{prob}\big(A(p,q,g_1,\ldots,g_r) = ((a_1,\ldots,a_r),(a'_1,\ldots,a'_r)),$$

$$(a_1,\ldots,a_r) \neq (a'_1,\ldots,a'_r), \quad \prod_{j=1}^{r} g_j^{a_j} = \prod_{j=1}^{r} g_j^{a'_j} :$$

$$g_j \overset{u}{\leftarrow} G_q \setminus \{1\}, 1 \leq j \leq r)$$

$$\geq 1/P(|p|).$$

Show that for every positive polynomial Q there is a probabilistic polynomial algorithm $\tilde{A} = \tilde{A}(p,q,g,y)$ such that

$$\text{prob}(\tilde{A}(p,q,g,y) = \text{Log}_{p,g}(y)) \geq 1 - 2^{-Q(|p|)},$$

for all $(p,q) \in \mathcal{K}$, $g \in G_q \setminus \{1\}$ and $y \in G_q$.

5. Let $J_k := \{n \in \mathbb{N} \mid n = pq, p, q \text{ distinct primes}, |p| = |q| = k\}$ be the set of RSA moduli with security parameter k. For a prime $\tilde{p}$, we denote by $J_{k,\tilde{p}} := \{n \in J_k \mid \tilde{p} \text{ does not divide } \varphi(n)\}$ the set of those moduli for which $\tilde{p}$ may serve as an RSA exponent. Let $\text{Primes}_{\leq 2k}$ be the primes of binary length $\leq 2k$.

Show that the RSA assumption remains valid if we first choose a prime exponent and then a suitable modulus. More precisely, show that the classical RSA assumption (Definition 6.7) implies that:

For every probabilistic polynomial algorithm A and every positive polynomial P there is a $k_0 \in \mathbb{N}$ such that for all $k \geq k_0$

$$\text{prob}(A(n,\tilde{p},x^{\tilde{p}}) = x \mid \tilde{p} \overset{u}{\leftarrow} \text{Primes}_{\leq 2k}, n \overset{u}{\leftarrow} J_{k,\tilde{p}}, x \overset{u}{\leftarrow} \mathbb{Z}_n^*) \leq \frac{1}{P(k)}.$$

6. Let $f = (f_i : D_i \longrightarrow R_i)_{i \in I}$, $I = (I_k)_{k \in \mathbb{N}}$, be a family of one-way functions with key generator K, and let $B = (B_i : D_i \longrightarrow \{0,1\})_{i \in I}$ be a hard-core predicate for f.

Show that for every positive polynomial P, there is a $k_0 \in \mathbb{N}$ such that for all $k \geq k_0$

$$\left| \text{prob}(B_i(x) = 0 : i \leftarrow K(1^k), x \overset{u}{\leftarrow} D_i) - \frac{1}{2} \right| \leq \frac{1}{P(k)}.$$

7. Let $f = (f_i : D_i \longrightarrow R_i)_{i \in I}$, $I = (I_k)_{k \in \mathbb{N}}$, be a family of one-way functions with key generator K, and let $B = (B_i : D_i \longrightarrow \{0,1\})_{i \in I}$ be a family of predicates which is computable by a Monte Carlo algorithm. Show that B is a hard-core predicate of f if and only if for every probabilistic polynomial algorithm $A(i,x,y)$ and every positive polynomial P there is a k_0 such that for all $k \geq k_0$

$$| \text{prob}(A(i,f_i(x),B_i(x)) = 1 : i \leftarrow K(1^k), x \overset{u}{\leftarrow} D_i)$$

$$- \text{prob}(A(i,f_i(x),z) = 1 : i \leftarrow K(1^k), x \overset{u}{\leftarrow} D_i, z \overset{u}{\leftarrow} \{0,1\}) | \leq \frac{1}{P(k)}.$$

If the functions f_i are bijective, then the latter statement means that the family $(\{(f_i(x), B_i(x)) : x \overset{u}{\leftarrow} D_i\})_{i \in I}$ of distributions cannot be distinguished from the uniform distributions on $(R_i \times \{0,1\})_{i \in I}$ by a probabilistic polynomial algorithm.

8. Let $I = (I_k)_{k \in \mathbb{N}}$ be a key set with security parameter k. Consider probabilistic polynomial algorithms A which on input $i \in I$ and $x \in X_i$ compute a Boolean value $A(i, x) \in \{0,1\}$. Assume that one family of probability distributions is given on $(I_k)_{k \in \mathbb{N}}$, and two families of probability distributions $p := (p_i)_{i \in I}$ and $q := (q_i)_{i \in I}$ are given on $X := (X_i)_{i \in I}$. Further assume that there are probabilistic polynomial sampling algorithms $S_1(i)$ and $S_2(i)$ which randomly choose an $x \in X_i$ with $\mathrm{prob}(S_1(i) = x) = p_i(x)$ and $\mathrm{prob}(S_2(i) = x) = q_i(x)$, for all $i \in I$ and $x \in X_i$.

Prove a result that is analogous to the statements of Propositions 6.3 and 6.17 for

$$|\, \mathrm{prob}(A(i, x) = 1 : i \leftarrow I_k, x \overset{p_i}{\leftarrow} X_i)$$
$$- \mathrm{prob}(A(i, x) = 1 : i \leftarrow I_k, x \overset{q_i}{\leftarrow} X_i)\,| \le \frac{1}{P(k)}.$$

9. Let $n := pq$, with distinct primes p and q. The quadratic residuosity assumption (Definition 6.11) states that it is infeasible to decide whether a given $x \in \mathrm{J}_n^{+1}$ is a quadratic residue or not. If p and q are kept secret, no efficient algorithm for selecting a quadratic non-residue modulo n is known. Thus, it might be easier to decide quadratic residuosity if, additionally, a random quadratic non-residue with Jacobi symbol 1 is revealed. In [GolMic84] (Section 6.2) it is shown that this is not the case. More precisely, let $I_k := \{n \mid n = pq, p, q \text{ distinct primes }, |p| = |q| = k\}$ and $I := (I_k)_{k \in \mathbb{N}}$. Let $\mathrm{QNR}_n := \mathbb{Z}_n^* \setminus \mathrm{QR}_n$ be the set of quadratic non-residues (see Definition A.54) and $\mathrm{QNR}_n^{+1} := \mathrm{QNR}_n \cap \mathrm{J}_n^{+1}$. Then the following statements are equivalent:

   a. For all probabilistic polynomial algorithms A, with inputs $n \in I$ and $x \in \mathrm{J}_n^{+1}$ and output in $\{0,1\}$, and every positive polynomial P, there exists a $k_0 \in \mathbb{N}$ such that for all $k \ge k_0$

   $$\mathrm{prob}(A(n, x) = \mathrm{PQR}_n(x) : n \overset{u}{\leftarrow} I_k, x \overset{u}{\leftarrow} \mathrm{J}_n^{+1}) \le \frac{1}{2} + \frac{1}{P(k)}.$$

   b. For all probabilistic polynomial algorithms A, with inputs $n \in I$ and $z, x \in \mathrm{J}_n^{+1}$ and output in $\{0,1\}$, and every positive polynomial P, there exists a $k_0 \in \mathbb{N}$ such that for all $k \ge k_0$

   $$\mathrm{prob}(A(n, z, x) = \mathrm{PQR}_n(x) : n \overset{u}{\leftarrow} I_k, z \overset{u}{\leftarrow} \mathrm{QNR}_n^{+1}, x \overset{u}{\leftarrow} \mathrm{J}_n^{+1})$$
   $$\le \frac{1}{2} + \frac{1}{P(k)}.$$

Study [GolMic84] and give a proof.

7. Bit Security of One-Way Functions

Let $f : X \longrightarrow Y$ be a bijective one-way function and let $x \in X$. Sometimes it is possible to compute some bits of x from $f(x)$ without inverting f. A function f does not necessarily hide everything about x, even if f is one way. Let b be a bit of x. We call b a *secure bit* of f if it is as difficult to compute b from $f(x)$ as it is to compute x from $f(x)$. We prove that the most-significant bit of x is a secure bit of Exp, and that the least-significant bit is a secure bit of RSA and Square.

We show how to compute x from $\mathrm{Exp}(x)$, assuming that we can compute the most-significant bit of x from $\mathrm{Exp}(x)$. Then we show the same for the least-significant bit and RSA or Square. First we assume that deterministic algorithms are used to compute the most- or least-significant bit. In this much easier case, we demonstrate the basic ideas. Then we study the probabilistic case. We assume that we can compute the most- or least-significant bit with a probability $p \geq 1/2 + 1/P(|x|)$, for some positive polynomial P, and derive that then x can be computed with an overwhelmingly high probability.

As a consequence, the discrete logarithm assumption implies that the most-significant bit is a hard-core predicate for the Exp family. Given the RSA or the factoring assumption, the least-significant bit yields a hard-core predicate for the RSA or the Square family.

Bit security is not only of theoretical interest. Bleichenbacher's 1-Million-Chosen-Ciphertext Attack against PKCS#1-based schemes shows that a leaking secure bit can lead to dangerous practical attacks (see Section 3.3.2).

Let $n, r \in \mathbb{N}$ such that $2^{r-1} \leq n < 2^r$. As usual, the *binary encoding* of $x \in \mathbb{Z}_n$ is the binary encoding in $\{0,1\}^r$ of the representative of x between 0 and $n - 1$ as an unsigned number (see Appendix A.2). This defines an embedding $\mathbb{Z}_n \subset \{0,1\}^r$. Bits of x and properties of x that depend on a representative of x are defined relative to this embedding. The property "x is even", for example, is such a property.

7.1 Bit Security of the Exp Family

Let p be an odd prime number and let g be a primitive root in $\mathbb{Z}_p^*$. We consider the discrete exponential function

$$\mathrm{Exp}_{p,g} : \mathbb{Z}_{p-1} \longrightarrow \mathbb{Z}_p^*, \ x \longmapsto g^x,$$

and its inverse

$$\mathrm{Log}_{p,g} : \mathbb{Z}_p^* \longrightarrow \mathbb{Z}_{p-1},$$

which is the discrete logarithm function.

$\mathrm{Log}_{p,g}(x)$ is even if and only if x is a square (Lemma A.55). The square property modulo a prime can be computed efficiently by Euler's criterion (Proposition A.58). Hence, we have an efficient algorithm that computes the least-significant bit of x from $\mathrm{Exp}_{p,g}(x)$. The most-significant bit, however, is as difficult to compute as the discrete logarithm.

Definition 7.1. Let p be an odd prime, and let g be a primitive root in $\mathbb{Z}_p^*$:

1. The predicate Msb_p is defined by

$$\mathrm{Msb}_p : \mathbb{Z}_{p-1} \longrightarrow \{0,1\}, x \longmapsto \begin{cases} 0 \text{ if } 0 \le x < \frac{p-1}{2}, \\ 1 \text{ if } \frac{p-1}{2} \le x < p-1. \end{cases}$$

2. The predicate $\mathrm{B}_{p,g}$ is defined by

$$\mathrm{B}_{p,g} : \mathbb{Z}_p^* \longrightarrow \{0,1\}, \mathrm{B}_{p,g}(x) = \mathrm{Msb}_p(\mathrm{Log}_{p,g}(x)).$$

Remark. If $p-1$ is a power of 2, then the predicate Msb_p is the most-significant bit of the binary encoding of x.

Let $x \in \mathrm{QR}_p$, $x \ne 1$. There is a probabilistic polynomial algorithm which computes the square roots of x (Algorithm A.67). Let $y := y_n \ldots y_0, y_i \in \{0,1\}$, be the binary representation of $y := \mathrm{Log}_{p,g}(x)$. As we observed before, $y_0 = 0$. Therefore, $w_1 := g^{\tilde{y}}$ with $\tilde{y} := y_n \ldots y_1$ is a root of x with $\mathrm{B}_{p,g}(w_1) = 0$. The other root w_2 is $g^{\tilde{y}+(p-1)/2}$, and obviously $\mathrm{B}_{p,g}(w_2) = 1$. Thus, exactly one of the two square roots w satisfies the condition $\mathrm{B}_{p,g}(w) = 0$.

Definition 7.2. Let $x \in \mathrm{QR}_p$. The square root w of x which satisfies the condition $\mathrm{B}_{p,g}(w) = 0$ is called the *principal square root* of x. The map $\mathrm{PSqrt}_{p,g}$ is defined by

$$\mathrm{PSqrt}_{p,g} : \mathrm{QR}_p \longrightarrow \mathbb{Z}_p^*, \ x \longmapsto \text{principal square root of } x.$$

Remark. The ability to compute the principal square root in polynomial time is equivalent to the ability to compute the predicate $\mathrm{B}_{p,g}$ in polynomial time. Namely, let A be a deterministic polynomial algorithm for the computation of $\mathrm{PSqrt}_{p,g}$. Then the algorithm

Algorithm 7.3.

```
int B(int p, g, x)
1   if A(p, g, x²) = x
2       then return 0
3       else return 1
```

computes $B_{p,g}$ and is deterministic polynomial.

Conversely, let B be a deterministic polynomial algorithm which computes $B_{p,g}$. If Sqrt is a polynomial algorithm for the computation of square roots modulo p, then the polynomial algorithm

Algorithm 7.4.

```
int A(int p, g, x)
 1   {u, v} ← Sqrt(x, p)
 2   if B(p, g, u) = 0
 3       then return u
 4       else return v
```

computes $\mathrm{PSqrt}_{p,g}$. It is deterministic if Sqrt is deterministic. We have a polynomial algorithm Sqrt (Algorithm A.67). It is deterministic for $p \equiv 3 \bmod 4$. In the other case, the only non-deterministic step is to find a quadratic non-residue modulo p, which is an easy task. If we select t numbers in $\{1, p-1\}$ at random, then the probability of finding a quadratic non-residue is $1 - 1/2^t$. Thus, the probability of success of A can be made almost 1, independent of the size of the input x.

In the following proposition and theorem, we show how to reduce the computation of the discrete logarithm function to the computation of $\mathrm{PSqrt}_{p,g}$. Given an algorithm A_1 for computing $\mathrm{PSqrt}_{p,g}$, we develop an algorithm A_2 for the discrete logarithm which calls A_1 as a subroutine.[1] The resulting algorithm A_2 has the same complexity as A_1. Therefore, $\mathrm{PSqrt}_{p,g}$ is also believed to be not efficiently computable.

Proposition 7.5. *Let A_1 be a deterministic polynomial algorithm such that*

$$A_1(p, g, x) = \mathrm{PSqrt}_{p,g}(x) \text{ for all } x \in \mathrm{QR}_p,$$

with p an odd prime and g a primitive root in $\mathbb{Z}_p^*$. Then there is a deterministic polynomial algorithm A_2 such that

$$A_2(p, g, x) = \mathrm{Log}_{p,g}(x) \text{ for all } x \in \mathrm{QR}_p.$$

Proof. The basic idea of the reduction is the following:
1. Let $x = g^y \in \mathbb{Z}_p^*$ and $y = y_k \ldots y_0$, $y_i \in \{0, 1\}$, $i = 0, \ldots, k$, be the binary encoding of y. We compute the bits of y from right to left. Bit y_0 is 0 if and only if $x \in \mathrm{QR}_p$ (Lemma A.55). This condition can be tested by use of Euler's criterion for quadratic residuosity (Proposition A.58).
2. To get the next bit y_1, we replace x by $xg^{-1} = g^{y_k \ldots y_1 0}$ if $y_0 = 1$. Then bit y_1 can be obtained from $\mathrm{PSqrt}_{p,g}(x) = g^{y_k \ldots y_1}$ as in step 1.

The following algorithm A_2, which calls A_1 as a subroutine, computes $\mathrm{Log}_{p,g}$.

---

[1] In our construction we use A_1 as an "oracle" for $\mathrm{PSqrt}_{p,g}$. Therefore, algorithms such as A_1 are sometimes called "*oracle algorithms*".

Algorithm 7.6.

```
int A₂(int p, g, x)
1   y ← empty word, k ← |p|
2   for c ← 0 to k − 1 do
3       if x ∈ QR_p
4           then y ← y‖0
5           else  y ← y‖1
6               x ← xg⁻¹
7       x ← A₁(p, g, x)
8   return y
```

This completes the proof.   □

Theorem 7.7. *Let $P, Q \in \mathbb{Z}[X]$ be positive polynomials and A_1 be a probabilistic polynomial algorithm such that*

$$\mathrm{prob}(A_1(p,g,x) = \mathrm{PSqrt}_{p,g}(x) : x \xleftarrow{u} \mathrm{QR}_n) \geq \frac{1}{2} + \frac{1}{P(k)},$$

where p is an odd prime number, g is a primitive root in $\mathbb{Z}_p^*$ and $k = |p|$ is the binary length of p. Then there is a probabilistic polynomial algorithm A_2 such that for every $x \in \mathbb{Z}_p^*$,

$$\mathrm{prob}(A_2(p,g,x) = \mathrm{Log}_{p,g}(x)) \geq 1 - 2^{-Q(k)}.$$

Proof. Let $\varepsilon := 1/P(k)$. In order to reduce $\mathrm{Log}_{p,g}$ to $\mathrm{PSqrt}_{p,g}$, we intend to proceed as in the deterministic case (Proposition 7.5). There, A_1 is applied k times by A_2. A_2 correctly yields the desired logarithm if PSqrt is correctly computed by A_1 in each step. Now the algorithm A_1 computes the function $\mathrm{PSqrt}_{p,g}$ with a probability of success of only $\geq 1/2 + \varepsilon$. Thus, the probability of success of A_2 is $\geq (1/2 + \varepsilon)^k$. This value is exponentially close to 0, and hence is too small.

The basic idea now is to replace A_1 by an algorithm B which computes $\mathrm{PSqrt}_{p,g}(x)$ with a high probability of success, for a polynomial fraction of inputs.

Lemma 7.8. *Under the assumptions of the theorem, let $t := k\varepsilon^{-2}$. Then there is a probabilistic polynomial algorithm B such that*

$$\mathrm{prob}(B(x) = \mathrm{PSqrt}_{p,g}(x)) \geq 1 - \frac{1}{k}, \text{ for } x = g^{2s}, 0 \leq s \leq \frac{p-1}{2t}.$$

Proof (of the Lemma). Let $x = g^{2s}, 0 \leq s \leq (p-1)/2t$. In our algorithm B we want to increase the probability of success on input x by computing $A_1(x)$ repeatedly. By assumption, we have

$$\mathrm{prob}(A_1(p,g,x) = \mathrm{PSqrt}_{p,g}(x) : x \xleftarrow{u} \mathrm{QR}_n) \geq \frac{1}{2} + \varepsilon.$$

Here the probability is also taken over the random choice of $x \in \text{QR}_n$. Therefore, we must modify x randomly each time we apply A_1. For this purpose, we iteratively select $r \in \left\{ 0, \ldots, \frac{(p-1)}{2} - 1 \right\}$ at random and compute $A_1(p, g, xg^{2r})$. If $A_1(p, g, xg^{2r})$ successfully computes $\text{PSqrt}_{p,g}(xg^{2r})$, then $\text{PSqrt}_{p,g}(x) = A_1(p, g, xg^{2r}) \cdot g^{-r}$, at least if $s + r < (p-1)/2$. The latter happens with a high probability, since s is assumed to be small. Since the points xg^{2r} are sampled randomly and independently, we can then compute the principal square root $\text{PSqrt}_{p,g}(x)$ with a high probability, using a majority decision on the values $A_1(p, g, xg^{2r}) \cdot g^{-r}$. The probability of success increases linearly with the number of sampled points, and it can be computed by Corollary B.26, which is a consequence of the Weak Law of Large Numbers.

Algorithm 7.9.

$\quad$ int $B(\text{int } x)$

$\quad$ 1 $\quad C_0 \leftarrow 0, C_1 \leftarrow 0$

$\quad$ 2 $\quad \{u, v\} \leftarrow \text{Sqrt}(x)$

$\quad$ 3 $\quad$ for $i \leftarrow 1$ to t do

$\quad$ 4 $\quad\quad$ select $r \in \left\{ 0, \ldots, \frac{p-1}{2} - 1 \right\}$ at random

$\quad$ 5 $\quad\quad$ if $A_1(p, g, xg^{2r}) = ug^r$

$\quad$ 6 $\quad\quad\quad$ then $C_0 \leftarrow C_0 + 1$

$\quad$ 7 $\quad\quad\quad$ else $\quad C_1 \leftarrow C_1 + 1$

$\quad$ 8 $\quad$ if $C_0 > C_1$

$\quad$ 9 $\quad\quad$ then return u

$\quad$ 10 $\quad\quad$ else return v

We now show that

$$\text{prob}(B(x) = \text{PSqrt}_{p,g}(x)) \geq 1 - \frac{1}{k}, \text{ for } x = g^{2s}, 0 \leq s \leq \frac{p-1}{2t}.$$

Let r_i be the randomly selected elements r and $x_i = xg^{2r_i}$, $i = 1, \ldots, t$. Every element $z \in \text{QR}_p$ has a unique representation $z = xg^{2r}$, $0 \leq r < (p-1)/2$. Therefore, the uniform and random choice of r_i implies the uniform and random choice of x_i. Let $x = g^{2s}$, $0 \leq s \leq (p-1)/2t$, and $r < (t-1)(p-1)/2t$. Then

$$2s + 2r < \frac{p-1}{t} + \frac{(t-1)(p-1)}{t} = p - 1,$$

and hence

$$\text{PSqrt}_{p,g}(x) \cdot g^r = \text{PSqrt}_{p,g}(xg^{2r}).$$

Let E_1 be the event $r < (t-1)(p-1)/2t$ and E_2 be the event $A_1(p, g, xg^{2r}) = \text{PSqrt}_{p,g}(xg^{2r})$. We have $\text{prob}(E_1) \geq 1 - 1/t$ and $\text{prob}(E_2) \geq 1/2 + \varepsilon$, and we correctly compute $\text{PSqrt}_{p,g}(x) \cdot g^r$, if both events E_1 and E_2 occur. Thus, we have (denoting by $\overline{E_2}$ the complement of event E_2)

$$\text{prob}(A_1(p, g, xg^{2^r}) = \text{PSqrt}_{p,g}(x) \cdot g^r)$$
$$\geq \text{prob}(E_1 \text{ and } E_2) \geq \text{prob}(E_1) - \text{prob}(\overline{E_2})$$
$$\geq 1 - \frac{1}{t} - \left(\frac{1}{2} - \varepsilon\right) = \frac{1}{2} + \varepsilon - \frac{1}{k}\varepsilon^2 \geq \frac{1}{2} + \frac{1}{2}\varepsilon.$$

In each of the t iterations of B, $\text{PSqrt}_{p,g}(x) \cdot g^r$ is computed with probability $> 1/2$. Taking the most frequent result, we get $\text{PSqrt}_{p,g}(x)$ with a very high probability. More precisely, we can apply Corollary B.26 to the independent random variables S_i, $i = 1, \ldots, t$, defined by

$$S_i = \begin{cases} 1 \text{ if } A_1(p, g, xg^{2^{r_i}}) = \text{PSqrt}_{p,g}(x) \cdot g^{r_i}, \\ 0 \text{ otherwise.} \end{cases}$$

We have $E(S_i) = \text{prob}(S_i = 1) \geq 1/2 + \varepsilon$.

If $u = \text{PSqrt}_{p,g}(x)$, then we conclude by Corollary B.26 that

$$\text{prob}(B(x) = \text{PSqrt}_{p,g}(x)) = \text{prob}\left(C_0 > \frac{t}{2}\right) \geq 1 - \frac{4}{4t\varepsilon^2} = 1 - \frac{1}{k}.$$

The case $v = \text{PSqrt}_{p,g}(x)$ follows analogously. This completes the proof of the lemma. □

We continue with the proof of the theorem. The following algorithm computes the discrete logarithm.

Algorithm 7.10.
```
int A(int p, g, x)
 1   y ← empty word, k ← |p|, t ← kε⁻²
 2   guess j ∈ {0, ..., t − 1} satisfying j(p−1)/t ≤ Log_{p,g}(x) < (j + 1)(p−1)/t
 3   x = xg^{−⌈j(p−1)/t⌉}
 4   for c ← 1 to k do
 5       if x ∈ QR_p
 6           then y ← y‖0
 7           else  y ← y‖1
 8               x ← xg⁻¹
 9       x ← B(x)
10   return y + ⌈j(p−1)/t⌉
```

In algorithm A we use "to guess" as a basic operation. To guess the right alternative means to find out the right choice by computation.

Here, to *guess* the correct j means to carry out lines 3–9 and then test whether $y + \left\lceil \frac{j(p-1)}{t} \right\rceil$ is equal to $\text{Log}_{p,g}(x)$, for $j = 0, 1, \ldots$. The test is done by modular exponentiation. We stop if the test succeeds, i.e., if $\text{Log}_{p,g}(x)$ is computed. We have to consider at most $t = k\varepsilon^{-2} = kP^2(k)$ many intervals.

Hence, in this way we get a polynomial algorithm. This notion of "to guess" will also be used in the subsequent sections.

We have $A(p, g, x) = \text{Log}_{p,g}(x)$ if B correctly computes $\text{PSqrt}_{p,g}(x)$ in each iteration of the for loop. Thus, we get

$$\text{prob}(A(p, g, x) = \text{Log}_{p,g}(x)) \geq \left(1 - \frac{1}{k}\right)^k.$$

As $(1 - 1/k)^k$ increases monotonously (it converges to e^{-1}), this implies

$$\text{prob}(A(p, g, x) = \text{Log}_{p,g}(x)) \geq \left(1 - \frac{1}{2}\right)^2 = \frac{1}{4}.$$

By Proposition 5.7, we get, by repeating the computation $A(p, g, x)$ independently, a probabilistic polynomial algorithm $A_2(p, g, x)$ with

$$\text{prob}(A_2(p, g, x) = \text{Log}_{p,g}(x)) > 1 - 2^{-Q(k)}.$$

This concludes the proof of the theorem. □

Remarks:

1. The expectation is that A will compute $\text{Log}_{p,g}(x)$ after four repetitions (see Lemma B.12). Thus, we expect that we have to call A_1 at most $4k^3\varepsilon^{-4}$ times to compute $\text{Log}_{p,g}(x)$.
2. In [BluMic84], Blum and Micali introduced the idea to reduce the discrete logarithm problem to the problem of computing principal square roots. They developed the techniques we used to prove Theorem 7.7. In that paper they also constructed cryptographically strong pseudo-random bit generators using hard-core bits. They proved that the most-significant bit of the discrete logarithm is unpredictable and achieved as an application the discrete exponential generator (see Section 8.1).

Corollary 7.11. *Let $I = \{(p, g) \mid p \text{ prime}, g \in \mathbb{Z}_p^* \text{ a primitive root}\}$. Provided the discrete logarithm assumption is true,*

$$\text{Msb} := \left(\text{Msb}_p : \mathbb{Z}_{p-1} \longrightarrow \{0, 1\}, x \longmapsto \begin{cases} 0 \text{ if } 0 \leq x < \frac{p-1}{2}, \\ 1 \text{ if } \frac{p-1}{2} \leq x < p - 1 \end{cases}\right)_{(p,g)\in I}$$

is a family of hard-core predicates for the Exp family

$$\text{Exp} = (\text{Exp}_{p,g} : \mathbb{Z}_{p-1} \longrightarrow \mathbb{Z}_p^*, x \longmapsto g^x \bmod p)_{(p,g)\in I}.$$

Proof. Assume Msb is not a family of hard-core predicates for Exp. Then, there is a positive polynomial $P \in \mathbb{Z}[X]$ and an algorithm A_1 such that for infinitely many k

$$\text{prob}(A_1(p, g, g^x) = \text{Msb}_p(x) : (p, g) \leftarrow I_k, x \xleftarrow{u} \mathbb{Z}_{p-1}) > \frac{1}{2} + \frac{1}{P(k)}.$$

By Proposition 6.17, there are positive polynomials Q, R such that

$$\mathrm{prob}\left(\left\{(p,g) \in I_k \mid \mathrm{prob}(A_1(p,g,g^x) = \mathrm{Msb}_p(x) : x \overset{u}{\leftarrow} \mathbb{Z}_{p-1})\right.\right.$$
$$\left.\left. > \frac{1}{2} + \frac{1}{Q(k)}\right\}\right) > \frac{1}{R(k)},$$

for infinitely many k. From the theorem and the remark after Definition 7.2 above, we conclude that there is an algorithm A_2 and a positive polynomial S such that

$$\mathrm{prob}\left(\left\{(p,g) \in I_k \mid \mathrm{prob}(A_2(p,g,g^x) = x : x \overset{u}{\leftarrow} \mathbb{Z}_{p-1}) \geq 1 - \frac{1}{S(k)}\right\}\right)$$
$$> \frac{1}{R(k)}$$

for infinitely many k. By Proposition 6.3, there is a positive polynomial T such that

$$\mathrm{prob}(A_2(p,g,g^x) = x : (p,g) \leftarrow I_k, x \overset{u}{\leftarrow} \mathbb{Z}_{p-1}) > \frac{1}{T(k)}$$

for infinitely many k, a contradiction to the discrete logarithm assumption (Definition 6.1).   □

Remark. Suppose that $p - 1 = 2^t a$, where a is odd. The t least significant bits of x can be easily computed from $g^x = \mathrm{Exp}_{p,g}(x)$ (see Exercise 3 of this chapter). But all the other bits of x are secure bits, i.e., each of them yields a hard-core predicate, as shown by Håstad and Näslund ([HåsNäs98]; [HåsNäs99]).

7.2 Bit Security of the RSA Family

Let $\mathrm{Lsb}(x) := x \bmod 2$ be the least-significant bit of $x \in \mathbb{Z}, x \geq 0$, with respect to the binary encoding of x as an unsigned number. In order to compute $\mathrm{Lsb}(x)$ for an element x in $\mathbb{Z}_n$, we apply Lsb to the representative of x between 0 and $n - 1$.

In this section, we study the RSA function

$$\mathrm{RSA}_{n,e} : \mathbb{Z}_n^* \longrightarrow \mathbb{Z}_n^*, \ x \longmapsto x^e$$

and its inverse $\mathrm{RSA}_{n,e}^{-1}$, for $n = pq$, with p and q odd, distinct primes, and e prime to $\varphi(n)$, $1 < e < \varphi(n)$.

To compute $\mathrm{Lsb}(x)$ from $y = x^e$ is as difficult as to compute x from y. The following Proposition 7.12 and Theorem 7.14 make this statement more precise. The proofs show how to reduce the computation of x from y to the computation of $\mathrm{Lsb}(x)$ from y.

First, we study the deterministic case where $\mathrm{Lsb}(x)$ can be computed from y by a deterministic algorithm in polynomial time.

Proposition 7.12. *Let A_1 be a deterministic polynomial algorithm such that*

$$A_1(n, e, x^e) = \mathrm{Lsb}(x) \text{ for all } x \in \mathbb{Z}_n^*,$$

where $n := pq$, p and q are odd distinct prime numbers, and e is relatively prime to $\varphi(n)$. Then there is a deterministic polynomial algorithm A_2 such that

$$A_2(n, e, x^e) = x \text{ for all } x \in \mathbb{Z}_n^*.$$

Proof. Let $x \in \mathbb{Z}_n^*$ and $y = x^e$. The basic idea of the inversion algorithm is to compute $a \in \mathbb{Z}_n^*$ and a rational number $u \in \mathbb{Q}, 0 \leq u < 1$, with

$$|ax \bmod n - un| < \frac{1}{2}.$$

Then we have $ax \bmod n = \lfloor un + 1/2 \rfloor$, and hence $x = a^{-1} \lfloor un + 1/2 \rfloor \bmod n$. This method to invert the RSA function is called *rational approximation*. We approximate $ax \bmod n$ by the rational number un.

For $z \in \mathbb{Z}$, let $\bar{z} := z \bmod n$. Let 2^{-1} denote the inverse element of $2 \bmod n$ in $\mathbb{Z}_n^*$. We start with $u_0 = 0$ and $a_0 = 1$ to get an approximation for $\overline{a_0 x}$ with

$$|\overline{a_0 x} - u_0 n| < n.$$

We define

$$a_t := 2^{-1} a_{t-1} \text{ and}$$

$$u_t := \frac{1}{2}(u_{t-1} + \mathrm{Lsb}(\overline{a_{t-1} x}))$$

(the last computation is done in $\mathbb{Q}$). In each step, we replace a_{t-1} by a_t and u_{t-1} by u_t, and we observe that

$$\overline{a_t x} = \overline{2^{-1} a_{t-1} x} = \begin{cases} \frac{1}{2}\overline{a_{t-1}x} & \text{if } \overline{a_{t-1}x} \text{ is even,} \\ \frac{1}{2}(\overline{a_{t-1}x} + n) & \text{if } \overline{a_{t-1}x} \text{ is odd,} \end{cases}$$

and hence

$$|\overline{a_t x} - u_t n| = \frac{1}{2}|\overline{a_{t-1}x} - u_{t-1}n|.$$

After $r = |n| + 1$ steps, we reach

$$|\overline{a_r x} - u_r n| < \frac{n}{2^r} < \frac{1}{2}.$$

Since $\mathrm{Lsb}(\overline{a_t x}) = A_1(n, e, a_t^e y \bmod n)$, we can decide whether $\overline{a_t x}$ is even without knowing x. Thus, we can compute a_t and u_t in each step, and finally get x. The following algorithm inverts the RSA function.

Algorithm 7.13.

   int A_2(int n, e, y)
  1   $a \leftarrow 1, u \leftarrow 0, k \leftarrow |n|$
  2   for $t \leftarrow 0$ to k do
  3      $u \leftarrow \frac{1}{2}(u + A_1(n, e, a^e y \bmod n))$
  4      $a \leftarrow 2^{-1} a \bmod n$
  5   return $a^{-1} \lfloor un + \frac{1}{2} \rfloor \bmod n$

This completes the proof of the Proposition. $\qquad\qquad\qquad\qquad$ □

Next we study the probabilistic case. Now the algorithm A_1 does not compute the predicate $\mathrm{Lsb}(x)$ deterministically, but only with a probability slightly better than guessing it at random. Nevertheless, RSA can be inverted with a high probability.

Theorem 7.14. *Let $P, Q \in \mathbb{Z}[X]$ be positive polynomials and A_1 be a probabilistic polynomial algorithm such that*

$$\mathrm{prob}(A_1(n, e, x^e) = \mathrm{Lsb}(x) : x \overset{u}{\leftarrow} \mathbb{Z}_n^*) \geq \frac{1}{2} + \frac{1}{P(k)},$$

where $n := pq$, $k := |n|$, p and q are odd distinct primes and e is relatively prime to $\varphi(n)$. Then there is a probabilistic polynomial algorithm A_2 such that

$$\mathrm{prob}(A_2(n, e, x^e) = x) \geq 1 - 2^{-Q(k)} \text{ for all } x \in \mathbb{Z}_n^*.$$

Proof. Let $y := x^e$ and let $\varepsilon := 1/P(k)$. As in the deterministic case, we use rational approximation to invert the RSA function. We try to approximate $ax \bmod n$ by a rational number un. To invert RSA correctly, we have to compute $\mathrm{Lsb}(ax)$ correctly in each step. However, now we only know that the probability for k correct computations of Lsb is $\geq (1/2 + \varepsilon)^k$, which is exponentially close to 0 and thus too small. In order to increase the probability of success, we develop the algorithm L. The probability of success of L is sufficiently high. This is the statement of the following lemma.

Lemma 7.15. *Under the assumptions of the theorem, there is a probabilistic polynomial algorithm L with the following properties: given $y := x^e$, randomly chosen $a, b \in \mathbb{Z}_n^*$,[2] $\alpha := \mathrm{Lsb}(ax \bmod n)$, $\beta := \mathrm{Lsb}(bx \bmod n)$, $u \in \mathbb{Q}$ with $|ax \bmod n - un| \leq \varepsilon^3 n/8$ and $v \in \mathbb{Q}$ with $|bx \bmod n - vn| \leq \varepsilon n/8$, then L successively computes values $l_t, t = 0, 1, 2, \ldots, k$, such that*

---

[2] Actually we randomly choose $a, b \in \mathbb{Z}_n$. In the rare case that $a, b \notin \mathbb{Z}_n^*$, we can factor n using Euclid's algorithm and compute x from x^e.

$$\text{prob}(l_t = \text{Lsb}(a_t x \bmod n)| \wedge_{j=0}^{t-1} l_j = \text{Lsb}(a_j x \bmod n) : a, b \xleftarrow{u} \mathbb{Z}_n^*) \geq 1 - \frac{1}{2k},$$

where $a_0 := a, a_t := 2^{-1}a_{t-1}$.

Proof (of the Lemma). Let $m := \min(2^t \varepsilon^{-2}, 2k\varepsilon^{-2})$. We may assume that both primes p and q are $> m$. Namely, if one of the primes is $\leq m$, then we may factorize n in polynomial time (and then easily compute the inverse of the RSA function), simply by checking whether one of the polynomially many numbers $\leq m$ divide n.

To compute $\text{Lsb}(a_t x \bmod n)$, L will apply A_1 m times. In each step, a least-significant bit is computed and the return value of L is the more frequently occurring bit. In the theorem, we assume that

$$\text{prob}(A_1(n, e, x^e) = \text{Lsb}(x) : x \xleftarrow{u} \mathbb{Z}_n^*) \geq \frac{1}{2} + \varepsilon.$$

The probability is also taken over the random choice of $x \in \mathbb{Z}_n^*$. Thus, we cannot simply repeat the execution of A_1 with the same input $(a_t x)^e$, but have to modify the input randomly. We use the modifiers $a_t + ia_{t-1} + b$, $i \in \mathbb{Z}, -m/2 \leq i \leq m/2 - 1$, and compute $\text{Lsb}((a_t + ia_{t-1} + b)x \bmod n)$.[3] As we will see below, the assumptions of the lemma guarantee that we can infer $\text{Lsb}(a_t x \bmod n)$ from $\text{Lsb}((a_t + ia_{t-1} + b)x \bmod n)$ with high probability (here sufficiently good rational approximations u_t of $a_t x \bmod n$ and v of $bx \bmod n$ are needed). a and b are chosen independently and at random, because then the modifiers $a_t + ia_{t-1} + b$ are pairwise independent. Then Corollary B.26, which is a consequence of the Weak Law of Large Numbers, applies. This implies that the probability of success of L is as large as desired.

We now describe how L works on input of $y = x^e, a, b, \alpha, \beta, u$ and v to compute $l_t, t = 0, 1, \ldots, k$. In its computation, L uses the variable a_t to store the current a_t, and the variable a_{t-1} to store the a_t from the preceding iteration. Analogously, we use variables u_t and u_{t-1}. $u_t n$ is a rational approximation of $a_t x$. We have $u_0 = u$ and $u_t = 1/2 (u_{t-1} + l_{t-1})$.

It is the goal of L to return $l_t = \text{Lsb}(a_t x \bmod n)$ for $t = 0, 1, \ldots, k$. The first iteration $t = 0$ is easy: l_0 is the given α. From now on, the variable α is used to store the last computed l_t.

Before L starts to compute $l_1, l_2, \ldots, l_k$, its variables are initialized by

$$a_{t-1} := a_0 = a, u_{t-1} := u.$$

To compute $l_t, t \geq 1$, L repeats the following subroutine.

---

[3] If $a_t + ia_{t-1} + b \notin \mathbb{Z}_n^*$, we factor n using Euclid's algorithm and compute x from x^e.

Algorithm 7.16.

$L'(\ )$

```
 1   C_0 ← 0; C_1 ← 0
 2   a_t ← 2^{-1}a_{t-1}; u_t ← \frac{1}{2}(u_{t-1} + α)
 3   for i ← -\frac{m}{2} to \frac{m}{2} - 1 do
 4       A ← a_t + ia_{t-1} + b
 5       W ← ⌊u_t + iu_{t-1} + v⌋
 6       B ← (iα + β + W) mod 2
 7       if A_1(n, e, A^e y mod n) ⊕ B = 0
 8           then C_0 ← C_0 + 1
 9           else C_1 ← C_1 + 1
10   u_{t-1} ← u_t, a_{t-1} ← a_t
11   if C_0 > C_1
12       then α ← 0
13       else α ← 1
14   return α
```

For $z \in \mathbb{Z}$, we denote by $\overline{z}$ the remainder of z modulo n.
For $i \in \{-m/2, \ldots, m/2 - 1\}$, let
$A_{t,i} := a_t + ia_{t-1} + b$,
$W'_{t,i} := u_t + iu_{t-1} + v, W_{t,i} = \lfloor W'_{t,i} \rfloor$,
$B_{t,i} := (i \cdot \mathrm{Lsb}(\overline{a_{t-1}x}) + \mathrm{Lsb}(\overline{bx}) + \mathrm{Lsb}(W_{t,i})) \bmod 2$.

We want to compute $\mathrm{Lsb}(\overline{a_t x})$ from $\mathrm{Lsb}(\overline{A_{t,i}x})$, $\mathrm{Lsb}(\overline{a_{t-1}x})$ and $\mathrm{Lsb}(\overline{bx})$. For this purpose, let $\lambda_{t,i} := \overline{a_t x} + i \cdot \overline{a_{t-1}x} + \overline{bx} = qn + \overline{A_{t,i}x}$ with $q = \lfloor \lambda_{t,i}/n \rfloor$. Then $\mathrm{Lsb}(\lambda_{t,i}) = (\mathrm{Lsb}(\overline{a_t x}) + i \cdot \mathrm{Lsb}(\overline{a_{t-1}x}) + \mathrm{Lsb}(\overline{bx})) \bmod 2$ and

$$\mathrm{Lsb}(\overline{A_{t,i}x}) = (\mathrm{Lsb}(\lambda_{t,i}) + \mathrm{Lsb}(q)) \bmod 2$$
$$= (\mathrm{Lsb}(\overline{a_t x}) + i \cdot \mathrm{Lsb}(\overline{a_{t-1}x}) + \mathrm{Lsb}(\overline{bx}) + \mathrm{Lsb}(q)) \bmod 2,$$

and we obtain

$$\mathrm{Lsb}(\overline{a_t x}) = (\mathrm{Lsb}(\overline{A_{t,i}x}) + i \cdot \mathrm{Lsb}(\overline{a_{t-1}x}) + \mathrm{Lsb}(\overline{bx}) + \mathrm{Lsb}(q)) \bmod 2.$$

The problem is to get q and $\mathrm{Lsb}(q)$. We will show that $W_{t,i}$ is equal to q with a high probability, and $W_{t,i}$ is easily computed from the rational approximations u_t of $\overline{a_t x}$, u_{t-1} of $\overline{a_{t-1}x}$ and v of $\overline{bx}$. If $W_{t,i} = q$, we have

$$\mathrm{Lsb}(\overline{a_t x}) = \mathrm{Lsb}(\overline{A_{t,i}x}) \oplus B_{t,i}.$$

We assume from now on that L computed the least-significant bit correctly in the preceding steps:

$$\mathrm{Lsb}(\overline{a_j x}) = l_j, 0 \le j \le t - 1.$$

Next, we give a lower bound for the probability that $W_{t,i} = q$. Let $Z = |\lambda_{t,i} - W'_{t,i}n|$.

$$Z = |\overline{a_t x} - u_t n + i(\overline{a_{t-1} x} - u_{t-1} n) + \overline{bx} - vn|$$

$$\leq \left|\frac{1}{2}(\overline{a_{t-1} x} - u_{t-1} n)(1 + 2i)\right| + |\overline{bx} - vn|$$

$$\leq \frac{n}{2^t}\frac{\varepsilon^3}{8}|1 + 2i| + \frac{\varepsilon}{8}n \leq \frac{\varepsilon}{8}n\left(\frac{\varepsilon^2 m}{2^t} + 1\right) \leq \frac{\varepsilon}{4}n.$$

Note that $|\overline{a_j x} - u_j n| = 1/2\,|(\overline{a_{j-1} x} - u_{j-1} n)|$ for $1 \leq j \leq t$ under our assumption $l_j = \mathrm{Lsb}(\overline{a_j x})$, $0 \leq j \leq t - 1$ (see the proof of Proposition 7.12). Moreover, $|1 + 2i| \leq m$, because $-m/2 \leq i \leq m/2 - 1$, and $m = \min(2^t \varepsilon^{-2}, 2k\varepsilon^{-2})$.

Now $W_{t,i} \neq q$ if and only if there is a multiple of n between $\lambda_{t,i}$ and $W'_{t,i} n$. There is no multiple of n between $\lambda_{t,i}$ and $W'_{t,i} n$ if

$$\frac{\varepsilon}{4}n < \lambda_{t,i} = \overline{A_{t,i} x} < n - \frac{\varepsilon}{4}n,$$

because $Z \leq \varepsilon/4\,n$.

Since $a, b \in \mathbb{Z}_n^*$ are selected uniformly and at random, the remainders $\overline{\lambda_{t,i}} = (a_t + i a_{t-1} + b)x \bmod n = ((2^{-1} + i)a_{t-1} + bx) \bmod n$ are also selected uniformly and at random.

This implies

$$\mathrm{prob}(W_{t,i} = q) \geq \mathrm{prob}\left(\frac{\varepsilon}{4}n < \overline{A_{t,i} x} < n - \frac{\varepsilon}{4}n\right) \geq 1 - \frac{\varepsilon}{2}.$$

We are now ready to show

$$\mathrm{prob}(l_t = \mathrm{Lsb}(\overline{a_t x})|\wedge_{j=0}^{t-1} l_j = \mathrm{Lsb}(\overline{a_j x})\}) \geq 1 - \frac{1}{2k}.$$

Let $E_{1,i}$ be the event $A_1(n, e, \overline{A_{t,i}^e y}) = \mathrm{Lsb}(\overline{A_{t,i} x})$ (here recall that $y = x^e$), and let $E_{2,i}$ be the event that $\overline{A_{t,i} x}$ satisfies the condition

$$\frac{\varepsilon}{4}n < \overline{A_{t,i} x} < n - \frac{\varepsilon}{4}n.$$

We have $\mathrm{prob}(E_{1,i}) \geq 1/2 + \varepsilon$ and $\mathrm{prob}(E_{2,i}) = 1 - \varepsilon/2$. We define random variables

$$S_i := \begin{cases} 1 \text{ if } E_{1,i} \text{ and } E_{2,i} \text{ occur,} \\ 0 \text{ otherwise.} \end{cases}$$

We do not err in computing $\mathrm{Lsb}(a_t x)$ in the i-th step of our algorithm L if both events $E_{1,i}$ and $E_{2,i}$ occur, i.e., if $S_i = 1$.

We have (denoting by $\overline{E_{1,i}}$ the complement of event $E_{1,i}$)

$$\mathrm{prob}(S_i = 1) = \mathrm{prob}(E_{1,i} \text{ and } E_{2,i}) \geq \mathrm{prob}(E_{2,i}) - \mathrm{prob}(\overline{E_{1,i}})$$

$$> \left(1 - \frac{\varepsilon}{2}\right) - \left(\frac{1}{2} - \varepsilon\right) = \frac{1}{2} + \frac{\varepsilon}{2}.$$

Let $i \neq j$. The probabilities $\text{prob}(S_i = d)$ and $\text{prob}(S_j = d)$ $(d \in \{0,1\})$ are taken over the random choice of $a, b \in \mathbb{Z}_n^*$ and the coin tosses of $A_1(n, e, \overline{A_{t,i}^e y})$ and $A_1(n, e, \overline{A_{t,j}^e y})$. The elements $a_0 = a$ and b are chosen independently (and uniformly), and we have $(\overline{A_{t,i}}, \overline{A_{t,j}}) = (a_{t-1}, b)\Delta = (2^{-t+1}a, b)\Delta$ with the invertible matrix

$$\Delta = \begin{pmatrix} 2^{-1}+i & 2^{-1}+j \\ 1 & 1 \end{pmatrix}$$

over $\mathbb{Z}_n^*$. Thus, $\overline{A_{t,i}}$ and $\overline{A_{t,j}}$ are also independent. This implies that the events $E_{2,i}$ and $E_{2,j}$ are independent and that the inputs $\overline{A_{t,i}^e y}$ and $\overline{A_{t,j}^e y}$ are independent random elements. Since the coin tosses during an execution of A_1 are independent of all other random events (see Chapter 5), the events $E_{1,i}$ and $E_{1,j}$ are also independent. We see that S_i and S_j are indeed independent. Note that the determinant $i - j$ of Δ is in $\mathbb{Z}_n^*$, since $|i - j| < m < \min\{p, q\}$.

The number of i, $-m/2 \leq i \leq m/2 - 1$, and hence the number of random variables S_i, is m. By Corollary B.26, we conclude

$$\text{prob}\left(\sum_i S_i > \frac{m}{2}\right) \geq 1 - \frac{1}{m\varepsilon^2} \geq 1 - \frac{1}{2k}.$$

Recall that we do not err in computing $\text{Lsb}(a_t x)$ in the i-th step of our algorithm L, if both events $E_{1,i}$ and $E_{2,i}$ occur, i.e., if $S_i = 1$. Thus, we have $C_0 \geq \sum_i S_i$ and hence $\text{prob}(C_0 > C_1) \geq 1 - 1/2k$ if $\text{Lsb}(\overline{a_t x}) = 0$, and $C_1 \geq \sum_i S_i$ and hence $\text{prob}(C_1 > C_0) \geq 1 - 1/2k$ if $\text{Lsb}(\overline{a_t x}) = 1$. Therefore, we have shown that

$$\text{prob}(l_t = \text{Lsb}(\overline{a_t x})| \wedge_{j=0}^{t-1} l_j = \text{Lsb}(\overline{a_j x})) \geq 1 - \frac{1}{2k}.$$

The proof of the lemma is complete.   □

We continue in the proof of the theorem. The following algorithm A (Algorithm 7.17) inverts the RSA function by the method of rational approximation. The basic structure of A is the same as that of Algorithm 7.13. We call L to compute $\text{Lsb}(ax)$. Therefore, we must meet the assumptions of Lemma 7.15. This is done in lines 1–4 of algorithm A. Algorithm L computes $l_0, \ldots, l_k$ in advance. Lines 7 and 8 of A also appear in L. In a real and efficient implementation, it is possible to avoid this redundancy.

As above in Algorithm 7.10, we can "guess" the right alternative. This means we can find out the right alternative in polynomial time. There are only a polynomial number of alternatives, and both the computation for each alternative as well as checking the result can be done in polynomial time. In order to guess u or v, we have to consider $8/\varepsilon^3 = 8P(k)^3$ and $8/\varepsilon = 8P(k)$ many intervals. There are only two alternatives for $\text{Lsb}(\overline{ax})$ and $\text{Lsb}(\overline{bx})$.

Algorithm 7.17.

int $A(\text{int } n, e, y)$

1   select $a, b \in \mathbb{Z}_n^*$ at random
2   guess $u, v \in \mathbb{Q} \cap [0, 1[$ satisfying
3   $|ax \bmod n - un| \leq \frac{\varepsilon^3 n}{8}, |bx \bmod n - vn| \leq \frac{\varepsilon n}{8}$
4   guess $\alpha \leftarrow \text{Lsb}(ax \bmod n)$, guess $\beta \leftarrow \text{Lsb}(bx \bmod n)$
5   Compute $l_0, l_1, \ldots, l_k$ by L
6   for $t \leftarrow 0$ to k do
7       $u \leftarrow \frac{1}{2}(u + l_t)$
8       $a \leftarrow 2^{-1}a \bmod n$
9   return $a^{-1} \lfloor un + \frac{1}{2} \rfloor \bmod n$

$A(n, e, y) = \text{RSA}_{n,e}^{-1}(y) = x$ for $y = x^e$ if L correctly computes $l_t = \text{Lsb}(\overline{a_t x})$ for $t = 1, \ldots, k$. Thus, we have

$$\text{prob}(A(n, e, y) = x) \geq \left(1 - \frac{1}{2k}\right)^k.$$

Since $(1 - 1/2k)^k$ increases monotonously (converging to $e^{-1/2}$), we conclude

$$\text{prob}(A(n, e, y) = x) \geq \frac{1}{2}.$$

Repeating the computation $A(n, e, y)$ independently, we get, by Proposition 5.7, a probabilistic polynomial algorithm $A_2(n, e, y)$ with

$$\text{prob}(A_2(n, e, y) = \text{RSA}_{n,e}^{-1}(y)) \geq 1 - 2^{-Q(k)},$$

and Theorem 7.14 is proven.   □

Remarks:

1. The expectation is that A will compute $\text{RSA}_{n,e}^{-1}(y)$ after two repetitions (see Lemma B.12). The input of A_1 does not depend on the guessed elements u, v, α and β (it only depends on a and b). Thus, we can also use the return values of A_1, computed for the first guess of u, v, α and β, for all subsequent guesses. Then we expect that we have to call A_1 at most $4k^2\varepsilon^{-2}$ times to compute $\text{RSA}_{n,e}^{-1}(y)$.
2. The bit security of the RSA family was first studied in [GolMicTon82], in which a method for inverting RSA by guessing the least-significant bit was introduced (see Exercise 11).
   The problem of inverting RSA, if the least-significant bit is predicted only with probability $\geq 1/2 + 1/P(k)$, is studied in [SchnAle84], [VazVaz84], [AleChoGolSch88] and [FisSch00]. The technique we used to prove Theorem 7.14 is from [FisSch00].

Corollary 7.18. *Let* $I := \{(n,e) \mid n = pq, p \text{ and } q \text{ odd distinct primes,}$
$|p| = |q|, 1 < e < \varphi(n), e \text{ prime to } \varphi(n)\}$. *Provided the RSA assumption is true, then*

$$\text{Lsb} = (\text{Lsb}_{n,e} : \mathbb{Z}_n^* \longrightarrow \{0,1\}, x \longmapsto \text{Lsb}(x))_{(n,e) \in I}$$

is a family of hard-core predicates for the RSA family

$$\text{RSA} = (\text{RSA}_{n,e} : \mathbb{Z}_n^* \longrightarrow \mathbb{Z}_n^*, x \longmapsto x^e)_{(n,e) \in I}.$$

Proof. The proof is analogous to the proof of Corollary 7.11.   □

Remark. Håstad and Näslund have shown that all the plaintext bits are secure bits of the RSA function, i.e., each of them yields a hard-core predicate ([HåsNäs98]; [HåsNäs99]).

7.3 Bit Security of the Square Family

Let $n := pq$, with p and q distinct primes, and $p, q \equiv 3 \bmod 4$. We consider the (bijective) modular squaring function

$$\text{Square}_n : \text{QR}_n \longrightarrow \text{QR}_n, x \longmapsto x^2$$

and its inverse, the modular square root function

$$\text{Sqrt}_n : \text{QR}_n \longrightarrow \text{QR}_n, y \longmapsto \text{Sqrt}_n(y)$$

(see Section 6.5). The computation of $\text{Sqrt}_n(y)$ can be reduced to the computation of the least-significant bit $\text{Lsb}(\text{Sqrt}_n(y))$ of the square root. This is shown in Proposition 7.19 and Theorem 7.22. In the proposition, the algorithm that computes the least-significant bit is assumed to be deterministic polynomial. Then the algorithm which we obtain by the reduction is also deterministic polynomial. It computes $\text{Sqrt}_n(y)$ for all $y \in \text{QR}_n$ (under the additional assumption $n \equiv 1 \bmod 8$).

Proposition 7.19. *Let* A_1 *be a deterministic polynomial algorithm such that*

$$A_1(n, x^2) = \text{Lsb}(x) \text{ for all } x \in \text{QR}_n,$$

where $n = pq$, p *and* q *are distinct primes,* $p, q \equiv 3 \bmod 4$ *and* $n \equiv 1 \bmod 8$. *Then there exists a deterministic polynomial algorithm* A_2 *such that*

$$A_2(n, y) = \text{Sqrt}_n(y) \text{ for all } y \in \text{QR}_n.$$

Proof. As in the RSA case, we use rational approximation to invert the Square function. Let $y = x^2, x \in QR_n$. Since n is assumed to be $\equiv 1 \bmod 8$, either $2 \in QR_n$ or $-2 \in QR_n$ (see the remark following this proof).

First, let $2 \in QR_n$. Then $2^{-1} \in QR_n$. We define $a_0 := 1$, $a_t := 2^{-1}a_{t-1} \bmod n$ for $t \geq 1$, and $u_0 := 1$, $u_t := 1/2\,(u_{t-1} + \mathrm{Lsb}(a_{t-1}x \bmod n))$ for $t \geq 1$. Since $2^{-1} \in QR_n$, we have $a_t \in QR_n$ and hence $a_t x \in QR_n$ for all $t \geq 1$. Thus, $\mathrm{Sqrt}_n(a_t^2 x^2) = a_t x$, and hence we can compute $\mathrm{Lsb}(a_t x \bmod n) = A_1(n, a_t^2 x^2) = A_1(n, a_t^2 y)$ by A_1 for all $t \geq 1$. The rational approximation works as in the RSA case.

Now, let $-2 \in QR_n$. Then $-2^{-1} \in QR_n$. We modify the method of rational approximation and define $a_0 = 1$, $a_t = -2^{-1}a_{t-1} \bmod n$ for $t \geq 1$, and $u_0 = 1$, $u_t = 1/2\,(2 - \mathrm{Lsb}(a_{t-1}x \bmod n) - u_{t-1})$ for $t \geq 1$. Then, we get

$$|a_t x \bmod n - u_t n| = \frac{1}{2}\,|(a_{t-1}x \bmod n - u_{t-1}n)|,$$

because

$$a_t x \bmod n = -2^{-1}a_{t-1}x \bmod n = 2^{-1}(n - a_{t-1}x \bmod n)$$

$$= \begin{cases} \frac{1}{2}(n - a_{t-1}x \bmod n) & \text{if } a_{t-1}x \bmod n \text{ is odd,} \\ \frac{1}{2}(n - a_{t-1}x \bmod n + n) & \text{otherwise.} \end{cases}$$

After $r = |n| + 1$ steps we reach

$$|a_r x - u_r n| \leq \frac{n}{2^r} < \frac{1}{2}.$$

Since $-2^{-1} \in QR_n$, we have $a_t \in QR_n$ and hence $a_t x \in QR_n$ for $t \geq 1$. Thus, $\mathrm{Sqrt}_n(a_t^2 x^2) = a_t x$, and hence we can compute $\mathrm{Lsb}(a_t x \bmod n) = A_1(n, a_t^2 x^2) = A_1(n, a_t^2 y)$ by A_1 for all $t \geq 1$.   □

Remarks:

1. As $p, q \equiv 3 \bmod 4$ is assumed, p and q are $\equiv 3 \bmod 8$ or $\equiv -1 \bmod 8$, and hence either $n \equiv 1 \bmod 8$ or $n \equiv 5 \bmod 8$. We do not consider the case $n \equiv 5 \bmod 8$ in the proposition.

2. The proof actually works if we have $2 \in QR_n$ or $-2 \in QR_n$. This is equivalent to $n \equiv 1 \bmod 8$, as follows from Theorem A.59. We have $2 \in QR_n$ if and only if $2 \in QR_p$ and $2 \in QR_q$, and this in turn is equivalent to $p \equiv q \equiv -1 \bmod 8$ (by Theorem A.59). On the other hand, $-2 \in QR_n$ if and only if $-2 \in QR_p$ and $-2 \in QR_q$, and this in turn is equivalent to $p \equiv q \equiv 3 \bmod 8$ (by Theorem A.59).

Studying the Square function, we face, in the probabilistic case, an additional difficulty compared to the reduction in the RSA case. Membership in the domain of the Square function – the set QR_n of quadratic residues – is not efficiently decidable without knowing the factorization of n. To overcome this difficulty, we develop a probabilistic polynomial reduction of the quadratic residuosity property to the predicate defined by the least-significant bit of Sqrt. Then the same reduction as in the RSA case also works for the Sqrt function.

Let $J_n^{+1} = \{x \in \mathbb{Z}_n^* \mid \left(\frac{x}{n}\right) = +1\}$. The predicate

$$\mathrm{PQR}_n : J_n^{+1} \longrightarrow \{0,1\}, \mathrm{PQR}_n(x) = \begin{cases} 1 \text{ if } x \in \mathrm{QR}_n, \\ 0 \text{ otherwise,} \end{cases}$$

is believed to be a trapdoor predicate (see Definition 6.11). In Proposition 7.20, we show how to reduce the computation of $\mathrm{PQR}_n(x)$ to the computation of the least-significant bit of a $\mathrm{Sqrt}_n(x)$.

Proposition 7.20. *Let $P, Q \in \mathbb{Z}[X]$ be positive polynomials, and let A_1 be a probabilistic polynomial algorithm such that*

$$\mathrm{prob}(A_1(n, x^2) = \mathrm{Lsb}(x) : x \xleftarrow{u} \mathrm{QR}_n) \geq \frac{1}{2} + \frac{1}{P(k)},$$

where $n = pq$, p and q are distinct primes, $p, q \equiv 3 \bmod 4$, and $k = |n|$. Then there exists a probabilistic polynomial algorithm A_2 such that

$$\mathrm{prob}(A_2(n, x) = \mathrm{PQR}_n(x)) \geq 1 - \frac{1}{Q(k)} \text{ for all } x \in J_n^{+1}.$$

Proof. Let $x \in J_n^{+1}$. If $x \in \mathrm{QR}_n$, then $x = \mathrm{Sqrt}_n(x^2)$ and therefore $\mathrm{Lsb}(x) = \mathrm{Lsb}(\mathrm{Sqrt}_n(x^2))$. If $x \notin \mathrm{QR}_n$, then $-x \bmod n = n - x = \mathrm{Sqrt}_n(x^2)$ and $\mathrm{Lsb}(x) \neq \mathrm{Lsb}(\mathrm{Sqrt}_n(x^2))$ (note that $-1 \in J_n^{+1} \setminus \mathrm{QR}_n$ for $p, q \equiv 3 \bmod 4$, by Theorem A.59). Consequently, we get

$$\mathrm{PQR}_n(x) = \mathrm{Lsb}(x) \oplus \mathrm{Lsb}(\mathrm{Sqrt}_n(x^2)) \oplus 1.$$

Since for each $y \in \mathrm{QR}_n$ there are exactly two elements $x \in J_n^{+1}$ with $x^2 = y$ and because $|J_n^{+1}| = 2|\mathrm{QR}_n|$, we conclude

$$\mathrm{prob}(A_1(n, x^2) = \mathrm{Lsb}(\mathrm{Sqrt}_n(x^2)) : x \xleftarrow{u} J_n^{+1})$$
$$= \mathrm{prob}(A_1(n, y) = \mathrm{Lsb}(\mathrm{Sqrt}_n(y)) : y \xleftarrow{u} \mathrm{QR}_n).$$

Hence, we get

$$\mathrm{prob}(\mathrm{PQR}_n(x) = A_1(n, x^2) \oplus \mathrm{Lsb}(x) \oplus 1 : x \xleftarrow{u} J_n^{+1}) \geq \frac{1}{2} + \frac{1}{P(k)}.$$

We construct an algorithm A_2 such that for every $x \in J_n^{+1}$

$$\mathrm{prob}(A_2(n, x) = \mathrm{PQR}_n(x)) \geq 1 - \frac{1}{Q(k)}.$$

Algorithm 7.21.

int A_2(int n, x)

1    $c \leftarrow 0, l \leftarrow \left\lceil \frac{Q(k)P^2(k)}{4} \right\rceil$

2   for $i \leftarrow 1$ to l do

3   select $r \in QR_n$ at random

4            $c \leftarrow c + (A_1(n, (rx)^2 \bmod n) \oplus \mathrm{Lsb}(rx \bmod n) \oplus 1)$

5   if $c > \frac{l}{2}$

6   then return 1

7   else return 0

Let $x \in J_n^{+1}$. $\mathrm{PQR}_n(x) = \mathrm{PQR}_n(rx)$ is computed l times by applying A_1 to the l independent random inputs rx. We compute $\mathrm{PQR}_n(rx)$ in each step with a probability $> 1/2$. The Weak Law of Large Numbers guarantees that, for sufficiently large l, we can compute $\mathrm{PQR}_n(x)$ by the majority of the results, with a high probability. More precisely, let the random variable S_i, $i = 1, \ldots, l$, be defined by

$$S_i = \begin{cases} 1 \text{ if } \mathrm{PQR}_n(rx) = A_1(n, (rx)^2) \oplus \mathrm{Lsb}(rx) \oplus 1, \\ 0 \text{ otherwise,} \end{cases}$$

with $r \in QR_n$ randomly chosen, as in the algorithm. Then we have $\mathrm{prob}(S_i = 1) \geq 1/2 + 1/P(k)$. The random variables S_i, $i = 1, \ldots, l$, are independent. If $\mathrm{PQR}_n(x) = 1$, we get by Corollary B.26

$$\mathrm{prob}(c > \frac{l}{2}) \geq 1 - \frac{P^2(k)}{4l} \geq 1 - \frac{1}{Q(k)} .$$

The case $\mathrm{PQR}_n(x) = 0$ follows analogously. Thus, we have shown

$$\mathrm{prob}(A_2(n, x) = \mathrm{PQR}_n(x)) \geq 1 - \frac{1}{Q(k)} ,$$

as desired.   □

Theorem 7.22. *Let* $P, Q \in \mathbb{Z}[X]$ *be positive polynomials and* A_1 *be a probabilistic polynomial algorithm such that*

$$\mathrm{prob}(A_1(n, x^2) = \mathrm{Lsb}(x) : x \xleftarrow{u} QR_n) \geq \frac{1}{2} + \frac{1}{P(k)} ,$$

where $n := pq$, p *and* q *are distinct primes,* $p, q \equiv 3 \bmod 4$, *and* $k := |n|$. *Then there is a probabilistic polynomial algorithm* A_2 *such that*

$$\mathrm{prob}(A_2(n, x) = \mathrm{Sqrt}_n(x)) \geq 1 - 2^{-Q(k)} \text{ for all } x \in QR_n .$$

Proof. The proof runs in the same way as the proof of Theorem 7.14. We only describe the differences to this proof. Here, the algorithm A_1 is only applicable to quadratic residues. However, it is easy to compute $\left(\frac{x}{n}\right)$ for $x \in \mathbb{Z}_n^*$, and we can use algorithm A_2 from Proposition 7.20 to check whether a given element $x \in J_n^{+1}$ is a quadratic residue. Assume that $\mathrm{prob}(A_2(n,x) = \mathrm{PQR}_n(x)) \geq 1 - 1/P^2(k)$.

If $p, q \equiv 3 \bmod 4$, we have $-1 \notin \mathrm{QR}_n$ (see Theorem A.59). Therefore, either a or $-a \in \mathrm{QR}_n$ for $\left(\frac{a}{n}\right) = 1$. We are looking for m multipliers $\tilde{a}$ of the form $a_t + ia_{t-1} + b$ with $\left(\frac{\tilde{a}}{n}\right) = 1$, where $m := \min(2^t \varepsilon^{-2}, 2k\varepsilon^{-2})$. If $\tilde{a} \in \mathrm{QR}_n$, $\mathrm{Lsb}(\tilde{a}x)$ can be computed with algorithm A_1, and if $-\tilde{a} = n - \tilde{a} \in \mathrm{QR}_n$, $\mathrm{Lsb}(\tilde{a}x) = 1 - \mathrm{Lsb}(-\tilde{a}x)$ and $\mathrm{Lsb}(-\tilde{a}x)$ can be computed with algorithm A_1. $\mathrm{Lsb}(\tilde{a}x)$ is correctly computed if A_1 correctly computes the predicate Lsb and A_2 from Proposition 7.20 correctly computes the predicate PQR_n. Both events are independent. Thus $\mathrm{Lsb}(\tilde{a}x)$ is computed correctly with a probability $> \left(1/2 + 1/P(k)\right)\left(1 - 1/P^2(k)\right) > \left(1/2 + 1/2P(k)\right)$. Thus we set $\varepsilon = 1/2P(k)$.

With i varying in an interval, the fraction of the multipliers $a_t + ia_{t-1} + b$ which are in J_n^{+1} differs from $1/2$ only negligibly, because J_n^{+1} is nearly uniformly distributed in $\mathbb{Z}_n^*$ (see [Peralta92]).

We double the range for i, take $i \in [-m, m-1]$, and halve the distances of the initial rational approximations:

$$|ax \bmod n - un| \leq \frac{\varepsilon^3 n}{16} \text{ and } |bx \bmod n - vn| \leq \frac{\varepsilon n}{16}.$$

Now we obtain the same estimates as in the proof of Theorem 7.14.   □

Corollary 7.23. *Let $I := \{n \mid n = pq, p \text{ and } q \text{ distinct primes}, |p| = |q|, p, q \equiv 3 \bmod 4\}$. Provided that the factorization assumption is true,*

$$\mathrm{QRLsb} = (\mathrm{QRLsb}_n : \mathrm{QR}_n \longrightarrow \{0,1\}, \ x \longmapsto \mathrm{Lsb}(x))_{n \in I}$$

is a family of hard-core predicates for

$$\mathrm{Square} = \left(\mathrm{Square}_n : \mathrm{QR}_n \longrightarrow \mathrm{QR}_n, \ x \longmapsto x^2\right)_{n \in I}.$$

Proof. The proof is analogous to the proof of Corollary 7.11. Observe that the ability to compute square roots modulo n is equivalent to the ability to compute the prime factors of n (Proposition A.70).   □

Exercises

1. Compute $\mathrm{Log}_{p,g}(17)$ using Algorithm 7.6, for $p := 19$ and $g := 2$ (note that 2, 4 and 13 are principal square roots).

2. Let p be an odd prime number, and g be a primitive root in $\mathbb{Z}_p^*$. For $y := \mathrm{Exp}_{p,g}(x)$, we have

$$B_{p,g}(y) = 0 \text{ if and only if } 0 \le x < \frac{p-1}{2},$$

$$B_{p,g}(y^2) = 0 \text{ if and only if } 0 \le x < \frac{p-1}{4} \text{ and } \frac{p-1}{2} \le x < \frac{3(p-1)}{4},$$

and so on. Let A_1 be a a deterministic polynomial algorithm such that $A_1(p,g,y) = B_{p,g}(y)$ for all $y \in \mathbb{Z}_p^*$.

By using a binary search technique, prove that there is a deterministic polynomial algorithm A_2 such that

$$A_2(p,g,y) = \mathrm{Log}_{p,g}(y)$$

for all $y \in \mathbb{Z}_p^*$.

3. Let p be an odd prime number. Suppose that $p-1 = 2^t a$, where a is odd. Let g be a primitive root in $\mathbb{Z}_p^*$:

   a. Show how the t least-significant bits (bits at the positions 0 to $t-1$) of $x \in \mathbb{Z}_{p-1}$ can be easily computed from $g^x = \mathrm{Exp}_{p,g}(x)$.

   b. Denote by $\mathrm{Lsb}_t(x)$ the t-th least-significant bit of x (bit at position t counted from the right, beginning with 0). Let A_1 be a deterministic polynomial algorithm such that

   $$A_1(p,g,g^x) = \mathrm{Lsb}_t(x)$$

   for all $x \in \mathbb{Z}_{p-1}$. By using A_1, construct a deterministic polynomial algorithm A_2 such that

   $$A_2(p,g,y) = \mathrm{Log}_{p,g}(y)$$

   for all $y \in \mathbb{Z}_p^*$. (Here assume that a deterministic algorithm for computing square roots exists.)

   c. Show that Lsb_t yields a hard-core predicate for the Exp family.

4. As in the preceding exercise, Lsb_j denotes the j-th least-significant bit.

   a. Let A_1 be a deterministic polynomial algorithm such that for all $x \in \mathbb{Z}_{p-1}$

   $$A_1(p,g,g^x,\mathrm{Lsb}_t(x),\dots,\mathrm{Lsb}_{t+j-1}(x)) = \mathrm{Lsb}_{t+j}(x),$$

   where p is an odd prime, $g \in \mathbb{Z}_p^*$ is a primitive root, $p-1 = 2^t a$, a is odd, $k = |p|$ and $j \in \{0,\dots,\lfloor \log_2(k) \rfloor\}$.

   Construct a deterministic polynomial algorithm A_2 such that

   $$A_2(p,g,y) = \mathrm{Log}_{p,g}(y)$$

   for all $y \in \mathbb{Z}_p^*$.

b. Let $P, Q \in \mathbb{Z}[X]$ be positive polynomials and A_1 be a probabilistic polynomial algorithm such that

$$\text{prob}(A_1(p, g, g^x, \text{Lsb}_t(x), \dots, \text{Lsb}_{t+j-1}(x))$$
$$= \text{Lsb}_{t+j}(x) : x \xleftarrow{u} \mathbb{Z}_{p-1}) \geq \frac{1}{2} + \frac{1}{P(k)},$$

where p is a an odd prime, $g \in \mathbb{Z}_p^*$ is a primitive root, $p - 1 = 2^t a$, a is odd, $k = |p|$ and $j \in \{0, \dots, \lfloor \log_2(k) \rfloor\}$.
Construct a probabilistic polynomial algorithm A_2 such that

$$\text{prob}(A_2(p, g, y) = \text{Log}_{p,g}(y)) \geq 1 - 2^{-Q(k)}$$

for all $y \in \mathbb{Z}_p^*$.

5. Let $I := \{(p, g) \mid p \text{ an odd prime}, g \in \mathbb{Z}_p^* \text{ a primitive root}\}$ and $I_k := \{(p, g) \in I \mid |p| = k\}$. Assume that the discrete logarithm assumption (see Definition 6.1) is true.
Show that for every probabilistic polynomial algorithm A, with inputs $p, g, y, b_0, \dots, b_{j-1}$, $1 \leq j \leq \lfloor \log_2(k) \rfloor$, and for every positive polynomial P, there is a $k_0 \in \mathbb{N}$ such that

$$\text{prob}(A(p, g, g^x, \text{Lsb}_t(x), \dots, \text{Lsb}_{t+j-1}(x))$$
$$= \text{Lsb}_{t+j}(x) : (p, g) \xleftarrow{u} I_k, x \xleftarrow{u} \mathbb{Z}_{p-1}) \leq \frac{1}{2} + \frac{1}{P(k)}$$

for $k \geq k_0$ and for all $j \in \{0, \dots, \lfloor \log_2(k) \rfloor\}$. t is defined by $p - 1 = 2^t a$, with a odd.
In particular, the predicates $\text{Lsb}_{t+j}, 0 \leq j \leq \lfloor \log_2(k) \rfloor$, are hard-core predicates for Exp.

6. Compute the rational approximation (a, u) for $13 \in \mathbb{Z}_{29}$.

7. Let $p := 17, q := 23, n := pq$ and $e := 3$. List the least-significant bits that A_1 will return if you compute $\text{RSA}_{n,e}^{-1}(49)$ using Algorithm 7.12 (note that $49 = 196^3 \bmod n$).

8. Let $n := pq$, with p and q distinct primes and e relatively prime to $\varphi(n)$, $x \in \mathbb{Z}_n^*$ and $y := \text{RSA}_{n,e}(x) = x^e \bmod n$. Msb is defined analogously to Definition 7.1. Show that you can compute $\text{Msb}(x)$ from y if and only if you can compute $\text{Lsb}(x)$ from y.

9. Show that the most-significant bit of x is a hard-core predicate for the RSA family, provided that the RSA assumption is true.

10. Use the most significant bit and prove Proposition 7.12 using a binary search technique (analogous to Exercise 2).

11. RSA inversion by binary division ([GolMicTon82]). Let $y := \text{RSA}_{n,e}(x) = x^e \bmod n$. Let A be an algorithm that on input y outputs $\text{Lsb}(x)$. Let

$k := |n|$, and let $2^{-e} \in \mathbb{Z}_n^*$ be the inverse of $2^e \in \mathbb{Z}_n^*$. Compute x from y by using the bit vector $(b_{k-1}, \ldots, b_0)$, defined by

$$y_0 = y,$$
$$b_0 = A(y_0),$$
$$y_i = \begin{cases} y_{i-1} 2^{-e} \bmod n & \text{if } b_{i-1} = 0, \\ (n - y_{i-1}) 2^{-e} \bmod n & \text{otherwise,} \end{cases}$$
$$b_i = A(y_i), \text{ for } 1 \le i \le k.$$

Describe the algorithm inverting RSA and show that it really does invert RSA.

12. a. Let A_1 be a deterministic polynomial algorithm such that

$$A_1(n, e, x^e, \mathrm{Lsb}_0(x), \ldots, \mathrm{Lsb}_{j-1}(x)) = \mathrm{Lsb}_j(x)$$

for all $x \in \mathbb{Z}_n^*$, where $n := pq$, p and q are odd distinct primes, e is relatively prime to $\varphi(n)$, $k := |n|$ and $j \in \{0, \ldots, \lfloor \log_2(k) \rfloor\}$. Construct a deterministic polynomial algorithm A_2 such that

$$A_2(n, e, x^e) = x$$

for all $x \in \mathbb{Z}_n^*$.

   b. Let $P, Q \in \mathbb{Z}[X]$ be positive polynomials and A_1 be a probabilistic polynomial algorithm such that

$$\mathrm{prob}(A_1(n, e, x^e, \mathrm{Lsb}_0(x), \ldots, \mathrm{Lsb}_{j-1}(x))$$
$$= \mathrm{Lsb}_j(x) : x \xleftarrow{u} \mathbb{Z}_n^*) \ge \frac{1}{2} + \frac{1}{P(k)},$$

where $n := pq$, p and q are odd distinct primes, e is relatively prime to $\varphi(n)$, $k := |n|$ and $j \in \{0, \ldots, \lfloor \log_2(k) \rfloor\}$. Construct a probabilistic polynomial algorithm A_2 such that

$$\mathrm{prob}(A_2(n, e, x^e) = x \ge 1 - 2^{-Q(k)}$$

for all $x \in \mathbb{Z}_n^*$.

13. Let $I := \{(n, e) \mid n = pq, p \ne q \text{ prime numbers}, |p| = |q|, e < \varphi(n), e \text{ prime to } \varphi(n)\}$ and $I_k := \{(n, e) \in I \mid n = pq, |p| = |q| = k\}$. Assume that the RSA assumption (see Definition 6.7) is true.

Show that for every probabilistic polynomial algorithm A, with inputs $n, e, y, b_0, \ldots, b_{j-1}, 0 \le j \le \lfloor \log_2(|n|) \rfloor$, and every positive polynomial P, there is a $k_0 \in \mathbb{N}$ such that

$$\mathrm{prob}(A(n, e, x^e, \mathrm{Lsb}_0(x), \ldots, \mathrm{Lsb}_{j-1}(x))$$
$$= \mathrm{Lsb}_j(x) : (n, e) \xleftarrow{u} I_k, x \xleftarrow{u} \mathbb{Z}_n^*) \le \frac{1}{2} + \frac{1}{P(k)}$$

for $k \geq k_0$ and for all $j \in \{0, \ldots, \lfloor \log_2(|n|) \rfloor\}$. In particular, the predicates $\mathrm{Lsb}_j, 0 \leq j \leq \lfloor \log_2(|n|) \rfloor$, are hard-core predicates for RSA.

8. One-Way Functions and Pseudorandomness

There is a close relationship between encryption and randomness. The security of encryption algorithms usually depends on the random choice of keys and bit sequences. A famous example is Shannon's result. Ciphers with perfect secrecy require randomly chosen key strings that are of the same length as the encrypted message. In Chapter 9, we will study the classical Shannon approach to provable security, together with more recent notions of security. One main problem is that truly random bit sequences of sufficient length are not available in most practical situations. Therefore, one works with pseudorandom bit sequences. They appear to be random, but actually they are generated by an algorithm. Such algorithms are called *pseudorandom bit generators*. They output, given a short random input value (called the *seed*), a long pseudorandom bit sequence. Classical techniques for the generation of pseudorandom bits or numbers (see [Knuth98]) yield well-distributed sequences. Therefore, they are well-suited for Monte Carlo simulations. However, they are often cryptographically insecure. For example, in linear congruential pseudorandom number generators or linear feedback shift registers (see, e.g., [MenOorVan96]), the secret parameters and hence the complete pseudorandom sequence can be efficiently computed from a small number of outputs.

It turns out that computationally perfect (hence cryptographically secure) pseudorandom bit generators can be derived from one-way permutations with hard-core predicates. We will discuss this close relation in this chapter. The pseudorandom bit generators G studied are families of functions whose indexes vary over a set of keys. Before we can use G, we have to select such a key with a sufficiently large security parameter.

Of course, even applying a perfect pseudorandom generator requires starting with a truly random seed. Thus, in any case you need some "natural" source of random bits, such as independent fair coin tosses (see Chapter 5).

8.1 Computationally Perfect Pseudorandom Bit Generators

In the definition of pseudorandom generators, we use the notion of polynomial functions.

Definition 8.1. We call a function $l : \mathbb{N} \longrightarrow \mathbb{N}$ a *polynomial function* if it is computable by a polynomial algorithm and if there is a polynomial $Q \in \mathbb{Z}[X]$ such that $l(k) \leq Q(k)$ for all $k \in \mathbb{N}$.

Now we are ready to define pseudorandom bit generators.

Definition 8.2. Let $I = (I_k)_{k \in \mathbb{N}}$ be a key set with security parameter k, and let K be a probabilistic polynomial sampling algorithm for I, which on input 1^k outputs an $i \in I_k$. Let l be a polynomial function.

A *pseudorandom bit generator* with *key generator* K and *stretch function* l is a family $G = (G_i)_{i \in I}$ of functions

$$G_i : X_i \longrightarrow \{0,1\}^{l(k)} \qquad (i \in I_k),$$

such that

1. G is computable by a deterministic polynomial algorithm G:
   $G(i, x) = G_i(x)$ for all $i \in I$ and $x \in X_i$.
2. There is a uniform sampling algorithm S for $X := (X_i)_{i \in I}$, which on input $i \in I$ outputs $x \in X_i$.

The generator G is *computationally perfect* (or *cryptographically secure*), if the pseudorandom sequences generated by G cannot be distinguished from true random sequences by an efficient algorithm; i.e., for every positive polynomial $P \in \mathbb{Z}[X]$ and every probabilistic polynomial algorithm A with inputs $i \in I_k, z \in \{0,1\}^{l(k)}$ and output in $\{0,1\}$, there is a $k_0 \in \mathbb{N}$ such that for all $k \geq k_0$

$$| \operatorname{prob}(A(i,z) = 1 : i \leftarrow K(1^k), z \xleftarrow{u} \{0,1\}^{l(k)})$$
$$- \operatorname{prob}(A(i, G_i(x)) = 1 : i \leftarrow K(1^k), x \xleftarrow{u} X_i) | \leq \frac{1}{P(k)}.$$

Remarks:

1. The probabilistic polynomial algorithm A in the definition may be considered as a statistical test trying to compute some property which distinguishes truly random sequences in $\{0,1\}^{l(k)}$ from the pseudorandom sequences generated by G. Classical statistical tests for randomness, such as the Chi-square test ([Knuth98], Chapter 3), can be considered as such tests and can be implemented as polynomial algorithms. Thus, "computationally perfect" means that no statistical test – which can be implemented as a probabilistic algorithm with polynomial running time – can significantly distinguish between true random sequences and sequences generated by G, provided a sufficiently large key i is chosen.
2. By condition 2, we can randomly generate uniformly distributed seeds $x \in X_i$ for the generator G. We could (but do not) generalize the definition and allow non-uniform seed generators S (see the analogous remark after

the formal definition of one-way functions – Definition 6.12, remark 2). The constructions and proofs given below also work in this more general case.

3. We study only computationally perfect pseudorandom generators. Therefore, we do not specify any other level of pseudorandomness for the sequences generated by G. In the literature, the term "pseudorandom generator" is sometimes only used for generators that are computationally perfect (see, e.g., [Goldreich99]; [Goldreich01]).

4. Our definition of computationally perfect pseudorandom generators is a definition in the "public-key model". The key i is an input to the statistical tests A (which are the adversaries). Thus, the key i is assumed to be public and available to everyone. Definition 8.2 can be adapted to the "private-key model", where the selected key i is kept secret, and hence is not known to the adversaries. The input i of the statistical tests A has to be omitted. We only discuss the public-key model.

5. *Admissible key generators* can be analogously defined as in Definition 6.13.

6. The probability in the definition is also taken over the random generation of a key i, with a given security parameter k. Even for very large k, there may be keys i such that A can successfully distinguish pseudorandom from truly random sequences. However, when generating a key i by K, the probability of obtaining one for which A has a significant chance of success is negligibly small (see Exercise 8 in Chapter 6).

7. As is common, we require that the pseudorandom generator can be implemented by a deterministic algorithm. However, if the sequences can be computed probabilistically in polynomial time, then we can also compute them almost deterministically: for every positive polynomial Q, there is a probabilistic polynomial algorithm $G(i, x)$ with

$$\text{prob}(G(i, x) = G_i(x)) \geq 1 - 2^{-Q(k)} \quad (i \in I_k)$$

(see Proposition 5.6 and Exercise 5 in Chapter 5). Thus, a modified definition, which relaxes condition 1 to "Monte Carlo computable", would also work. In all our examples, the pseudorandom sequences can be efficiently computed by deterministic algorithms.

We will now derive pseudorandom bit generators from one-way permutations with hard-core predicates (see Definition 6.15). These generators turn out to be computationally perfect. The construction was introduced by Blum and Micali ([BluMic84]).

Definition 8.3. Let $I = (I_k)_{k \in \mathbb{N}}$ be a key set with security parameter k, and let $Q \in \mathbb{Z}[X]$ be a positive polynomial.
Let $f = (f_i : D_i \longrightarrow D_i)_{i \in I}$ be a family of one-way permutations with hard-core predicate $B = (B_i : D_i \longrightarrow \{0, 1\})_{i \in I}$ and key generator K. Then we have the following pseudorandom bit generator with stretch function Q and key generator K:

$$G := G(f, B, Q) := (G_i : D_i \longrightarrow \{0,1\}^{Q(k)})_{k \in \mathbb{N}, i \in I_k},$$

$$x \in D_i \longmapsto (B_i(x), B_i(f_i(x)), B_i(f_i^2(x)), \ldots, B_i(f_i^{Q(k)-1}(x))).$$

We call this generator the *pseudorandom bit generator induced by* f, B *and* Q.

Remark. We obtain the pseudorandom bit sequence by a very simple construction: choose some random seed $x \xleftarrow{u} D_i$. Compute the first pseudorandom bit as $B_i(x)$, apply f_i and get $y := f_i(x)$. Compute the next pseudorandom bit as $B_i(y)$, apply f_i to y and get a new $y := f_i(y)$. Compute the next pseudorandom bit as $B_i(y)$, and so on.

Examples:

1. Provided that the discrete logarithm assumption is true, the discrete exponential function

$$\mathrm{Exp} = (\mathrm{Exp}_{p,g} : \mathbb{Z}_{p-1} \longrightarrow \mathbb{Z}_p^*, \ x \longmapsto g^x)_{(p,g) \in I}$$

   with $I := \{(p,g) \mid p \text{ prime}, g \in \mathbb{Z}_p^* \text{ a primitive root}\}$ is a bijective one-way function, and the most-significant bit $\mathrm{Msb}_p(x)$ defined by

$$\mathrm{Msb}_p(x) = \begin{cases} 0 \text{ for } 0 \leq x < \frac{p-1}{2}, \\ 1 \text{ for } \frac{p-1}{2} \leq x < p-1, \end{cases}$$

   is a hard-core predicate for Exp (see Section 7.1). Identifying $\mathbb{Z}_{p-1}$ with $\mathbb{Z}_p^*$ in the straightforward way,[1] we may consider Exp as a one-way permutation. The induced pseudorandom bit generator is called the *discrete exponential generator* (or *Blum–Micali generator*).

2. Provided that the RSA assumption is true, the RSA family

$$\mathrm{RSA} = (\mathrm{RSA}_{n,e} : \mathbb{Z}_n^* \longrightarrow \mathbb{Z}_n^*, \ x \longmapsto x^e)_{(n,e) \in I}$$

   with $I := \{(n,e) \mid n = pq, \ p,q \text{ distinct primes}, |p| = |q|, 1 < e < \varphi(n), e \text{ prime to } \varphi(n)\}$ is a one-way permutation, and the least-significant bit $\mathrm{Lsb}_n(x)$ is a hard-core predicate for RSA (see Section 7.2). The induced pseudorandom bit generator is called the *RSA generator*.

3. Provided that the factorization assumption is true, the modular squaring function

$$\mathrm{Square} = (\mathrm{Square}_n : \mathrm{QR}_n \longrightarrow \mathrm{QR}_n, \ x \longmapsto x^2)_{n \in I}$$

   with $I := \{n \mid n = pq, \ p,q \text{ distinct primes}, |p| = |q|, p,q \equiv 3 \bmod 4\}$ is a one-way permutation, and the least-significant bit $\mathrm{Lsb}_n(x)$ is a hard-core predicate for Square (see Section 7.3). The induced pseudorandom

---

[1] $\mathbb{Z}_{p-1} = \{0, \ldots, p-2\} \longrightarrow \mathbb{Z}_p^* = \{1, \ldots, p-1\}, 0 \mapsto p-1, x \mapsto x$ for $1 \leq x \leq p-2$.

bit generator is called the $(x^2 \bmod n)$ generator or *Blum–Blum–Shub generator*.

Here the computation of the pseudorandom bits is particularly simple: choose a random seed $x \in \mathbb{Z}_n^*$, repeatedly square x and reduce it by modulo n, take the least-significant bit after each step.

We will now show that the pseudorandom bit generators, induced by one-way permutations, are computationally perfect. For later applications, it is useful to prove a slightly more general statement that covers the pseudorandom bits and the seed, which is encrypted using $f_i^{Q(k)}$.

Theorem 8.4. *Let $I = (I_k)_{k \in \mathbb{N}}$ be a key set with security parameter k, and let $Q \in \mathbb{Z}[X]$ be a positive polynomial. Let $f = (f_i : D_i \longrightarrow D_i)_{i \in I}$ be a family of one-way permutations with hard-core predicate $B = (B_i : D_i \longrightarrow \{0,1\})_{i \in I}$ and key generator K. Let $G := G(f, B, Q)$ be the induced pseudorandom bit generator.*

Then, for every probabilistic polynomial algorithm A with inputs $i \in I_k, z \in \{0,1\}^{Q(k)}, y \in D_i$ and output in $\{0,1\}$, and every positive polynomial $P \in \mathbb{Z}[X]$, there is a $k_0 \in \mathbb{N}$ such that for all $k \geq k_0$

$$| \operatorname{prob}(A(i, G_i(x), f_i^{Q(k)}(x)) = 1 : i \leftarrow K(1^k), x \xleftarrow{u} D_i)$$
$$- \operatorname{prob}(A(i, z, y) = 1 : i \leftarrow K(1^k), z \xleftarrow{u} \{0,1\}^{Q(k)}, y \xleftarrow{u} D_i) | \leq \frac{1}{P(k)}.$$

Remark. The theorem states that for sufficiently large keys, the probability of distinguishing successfully between truly random sequences and pseudorandom sequences – using a given efficient algorithm – is negligibly small, even if the encryption $f_i^{Q(k)}(x)$ of the seed x is known.

Proof. Assume that there is a probabilistic polynomial algorithm A such that the inequality is false for infinitely many k. Replacing A by $1 - A$ if necessary, we may drop the absolute value and assume that

$$\operatorname{prob}(A(i, G_i(x), f_i^{Q(k)}(x)) = 1 : i \leftarrow K(1^k), x \xleftarrow{u} D_i)$$
$$- \operatorname{prob}(A(i, z, y) = 1 : i \leftarrow K(1^k), z \xleftarrow{u} \{0,1\}^{Q(k)}, y \xleftarrow{u} D_i) > \frac{1}{P(k)},$$

for k in an infinite subset $\mathcal{K}$ of $\mathbb{N}$.

For $k \in \mathcal{K}$ and $i \in I_k$, we consider the following sequence of distributions $p_{i,0}, p_{i,1}, \ldots, p_{i,Q(k)}$ on $Z_i := \{0,1\}^{Q(k)} \times D_i$:[2]

---

[2] We use the following notation: $\{S(x) : x \leftarrow X\}$ denotes the image of the distribution on X under $S : X \longrightarrow Z$, i.e., the probability of $z \in Z$ is given by the probability for z appearing as $S(x)$, if x is randomly selected from X.

$$p_{i,0} := \{(b_1, \ldots, b_{Q(k)}, y) : (b_1, \ldots, b_{Q(k)}) \overset{u}{\leftarrow} \{0,1\}^{Q(k)}, y \overset{u}{\leftarrow} D_i\}$$

$$p_{i,1} := \{(b_1, \ldots, b_{Q(k)-1}, B_i(x), f_i(x)) :$$
$$(b_1, \ldots, b_{Q(k)-1}) \overset{u}{\leftarrow} \{0,1\}^{Q(k)-1}, x \overset{u}{\leftarrow} D_i\}$$

$$p_{i,2} := \{(b_1, \ldots, b_{Q(k)-2}, B_i(x), B_i(f_i(x)), f_i^2(x)) :$$
$$(b_1, \ldots, b_{Q(k)-2}) \overset{u}{\leftarrow} \{0,1\}^{Q(k)-2}, x \overset{u}{\leftarrow} D_i\}$$

$$\vdots$$

$$p_{i,r} := \{(b_1, \ldots, b_{Q(k)-r}, B_i(x), B_i(f_i(x)), \ldots, B_i(f_i^{r-1}(x)), f_i^r(x)) :$$
$$(b_1, \ldots, b_{Q(k)-r}) \overset{u}{\leftarrow} \{0,1\}^{Q(k)-r}, x \overset{u}{\leftarrow} D_i\}$$

$$\vdots$$

$$p_{i,Q(k)} := \{(B_i(x), B_i(f_i(x)), \ldots, B_i(f_i^{Q(k)-1}(x)), f_i^{Q(k)}(x)) : x \overset{u}{\leftarrow} D_i\}.$$

We start with truly random bit sequences. In each step, we replace one more truly random bit from the right with a pseudorandom bit. The seed x encrypted by f_i^r is always appended on the right. Note that the image $\{f_i(x) : x \overset{u}{\leftarrow} D_i\}$ of the uniform distribution under f_i is again the uniform distribution, since f_i is bijective. Finally, in $p_{i,Q(k)}$ we have the distribution of the pseudorandom sequences supplemented by the encrypted seed. We observe that

$$\mathrm{prob}(A(i, z, y) = 1 : i \leftarrow K(1^k), z \overset{u}{\leftarrow} \{0,1\}^{Q(k)}, y \overset{u}{\leftarrow} D_i)$$
$$= \mathrm{prob}(A(i, z, y) = 1 : i \leftarrow K(1^k), (z, y) \overset{p_{i,0}}{\leftarrow} Z_i)$$

and

$$\mathrm{prob}(A(i, G_i(x), f_i^{Q(k)}(x)) = 1 : i \leftarrow K(1^k), x \overset{u}{\leftarrow} D_i)$$
$$= \mathrm{prob}(A(i, z, y) = 1 : i \leftarrow K(1^k), (z, y) \overset{p_{i,Q(k)}}{\leftarrow} Z_i).$$

Thus, our assumption says that for $k \in \mathcal{K}$, the algorithm A is able to distinguish between the distribution $p_{i,Q(k)}$ (of pseudorandom sequences) and the (uniform) distribution $p_{i,0}$. Hence, A must be able to distinguish between two subsequent distributions $p_{i,r}$ and $p_{i,r+1}$, for some r.

Since f_i is bijective, we have the following equation (8.1):

$$p_{i,r} = \{(b_1, \ldots, b_{Q(k)-r}, B_i(x), B_i(f_i(x)), \ldots, B_i(f_i^{r-1}(x)), f_i^r(x)) :$$
$$(b_1, \ldots, b_{Q(k)-r}) \overset{u}{\leftarrow} \{0,1\}^{Q(k)-r}, x \overset{u}{\leftarrow} D_i\}$$
$$= \{(b_1, \ldots, b_{Q(k)-r}, B_i(f_i(x)), B_i(f_i^2(x)), \ldots, B_i(f_i^r(x)), f_i^{r+1}(x)) :$$
$$(b_1, \ldots, b_{Q(k)-r}) \overset{u}{\leftarrow} \{0,1\}^{Q(k)-r}, x \overset{u}{\leftarrow} D_i\}.$$

We see that $p_{i,r}$ differs from $p_{i,r+1}$ only at one position, namely at position $Q(k) - r$. There, the hard-core bit $B_i(x)$ is replaced by a truly random bit.

Therefore, algorithm A, which distinguishes between $p_{i,r}$ and $p_{i,r+1}$, can also be used to compute $B_i(x)$ from $f_i(x)$.

More precisely, we will derive a probabilistic polynomial algorithm $\tilde{A}(i, y)$ from A that on inputs $i \in I_k$ and $y := f_i(x)$ computes $B_i(x)$ with probability $> 1/2 + 1/P(k)Q(k)$, for the infinitely many $k \in \mathcal{K}$. This contradiction to the hard-core property of B will finish the proof of the theorem.

For $k \in \mathcal{K}$, we have

$$
\frac{1}{P(k)} < \mathrm{prob}(A(i, z, y) = 1 : i \leftarrow K(1^k), (z, y) \overset{p_{i,Q(k)}}{\leftarrow} Z_i)
$$

$$
- \mathrm{prob}(A(i, z, y) = 1 : i \leftarrow K(1^k), (z, y) \overset{p_{i,0}}{\leftarrow} Z_i)
$$

$$
= \sum_{r=0}^{Q(k)-1} (\mathrm{prob}(A(i, z, y) = 1 : i \leftarrow K(1^k), (z, y) \overset{p_{i,r+1}}{\leftarrow} Z_i)
$$

$$
- \mathrm{prob}(A(i, z, y) = 1 : i \leftarrow K(1^k), (z, y) \overset{p_{i,r}}{\leftarrow} Z_i)).
$$

Randomly choosing r, we expect that the r-th term in the sum is $> 1/P(k)Q(k)$.

On inputs $i \in I_k, y \in D_i$, the algorithm $\tilde{A}$ works as follows:

1. Choose r, with $0 \leq r < Q(k)$, uniformly at random.
2. Independently choose random bits $b_1, b_2, \ldots, b_{Q(k)-r-1}$ and another random bit b.
3. For $y = f_i(x) \in D_i$, let

$$
\tilde{A}(i, y) = \tilde{A}(i, f_i(x))
$$

$$
:= \begin{cases} b & \text{if } A(i, b_1, \ldots, b_{Q(k)-r-1}, b, \\ & \qquad B_i(f_i(x)), \ldots, B_i(f_i^r(x)), f_i^{r+1}(x)) = 1, \\ 1-b & \text{otherwise.} \end{cases}
$$

If A distinguishes between $p_{i,r}$ and $p_{i,r+1}$, it yields 1 with higher probability if the $(Q(k) - r)$-th bit of its input is $B_i(x)$ and not a random bit. Therefore, we guess in our algorithm that the randomly chosen b is the desired hard-core bit if A outputs 1.

We now check that $\tilde{A}$ indeed computes the hard-core bit with a non-negligible probability. Let R be the random variable describing the choice of r in the first step of the algorithm. Since r is selected with respect to the uniform distribution, we have $\mathrm{prob}(R = r) = 1/Q(k)$ for all r. Applying Lemma B.13, we get

$$\text{prob}(\tilde{A}(i, f_i(x)) = B_i(x) : i \leftarrow K(1^k), x \stackrel{u}{\leftarrow} D_i)$$

$$= \frac{1}{2} + \text{prob}(\tilde{A}(i, f_i(x)) = b \,|\, B_i(x) = b) - \text{prob}(\tilde{A}(i, f_i(x)) = b)$$

$$= \frac{1}{2} + \sum_{r=0}^{Q(k)-1} \text{prob}(R = r) \cdot (\text{prob}(\tilde{A}(i, f_i(x)) = b \,|\, B_i(x) = b, R = r)$$

$$- \text{prob}(\tilde{A}(i, f_i(x)) = b \,|\, R = r))$$

$$= \frac{1}{2} + \frac{1}{Q(k)} \sum_{r=0}^{Q(k)-1} (\text{prob}(\tilde{A}(i, f_i(x)) = b \,|\, B_i(x) = b : i \leftarrow K(1^k), x \stackrel{u}{\leftarrow} D_i)$$

$$- \text{prob}(\tilde{A}(i, f_i(x)) = b : i \leftarrow K(1^k), x \stackrel{u}{\leftarrow} D_i))$$

$$= \frac{1}{2} + \frac{1}{Q(k)} \sum_{r=0}^{Q(k)-1} (\text{prob}(A(i, z, y) = 1 : i \leftarrow K(1^k), (z, y) \stackrel{p_{i,r+1}}{\leftarrow} Z_i)$$

$$- \text{prob}(A(i, z, y) = 1 : i \leftarrow K(1^k), (z, y) \stackrel{p_{i,r}}{\leftarrow} Z_i))$$

$$> \frac{1}{2} + \frac{1}{Q(k)P(k)},$$

for the infinitely many $k \in \mathcal{K}$. The probabilities in lines 2 and 3 are computed with respect to $i \leftarrow K(1^k)$ and $x \stackrel{u}{\leftarrow} D_i$ (and the random choice of the elements b_i, b, r). Since r is chosen independently, we can omit the conditions $R = r$. Taking the probability $\text{prob}(\tilde{A}(i, f_i(x)) = b \,|\, B_i(x) = b)$ conditional on $B_i(x) = b$ just means that the inputs to A in step 3 of the algorithm $\tilde{A}$ are distributed according to $p_{i,r+1}$. Finally, recall equation (8.1) for $p_{i,r}$ from above.

Since B is a hard-core predicate, our computation yields the desired contradiction, and the proof of Theorem 8.4 is complete. $\square$

Corollary 8.5 *(Theorem of Blum and Micali). Pseudorandom bit generators induced by one-way permutations with hard-core predicates are computationally perfect.*

Proof. Let $A(i, z)$ be a probabilistic polynomial algorithm with inputs $i \in I_k$, $z \in \{0,1\}^{Q(k)}$ and output in $\{0,1\}$. We define $\tilde{A}(i, z, y) := A(i, z)$, and observe that

$$\text{prob}(\tilde{A}(i, z, y) = 1 : i \leftarrow I_k, z \leftarrow \{0,1\}^{Q(k)}, y \leftarrow D_i)$$
$$= \text{prob}(A(i, z) = 1 : i \leftarrow I_k, z \leftarrow \{0,1\}^{Q(k)}),$$

and the corollary follows from Theorem 8.4 applied to $\tilde{A}$. $\square$

8.2 Yao's Theorem

Computationally perfect pseudorandom bit generators such as the ones induced by one-way permutations are characterized by another unique feature: it is not possible to predict the next bit in the pseudorandom sequence from the preceding bits.

Definition 8.6. Let $I = (I_k)_{k \in \mathbb{N}}$ be a key set with security parameter k, and let $G = (G_i : X_i \longrightarrow \{0,1\}^{l(k)})_{i \in I}$ be a pseudorandom bit generator with polynomial stretch function l and key generator K:

1. A *next-bit predictor* for G is a probabilistic polynomial algorithm $A(i, z_1 \ldots z_r)$ which, given $i \in I_k$, outputs a bit ("the next bit") from r input bits z_j $(0 \leq r < l(k))$.
2. G *passes all next-bit tests* if and only if for every next-bit predictor A and every positive polynomial $P \in \mathbb{Z}[X]$, there is a $k_0 \in \mathbb{N}$ such that for all $k \geq k_0$ and all $0 \leq r < l(k)$

$$\text{prob}(A(i, G_{i,1}(x) \ldots G_{i,r}(x)) = G_{i,r+1}(x) : i \leftarrow K(1^k), x \xleftarrow{u} X_i)$$

$$\leq \frac{1}{2} + \frac{1}{P(k)}.$$

Here and in what follows we denote by $G_{i,j}, 1 \leq j \leq l(k)$, the j-th bit generated by G_i:

$$G_i(x) = (G_{i,1}(x), G_{i,2}(x), \ldots, G_{i,l(k)}(x)).$$

Remarks:

1. A next-bit predictor has two inputs: the key i and a bit string $z_1 \ldots z_r$ of variable length.
2. As usual, the probability in the definition is also taken over the random choice of a key i with security parameter k. This means that when randomly generating a key i, the probability of obtaining one for which A has a significant chance of predicting a next bit is negligibly small (see Proposition 6.17).

Theorem 8.7 *(Yao's Theorem). Let $I = (I_k)_{k \in \mathbb{N}}$ be a key set with security parameter k, and let $G = (G_i : X_i \longrightarrow \{0,1\}^{l(k)})_{i \in I}$ be a pseudorandom bit generator with polynomial stretch function l and key generator K.*
Then G is computationally perfect if and only if G passes all next-bit tests.

Proof. Assume that G is computationally perfect and does not pass all next-bit tests. Then there is a next-bit predictor A and a positive polynomial P, such that for k in an infinite subset $\mathcal{K}$ of $\mathbb{N}$, we have a position r_k, $0 \leq r_k < l(k)$, with $q_{k,r_k} > 1/2 + 1/P(k)$, where

$q_{k,r} := \mathrm{prob}(A(i, G_{i,1}(x) \ldots G_{i,r}(x)) = G_{i,r+1}(x) : i \leftarrow K(1^k), x \xleftarrow{u} X_i)$.

By Proposition 6.18, we can compute the probabilities $q_{k,r}$, $r = 0, 1, 2, \ldots$, approximately with high probability, and we conclude that there is a probabilistic polynomial algorithm R which on input 1^k finds a position where the next-bit predictor is successful:

$$\mathrm{prob}\left(q_{k,R(1^k)} > \frac{1}{2} + \frac{1}{2P(k)}\right) \geq 1 - \frac{1}{4P(k)}.$$

We define a probabilistic polynomial algorithm $\tilde{A}$ (a statistical test) for inputs $i \in I$ and $z = (z_1, \ldots, z_{l(k)}) \in \{0,1\}^{l(k)}$ as follows: Let $r := R(1^k)$ and set

$$\tilde{A}(i, z) := \begin{cases} 1 & \text{if } z_{r+1} = A(i, z_1 \ldots z_r), \\ 0 & \text{otherwise.} \end{cases}$$

For truly random sequences, it is not possible to predict a next bit with probability $> 1/2$. Thus, we have for the uniform distribution on $\{0,1\}^{l(k)}$ that

$$\mathrm{prob}(\tilde{A}(i,z) = 1 : i \leftarrow K(1^k), z \xleftarrow{u} \{0,1\}^{l(k)}) \leq \frac{1}{2}.$$

We obtain, for the infinitely many $k \in \mathcal{K}$, that

$| \mathrm{prob}(\tilde{A}(i, G_i(x)) = 1 : i \leftarrow K(1^k), x \xleftarrow{u} X_i) -$

$\qquad\qquad \mathrm{prob}(\tilde{A}(i,z) = 1 : i \leftarrow K(1^k), z \xleftarrow{u} \{0,1\}^{l(k)}) |$

$$> \left(1 - \frac{1}{4P(k)}\right) \cdot \left(\frac{1}{2} + \frac{1}{2P(k)} - \frac{1}{2}\right) - \frac{1}{4P(k)}$$

$$= \frac{1}{4P(k)} - \frac{1}{8P^2(k)} \geq \frac{1}{8P(k)},$$

which is a contradiction to the assumption that G is computationally perfect.

Conversely, assume that the sequences generated by G pass all next-bit tests, but can be distinguished from truly random sequences by a statistical test A. This means that

$| \mathrm{prob}(A(i, G_i(x)) = 1 : i \leftarrow K(1^k), x \xleftarrow{u} X_i)$

$$- \mathrm{prob}(A(i,z) = 1 : i \leftarrow K(1^k), z \xleftarrow{u} \{0,1\}^{l(k)}) | > \frac{1}{P(k)},$$

for some positive polynomial P and k in an infinite subset $\mathcal{K}$ of $\mathbb{N}$. Replacing A by $1 - A$, if necessary, we may drop the absolute value.

The proof now runs in a similar way to the proof of Theorem 8.4. For $k \in \mathcal{K}$ and $i \in I_k$, we consider a sequence $p_{i,0}, p_{i,1}, \ldots, p_{i,l(k)}$ of distributions on $\{0,1\}^{l(k)}$:

$$p_{i,0} := \{(b_1, \ldots, b_{l(k)}) : (b_1, \ldots, b_{l(k)}) \stackrel{u}{\leftarrow} \{0,1\}^{l(k)}\}$$

$$p_{i,1} := \{(G_{i,1}(x), b_2, \ldots, b_{l(k)}) : (b_2, \ldots, b_{l(k)}) \stackrel{u}{\leftarrow} \{0,1\}^{l(k)-1}, x \stackrel{u}{\leftarrow} X_i\}$$

$$p_{i,2} := \{(G_{i,1}(x), G_{i,2}(x), b_3, \ldots, b_{l(k)}) :$$
$$(b_3, \ldots, b_{l(k)}) \stackrel{u}{\leftarrow} \{0,1\}^{l(k)-2}, x \stackrel{u}{\leftarrow} X_i\}$$

$$\vdots$$

$$p_{i,r} := \{(G_{i,1}(x), G_{i,2}(x), \ldots, G_{i,r}(x), b_{r+1}, \ldots, b_{l(k)}) :$$
$$(b_{r+1}, \ldots, b_{l(k)}) \stackrel{u}{\leftarrow} \{0,1\}^{l(k)-r}, x \stackrel{u}{\leftarrow} X_i\}$$

$$\vdots$$

$$p_{i,l(k)} := \{(G_{i,1}(x), G_{i,2}(x), \ldots, G_{i,l(k)}(x)) : x \stackrel{u}{\leftarrow} X_i\}.$$

We start with truly random bit sequences, and in each step we replace one more truly random bit from the left by a pseudorandom bit. Finally, in $p_{i,l(k)}$ we have the distribution of the pseudorandom sequences.

Our assumption says that for $k \in \mathcal{K}$, algorithm A is able to distinguish between the distribution $p_{i,l(k)}$ (of pseudorandom sequences) and the (uniform) distribution $p_{i,0}$. Again, the basic idea of the proof now is that A must be able to distinguish between two subsequent distributions $p_{i,r}$ and $p_{i,r+1}$, for some r. However, $p_{i,r+1}$ differs from $p_{i,r}$ in one position only, and there a truly random bit is replaced by the next bit $G_{i,r+1}(x)$ of the pseudorandom sequence. Therefore, algorithm A can also be used to predict $G_{i,r+1}(x)$.

More precisely, we will derive a probabilistic polynomial algorithm $\tilde{A}(i, z_1, \ldots, z_r)$ that successfully predicts the next bit $G_{i,r+1}(x)$ from $G_{i,1}(x), G_{i,2}(x), \ldots, G_{i,r}(x)$ for some $r = r_k$, for the infinitely many $k \in \mathcal{K}$. This contradiction to the assumption that G passes all next-bit tests will finish the proof of the theorem.

Since A is able to distinguish between the uniform distribution and the distribution induced by G, we get for $k \in \mathcal{K}$ that

$$\frac{1}{P(k)} < \text{prob}(A(i,z) = 1 : i \leftarrow K(1^k), z \stackrel{p_{i,l(k)}}{\leftarrow} \{0,1\}^{l(k)})$$

$$- \text{prob}(A(i,z) = 1 : i \leftarrow K(1^k), z \stackrel{p_{i,0}}{\leftarrow} \{0,1\}^{l(k)})$$

$$= \sum_{r=0}^{l(k)-1} (\text{prob}(A(i,z) = 1 : i \leftarrow K(1^k), z \stackrel{p_{i,r+1}}{\leftarrow} \{0,1\}^{l(k)})$$

$$- \text{prob}(A(i,z) = 1 : i \leftarrow K(1^k), z \stackrel{p_{i,r}}{\leftarrow} \{0,1\}^{l(k)})).$$

We conclude that for $k \in \mathcal{K}$, there is some r_k, $0 \leq r_k < l(k)$, with

$$\frac{1}{P(k)l(k)} < \text{prob}(A(i,z) = 1 : i \leftarrow K(1^k), z \overset{p_{i,r_k+1}}{\leftarrow} \{0,1\}^{l(k)})$$

$$- \text{prob}(A(i,z) = 1 : i \leftarrow K(1^k), z \overset{p_{i,r_k}}{\leftarrow} \{0,1\}^{l(k)}).$$

This means that $A(i,z)$ yields 1 for

$$z = (G_{i,1}(x), G_{i,2}(x), \ldots, G_{i,r_k}(x), b, b_{r_k+2}, \ldots, b_{l(k)})$$

with higher probability if b is equal to $G_{i,r_k+1}(x)$ and not a truly random bit.

On inputs $i \in I_k, z_1 \ldots z_r$ $(0 \le r < l(k))$, algorithm $\tilde{A}$ is defined as follows:

1. Choose truly random bits $b, b_{r+2}, \ldots, b_{l(k)}$, and set

$$z := (z_1, \ldots, z_r, b, b_{r+2}, \ldots, b_{l(k)}).$$

2. Let

$$\tilde{A}(i, z_1 \ldots z_r) := \begin{cases} b & \text{if } A(i,z) = 1, \\ 1 - b & \text{if } A(i,z) = 0. \end{cases}$$

Applying Lemma B.13, we get

$$\text{prob}(\tilde{A}(i, G_{i,1}(x) \ldots G_{i,r_k}(x)) = G_{i,r_k+1}(x) : i \leftarrow K(1^k), x \overset{u}{\leftarrow} X_i)$$

$$= \frac{1}{2} + \text{prob}(\tilde{A}(i, G_{i,1}(x) \ldots G_{i,r_k}(x)) = b \mid$$

$$G_{i,r_k+1}(x) = b : i \leftarrow K(1^k), x \overset{u}{\leftarrow} X_i)$$

$$- \text{prob}(\tilde{A}(i, G_{i,1}(x) \ldots G_{i,r_k}(x)) = b : i \leftarrow K(1^k), x \overset{u}{\leftarrow} X_i)$$

$$= \frac{1}{2} + \text{prob}(A(i,z) = 1 : i \leftarrow K(1^k), z \overset{p_{i,r_k+1}}{\leftarrow} \{0,1\}^{l(k)})$$

$$- \text{prob}(A(i,z) = 1 : i \leftarrow K(1^k), z \overset{p_{i,r_k}}{\leftarrow} \{0,1\}^{l(k)})$$

$$> \frac{1}{2} + \frac{1}{P(k)l(k)},$$

for the infinitely many $k \in \mathcal{K}$. This is the desired contradiction and completes the proof of Yao's Theorem. $\qquad \square$

Exercises

1. Let $I = (I_k)_{k \in \mathbb{N}}$ be a key set with security parameter k, and let $G = (G_i)_{i \in I}$ be a computationally perfect pseudorandom bit generator with polynomial stretch function l. Let $\pi = (\pi_i)_{i \in I}$ be a family of permutations, where π_i is a permutation of $\{0,1\}^{l(k)}$ for $i \in I_k$. Assume that π can be computed by a polynomial algorithm Π, i.e., $\pi_i(y) = \Pi(i,y)$. Let $\pi \circ G$ be the composition of π and G: $x \mapsto \pi_i(G_i(x))$.

   Show that $\pi \circ G$ is also a computationally perfect pseudorandom bit generator.

2. Give an example of a computationally perfect pseudorandom bit generator $G = (G_i)_{i \in I}$ and a family of permutations π such that $\pi \circ G$ is not computationally perfect.

   (According to Exercise 1, π cannot be computable in polynomial time.)

3. Let $G = (G_i)_{i \in I}$ be a pseudorandom bit generator with polynomial stretch function l and key generator K.

   Show that G is computationally perfect if and only if next bits in the past cannot be predicted; i.e., for every probabilistic polynomial algorithm $A(i, z_{r+1} \ldots z_{l(k)})$ which, given $i \in I_k$, outputs a bit ("the next bit in the past") from $l(k) - r$ input bits z_j, there is a $k_0 \in \mathbb{N}$ such that for all $k \geq k_0$ and all $1 \leq r \leq l(k)$

   $$\mathrm{prob}(G_{i,r}(x) = A(i, G_{i,r+1}(x) \ldots G_{i,l(k)}(x)) : i \leftarrow K(1^k), x \xleftarrow{u} X_i)$$
   $$\leq \frac{1}{2} + \frac{1}{P(k)}.$$

4. Let Q be a positive polynomial, and let $G = (G_i)_{i \in I}$ be a computationally perfect pseudorandom bit generator with

   $$G_i : \{0,1\}^{Q(k)} \longrightarrow \{0,1\}^{Q(k)+1} \quad (i \in I_k),$$

   i.e., G extends the binary length of the seeds by 1. Recursively, define the pseudorandom bit generators G^l by

   $$G^1 := G, \quad G_i^l(x) := (G_{i,1}(x), G_i^{l-1}(G_{i,2}(x), \ldots, G_{i,Q(k)+1}(x))).$$

   As before, we denote by $G_{i,j}^l(x)$ the j-th bit of $G_i^l(x)$. Let l vary with the security parameter k, i.e., $l = l(k)$, and assume that $l : \mathbb{N} \longrightarrow \mathbb{N}$ is a polynomial function.

   Show that G^l is computationally perfect.

5. Prove the following stronger version of Yao's Theorem (Theorem 8.7).

   Let $I = (I_k)_{k \in \mathbb{N}}$ be a key set with security parameter k, and let $G = (G_i : X_i \longrightarrow \{0,1\}^{l(k)})_{i \in I}$ be a pseudorandom bit generator with polynomial stretch function l and key generator K.

   Let $f = (f_i : X_i \longrightarrow Y_i)_{i \in I}$ be a Monte Carlo computable family of maps. Then, the following statements are equivalent:

   a. For every probabilistic polynomial algorithm $A(i, y, z)$ and every positive polynomial P, there is a $k_0 \in \mathbb{N}$ such that for all $k \geq k_0$ and all $0 \leq r < l(k)$

   $$\mathrm{prob}(G_{i,r+1}(x) = A(i, f_i(x), G_{i,1}(x) \ldots G_{i,r}(x)) :$$
   $$i \leftarrow K(1^k), x \xleftarrow{u} X_i)$$
   $$\leq \frac{1}{2} + \frac{1}{P(k)}.$$

b. For every probabilistic polynomial algorithm $A(i, y, z)$ and every positive polynomial P, there is a $k_0 \in \mathbb{N}$ such that for all $k \geq k_0$

$$| \operatorname{prob}(A(i, f_i(x), z) = 1 : i \leftarrow K(1^k), x \overset{u}{\leftarrow} X_i, z \overset{u}{\leftarrow} \{0,1\}^{l(k)})$$
$$- \operatorname{prob}(A(i, f_i(x), G_i(x)) = 1 : i \leftarrow K(1^k), x \overset{u}{\leftarrow} X_i) |$$
$$\leq \frac{1}{P(k)}.$$

In this exercise a setting is modeled in which some information $f_i(x)$ about the seed x is known to the adversary.

6. Let $f = (f_i : D_i \longrightarrow R_i)_{i \in I}$ be a family of one-way functions with key generator K, and let $B = (B_i : D_i \longrightarrow \{0,1\}^{l(k)})_{i \in I}$ be a family of l-bit predicates which is computable by a Monte Carlo algorithm (l a polynomial function). Let $B_i = (B_{i,1}, B_{i,2}, \ldots, B_{i,l(k)})$. We call B an l-bit hard-core predicate for f (or simultaneously secure bits of f), if for every probabilistic polynomial algorithm $A(i, y, z_1, \ldots, z_l)$ and every positive polynomial P, there is a $k_0 \in \mathbb{N}$ such that for all $k \geq k_0$

$$| \operatorname{prob}(A(i, f_i(x), B_{i,1}(x), \ldots, B_{i,l(k)}(x)) = 1 : i \leftarrow K(1^k), x \overset{u}{\leftarrow} D_i)$$
$$- \operatorname{prob}(A(i, f_i(x), z) = 1 : i \leftarrow K(1^k), x \overset{u}{\leftarrow} D_i, z \overset{u}{\leftarrow} \{0,1\}^{l(k)}) |$$
$$\leq \frac{1}{P(k)}.$$

For $l = 1$, the definition is equivalent to our previous Definition 6.15 of hard-core bits (see Exercise 7 in Chapter 6).

Now assume that B is an l-bit hard-core predicate, and let $C = (C_i : \{0,1\}^{l(k)} \longrightarrow \{0,1\})_{i \in I}$ be a Monte Carlo computable family of predicates with $\operatorname{prob}(C_i(x) = 0 : x \overset{u}{\leftarrow} \{0,1\}^{l(k)}) = 1/2$ for all $i \in I$.

Show that the composition $C \circ B$, $x \in D_i \mapsto C_i(B_i(x))$, is a hard-core predicate for f.

7. Let $f = (f_i : D_i \longrightarrow R_i)_{i \in I}$ be a family of one-way functions with key generator K, and let $B = (B_i : D_i \longrightarrow \{0,1\}^{l(k)})_{i \in I}$ be a family of l-bit predicates for f. Let $B_i = (B_{i,1}, B_{i,2}, \ldots, B_{i,l(k)})$. Assume that knowing $f_i(x)$ and $B_{i,1}(x), \ldots, B_{i,j-1}(x)$ does not help in the computation of $B_{i,j}(x)$. More precisely, assume that for every probabilistic polynomial algorithm $A(i, y, z)$ and every positive polynomial P, there is a $k_0 \in \mathbb{N}$ such that for all $1 \leq j \leq l(k)$

$$\operatorname{prob}(A(i, f_i(x), B_{i,1}(x) \ldots B_{i,j-1}(x)) = B_{i,j}(x) : i \leftarrow K(1^k), x \overset{u}{\leftarrow} X_i)$$
$$\leq \frac{1}{2} + \frac{1}{P(k)}.$$

(In particular, the $(B_{i,j})_{i \in I}$ are hard-core predicates for f.)
Show that the bits $B_{i,1}, \ldots, B_{i,l}$ are simultaneously secure bits for f.

Examples:
The $\lfloor \log_2(|n|) \rfloor$ least-significant bits are simultaneously secure for the RSA (and the Square) one-way function (Exercise 12 in Chapter 7). If p is a prime, $p - 1 = 2^t a$ and a is odd, then the bits at the positions $t, t + 1, \ldots, t + \lfloor \log_2(|p|) \rfloor$ (counted from the right, starting with 0) are simultaneously secure for the discrete exponential one-way function (Exercise 5 in Chapter 7).

8. Let $f = (f_i : D_i \longrightarrow D_i)_{i \in I}$ be a family of one-way permutations with key generator K, and let $B = (B_i)_{i \in I}$ be an l-bit hard-core predicate for f. Let G be the following pseudorandom bit generator with stretch function lQ (Q a positive polynomial):

$$G := \left(G_i : D_i \longrightarrow \{0, 1\}^{l(k)Q(k)} \right)_{k \in \mathbb{N}, i \in I_k}$$

$$x \in D_i \longmapsto (B_i(x), B_i(f_i(x)), B_i(f_i^2(x)), \ldots, B_i(f_i^{Q(k)-1}(x))).$$

Prove a statement that is analogous to Theorem 8.4. In particular, prove that G is computationally perfect.

Example:
Taking the Square one-way permutation $x \mapsto x^2$ ($x \in QR_n$, with n a product of distinct primes) and the $\lfloor \log_2(|n|) \rfloor$ least-significant bits, we get the *generalized Blum–Blum–Shub generator*. It is used in the Blum–Goldwasser probabilistic encryption scheme (see Chapter 9).

9. Provably Secure Encryption

This chapter deals with provable security. It is desirable that mathematical proofs show that a given cryptosystem resists certain types of attacks. The security of cryptographic schemes and randomness are closely related. An encryption method provides secrecy only if the ciphertexts appear sufficiently random to the adversary. Therefore, probabilistic encryption algorithms are required. The pioneering work of Shannon on provable security, based on his information theory, is discussed in Section 9.1. For example, we prove that Vernam's one-time pad is a perfectly secret encryption. Shannon's notion of perfect secrecy may be interpreted in terms of probabilistic attacking algorithms that try to distinguish between two candidate plaintexts (Section 9.2). Unfortunately, Vernam's one-time pad is not practical in most situations. In Section 9.3, we give important examples of probabilistic encryption algorithms that are practical. One-way permutations with hard-core predicates yield computationally perfect pseudorandom bit generators (Chapter 8), and these can be used to define "public-key pseudorandom one-time pads", by analogy to Vernam's one-time pad: the plaintext bits are XORed with pseudorandom bits generated from a short, truly random (one-time) seed. More recent notions of provable security, which include the computational complexity of attacking algorithms, are considered in Section 9.4. The computational analogue of Shannon's perfect secrecy, ciphertext-indistinguishability, is defined. A typical security proof for probabilistic public-key encryption schemes is given. We show that the public-key one-time pads, introduced in Section 9.3, provide computationally perfect secrecy against passive eavesdroppers, who perform ciphertext-only or chosen-plaintext attacks. Encryption schemes that are secure against adaptively-chosen-ciphertext attacks, are considered in Section 9.5. The security proof for Boneh's SAEP is a typical proof in the random oracle model, the proof for Cramer–Shoup's public-key encryption scheme is based solely on a standard number-theoretic assumption and the collision-resistance of the hash function used. An introduction to the "unconditional security approach" is given in Chapter 10. In this approach, the goal is to design practical cryptosystems which provably come close to perfect information-theoretic security, without relying on unproven assumptions about problems from computational number theory.

9.1 Classical Information-Theoretic Security

A deterministic public-key encryption algorithm E necessarily leaks information to an adversary. For example, recall the small-message-space attack on RSA (Section 3.3.2). An adversary intercepts a ciphertext c and knows that the transmitted message m is from a small set $\{m_1, \ldots, m_r\}$ of possible messages. Then he easily finds out m by computing the ciphertexts $E(m_1), \ldots, E(m_r)$ and comparing them with c. This example shows that randomness in encryption is necessary to ensure real secrecy. Learning the encryption $c \doteq E(m)$ of a message m, an adversary should not be able to predict the ciphertext the next time when m is encrypted by E. This observation applies also to symmetric-key encryption schemes. Thus, to obtain a provably secure encryption scheme, we have to study randomized encryption algorithms.

Definition 9.1. An encryption algorithm E, which on input $m \in M$ outputs a ciphertext $c \in C$, is called a *randomized encryption* if E is a non-deterministic probabilistic algorithm.

The random behavior of a randomized encryption E is caused by its coin tosses. These coin tosses may be considered as the random choice of a one-time key (for each message to be encrypted a new random key is chosen, independently of the previous choices). Take, for example, Vernam's one-time pad which is the classical example of a randomized (and provably secure) cipher. We recall its definition.

Definition 9.2. Let $n \in \mathbb{N}$ and $M := C := \{0,1\}^n$. The randomized encryption E which encrypts a message $m \in M$ by XORing it bitwise with a randomly and uniformly chosen bit sequence $k \overset{u}{\leftarrow} \{0,1\}^n$ of the same length, $E(m) := m \oplus k$, is called *Vernam's one-time pad*.

As the name indicates, key k is used only once: each time a message m is encrypted, a new bit sequence is randomly chosen as the encryption key. This choice of the key is viewed as the coin tosses of a probabilistic algorithm. The security of a randomized encryption algorithm is related to the level of randomness caused by its coin tosses. More randomness means more security. Vernam's one-time pad includes a maximum of randomness and hence, provably, provides a maximum of security, as we will see below (Theorem 9.5).

The problem with Vernam's one-time pad is that truly random keys of the same length as the message have to be generated and securely transmitted to the recipient. This is rarely a practical operation (for an example, see Section 2.1.1). Later (in Section 9.4), we will see how to obtain practical, but still provably secure probabilistic encryption methods, by using high quality pseudorandom bit sequences as keys.

The classical notion of security of an encryption algorithm is based on Shannon's information theory and his famous papers [Shannon48] and

[Shannon49]. Appendix B.5 gives an introduction to information theory and its basic notions, such as entropy, uncertainty and mutual information.

We consider a randomized encryption algorithm E mapping plaintexts $m \in M$ to ciphertexts $c \in C$. We assume that the messages to be encrypted are generated according to some probability distribution, i.e., M is assumed to be a probability space. The distribution on M and the algorithm E induce probability distributions on $M \times C$ and C (see Section 5.1). As usual, the probability space induced on $M \times C$ is denoted by MC and, for $m \in M$ and $c \in C$, $\mathrm{prob}(c|m)$ denotes the probability that c is the ciphertext if the plaintext is m. Analogously, $\mathrm{prob}(m|c)$ is the probability that m is the plaintext if c is the ciphertext.[1]

Without loss of generality, we assume that $\mathrm{prob}(m) > 0$ for all $m \in M$ and that $\mathrm{prob}(c) > 0$ for all $c \in C$.

Definition 9.3 *(Shannon).* The encryption E is *perfectly secret* if C and M are independent, i.e., the distribution of MC is the product of the distributions on M and C:

$$\mathrm{prob}(m, c) = \mathrm{prob}(m) \cdot \mathrm{prob}(c), \text{ for all } m \in M, c \in C.$$

Perfect secrecy can be characterized in different ways.

Proposition 9.4. *The following statements are equivalent:*

1. *E is perfectly secret.*
2. *The mutual information* $\mathrm{I}(M; C) = 0$.
3. $\mathrm{prob}(m|c) = \mathrm{prob}(m)$, *for all* $m \in M$ *and* $c \in C$.
4. $\mathrm{prob}(c|m) = \mathrm{prob}(c)$, *for all* $m \in M$ *and* $c \in C$.
5. $\mathrm{prob}(c|m) = \mathrm{prob}(c|m')$, *for all* $m, m' \in M$ *and* $c \in C$.
6. $\mathrm{prob}(E(m) = c) = \mathrm{prob}(c)$, *for all* $m \in M$ *and* $c \in C$.
7. $\mathrm{prob}(E(m) = c) = \mathrm{prob}(E(m') = c)$, *for all* $m, m' \in M$ *and* $c \in C$;
   i.e., the distribution of $E(m)$ does not depend on m.

Proof. All statements of Proposition 9.4 are contained in, or immediately follow from Proposition B.39. For the latter two statements, observe that

$$\mathrm{prob}(c|m) = \mathrm{prob}(E(m) = c),$$

by the definition of $\mathrm{prob}(E(m) = c)$ (see Chapter 5).   □

Remarks:

1. The probabilities in statement 7 only depend on the coin tosses of E. This means, in particular, that the perfect secrecy of an encryption algorithm E does not depend on the distribution of the plaintexts.

---

[1] The notation is introduced on p. 462 in Appendix B.1.

2. Let Eve be an attacker trying to discover information about the plaintexts from the ciphertexts that she is able to intercept. Assume that Eve is well informed and knows the distribution of the plaintexts. Then perfect secrecy means that her uncertainty about the plaintext (as precisely defined in information theory, see Appendix B.5, Definition B.35) is the same whether or not she observes the ciphertext c: learning the ciphertext does not increase her information about the plaintext m. Thus, perfect secrecy really means unconditional security against ciphertext-only attacks.

   A perfectly secret randomized encryption E also withstands the other types of attacks, such as the known-plaintext attacks and adaptively-chosen-plaintext/ciphertext attacks discussed in Section 1.3. Namely, the security is guaranteed by the randomness caused by the coin tosses of E. Encrypting, say, r messages, means applying E r times. The coin tosses within one of these executions of E are independent of the coin tosses in the other executions (in Vernam's one-time pad, this corresponds to the fact that an individual key is chosen independently for each message). Knowing details about previous encryptions does not help the adversary. Each encryption is a new and independent random experiment and, hence, the probabilities $\text{prob}(c|m)$ are the same, whether we take them conditional on other plaintext–ciphertext pairs (m', c') or not. Note that additional knowledge of the adversary is included by conditioning the probabilities on this knowledge.

3. The mutual information is a typical measure defined in information theory (see Definition B.37). It measures the average amount of information Eve obtains about the plaintext m when learning the ciphertext c.

Vernam's one-time pad is a perfectly secret encryption. More generally, we prove the following theorem.

Theorem 9.5 *(Shannon). Let $M := C := K := \{0,1\}^n$, and let E be a one-time pad, which encrypts $m := (m_1, \ldots, m_n) \in M$ by XORing it with a random key string $k := (k_1, \ldots, k_n) \in K$, chosen independently from m:*

$$E(m) := m \oplus k := (m_1 \oplus k_1, \ldots, m_n \oplus k_n).$$

Then E is perfectly secret if and only if K is uniformly distributed.

Proof. We have

$$\text{prob}_{MC}(m, c) = \text{prob}_{MK}(m, m \oplus c) = \text{prob}_M(m) \cdot \text{prob}_K(m \oplus c).$$

If M and C are independent, then

$$\text{prob}_M(m) \cdot \text{prob}_C(c) = \text{prob}_{MC}(m, c) = \text{prob}_M(m) \cdot \text{prob}_K(m \oplus c).$$

Hence

$$\text{prob}_K(m \oplus c) = \text{prob}_C(c), \text{ for all } m \in M.$$

This means that $\text{prob}_K(k)$ is the same for all $k \in K$. Thus, K is uniformly distributed. Conversely, if K is uniformly distributed, then

$$\text{prob}_C(c) = \sum_{m \in M} \text{prob}_{MK}(m, m \oplus c) = \sum_{m \in M} \text{prob}_M(m) \cdot \text{prob}_K(m \oplus c)$$

$$= \sum_{m \in M} \text{prob}_M(m) \cdot \frac{1}{2^n}$$

$$= \frac{1}{2^n}.$$

Hence, C is also distributed uniformly, and we obtain:

$$\text{prob}_{MC}(m, c) = \text{prob}_{MK}(m, m \oplus c)$$

$$= \text{prob}_M(m) \cdot \text{prob}_K(m \oplus c) = \text{prob}_M(m) \cdot \frac{1}{2^n}$$

$$= \text{prob}_M(m) \cdot \text{prob}_C(c).$$

Thus, M and C are independent. □

Remarks:

1. Note that we do not consider the one-time pad as a cipher for plaintexts of varying length: we have to assume that all plaintexts have the same length n. Otherwise some information, namely the length of the plaintext, leaks to an adversary Eve, and the encryption could not be perfectly secret.

2. There is a high price to pay for the perfect secrecy of Vernam's one-time pad. For each message to be encrypted, of length n, n independent random bits have to be chosen for the key. One might hope to find a more sophisticated, perfectly secret encryption method requiring less randomness. Unfortunately, this hope is destroyed by the following result which was proven by Shannon ([Shannon49]).

Theorem 9.6. *Let E be a randomized encryption algorithm with the deterministic extension $E_D : M \times K \longrightarrow C$. Each time a message $m \in M$ is encrypted, a one-time key k is chosen randomly from K (according to some probability distribution on K), independently from the choice of m. Assume that the plaintext m can be recovered from the ciphertext c and the one-time key k (no other information is necessary for decryption). Then, if E is perfectly secret, the uncertainty of the keys cannot be smaller than the uncertainty of the messages:*

$$H(K) \geq H(M).$$

Remark. The uncertainty of a probability space M (see Definition B.35) is maximal and equal to $log_2(|M|)$ if the distribution of M is uniform (Proposition B.36). Hence, if $M = \{0, 1\}^n$ as in Theorem 9.5, then the entropy of any key set K – yielding a perfectly secret encryption – is at least n. Thus, the random choice of $k \in K$ requires the choice of at least n truly random bits.

Note at this point that the perfect secrecy of an encryption does not depend on the distribution of the plaintexts (Proposition 9.4). Therefore, we may assume that M is uniformly distributed and, as a consequence, that $H(M) = n$.

Proof. The plaintext m can be recovered from the ciphertext c and the one-time key k. This means that there is no uncertainty about the plaintext if both the ciphertext and the key are known, i.e., the conditional entropy $H(M|KC) = 0$ (see Definition B.37). Perfect secrecy means $I(M;C) = 0$ (Proposition 9.4), or equivalently, $H(C) = H(C|M)$ (Proposition B.39). Since M and K are assumed to be independent, $I(K;M) = I(M;K) = 0$ (Proposition B.39). We compute by use of Proposition B.38 and Definition B.40 the following:

$$
\begin{aligned}
H(K) - H(M) &= I(K;M) + H(K|M) - I(M;K) - H(M|K) \\
&= H(K|M) - H(M|K) \\
&= I(K;C|M) + H(K|CM) - I(M;C|K) - H(M|KC) \\
&= I(K;C|M) + H(K|CM) - I(M;C|K) \\
&\geq I(K;C|M) - I(M;C|K) \\
&= H(C|M) - H(C|KM) - H(C|K) + H(C|KM) \\
&= H(C|M) - H(C|K) \\
&= H(C) - H(C) + I(K;C) = I(K;C) \\
&\geq 0.
\end{aligned}
$$

The proof of Theorem 9.6 is finished.   □

Remark. In Vernam's one-time pad it is not possible, without destroying perfect secrecy, to use the same randomly chosen key for the encryption of two messages. This immediately follows, for example, from Theorem 9.6. Namely, such a modified Vernam one-time pad may be described as a probabilistic algorithm from $M \times M$ to $C \times C$, with the deterministic extension

$$
M \times M \times K \longrightarrow C \times C, \quad (m, m', k) \mapsto (m \oplus k, m' \oplus k),
$$

where $M = K = C = \{0,1\}^n$. Assuming the uniform distribution on M, we have

$$
H(K) = n < H(M \times M) = 2n.
$$

9.2 Perfect Secrecy and Probabilistic Attacks

We model the behavior of an adversary Eve by probabilistic algorithms, and show the relation between the failure of such algorithms and perfect secrecy.

In Section 9.3, we will slightly modify this model by restricting the computing power of the adversary to polynomial resources.

As in Section 9.1, let E be a randomized encryption algorithm that maps plaintexts $m \in M$ to ciphertexts $c \in C$ and is used by Alice to encrypt her messages. As before, Alice chooses the messages $m \in M$ according to some probability distribution. The distribution on M and the algorithm E induce probability distributions on $M \times C$ and C. $\mathrm{prob}(m, c)$ is the probability that m is the chosen message and that the probabilistic encryption of m yields c.

We first consider a probabilistic algorithm A which on input $c \in C$ outputs a plaintext $m \in M$. Algorithm A models an adversary Eve performing a ciphertext-only attack and trying to decrypt ciphertexts. Recall that the coin tosses of a probabilistic algorithm are independent of any other random events in the given setting (see Chapter 5). Thus, the coin tosses of A are independent of the choice of the message and the coin tosses of E. This is a reasonable model, because sender Alice, generating and encrypting messages, and adversary Eve operate independently. We have

$$\mathrm{prob}(m, c, A(c) = m) = \mathrm{prob}(m, c) \cdot \mathrm{prob}(A(c) = m),$$

for $m \in M$ and $c \in C$ (see Chapter 5). $\mathrm{prob}(A(c) = m)$ is the conditional probability that $A(c)$ yields m, assuming that m and c are fixed. It is determined by the coin tosses of A. The *probability of success* of A is given by

$$\mathrm{prob}_{\mathrm{success}}(A) := \sum_{m,c} \mathrm{prob}(m, c) \cdot \mathrm{prob}(A(c) = m)$$
$$= \sum_{m,c} \mathrm{prob}(m) \cdot \mathrm{prob}(E(m) = c) \cdot \mathrm{prob}(A(c) = m)$$
$$= \mathrm{prob}(A(c) = m : m \leftarrow M, c \leftarrow E(m)).$$

Proposition 9.7. *If E is perfectly secret, then for every probabilistic algorithm A which on input $c \in C$ outputs a plaintext $m \in M$*

$$\mathrm{prob}_{\mathrm{success}}(A) \leq \max_{m \in M} \mathrm{prob}(m).$$

Proof.

$$\mathrm{prob}_{\mathrm{success}}(A) = \sum_{m,c} \mathrm{prob}(m, c) \cdot \mathrm{prob}(A(c) = m)$$
$$= \sum_c \mathrm{prob}(c) \cdot \sum_m \mathrm{prob}(m|c) \cdot \mathrm{prob}(A(c) = m)$$
$$= \sum_c \mathrm{prob}(c) \cdot \sum_m \mathrm{prob}(m) \cdot \mathrm{prob}(A(c) = m) \quad \text{(by Proposition 9.4)}$$

$$\leq \max_{m \in M} \mathrm{prob}(m) \cdot \sum_c \mathrm{prob}(c) \cdot \sum_m \mathrm{prob}(A(c) = m)$$

$$= \max_{m \in M} \mathrm{prob}(m),$$

and the proposition follows.   □

Remarks:

1. In Proposition 9.7, as in the whole of Section 9.2, we do not assume any limits for the resources of the algorithms. The running time and the memory requirements may be exponential.
2. Proposition 9.7 says that for a perfectly secret encryption, selecting a plaintext with maximal probability from M, without looking at the ciphertext, is optimum under all attacks that try to derive the plaintext from the ciphertext. If M is uniformly distributed, then randomly selecting a plaintext is an optimal strategy.

Perfect secrecy may also be described in terms of distinguishing algorithms.

Definition 9.8. A *distinguishing algorithm* for E is a probabilistic algorithm A which on inputs $m_0, m_1 \in M$ and $c \in C$ outputs an $m \in \{m_0, m_1\}$.

Remark. A distinguishing algorithm A models an adversary Eve, who, given a ciphertext c and two plaintext candidates m_0 and m_1, tries to find out which one of the both is the correct plaintext, i.e., which one is encrypted as c. Again, recall that the coin tosses of A are independent of a random choice of the messages and the coin tosses of the encryption algorithm (see Chapter 5). Thus, the adversary Eve and the sender Alice, generating and encrypting messages, are modeled as working independently.

Proposition 9.9. E *is perfectly secret if and only if for every probabilistic distinguishing algorithm A and all $m_0, m_1 \in M$,*

$$\mathrm{prob}(A(m_0, m_1, c) = m_0 : c \leftarrow E(m_0))$$
$$= \mathrm{prob}(A(m_0, m_1, c) = m_0 : c \leftarrow E(m_1)).$$

Proof. E is perfectly secret if and only if the distribution of $E(m)$ does not depend on m (Proposition 9.4). Thus, the equality obviously holds if E is perfectly secret.

Conversely, assume that E is not perfectly secret. There are no limits for the running time of our algorithms. Then there is an algorithm P which starts with a description of the encryption algorithm E and analyzes the paths and coin tosses of E and, in this way, computes the probabilities $\mathrm{prob}(c|m)$:

$$P(c, m) := \mathrm{prob}(c|m), \text{ for all } c \in C, m \in M.$$

We define the following distinguishing algorithm:

$$A(m_0, m_1, c) := \begin{cases} m_0 \text{ if } P(c, m_0) > P(c, m_1), \\ m_1 \text{ otherwise.} \end{cases}$$

Since E is not perfectly secret, there are $m_0, m_1 \in M$ and $c_0 \in C$ such that $P(c_0, m_0) = \mathrm{prob}(c_0 \mid m_0) > P(c_0, m_1) = \mathrm{prob}(c_0 \mid m_1)$ (Proposition 9.4). Let

$$C_0 := \{c \in C \mid \mathrm{prob}(c \mid m_0) > \mathrm{prob}(c \mid m_1)\} \text{ and}$$

$$C_1 := \{c \in C \mid \mathrm{prob}(c \mid m_0) \leq \mathrm{prob}(c \mid m_1)\}.$$

Then $A(m_0, m_1, c) = m_0$ for $c \in C_0$, and $A(m_0, m_1, c) = m_1$ for $c \in C_1$. We compute

$$\begin{aligned} \mathrm{prob}(A(m_0, &m_1, c) = m_0 : c \leftarrow E(m_0)) \\ &- \mathrm{prob}(A(m_0, m_1, c) = m_0 : c \leftarrow E(m_1)) \\ &= \sum_{c \in C} \mathrm{prob}(c \mid m_0) \cdot \mathrm{prob}(A(m_0, m_1, c) = m_0) \\ &\quad - \sum_{c \in C} \mathrm{prob}(c \mid m_1) \cdot \mathrm{prob}(A(m_0, m_1, c) = m_0) \\ &= \sum_{c \in C_0} \mathrm{prob}(c \mid m_0) - \mathrm{prob}(c \mid m_1) \\ &\geq \mathrm{prob}(c_0 \mid m_0) - \mathrm{prob}(c_0 \mid m_1) \\ &> 0, \end{aligned}$$

and see that a violation of perfect secrecy causes a violation of the equality condition. The proof of the proposition is finished. $\qquad\square$

Proposition 9.10. *E is perfectly secret if and only if for every probabilistic distinguishing algorithm A and all $m_0, m_1 \in M$, with $m_0 \neq m_1$,*

$$\mathrm{prob}(A(m_0, m_1, c) = m : m \xleftarrow{u} \{m_0, m_1\}, c \leftarrow E(m)) = \frac{1}{2}.$$

Proof.

$$\begin{aligned} \mathrm{prob}(A(m_0, &m_1, c) = m : m \xleftarrow{u} \{m_0, m_1\}, c \leftarrow E(m)) \\ &= \frac{1}{2} \cdot \mathrm{prob}(A(m_0, m_1, c) = m_0 : c \leftarrow E(m_0)) \\ &\quad + \frac{1}{2} \cdot \mathrm{prob}(A(m_0, m_1, c) = m_1 : c \leftarrow E(m_1)) \\ &= \frac{1}{2} + \frac{1}{2} \cdot (\mathrm{prob}(A(m_0, m_1, c) = m_0 : c \leftarrow E(m_0)) \\ &\quad - \mathrm{prob}(A(m_0, m_1, c) = m_0 : c \leftarrow E(m_1)), \end{aligned}$$

and the proposition follows from Proposition 9.9. $\qquad\square$

Remark. Proposition 9.10 characterizes a perfectly secret encryption scheme in terms of a passive eavesdropper A, who performs a ciphertext-only attack. But, as we observed before, the statement would remain true, if we model an (adaptively-)chosen-plaintext/ciphertext attacker by algorithm A (see the remark after Proposition 9.4).

9.3 Public-Key One-Time Pads

Vernam's one-time pad is provably secure (Section 9.1) and thus appears to be a very attractive encryption method. However, there is the problem that truly random keys of the same length as the message have to be generated and securely transmitted to the recipient. The idea now is to use high quality ("cryptographically secure") pseudorandom bit sequences as keys and to obtain in this way practical, but still provably secure randomized encryption methods.

Definition 9.11. Let $I = (I_k)_{k \in \mathbb{N}}$ be a key set with security parameter k, and let $G = (G_i)_{i \in I}$, $G_i : X_i \longrightarrow \{0,1\}^{l(k)}$ $(i \in I_k)$, be a pseudorandom bit generator with polynomial stretch function l and key generator K (see Definition 8.2).

The probabilistic polynomial encryption algorithm $E(i, m)$ which, given (a public key) $i \in I_k$, encrypts a message $m \in \{0,1\}^{l(k)}$ by bitwise XORing it with the pseudorandom sequence $G_i(x)$, generated by G_i from a randomly and uniformly chosen seed $x \in X_i$,

$$E(i, m) := m \oplus G_i(x), \quad x \xleftarrow{u} X_i,$$

is called the *pseudorandom one-time pad induced by* G. Keys i are assumed to be generated by K.

Example. Let $f = (f_i : D_i \longrightarrow D_i)_{i \in I}$ be a family of one-way permutations with hard-core predicate $B = (B_i : D_i \longrightarrow \{0,1\})_{i \in I}$, and let Q be a polynomial. f, B and Q induce a pseudorandom bit generator $G(f, B, Q)$ with stretch function Q (see Definition 8.3), and hence a pseudorandom one-time pad.

We will see in Section 9.4 that computationally perfect pseudorandom generators (Definition 8.2), such as the $G(f, B, Q)$s, lead to provably secure encryption schemes. Nevertheless, one important problem remains unsolved: how to transmit the secret one-time key – the randomly chosen seed x – to the recipient of the message?

If G is induced by a family of trapdoor permutations with hard-core predicate, there is an easy answer. We send x hidden by the one-way function together with the encrypted message. Knowing the trapdoor, the recipient is able to determine x.

Definition 9.12. Let $I = (I_k)_{k\in\mathbb{N}}$ be a key set with security parameter k, and let Q be a positive polynomial. Let $f = (f_i : D_i \longrightarrow D_i)_{i\in I}$ be a family of trapdoor permutations with hard-core predicate B and key generator K. Let $G(f, B, Q)$ be the induced pseudorandom bit generator with stretch function Q. For every recipient of messages, a public key $i \in I_k$ (and the associated trapdoor information) is generated by K.

The probabilistic polynomial encryption algorithm $E(i, m)$ which encrypts a message $m \in \{0, 1\}^{Q(k)}$ as

$$E(i, m) := (m \oplus G(f, B, Q)_i(x), f_i^{Q(k)}(x)),$$

with x chosen randomly and uniformly from D_i (for each message m), is called the *public-key one-time pad* induced by f, B and Q.

Remarks:

1. Recall that we get $G(f, B, Q)_i(x)$ by repeatedly applying f_i to x and taking the hard-core bits $B_i(f_i^j(x))$ of the sequence

$$x, f_i(x), f_i^2(x), f_i^3(x), \ldots, f_i^{Q(k)-1}(x)$$

   (see Definition 8.3). In order to encrypt the seed x, we then apply f_i once more. Note that we cannot take $f_i^j(x)$ with $j < Q(k)$ as the encryption of x, because this would reveal bits from the sequence $G(f, B, Q)_i(x)$.
2. Since f_i is a permutation of D_i and the recipient Bob knows the trapdoor, he has an efficient algorithm for f_i^{-1}. He is able to compute the sequence

$$x, f_i(x), f_i^2(x), f_i^3(x), \ldots, f_i^{Q(k)-1}(x)$$

   from $f_i^{Q(k)}(x)$, by repeatedly applying f_i^{-1}. In this way, he can easily decrypt the ciphertext.
3. In the public-key one-time pad, the basic pseudorandom one-time pad is augmented by an asymmetric (i.e., public-key) way of transmitting the one-time symmetric encryption key x.
4. Like the basic pseudorandom one-time pad, the augmented version is provably secure against passive attacks (see Theorem 9.16). Supplying the encrypted key $f_i^{Q(k)}(x)$ does not diminish the secrecy of the encryption scheme.
5. Pseudorandom one-time pads and public-key one-time pads are straightforward analogies to the classical probabilistic encryption method, Vernam's one-time pad. We will see in Section 9.4 that more recent notions of secrecy, such as indistinguishability (introduced in [GolMic84]), are also analogous to the classical notion in Shannon's work. The statements on the secrecy of pseudorandom and public-key one-time pads (see Section 9.4) are analogous to the classical results by Shannon.

6. The notion of *probabilistic public-key encryption*, whose security may be rigorously proven in a complexity theoretic model, was suggested by Goldwasser and Micali ([GolMic84]). They introduced the hard-core predicates of trapdoor functions (or, more generally, trapdoor predicates) as the basic building blocks of such schemes. The implementation of probabilistic public-key encryption, given in [GolMic84] and known as *Goldwasser–Micali probabilistic encryption* (see Exercise 7), is based on the quadratic residuosity assumption (Definition 6.11). During encryption, messages are expanded by a factor proportional to the security parameter k. Thus, this implementation is quite wasteful in space and bandwidth and is therefore not really practical. The public-key one-time pads, introduced in [BluGol85] and [BluBluShu86], avoid this large message expansion. They are the *efficient implementations of (asymmetric) probabilistic encryption*.

9.4 Passive Eavesdroppers

We study the security of public-key encryption schemes against passive eavesdroppers, who perform ciphertext-only attacks. In a public-key encryption scheme, a ciphertext-only attacker (as everybody) can encrypt messages of his choice at any time by using the publicly known key. Therefore, security against ciphertext-only attacks in a public-key encryption scheme also includes security against adaptively-chosen-plaintext attacks.

The stronger notion of security against adaptively-chosen-ciphertext attacks is considered in the subsequent Section 9.5.

Throughout this section we consider a probabilistic polynomial encryption algorithm $E(i, m)$, such as the pseudorandom or public-key one-time pads defined in Section 9.3. Here, $I = (I_k)_{k \in \mathbb{N}}$ is a key set with security parameter k and, for every $i \in I$, E maps plaintexts $m \in M_i$ to ciphertexts $c := E(i, m) \in C_i$. The keys i are generated by a probabilistic polynomial algorithm K and are assumed to be public. In our examples, the encryption E is derived from a family $f = (f_i)_{i \in I}$ of one-way permutations, and the index i is the public key of the recipient.

We define distinguishing algorithms A for E completely analogous to Definition 9.8. Now the computational resources of the adversary, modeled by A, are limited. A is required to be polynomial.

Definition 9.13. A *probabilistic polynomial distinguishing algorithm* for E is a probabilistic polynomial algorithm $A(i, m_0, m_1, c)$ which on inputs $i \in I, m_0, m_1 \in M_i$ and $c \in C_i$ outputs an $m \in \{m_0, m_1\}$.

Below we show that pseudorandom one-time pads induced by computationally perfect pseudorandom bit generators have computationally perfect secrecy. This result is analogous to the classical result by Shannon that Vernam's one-time pad is perfectly secret. Using truly random bit sequences as the key in the one-time pad, the probability of success of an attack with unlimited resources – which tries to distinguish between two candidate plaintexts – is equal to $1/2$; so there is no use in observing the ciphertext. Using computationally perfect pseudorandom bit sequences, the probability of success of an attack with polynomial resources is, at most, negligibly more than $1/2$.

Definition 9.14. The encryption E is called *ciphertext-indistinguishable* or (for short) *indistinguishable*, if for every probabilistic polynomial distinguishing algorithm $A(i, m_0, m_1, c)$ and every probabilistic polynomial sampling algorithm S, which on input $i \in I$ yields $S(i) = \{m_0, m_1\} \subset M_i$, and every positive polynomial $P \in \mathbb{Z}[X]$, there is a $k_0 \in \mathbb{N}$ such that for all $k \geq k_0$:

$$\mathrm{prob}(A(i, m_0, m_1, c) = m : i \leftarrow K(1^k), \{m_0, m_1\} \leftarrow S(i),$$
$$m \xleftarrow{u} \{m_0, m_1\}, c \leftarrow E(i, m)) \leq \frac{1}{2} + \frac{1}{P(k)}.$$

Remarks:

1. The definition is a definition in the "public-key model": the keys i are public and hence available to the distinguishing algorithms. It can be adapted to a private-key setting. See the analogous remark after the definition of pseudorandom generators (Definition 8.2).
2. Algorithm A models a passive adversary, who performs a ciphertext-only attack. But, everybody knows the public key i and can encrypt messages of his choice at any time. This implies that the adversary algorithm A may include the encryption of messages of its choice. We see that ciphertext-indistinguishability, as defined here, means security against adaptively-chosen-plaintext attacks. Chosen-ciphertext attacks are considered in Section 9.5.
3. The output of the sampling algorithm S is a subset $\{m_0, m_1\}$ with two members, and therefore $m_0 \neq m_1$.
   If the message spaces M_i are unrestricted, like $\{0, 1\}^*$, then it is usually required that the two candidate plaintexts m_0, m_1, generated by S, have the same bit length. This additional requirement is reasonable. Typically, the length of the plaintext and the length of the ciphertext are closely related. Hence, information about the length of the plaintexts necessarily leaks to an adversary, and plaintexts of different length can easily be distinguished. For the same reason, we considered Vernam's one-time pad as a cipher for plaintexts of a fixed length in Section 9.1 (see the remark after Theorem 9.5).
   In this section, we consider only schemes where all plaintexts are of the same bit length.
4. In view of Proposition 9.10, the definition is the computational analogy to the notion of perfect secrecy, as defined by Shannon.

Non-perfect secrecy means that some algorithm A is able to distinguish between distinct plaintexts m_0 and m_1 (given the ciphertext) with a probability $> 1/2$. The running time of A may be exponential. If an encryption scheme is not ciphertext-indistinguishable, then an algorithm with polynomial running time is able to distinguish between distinct plaintexts m_0 and m_1 (given the ciphertext) with a probability significantly larger than $1/2$. In addition, the plaintexts m_0 and m_1 can be found in polynomial time by some probabilistic algorithm S. This additional requirement is adequate. A secrecy problem can only exist for messages which can be generated in practice by using a probabilistic polynomial algorithm. The message generation is modeled uniformly by S for all keys i.[2]

5. The notion of ciphertext-indistinguishability was introduced by Goldwasser and Micali ([GolMic84]). They call it *polynomial security* or *polynomial-time indistinguishability*. Ciphertext-indistinguishable encryption schemes are also called *schemes with indistinguishable encryptions*. Another notion of security was introduced in [GolMic84]. An encryption scheme is called *semantically secure*, if it has the following property: Whatever a passive adversary Eve is able to compute about the plaintext in polynomial time given the ciphertext, she is also able to compute in polynomial time without the ciphertext. The messages to be encrypted are assumed to be generated by a probabilistic polynomial algorithm. Semantic security is equivalent to indistinguishability ([GolMic84]; [MicRacSlo88]; [WatShiIma03]; [Goldreich04]).

6. Recall that the execution of a probabilistic algorithm is an independent random experiment (see Chapter 5). Thus, the coin tosses of the distinguishing algorithm A are independent of the coin tosses of the sampling algorithm S and the coin tosses of the encryption algorithm E. This reflects the fact that the sender and the adversary operate independently.

7. The probability in our present definition is also taken over the random generation of a key i, with a given security parameter k. Even for very large k, there may be insecure keys i such that A is able to distinguish successfully between two plaintext candidates. However, when randomly generating keys by the key generator, the probability of obtaining an insecure one is negligibly small (see Proposition 6.17 for a precise statement).

8. Ciphertext-indistinguishable encryption algorithms are necessarily randomized encryptions. When encrypting a plaintext m twice, the probability that we get the same ciphertext must be negligibly small. Otherwise it would be easy to distinguish between two messages m_0 and m_1 by comparing the ciphertext with encryptions of m_0 and m_1.

---

[2] By applying a (more general) non-uniform model of computation (non-uniform polynomial-time algorithms instead of probabilistic polynomial algorithms, see, for example, [Goldreich99], [Goldreich04]), one can dispense with the sampling algorithm S.

Theorem 9.15. *Let E be the pseudorandom one-time pad induced by a computationally perfect pseudorandom bit generator G. Then E is ciphertext-indistinguishable.*

Proof. The proof runs in exactly the same way as the proof of Theorem 9.16, yielding a contradiction to the assumption that G is computationally perfect. See below for the details.   □

General pseudorandom one-time pads leave open how the secret encryption key – the randomly chosen seed – is securely transmitted to the receiver of the message. Public-key one-time pads provide an answer. The key is encrypted by the underlying one-way permutation and becomes part of the encrypted message (see Definition 9.12). The indistinguishability is preserved.

Theorem 9.16. *Let E be the public-key one-time pad induced by a family $f = (f_i : D_i \longrightarrow D_i)_{i \in I}$ of trapdoor permutations with hard-core predicate B. Then E is ciphertext-indistinguishable.*

Remark. Theorem 9.16 states that public-key cryptography provides variants of the one-time pad which are provably secure and practical. XORing the plaintext with a pseudorandom bit sequence generated from a short random seed by a trapdoor permutation with hard-core predicate (e.g. use the Blum–Blum–Shub generator or the RSA generator, see Section 8.1,) yields an encryption with indistinguishability. Given the ciphertext, an adversary is provably not able to distinguish between two plaintexts. In addition, it is possible in this public-key one-time pad to securely transmit the key string (more precisely, the seed of the pseudorandom sequence) to the recipient, simply by encrypting it by means of the one-way function.

Of course, the security proof for a public-key one-time pad, such as the RSA- or Blum–Blum–Shub-based one-time pad, is conditional. It depends on the validity of basic unproven (though widely believed) assumptions, such as the RSA assumption (Definition 6.7), or the factoring assumption (Definition 6.9).

Computing the pseudorandom bit sequences using a one-way permutation requires complex computations, such as exponentiation and modular reductions. Thus, the classical private-key symmetric encryption methods, like the DES (see Chapter 2.1) or stream ciphers, using shift registers to generate pseudorandom sequences (see, e.g., [MenOorVan96], Chapter 6), are much more efficient than public-key one-time pads, and hence are better suited for large amounts of data.

However, notice that the one-way function of the Blum–Blum–Shub generator (see Chapter 8) is a quite simple one. Quadratic residues $x \bmod n$ are squared: $x \mapsto x^2 \bmod n$. A public-key one-time pad, whose efficiency is comparable to standard RSA encryption, can be implemented based on this generator.

Namely, suppose $n = pq$, with distinct primes $p, q \equiv 3 \bmod 4$ of binary length k. Messages m of length l are to be encrypted. In order to encrypt m, we randomly choose an element from $\mathbb{Z}_n^*$ and square it to get a random x in QR_n. This requires $O(k^2)$ steps. To get the pseudorandom sequence and to encrypt the random seed x we have to compute l squares modulo n, which comes out to $O(k^2 l)$ steps. XORing requires $O(l)$ steps. Thus, encryption is finished in $O(k^2 l)$ steps.

To decrypt, we first compute the seed x from x^{2^l} by drawing the square root l times in QR_n. We can do this by drawing the square roots modulo p and modulo q, and applying the Chinese Remainder Theorem (see Proposition A.68).

Assume, in addition, that we have even chosen $p, q \equiv 7 \bmod 8$. Then $p + 1/8$ is in $\mathbb{N}$, and for every quadratic residue $a \in QR_p$ we get a square root b which again is an element of QR_p by setting

$$b = a^{(p+1)/4} = \left(a^{(p+1)/8} \right)^2.$$

Here, note that

$$b^2 = a^{(p+1)/2} = a \cdot a^{(p-1)/2} = a,$$

since $a^{(p-1)/2} = \left(\frac{a}{p} \right) = 1$ for quadratic residues $a \in QR_p$ (Proposition A.58). Thus, we can derive $x \bmod p$ from $y = x^{2^l}$ by

$$x \bmod p = y^u \bmod p, \quad \text{with } u = \left(\frac{p+1}{4} \right)^l \bmod (p-1).$$

The exponent u can be reduced modulo $p-1$, since $a^{p-1} = 1$ for all $a \in \mathbb{Z}_p^*$. We assume that the message length l is fixed. Then the exponent u can be computed in advance, and we see that figuring out $x \bmod p$ (or $x \bmod q$) requires at most k squarings applying the square-and-multiply algorithm (Algorithm A.27). Thus, it can be done in $O(k^3)$ steps.

Reducing $x^{2^l} \bmod p$ (and $x^{2^l} \bmod q$) at the beginning and applying the Chinese Remainder Theorem at the end requires at most $O(k^2)$ steps. Summarizing, we see that computing x requires $O(k^3)$ steps. Now, completing the decryption essentially means performing an encryption whose cost is $O(k^2 l)$, as we saw above. Hence, the complete decryption procedure takes $O(k^3 + k^2 l)$ steps. If $l = O(k)$, this is equal to $O(k^3)$ and thus of the same order as the running time of an RSA encryption.

The efficiency of the Blum–Blum–Shub-based public-key one-time pad (as well as that of the RSA-based one) can be increased even further, by modifying the generation of the pseudorandom bit sequence. Instead of taking only the least-significant bit of $x^{2^j} \bmod n$, you may take the $\lfloor \log_2(|n|) \rfloor$ least-significant bits after each squaring. These bits form a $\lfloor \log_2(|n|) \rfloor$-bit hard-core predicate of the modular squaring function and are simultaneously secure (see Exercise 7 in Chapter 8). The resulting public-key one-time pad is called *Blum–Goldwasser probabilistic encryption* ([BluGol85]). It is also ciphertext-indistinguishable (see Exercise 8 in Chapter 8).

Our considerations are not only valid asymptotically, as the O notation might suggest. Take for example $k = 512$ and $|n| = 1024$, and encrypt 1024-bit messages m. In the x^2 mod n public-key one-time pad, always use the $\log_2(|n|) = 10$ least-significant bits. To encrypt a message m, about 100 modular squarings of 1024-bit numbers are necessary. To decrypt a ciphertext, we first determine the seed x by at most $1024 = 512 + 512$ modular squarings and multiplications of 512-bit numbers, and then compute the plaintext using about 100 modular squarings, as in encryption. Encrypting and decrypting messages $m \in \mathbb{Z}_n$ by RSA requires up to 1024 modular squarings and multiplications of 1024-bit (encryption) or 512-bit (decryption) numbers, with the actual number depending on the size of the encryption and decryption exponents. In the estimates, we did not count the (few) operations associated with applying the Chinese Remainder Theorem during decryption.

Proof (of Theorem 9.16). Let K be the key generator of f and $G :=$ $G(f, B, Q)$ be the pseudorandom bit generator (Chapter 8). Recall that

$$E(i, m) = (m \oplus G_i(x), f_i^{Q(k)}(x)),$$

where $i \in I = (I_k)_{k \in \mathbb{N}}$, $m \in \{0,1\}^{Q(k)}$ (for $i \in I_k$) and the seed x is randomly and uniformly chosen from D_i.

The image of the uniform distribution on D_i under $f_i^{Q(k)}$ is again the uniform distribution on D_i, because f_i is bijective.

You can obtain a proof of Theorem 9.15 simply by omitting "$\times D_i$", $y, f_i^{Q(k)}(x)$ everywhere (and by replacing $Q(k)$ by $l(k)$). Our proof yields a contradiction to Theorem 8.4. In the proof of Theorem 9.15, you get the completely analogous contradiction to the assumption that the pseudorandom bit generator G is computationally perfect.

Now assume that there is a probabilistic polynomial distinguishing algorithm $A(i, m_0, m_1, c, y)$, with inputs $i \in I, m_0, m_1 \in M_i, c \in \{0,1\}^{Q(k)}$ (if $i \in I_k$), $y \in D_i$, a probabilistic polynomial sampling algorithm $S(i)$ and a positive polynomial P such that

$$\text{prob}(A(i, m_0, m_1, c, y) = m :$$
$$i \leftarrow K(1^k), \{m_0, m_1\} \leftarrow S(i), m \xleftarrow{u} \{m_0, m_1\}, (c, y) \leftarrow E(i, m))$$
$$= \text{prob}(A(i, m_0, m_1, m \oplus G_i(x), f_i^{Q(k)}(x)) = m :$$
$$i \leftarrow K(1^k), \{m_0, m_1\} \leftarrow S(i), m \xleftarrow{u} \{m_0, m_1\}, x \xleftarrow{u} D_i)$$
$$> \frac{1}{2} + \frac{1}{P(k)},$$

for infinitely many k. We define a probabilistic polynomial statistical test $\tilde{A} = \tilde{A}(i, z, y)$, with inputs $i \in I, z \in \{0,1\}^{Q(k)}$ (if $i \in I_k$) and $y \in D_i$ and output in $\{0,1\}$:

1. Apply $S(i)$ and get $\{m_0, m_1\} := S(i)$.

2. Randomly choose m in $\{m_0, m_1\} : m \xleftarrow{u} \{m_0, m_1\}$.
3. Let

$$\tilde{A}(i, z, y) := \begin{cases} 1 \text{ if } A(i, m_0, m_1, m \oplus z, y) = m, \\ 0 \text{ otherwise.} \end{cases}$$

The statistical test $\tilde{A}$ will be able to distinguish between the pseudorandom sequences produced by G and truly random, uniformly distributed sequences, thus yielding the desired contradiction. We have to compute the probability

$$\text{prob}(\tilde{A}(i, z, y) = 1 : i \leftarrow K(1^k), (z, y) \leftarrow \{0, 1\}^{Q(k)} \times D_i)$$

for both the uniform distribution on $\{0, 1\}^{Q(k)} \times D_i$ and the distribution induced by $(G, f^{Q(k)})$. More precisely, we have to compare the probabilities

$$p_{k,G} := \text{prob}(\tilde{A}(i, G_i(x), f_i^{Q(k)}(x)) = 1 : i \leftarrow K(1^k), x \xleftarrow{u} D_i) \quad \text{and}$$

$$p_{k,\text{uni}} := \text{prob}(\tilde{A}(i, z, y) = 1 : i \leftarrow K(1^k), z \xleftarrow{u} \{0, 1\}^{Q(k)}, y \xleftarrow{u} D_i).$$

Our goal is to prove that

$$p_{k,G} - p_{k,\text{uni}} > \frac{1}{P(k)},$$

for infinitely many k. This contradicts Theorem 8.4 and finishes the proof. From the definition of $\tilde{A}$ we get

$$\text{prob}(\tilde{A}(i, z, y) = 1 : i \leftarrow K(1^k), (z, y) \leftarrow \{0, 1\}^{Q(k)} \times D_i)$$
$$= \text{prob}(A(i, m_0, m_1, m \oplus z, y) = m : i \leftarrow K(1^k),$$
$$(z, y) \leftarrow \{0, 1\}^{Q(k)} \times D_i, \{m_0, m_1\} \leftarrow S(i), m \xleftarrow{u} \{m_0, m_1\}).$$

Since the random choice of (z, y) and the random choice of m in the probabilistically computed pair $S(i)$ are independent, we may switch them in the probability and obtain

$$\text{prob}(\tilde{A}(i, z, y) = 1 : i \leftarrow K(1^k), (z, y) \leftarrow \{0, 1\}^{Q(k)} \times D_i)$$
$$= \text{prob}(A(i, m_0, m_1, m \oplus z, y) = m : i \leftarrow K(1^k),$$
$$\{m_0, m_1\} \leftarrow S(i), m \xleftarrow{u} \{m_0, m_1\}, (z, y) \leftarrow \{0, 1\}^{Q(k)} \times D_i).$$

Now consider $p_{k,G}$:

$$p_{k,G} = \text{prob}(A(i, m_0, m_1, m \oplus G_i(x), f_i^{Q(k)}(x)) = m :$$
$$i \leftarrow K(1^k), \{m_0, m_1\} \leftarrow S(i), m \xleftarrow{u} \{m_0, m_1\}, x \xleftarrow{u} D_i).$$

We assumed that this probability is $> 1/2 + 1/P(k)$, for infinitely many k.

The following computation shows that $p_{k,\text{uni}} = 1/2$ for all k, thus completing the proof. We have

$$p_{k,\text{uni}} = \text{prob}(A(i, m_0, m_1, m \oplus z, y) = m : i \leftarrow K(1^k), \{m_0, m_1\} \leftarrow S(i),$$
$$m \overset{u}{\leftarrow} \{m_0, m_1\}, z \overset{u}{\leftarrow} \{0,1\}^{Q(k)}, y \overset{u}{\leftarrow} D_i)$$
$$= \sum_i \text{prob}(K(1^k) = i) \cdot \sum_{m_0, m_1} \text{prob}(S(i) = \{m_0, m_1\}) \cdot p_{i, m_0, m_1},$$

with

$$p_{i, m_0, m_1} = \text{prob}(A(i, m_0, m_1, m \oplus z, y) = m :$$
$$m \overset{u}{\leftarrow} \{m_0, m_1\}, z \overset{u}{\leftarrow} \{0,1\}^{Q(k)}, y \overset{u}{\leftarrow} D_i).$$

Vernam's one-time pad is perfectly secret (Theorem 9.5). Thus, the latter probability is equal to $1/2$ by Proposition 9.10. Note that the additional input $y \overset{u}{\leftarrow} D_i$ of A, not appearing in Proposition 9.10, can be viewed as additional coin tosses of A. So, we finally get the desired equation

$$p_{k,\text{uni}} = \sum_i \text{prob}(K(1^k) = i) \sum_{m_0, m_1} \text{prob}(S(i) = \{m_0, m_1\}) \cdot \frac{1}{2} = \frac{1}{2},$$

and the proof of the theorem is finished.   □

9.5 Chosen-Ciphertext Attacks

In the preceding section, we studied encryption schemes which are ciphertext-indistinguishable and provide security against a passive eavesdropper, who performs a ciphertext-only or an adaptively-chosen-plaintext attack. The schemes may still be insecure against an attacker, who manages to get temporary access to the decryption device and who executes a chosen-ciphertext or an adaptively-chosen-ciphertext attack (see Section 1.3 for a classification of attacks). Given a ciphertext c, such an attacker Eve tries to get information about the plaintext. In the course of her attack, Eve can get decryptions of ciphertexts c' from the decryption device, with the only restriction that $c' \neq c$. She has temporary access to a "decryption oracle".

Consider, for example, the efficient implementation of Goldwasser–Micali's probabilistic encryption, which we called public-key one-time pad and which we studied in the preceding sections. A public-key one-time pad encrypts a message m by XORing it bitwise with a pseudorandom key stream $G(x)$, where x is a secret random seed. The pseudorandom bit generator G is induced by a family $f = (f_i : D_i \longrightarrow D_i)_{i \in I}$ of trapdoor permutations with hard-core predicate B. The public-key-encrypted seed $f^l(x)$ (where l is the bit length of the messages m) is transmitted together with the encrypted message $m \oplus G(x)$ (see Section 9.4 above).

These encryption schemes are provably secure against passive eavesdroppers (Theorem 9.16), but they are insecure against a chosen-ciphertext attacker Eve. Eve submits ciphertexts (c, y) for decryption, where y is any element in the domain of f and $c = c_1 c_2 \ldots c_l$ is any bit string of length l. If Eve only obtains the last bit m_l of the plaintext $m = m_1 m_2 \ldots m_l$ from the decryption device, then she immediately derives the hard-core bit $B(f^{-1}(y))$ of $f^{-1}(y)$, since $B(f^{-1}(y)) = c_l \oplus m_l$. Therefore, Eve has an oracle that provides her with the hard-core bit $B(f^{-1}(y))$ for every y. Now assume that f is the RSA function modulo a composite $n = pq$ or the Rabin function, which squares quadratic residues modulo $n = pq$, as in Blum–Blum–Shub-encryption, and B is the least significant bit. The hard-core-bit oracle enables Eve to compute the inverse of the RSA or Rabin function modulo n by using an efficient algorithm, which calls the oracle as a subroutine. We constructed these algorithms in Sections 7.2 and 7.3. Then, of course, Eve can also compute the seed x from $f^l(x)$ and derive the plaintext $m = c \oplus G(x)$ for every ciphertext c.

We have to worry about adaptively-chosen-ciphertext attacks. One can imagine scenarios where Bob, the owner of the secret decryption key, might think that decryption requests are reasonable – for example, if an incomplete configuration or control of privileges enables the attacker Eve from time to time to get access to the decryption device. If a system is secure against chosen-ciphertext attacks, then it also resists partial chosen-ciphertext attacks. In such an attack, adversary Eve does not get the full plaintext in response to her decryption requests, but only some partial information. Partial-chosen-ciphertext attacks are a real danger in practice. We just discussed a partial-chosen-ciphertext attack against public-key one-time pads. In Section 3.3.2, we described Bleichenbacher's 1-Million-Chosen-Ciphertext Attack against PKCS#1(v1.5)-based schemes, which is a practical example of a partial-chosen-ciphertext attack.

Therefore, it is desirable to have encryption schemes which are provably secure against adaptively-chosen-ciphertext attacks. We give two examples of such schemes. The security proof of the first scheme, Boneh's SAEP, relies on the random oracle model, which we described in Section 3.3.5, and the factoring assumption (Definition 6.9). The security proof of the second scheme, Cramer–Shoup's public-key encryption scheme, is based solely on a standard number-theoretic assumption of the hardness of a computational problem and on a standard hash function assumption (collision-resistance). The stronger random oracle model is not needed.

We start with a definition of the security notion. The definition includes a precise description of the attack model. As before, we strive for ciphertext-indistinguishability. This notion can be extended to cover adaptively-chosen-ciphertext attacks ([NaoYun90]; [RacSim91]).

Definition 9.17. Let E be a public-key encryption scheme.

1. An *adaptively-chosen-ciphertext attack algorithm A* against E is a probabilistic polynomial algorithm that interacts with its environment, called the challenger, as follows:

a. **Setup:** The challenger C randomly generates a public-secret key pair (pk, sk) for E and calls A with the public key pk as input. The secret key sk is kept secret.

b. **Phase I:** The adversary A issues a sequence of decryption requests for various ciphertexts c'. The challenger responds with the decryption of the valid ciphertexts c'.

c. **Challenge:** At some point, algorithm A outputs two distinct messages m_0, m_1. The challenger selects a message $m \in \{m_0, m_1\}$ at random and responds with the "challenge" ciphertext c, which is an encryption $E(pk, m)$ of m.

d. **Phase II:** The adversary A continues to request the decryption of ciphertexts c', with the only constraint that $c' \neq c$. The challenger decrypts c', if c' is valid, and sends the plaintext to A. Finally, A terminates and outputs $m' \in \{m_0, m_1\}$.

The attacker A is successful, if $m' = m$.

We also call A, more precisely, an *adaptively-chosen-ciphertext distinguishing algorithm*.

2. E is *ciphertext-indistinguishable against adaptively-chosen-ciphertext attacks*, if for every adaptively-chosen-ciphertext distinguishing algorithm A, the probability of success is $\leq 1/2 + \varepsilon$, with ε negligible.

We also call such an encryption scheme E *indistinguishability-secure*, or *secure*, for short, *against adaptively-chosen-ciphertext attacks*.

Remarks:

1. In the real attack, the challenger is Bob, the legitimate owner of a public-secret key pair (pk, sk). He is attacked by the adversary A.

2. If the message space is unrestricted, like $\{0, 1\}^*$, then it is usually required that the two candidate plaintexts m_0, m_1 have the same bit length. This additional requirement is reasonable. Typically, the length of the plaintext and the length of the ciphertext are closely related. Hence, information about the length of the plaintexts necessarily leaks to an adversary, and plaintexts of different length can easily be distinguished (also see the remark on Vernam's one-time pad after Theorem 9.5).

   If the message space is a number-theoretic or geometric domain, as in the typical public-key encryption scheme, then, usually, all messages have the same bit length. Take, for example, $\mathbb{Z}_n$. The messages $m \in \mathbb{Z}_n$ are all encoded as bit strings of length $\lfloor \log_2(n) \rfloor + 1$; they are padded out with the appropriate number of leading zeros, if necessary.

3. In the literature, adaptively-chosen-ciphertext attacks are sometimes denoted by the acronym CCA2, and sometimes they are simply called chosen-ciphertext attacks. Non-adaptive chosen-ciphertext attacks, which can request the decryption of ciphertexts only in phase I of the attack, are often called *lunchtime attacks* or *midnight attacks* and denoted by the acronym CCA1.

4. The notion of semantic security (see the remarks after Definition 9.14) can also be carried over to the adaptively-chosen-ciphertext setting. It means that whatever an adversary Eve is able to compute about the plaintext m in polynomial time given the ciphertext c, she is also able to compute in polynomial time without the ciphertext, even if Eve gets the decryption of any adaptively chosen ciphertexts $c' \neq c$. As shown recently, semantic security and indistinguishability are also equivalent in the case of adaptively-chosen-ciphertext attacks ([WatShiIma03]; [Goldreich04]).

9.5.1 A Security Proof in the Random Oracle Model

Boneh's Simplified OAEP – SAEP. As an example, we study Boneh's Simple-OAEP encryption, or SAEP for short ([Boneh01]). In the encryption scheme, a collision-resistant hash function h is used. The security proof for SAEP is a proof in the random oracle model. Basically, this means that h is assumed to be a truly random function (see page 73 for a more precise description of the random oracle model).

In contrast to Bellare's OAEP, which we studied in Section 3.3.3, SAEP applies Rabin encryption (Section 3.5.1) and not the RSA function. Compared to OAEP, the padding scheme is considerably simplified. It requires only one cryptographic hash function. The slightly more complex padding scheme SAEP+ is provably secure and can be applied to the RSA and the Rabin function (see [Boneh01]). As OAEP, it requires an additional hash function G. We do not discuss SAEP+ here.

Key Generation. Let $k \in \mathbb{N}$ be an even security parameter (e.g. $k = 1024$). Bob generates a $(k+2)$-bit modulus $n = pq$, with $2^{k+1} < n < 2^{k+1} + 2^k$ (i.e., the two most significant bits of n are 10), where p and q are $(k/2 + 1)$-bit primes, with $p \equiv q \equiv 3 \bmod 4$. The primes p, q are chosen randomly. The public key is n, the private key is (p, q).

The security parameter is split into 3 parts, $k = l + s_0 + s_1$, with $l \leq k/4$ and $l + s_0 \leq k/2$. In practice, typical values for these parameters are $k = 1024, l = 256, s_0 = 128, s_1 = 640$. The constraints on the lengths of the security parameters are necessary for the security proof (see Theorem 9.19 below).

We make use of a collision-resistant hash function

$$h : \{0,1\}^{s_1} \longrightarrow \{0,1\}^{l+s_0}.$$

Notation. As usual, we denote by 0^r (or 1^r) the constant bit string $000\ldots0$ (or $111\ldots1$) of length r ($r \in \mathbb{N}$). As always, let $\|$ denote the concatenation of strings and $\oplus$ be the bitwise XOR operator.

Encryption. To encrypt an l-bit message $m \in \{0,1\}^l$ for Bob, Alice proceeds in the following steps:

1. She chooses a random bit string $r \in \{0,1\}^{s_1}$.

2. She appends s_0 0-bits to m to obtain the $(l + s_0)$-bit string $x := m\|0^{s_0}$.
3. She sets $y = (x \oplus h(r))\|r$.
4. She views the k-bit string y as a k-bit integer and applies the Rabin trapdoor function modulo n to obtain the ciphertext c:

$$c := y^2 \bmod n.$$

Note that $y < 2^k < n/2$.

Remarks:

1. At a first glance, the length l of the plaintexts might appear small (recall that a typical value is $l = 256$). But usually we encrypt only short data by using a public-key method, for example, session keys for a symmetric cipher, such as Triple-DES or AES, and for this purpose 256 bits are really sufficient.
2. All ciphertexts are quadratic residues modulo n. But not every quadratic residue appears as a ciphertext. Let c be a quadratic residue modulo n. Then c is the encryption of some plaintext in $\{0,1\}^l$, if there is a square root y of c modulo n, such that
   a. $y < 2^k$, i.e., we may consider y as a bit string of length k, and
   b. the s_0 least-significant bits of $v \oplus h(r)$ are all 0, where $y = v\|r, v \in \{0,1\}^{l+s_0}, r \in \{0,1\}^{s_1}$.
   In this case, the plaintext m, whose encryption is c, consists of the l most-significant bits of $v \oplus h(r)$, i.e., $v \oplus h(r) = m\|0^{s_0}$, and we call c a *valid ciphertext* (or a *valid encryption of* m) *with respect to* y.

The security of SAEP is based on the factoring assumption: it is practically infeasible to compute the prime factors p and q of n (see Section 6.5 for a precise statement of the assumption). Decrypting ciphertexts requires drawing square roots. Without knowing the prime factors of n, it is infeasible to compute square roots modulo n. The ability to compute modular square roots is equivalent to the ability to factorize the modulus n (Proposition A.68, Lemma A.69). Bob can compute square roots modulo n because he knows the secret factors p and q.

We recall from Proposition A.68 some basics on computing square roots.

Let $c \in \mathbb{Z}_n$ be a quadratic residue, $c \neq 0$. If c is prime to n, then c has 4 distinct square roots modulo n. If c is not prime to n, i.e., if c is a multiple of p or q, then c has only 2 square roots modulo n. In the latter case, the factors p, q of n can be easily derived by computing $\gcd(c, n)$ with the Euclidean algorithm. The probability for this case is negligibly small, if c is a random quadratic residue.

If $y^2 \bmod n = c$, then also $(n - y)^2 \bmod n = c$. Hence, exactly two of the 4 roots (or one of the two roots) are $< n/2$.

The residue $[0] \in \mathbb{Z}_n$ has the only square root $[0]$.

If both primes p and q are $\equiv 3 \bmod 4$, as here in SAEP, and if the factors p and q are known, then the square roots of c can be easily and efficiently computed as follows:

1. $p + 1/4$ and $q + 1/4$ are integers, and the powers $z_p = c^{(p+1)/4} \bmod p$ of $c \bmod p$ and $z_q = c^{(q+1)/4} \bmod q$ of $c \bmod q$ are square roots of c modulo p and modulo q (see Proposition A.66). Recall that $z_p^2 = c^{(p+1)/2} = c \cdot c^{(p-1)/2}$ and $c^{(p-1)/2} \equiv 1 \bmod p$, if $c \bmod p \neq 0$ (see Euler's criterion, Proposition A.58). If $c \bmod p \neq 0$, then $\pm z_p$ are the two distinct square roots of $c \bmod p$. If $c \bmod p = 0$, then there is the single root $z_p = 0$ of c modulo p.

2. To get the square roots of c modulo n, we map $(\pm z_p, \pm z_q)$ to $\mathbb{Z}_n$ by applying the inverse Chinese remainder map. If c is prime to n, then we obtain 4 distinct roots. If $z_p = 0$ or $z_q = 0$, we get 2 distinct roots, and if both z_p and z_q are 0, then there is the only root 0 (see Proposition A.68).

Decryption. Bob decrypts a ciphertext c by using his secret key (p, q) as follows:

1. He computes the square roots of c modulo n, as just described.
   In this computation, he tests that $z_p^2 \equiv c \bmod p$ and $z_q^2 \equiv c \bmod q$. If either test fails, then c is not a quadratic residue and Bob rejects c.

2. Two of the four roots (or one of the two roots) are $> n/2$ and hence can be discarded. Bob is left with 2 square roots y_1, y_2 (let $y_1 = y_2$, if there is only one left). If neither of y_1, y_2 is $< 2^k$, then Bob rejects c.
   From now on, we assume that $y_1 < 2^k$. If $y_2 \geq 2^k$, then Bob does not have to distinguish between two candidates, and thus he can simplify the following steps (omit the processing of y_2). So, assume now that both y_1 and y_2 are $< 2^k$. Then, we may view them as strings in $\{0, 1\}^k$.

3. Bob writes $y_1 = v_1 \| r_1$ and $y_2 = v_2 \| r_2$, with $v_1, v_2 \in \{0, 1\}^{l+s_0}$ and $r_1, r_2 \in \{0, 1\}^{s_1}$, and computes $x_1 = v_1 \oplus h(r_1)$ and $x_2 = v_2 \oplus h(r_2)$.

4. He writes $x_1 = m_1 \| t_1$ and $x_2 = m_2 \| t_2$, with $m_1, m_2 \in \{0, 1\}^l$ and $t_1, t_2 \in \{0, 1\}^{s_0}$. If either none or both of t_1, t_2 are $00 \ldots 0$, then Bob rejects c. Otherwise, let $i \in \{1, 2\}$ be the unique i with $t_i = 00 \ldots 0$. Then, m_i is the decryption of c.

Remark. It might happen that Bob rejects a valid encryption c of a message m, because c is valid with respect to both roots y_1 and y_2 and hence both t_1 and t_2 are 00...0. But the probability for this event is negligibly small, at least if h comes close to a random function.

Namely, assume $y_1 \neq y_2$ and c is a valid encryption of m with respect to y_1, i.e., $y_1 = ((m \| 0^{s_0}) \oplus h(r_1)) \| r_1$. If h is assumed to be a random function, then the probability that c is also a valid ciphertext with respect to y_2 is about $\cdot 1/2^{s_0}$. There are 2 cases. If $r_1 \neq r_2$, then $h(r_2)$ is randomly generated independently from $h(r_1)$. Hence, the probability that the s_0 least-significant bits of $h(r_2)$ are just the s_0 least-significant bits of v_2 is $1/2^{s_0}$. If $r_1 = r_2$ and c is valid with respect to both roots, then the $(s_0 + s_1)$ least-significant bits of y_1 and y_2 coincide, i.e., $y_2 = y_1 + 2^{s_0+s_1}\delta$, with absolute value $|\delta| < 2^l$. Since $|\delta| < 2^l < 2^{k/2} < p, q$, we know that δ is a unit modulo n. From

$$y_2^2 = (y_1 + 2^{s_0+s_1}\delta)^2 = y_1^2 + 2^{s_0+s_1+1}\delta y_1 + 2^{2(s_0+s_1)}\delta^2 \equiv c \equiv y_1^2 \bmod n,$$

we conclude that $y_1 \equiv -2^{s_0+s_1-1}\delta \bmod n$, and the probability for that is $1/2^{s_0+s_1-1}$.

The rejection of valid ciphertexts can be completely avoided by a slight modification of the encryption algorithm. Alice repeats the random generation of r, until y has Jacobi symbol 1. Then Bob can always select the correct square root by taking the unique square root $< 2^k$ with Jacobi symbol 1. We described a similar approach in Section 3.5.1 on Rabin's encryption. However, this makes the encryption scheme less efficient, and it is not necessary.

The proof of security for SAEP is based on an important result due to Coppersmith ([Coppersmith97]).

Theorem 9.18. *Let n be an integer, and let $f(X) \in \mathbb{Z}_n[X]$ be a monic polynomial of degree d. Then there is an efficient algorithm which finds all $x \in \mathbb{Z}$ such that the absolute value $|x| < n^{1/d}$ and $f(x) \equiv 0 \bmod n$.*

For a proof, see [Coppersmith97]. The special case $f(X) = X^d - c, c \in \mathbb{Z}_n$, is easy. To get the solutions x with $|x| < n^{1/d}$, you can compute the ordinary d-th roots of c, because for $0 \leq x < n^{1/d}$ we have $x^d \bmod n = x^d$. To compute the ordinary d-th roots is easy, since $x \mapsto x^d$ (without taking residues) is strictly monotonic (take, for example, the simple Algorithm 3.5).

Theorem 9.19. *Assume that the hash function h in SAEP is a random oracle. Let $n = pq$ be a key for SAEP with security parameter $k = l + s_0 + s_1$ (i.e., $2^{k+1} < n < 2^{k+1} + 2^k$). Assume $l \leq k/4$ and $l + s_0 \leq k/2$. Let $A(n, l, s_0, s_1)$ be a probabilistic distinguishing algorithm with running time t that performs an adaptively-chosen-ciphertext attack against SAEP and has a probability of success $\geq 1/2 + \varepsilon$. Let q_d be the number of A's decryption queries, and let q_h be the number of A's queries of the random oracle h.*

Then there is a probabilistic algorithm B for factoring the modulus n with

$$\text{running time} \quad t + O(q_d q_h t_C + q_d t'_C) \quad \text{and}$$

$$\text{probability of success} \geq \frac{1}{6} \cdot \varepsilon \cdot \left(1 - \frac{2q_d}{2^{s_0}} - \frac{2q_d}{2^{s_1}}\right).$$

Here, t_C (resp. t'_C) is the running time of Coppersmith's algorithm for finding "small-size" roots of polynomials of degree 2 (resp. 4) modulo n (i.e., roots with absolute value $\leq n^{1/2}$ resp. $\leq n^{1/4}$).

Remarks:

1. Recall that typical values of the parameters are $k = 1024, l = 256, s_0 = 128, s_1 = 640$. The number q_d of decryption queries that an adversary can issue in practice should be limited by 2^{40}. We see that the fractions $2q_d/2^{s_0}$ and $2q_d/2^{s_1}$ are negligibly small.

2. Provided the factoring assumption is true, we conclude from Boneh's theorem that a probabilistic polynomial attacking algorithm like A, whose probability of success is $\geq 1/2 + 1/P(k)$, P a positive polynomial (i.e., A has a non-negligible "advantage"), cannot exist. Otherwise, B would factor moduli n in polynomial time. Thus, SAEP is indistinguishability-secure against adaptively-chosen-ciphertext attacks (in the random oracle model).

   But Boneh's result is more precise: It gives a *tight reduction* from the problem of attacking SAEP to the problem of factoring. If we had a successful attacking algorithm A against SAEP, we could give precise estimates for the running time and the probability of successfully factoring n. Or, conversely, if we can state reasonable bounds for the running time and the probability of success in factoring numbers n of a given bit length, we can derive a concrete lower bound for the running time of algorithms attacking SAEP with a given probability of success.

Proof. We sketch the basic ideas of the proof. In particular, we illustrate how the random oracle assumption is applied. For more details we refer to [Boneh01].

It suffices to construct an algorithm S which on inputs n, l, s_0, s_1 and $c = \alpha^2 \bmod n$, for a randomly chosen α with $0 \leq \alpha < 2^k$, outputs a square root α', $0 \leq \alpha' < 2^k$, of c with probability $\geq \varepsilon \cdot (1 - 2q_d/2^{s_0} - 2q_d/2^{s_1})$. Namely, with probability $> 1/3$, a number c with $0 \leq c < n$ has two distinct square roots modulo n in $[0, 2^k[$ [3]. Hence $\alpha \neq \alpha'$ with probability $\geq 1/6$ and then n can be factored by computing $\gcd(n, \alpha - \alpha')$ (see Lemma A.69 and Section 3.5.1).

In the following we describe the algorithm S. The algorithm efficiently computes a square root $< 2^k$ of c, without knowing p, q, in two cases, which we study first.

1. Let $y = v\|r$, where v and r are bit strings of lengths $(l + s_0)$ and s_1. If y is a root of c, then v is a root of the quadratic polynomial $(2^{s_1}X + r)^2 - c$ modulo n, with $0 \leq v < 2^{l+s_0} \leq 2^{k/2} < n^{1/2}$.

   Thus, if S happens to know or guess correctly the s_1 lower significant bits r of a root y, then it efficiently finds y by applying the following algorithm

   $CompRoot_1(c, r)$:

   Compute the roots $v < 2^{k/2}$ of $(2^{s_1}X + r)^2 - c$ modulo n by using Coppersmith's algorithm. If such a v is found, return the root $y = v\|r$.

2. Let $y = m\|w$ be a square root modulo n of c, with $m \in \{0, 1\}^l$ and $w \in \{0, 1\}^{s_0+s_1}$. Let $c' \in \mathbb{Z}_n$ be a further quadratic residue, $c' \neq c$, and assume that c' has a k-bit square root $y' = m'\|w$ modulo n, whose $(s_0 + s_1)$ lower significant bits w are the same as those of y. Then we

---

[3] See Fact 2 in [Boneh01]; n is chosen between 2^{k+1} and $2^{k+1} + 2^k$ to get this estimate.

may write $y' = y + 2^{s_0+s_1}\delta$, where $\delta = m' - m$ is an integer with absolute value $|\delta| < 2^l \leq 2^{k/4} < n^{1/4}$. y is a common root of the polynomials $f(X) = X^2 - c$ and $g(X,\delta) = (X + 2^{s_0+s_1}\delta)^2 - c'$ modulo n. Therefore, δ is a root of the resultant $\mathcal{R}$ of $f(X)$ and $g(X,\Delta) = (X + 2^{s_0+s_1}\Delta)^2 - c'$. The resultant $\mathcal{R}$ is a polynomial modulo n in Δ of degree 4 (see, for example, [Lang05], Chapter IV). Since $|\delta| < 2^l \leq 2^{k/4} < n^{1/4}$, we can compute δ efficiently by using Coppersmith's algorithm for polynomials of degree 4. The greatest common divisor $\gcd(f(X), g(X,\delta))$ is $X - y$. Thus, if S happens to get such an element c', then it efficiently finds a square root y of c by applying the following algorithm
$CompRoot_2(c, c')$:
Compute the roots δ with absolute value $|\delta| < 2^l$ of the resultant $\mathcal{R}$ modulo n by using Coppersmith's algorithm. Compute the greatest common divisor $X - y$ of $X^2 - c$ and $(X + 2^{s_0+s_1}\delta)^2 - c'$ modulo n. If such a δ (and then y) is found, return the root y.

Algorithm S interacts with the attacking algorithm A. In the real attack, A interacts with Bob, the legitimate owner of the public-secret key pair $(n, (p, q))$, to obtain decryptions of ciphertexts of its choice, and with the random oracle h to get hash values $h(m)$ (in practice, interacting typically means to communicate with another computer program). Now S is constructed to replace both Bob and the random oracle h in the attack. It "simulates" Bob and h.

Each time, when A issues a query, S has a chance to compute a square root of c, and of course S terminates when a root is found.

S has no problems answering the hash queries. Since h is a random oracle, S can assign a randomly generated $v \in \{0,1\}^{l+s_0}$ as hash value $h(r)$. The only restriction is: If the hash value of m is queried more than once, then always the same value has to be provided. Therefore, S has to store the list $\mathcal{H}$ of hash value pairs $(r, h(r))$ that it has given to A.

The structure of S is the following.

S calls A with the public key n and the security parameters l, s_0, s_1 as input. Then it waits for the queries of A.

1. If A queries the hash value for r, then
   a. if $(r, h(r))$ is on the list $\mathcal{H}$ of previous responses to hash queries, then S again sends $h(r)$ as hash value to A;
   b. else S applies algorithm $CompRoot_1(c, r)$; if S finds a root of c modulo n in this way, then it returns the root and terminates;
   c. else S picks a random bit string v of length $(l + s_0)$, puts (r, v) on its list $\mathcal{H}$ of hash values and sends $v = h(r)$ as hash value to A.
2. if A queries the decryption of a ciphertext c', then
   a. S applies algorithm $CompRoot_2(c, c')$; if S finds a root of c modulo n in this way, it returns the root and terminates;
   b. else S applies, for each $(r', h(r'))$ on the list $\mathcal{H}$ of previous hash values, algorithm $CompRoot_1(c', r')$; if S finds a square root $v'\|r'$ of c' in this

way, then it computes $w' = v' \oplus h(r')$; if the s_0 lower significant bits of w' are all 0, i.e., $w' = m'\|0^{s_0}$, then S sends m' as plaintext to A, else S rejects c' as an "invalid ciphertext";

c. else S could not find a square root of c' in the previous step b and rejects c' as an "invalid ciphertext";

3. if A produces the two candidate plaintexts $m_0, m_1 \in \{0, 1\}^l$, then S sends c as encryption of m_0 or m_1 to A.

If A terminates, then S also terminates (if it has not terminated before).

To analyze algorithm S, let y_1 and y_2 be the two square roots of c with $0 \le y_1, y_2 < 2^k$ (take $y_2 = y_1$, if there is only one). The goal of S is to compute y_1 or y_2. We decompose $y_i = v_i\|r_i$, with $r_i \in \{0,1\}^{s_1}$.

We consider several cases.

1. If A happens to query the hash value for r_1 or r_2, then S successfully finds one of the roots y_1, y_2 in step 1b.

2. If A happens to query the decryption of a ciphertext c', and if c' has a k-bit square root y' modulo n, whose $(s_0 + s_1)$ lower significant bits are the same as those of y_1 or y_2, then S successfully finds one of the roots y_1, y_2 in step 2a.

We observed above that S can easily play the role of the random oracle h and answer hash queries. Sometimes, S can answer decryption requests correctly.

3. If A queries the decryption of a ciphertext c' that is valid with respect to its square root y' modulo n, and if A has previously asked for the hash value $h(r')$ of the s_1 rightmost bits r' of y' ($c' = y'^2 \bmod n; y' = v'\|r', r' \in \{0,1\}^{s_1}$), then S responds in step 2b to A with the correct plaintext.

But S does not know the secret key, and so S cannot perfectly simulate Bob. We now study the two cases where, from A's point of view, the behavior of S might appear different from Bob's behavior.

In the real attack, Bob sends a valid encryption of m_0 or m_1 as challenge ciphertext to A. The choice of the number c, which S sends as challenge ciphertext, has nothing to do with m_0 or m_1. Therefore, we have to study the next case 4.

4. The number c, which S presents as challenge ciphertext at the end of phase I, is, from A's point of view, not a valid encryption of m_0 or m_1 with respect to y_i, where $i = 1$ or $i = 2$. This means that $v_i \oplus h(r_i) \ne m_b\|0^{s_0}$ or, equivalently, $h(r_i) \ne (m_b\|0^{s_0}) \oplus v_i$ for $b = 0, 1$.

   This can only happen either

   • if A has asked for $h(r_i)$ in phase I, and S has responded with a non-appropriate hash value for r_i, or

- if A has asked in phase I for the decryption of a ciphertext c' with $c' = y'^2 \bmod n, y' = v' \| r', r' \in \{0,1\}^{s_1}, v' \in \{0,1\}^{l+s_0}$ and $r' = r_i$, i.e., the s_0 rightmost bits of y' and y_i are equal[4]; in this case the answer of S might have put some restriction on $h(r_i)$.

Otherwise, the hash value $h(r_i)$, which is randomly generated by the oracle, is independent of A's point of view at the end of phase I. This implies that, from A's point of view, $h(r_i) = (m_b \| 0^{s_0}) \oplus v_i$ has the same probability as $h(r_i) = w$, for any other $(l + s_0)$-bit string w. Therefore c is a valid encryption of m_0 or m_1 with respect to y_i.

That A has asked for $h(r_i)$ before can be excluded, because then S has already successfully terminated in step 1b.

The number c is generated by squaring a randomly chosen α, and c is not known to A in phase I. Therefore, the random choice of c is independent from $A's$ decryption queries in phase I. Hence, the probability that a particular decryption query in phase I involves r_1 or r_2 is $2 \cdot 1/2^{s_1}$. There are at most q_d decryption queries. Thus, the probability of case 4 is $\leq q_d \cdot 2/2^{s_1}$.

5. The attacker A asks for the decryption of a ciphertext c' and S rejects it (in step 2b or step 3), whereas Bob, by using his secret key, accepts c' and provides A with the plaintext m'. Then, we are not in case 2, because, in case 2, S successfully terminates in step 2a before giving an answer to A.

We assume that c is a valid ciphertext for m_0 or m_1 with respect to both square roots y_1 and y_2, i.e., that we are not in case 4. This means that for $i = 1$ and $i = 2$, we have $v \oplus h(r_i) = m_b \| 0^{s_0}$ for $b = 0$ or $b = 1$.

Decrypting c', Bob finds:

$$c' = y'^2 \bmod n, y' = v' \| r', v' \in \{0,1\}^{l+s_0}, r' \in \{0,1\}^{s_1}, v' \oplus h(r') = m' \| 0^{s_0}.$$

If $r' = r_i$ for $i = 1$ or $i = 2$, then the s_0 rightmost bits of v' and v are the same – they are equal to the s_0 rightmost bits of $h(r') = h(r_i)$. Then the $(s_0 + s_1)$ rightmost bits of y' and y_i coincide and we are in case 2, which we have excluded. Hence $r' \neq r_1$ and $r' \neq r_2$.

The hash value $h(r')$ has not been queried before, because otherwise we would be in case 3, and S would have accepted c' and responded with the plaintext m'.

Hence, the hash value $h(r')$ is a random value which is independent from preceding hash queries and our assumption that c is a valid encryption of m_0 or m_1 (which constrains $h(r_1)$ and $h(r_2)$, see above). But then the probability that the s_0 rightmost bits of $h(r') \oplus v'$ are all 0 (which is necessary for Bob to accept) is $1/2^{s_0}$. Since there are at most 2 square roots y', and A queries the decryption of q_d ciphertexts, the probability that case 5 occurs is $\leq q_d \cdot 2/2^{s_0}$.

---

[4] Note that S successfully terminates in step 2a only if there are $(s_0 + s_1)$ common rightmost bits.

Now assume that neither case 4 nor case 5 occurs. Following Boneh, we call this assumption $\mathcal{G}oodSim$.

Under this assumption $\mathcal{G}oodSim$, S behaves exactly like Bob and the random oracle h. Thus, A operates in the same probabilistic setting, and it therefore has the same probability of success $1/2 + \varepsilon$ as in the original attack.

If, in addition, cases 1 and 2 do not occur, then the randomly generated c (and r_1, r_2), and the hash values $h(r_1)$ and $h(r_2)$, which are generated by the random oracle, are independent of all hash and decryption queries issued by A and the responses given by S. Therefore A has not collected any information about $h(r_1)$ and $h(r_2)$. This means that the ciphertext c hides the plaintext m_b perfectly in the information-theoretic sense – the encryption includes a bitwise XOR with a truly random bit string $h(r_i)$. We have seen in Section 9.2 that then A's probability of correctly distinguishing between m_0 and m_1 is $1/2$.

Therefore, the advantage ε in A's probability of success necessarily results from the cases 1, 2 (we still assume $\mathcal{G}oodSim$). In the cases 1 and 2, the algorithm S successfully computes a square root of c. Hence the probability of success $\mathrm{prob}_{success}(S \,|\, \mathcal{G}oodSim)$ of algorithm S assuming $\mathcal{G}oodSim$ is $\geq \varepsilon$. The probability of the cases 4 and 5 is $\leq 2q_d/2^{s_1} + 2q_d/2^{s_0}$ and hence $\mathrm{prob}(\mathcal{G}oodSim) \geq 1 - 2q_d/2^{s_1} - 2q_d/2^{s_0}$, and we conclude that the probability of success of S is

$$\geq \mathrm{prob}(\mathcal{G}oodSim) \cdot \mathrm{prob}_{success}(S \,|\, \mathcal{G}oodSim) \geq \varepsilon \cdot \left(1 - \frac{2q_d}{2^{s_1}} - \frac{2q_d}{2^{s_0}}\right).$$

The running time of S is essentially the running time of A plus the running times of the Coppersmith algorithms. The algorithm for quadratic polynomials is called after each of the q_h hash queries (step 1b). After each of the q_d decryption queries, it may be called for all of the $\leq q_h$ known hash pairs $(r, h(r))$ (step 2b). The algorithm for polynomials of degree 4 is called after each of the q_d decryption queries (step 2a). Thus we can estimate the running time of S by

$$t + O(q_d q_h t_C + q_d t'_C),$$

and we see that the algorithm S gives a tight reduction from the problem of attacking SAEP to the problem of factoring.   □

Remark. True random functions cannot be implemented in practice. Therefore, a proof in the random oracle model – treating hash functions as equivalent to random functions – can never be a complete proof of security for a cryptographic scheme. But, intuitively, the random oracle model seems to be reasonable. In practice, a well-designed hash function should never have any features that distinguish it from a random function and that an attacker could exploit (see Section 2.2.4).

In recent years, doubts about the random oracle model have been expressed. Examples of cryptographic schemes were constructed which are provably secure in the random oracle model, but are insecure in any real-world implementation, where the random oracle is replaced by a real hash function ([CanGolHal98]; [GolTau03]; [GolTau03a]; [MauRenHol04]; [CanGolHal04]). However, the examples appear contrived and far from systems that would be designed in the real world. The confidence in the soundness of the random oracle assumption is still high among cryptographers, and the random oracle model is still considered a useful tool for validating cryptographic constructions. See, for example, [KobMen05] for a discussion of this point.

Nevertheless, it is desirable to have encryption schemes whose security can be proven solely under a standard assumption that some computational problem in number theory cannot be efficiently solved. An example of such a scheme is given in the next section.

9.5.2 Security Under Standard Assumptions

The Cramer–Shoup Public-Key Encryption Scheme. The security of the Cramer–Shoup public-key encryption scheme ([CraSho98]) against adaptively-chosen-ciphertext attacks can be proven assuming that the decision Diffie–Hellman problem (Section 4.8.3) cannot be solved efficiently and that the hash function used is collision-resistant. The random oracle model is not needed. In Chapter 11, we will give examples of signature schemes whose security can be proven solely under the assumed difficulty of computational problems (for example, Cramer–Shoup's signature scheme).

First, we recall the decision Diffie–Hellman problem (see Section 4.8.3).

Let p and q be large prime numbers such that q is a divisor of $p-1$, and let G_q be the (unique) subgroup of order q of $\mathbb{Z}_p^*$. G_q is a cyclic group, and every element $g \in \mathbb{Z}_p^*$ of order q is a generator of G_q (see Lemma A.41).

Given $g_1, u_1 = g_1^x, g_2 = g_1^y, u_2$ with random elements $g_1, u_2 \in G_q$ and randomly chosen exponents $x, y \in \mathbb{Z}_q^*$, decide if $u_2 = g_1^{xy}$. This is equivalent to decide, for randomly (and independently) chosen elements $g_1, u_1, g_2, u_2 \in G_q$, if

$$\log_{g_2}(u_2) = \log_{g_1}(u_1).$$

If the equality holds, we say that (g_1, u_1, g_2, u_2) has the Diffie–Hellman property.

The *decision Diffie–Hellman assumption* says that no probabilistic polynomial algorithm exists to solve the decision Diffie–Hellman problem.

Notation. For pairs $u = (u_1, u_2)$, $x = (x_1, x_2)$ and a scalar value r we shortly write $u^x := u_1^{x_1} u_2^{x_2}$, $u^{rx} := u_1^{rx_1} u_2^{rx_2}$.

Key Generation. Bob randomly generates large prime numbers p and q such that q is a divisor of $p-1$. He randomly chooses a pair $g = (g_1, g_2)$ of elements $g_1, g_2 \in G_q$.[5]

---

[5] Compare, for example, the key generation in the Digital Signature Standard, Section 3.4.3.

Then Bob randomly chooses three pairs of exponents $x = (x_1, x_2), y = (y_1, y_2), z = (z_1, z_2)$, with $x_1, x_2, y_1, y_2, z_1, z_2 \in \mathbb{Z}_q^*$ and computes modulo p:

$$d = g^x = g_1^{x_1} g_2^{x_2}, e = g^y = g_1^{y_1} g_2^{y_2}, f = g^z = g_1^{z_1} g_2^{z_2}.$$

Bob's public key is (p, q, g, d, e, f), his private key is (x, y, z).

For encryption, we need a collision-resistant hash function

$$h : \{0, 1\}^* \longrightarrow \mathbb{Z}_q^* = \{0, 1, \ldots q - 1\}.$$

h outputs bit strings of length $|q| = \lfloor \log_2(q) \rfloor + 1$. In practice, as in DSS, we might have $|q| = 160$ and $h = \text{SHA-1}$, see Section 3.4.3.

Encryption. Alice can encrypt messages $m \in G_q$ for Bob, i.e., elements m of order q in $\mathbb{Z}_p^*$. To encrypt a message m for Bob, Alice chooses a random $r \in \mathbb{Z}_q^*$ and computes the ciphertext

$$c = (u_1, u_2, w, v) := (g_1^r, g_2^r, f^r \cdot m, v), \text{ with } v := d^r e^{r \cdot h(u_1, u_2, w)}.$$

Note that the Diffie–Hellman property holds for (g_1, u_1, g_2, u_2).

Decryption. Bob decrypts a ciphertext $c = (u_1, u_2, w, v)$ by using his private key (x, y, z) as follows:

1. Bob checks the verification code v. He checks if $u^{x + y \cdot h(u_1, u_2, w)} = v$. If this equation does not hold, he rejects c.
2. He recovers the plaintext by computing $w \cdot u^{-z}$.

Remarks:

1. We follow here the description of Cramer–Shoup's encryption scheme in [KobMen05]. The actual scheme in [CraSho98] is the special case with $z_2 = 0$. Cramer and Shoup prove the security of the scheme using a slightly weaker assumption on the hash function. They assume that h is a member of a universal one-way family of hash functions (as defined in Section 11.4).
2. If $c = (u_1, u_2, w, v)$ is the correct encryption of a message m, then the verification code v passes the check in step 1,

$$u^{x + y \cdot h(u_1, u_2, w)} = (g^r)^x \cdot (g^r)^{y \cdot h(u_1, u_2, w)}$$

$$= (g^x)^r \cdot (g^y)^{r \cdot h(u_1, u_2, w)} = d^r \cdot e^{r \cdot h(u_1, u_2, w)} = v,$$

and the plaintext m is recovered in step 2,

$$f^r = (g^z)^r = (g^r)^z = u^z, \text{ hence } w \cdot u^{-z} = m \cdot f^r \cdot f^{-r} = m.$$

3. The private key $(x, y, z) = ((x_1, x_2), (y_1, y_2), (z_1, z_2))$ is an element in the $\mathbb{Z}_q$-vectorspace $\mathbb{Z}_q^6$. Publishing the public key (d, e, f) means to publish the following conditions on (x, y, z):

$$(1)\ d = g^x\ ,\quad (2)\ e = g^y\ ,\quad (3)\ f = g^z.$$

These are linear equations for $x_1, x_2, y_1, y_2, z_1, z_2$. They can also be written as (with $\lambda := \log_{g_1}(g_2)$)

$$(1)\ \log_{g_1}(d) = x_1 + \lambda \cdot x_2$$
$$(2)\ \log_{g_1}(e) = \qquad\qquad y_1 + \lambda \cdot y_2$$
$$(3)\ \log_{g_1}(f) = \qquad\qquad z_1 + \lambda \cdot z_2.$$

The equations define lines L_x, L_y, L_z in the plane $\mathbb{Z}_q^2$. For a given public key (d, e, f), the private key element x (resp. y, z) is a (uniformly distributed) random element of L_x (resp. L_y, L_z); each of the elements in L_x (resp. L_y, L_z) has the same probability $1/q$ of being x (resp. y, z).

4. Let $c := (u_1, u_2, w, v)$ be a ciphertext-tuple. Then c is accepted for decryption only if the verification condition (4) $v = u^{x+y\cdot\alpha}$, with $\alpha = h(u_1, u_2, w)$, holds. The verification condition is a linear equation for x and y:

$$(4)\ \log_{g_1}(v) = \log_{g_1}(u_1)x_1 + \lambda\log_{g_2}(u_2)x_2 + \alpha\log_{g_1}(u_1)y_1 + \alpha\lambda\log_{g_2}(u_2)y_2.$$

Equations $(1), (2)$ and (4) are linearly independent, if and only if $\log_{g_1}(u_1) \neq \log_{g_2}(u_2)$.

The ciphertext-tuple c is the correct encryption of a message, if and only if c satisfies the verification condition and $\log_{g_1}(u_1) = \log_{g_2}(u_2)$, i.e., (g_1, u_1, g_2, u_2) has the Diffie–Hellman property. In this case, we call c a *valid ciphertext*.

Now assume that $\log_{g_1}(u_1) \neq \log_{g_2}(u_2)$. Then, the probability that c passes the check of the verification code and is not rejected in the decryption step 1 is $1/q$ and hence negligibly small. Namely, $(1), (2)$ and (4) are linearly independent. This implies that (x, y) is an element of the line in $\mathbb{Z}_q^4$ which is defined by $(1), (2), (4)$. The probability for that is $1/q$, since (x, y) is a random element of the 2-dimensional space $L_x \times L_y$.

Theorem 9.20. *Let the decision Diffie–Hellman assumption be true and let h be a collision-resistant hash function. Then the probability of success of any attacking algorithm A, which on input of a random public key (p, q, g, d, e, f) executes an adaptively-chosen-ciphertext attack and tries to distinguish between two plaintexts, is $\leq 1/2 + \varepsilon$, with ε negligible.*

Proof. We discuss the essential ideas of the proof. For more details, we refer to [CraSho98].

The proof runs by contradiction. Assume that there is an algorithm A which on input of a randomly generated public key successfully distinguishes between ciphertexts in an adaptively-chosen-ciphertext attack, with a probability of success of $1/2 + \varepsilon$, ε non-negligible.

Then we can construct a probabilistic polynomial algorithm S which answers the decision Diffie–Hellman problem with a probability close to 1, in contradiction to the decision Diffie–Hellman assumption. The algorithm S successfully finds out whether a random 4-tuple (g_1, u_1, g_2, u_2) has the Diffie–Hellman property or not.

As in the proof of Theorem 9.19, algorithm S interacts with the attacking algorithm A. In the real attack, A interacts with Bob, the legitimate owner of the secret key, to obtain decryptions of ciphertexts of its choice (in practice, interacting typically means to communicate with another computer program). Now S is constructed to replace Bob in the attack. S "simulates" Bob.

On input of $(p, q, g_1, u_1, g_2, u_2)$ the algorithm S repeatedly generates a random private key (x, y, z) and computes the corresponding public key $d = g^x, e = g^y, f = g^z$. Then S calls A with the public key, and A executes its attack. The difference to the real attack is that A communicates with S instead of Bob. At some point (end of phase I of the attack), A outputs two plaintexts m_0, m_1. S randomly selects a bit $b \in \{0, 1\}$, sets

$$ w := u^z m_b, \alpha := h(u_1, u_2, w), v := u^{x + \alpha \cdot y} $$

and sends $c := (u_1, u_2, w, v)$ to A, as an encryption of m_b. The verification code v is correct by construction, even if $\log_{g_1} u_1 \neq \log_{g_2} u_2$ and c is not a valid encryption of m_b.

At any time in phase I or phase II, A can request the decryption of a ciphertext $c' = (u'_1, u'_2, w', v')$, $c' \neq c$. If c' satisfies the verification condition, then equation (4) holds for c', i.e., we have, with $\alpha' = h(u'_1, u'_2, w')$:

$$ (5)\, \log_{g_1}(v') = \log_{g_1}(u'_1)x_1 + \lambda \log_{g_2}(u'_2)x_2 + \alpha' \log_{g_1}(u'_1)y_1 + \alpha' \lambda \log_{g_2}(u'_2)y_2. $$

We observe: If $\log_{g_1}(u_1) \neq \log_{g_2}(u_2)$ and $\log_{g_1}(u'_1) \neq \log_{g_2}(u'_2)$, then the equations $(1), (2), (4), (5)$ are linearly independent, if and only if $\alpha' \neq \alpha$.

If c' satisfies the verification condition, then, up to negligible probability, c' is a valid ciphertext (see Remark 4 above) and S answers with the correct plaintext. S can easily check the verification condition and decrypt the ciphertext, because it knows the secret key. Here, S behaves exactly like Bob in the real attack.

If (g_1, u_1, g_2, u_2) has the Diffie–Hellman property, then the ciphertext $c := (u_1, u_2, w, v)$, which S presents to A as an encryption of m_b, is a valid ciphertext, and the probability distribution of c is the same, as if Bob produced the ciphertext. Hence, in this case, A operates in the same setting as in the real attack.

If (g_1, u_1, g_2, u_2) does not have the Diffie–Hellman property, then A operates in a setting which does not occur in the real attack, since Bob only produces valid ciphertexts. We do not know how the attacker A responds to the modified setting. For example, A could fail to terminate in its expected running time or it could output an error message or it could produce the usual output and answer, whether m_0 or m_1 is encrypted. But fortunately the concrete behavior of A is not relevant in this case: if (g_1, u_1, g_2, u_2) does not have the Diffie–Hellman property, then it is guaranteed that up to a negligible probability, m_b is perfectly hidden in $w = u^z m_b$ to A and A cannot generate any advantage from knowing w in distinguishing between m_0 and m_1.

To understand this, assume that (g_1, u_1, g_2, u_2) does not have the Diffie–Hellman property. Then the probability that the attacker A does get some information about the private key z by asking for the decryption of ciphertexts c' of its choice, $c' \neq c$, is negligibly small.

Namely, let A ask for the decryption of a ciphertext $c' = (u_1', u_2', w', v')$. Let $\alpha' = h(u_1', u_2', w')$.

Since we assume that $\log_{g_1}(u_1) \neq \log_{g_2}(u_2)$, equations $(1), (2), (4)$ are linearly independent and define a line L. From A's point of view, the private key $(x, y) = (x_1, x_2, y_1, y_2)$ is a random point on the line L.

We have to distinguish between several cases.

1. $\alpha' = \alpha$. Since h is collision-resistant[6], the probability that A can generate a triple $(u_1', u_2', w') \neq (u_1, u_2, w)$, with $h(u_1', u_2', w') = \alpha' = \alpha = h(u_1, u_2, w)$, is negligibly small. Hence the case $\alpha' = \alpha$ can happen, up to a negligible probability, only if $(u_1', u_2', w') = (u_1, u_2, w)$. But then $v' \neq v$, because $c' \neq c$, and hence $v' \neq (u')^{x+\alpha' \cdot y} = u^{x+\alpha \cdot y} = v$, and c' is rejected.

2. Now assume that $\alpha' \neq \alpha$ and (g_1, u_1', g_2, u_2') does not have the Diffie–Hellman property. If c' is not rejected, then the following equation (5) holds (it is equation (4) stated for c'):

(5) $\log_{g_1}(v')$

$$= \log_{g_1}(u_1')x_1 + \lambda \log_{g_2}(u_2')x_2 + \alpha' \log_{g_1}(u_1')y_1 + \alpha' \lambda \log_{g_2}(u_2')y_2.$$

Since $\alpha' \neq \alpha$, equation (5) puts, in addition to $(1), (2), (4)$, another linearly independent condition on (x, y). Hence, at most one point $(\tilde{x}, \tilde{y})$ satisfies these equations. Since (x, y) is a randomly chosen element on the line L, the probability that $(x, y) = (\tilde{x}, \tilde{y})$ is $\leq 1/q$. Hence, up to a negligible probability, c' is rejected.

3. The remaining case is that $\alpha' \neq \alpha$ and (g_1, u_1', g_2, u_2') has the Diffie–Hellman property. Either the check of the verification code v' fails and c' is rejected or equation (5) holds. In the latter case, S decrypts c'

---

[6] Second-pre-image resistance would suffice.

and provides A with the plaintext m'. Thus, what A learns from the decryption is the equation (6) $m' = w' \cdot u'^{-z}$, which can be reformulated as

$$(6) \quad \log_{g_1}(w' \cdot m'^{-1}) = \log_{g_1}(u'_1)z_1 + \lambda \log_{g_2}(u'_2)z_2$$
$$= \log_{g_1}(u'_1)(z_1 + \lambda z_2).$$

But this equation is linearly dependent on (3), so it does not give A any additional information on z. A learns from $m' = w' \cdot u'^{-z}$ the same information that it already knew from the public-secret-key equation (3) $f = e^z$: the private key z is a random element on the line L_z.

Summarizing, if (g_1, u_1, g_2, u_2) does not have the Diffie–Hellman property, then up to a negligible probability, from the attacker A's point of view, the secret key z is a random point on a line in $\mathbb{Z}_q^2$. This means that from A's point of view, m_b is perfectly (in the information-theoretic sense) hidden in $w = f^z m_b$ by a random element of G_q. We learnt in Section 9.2 that therefore A's probability of distinguishing successfully between m_0 and m_1 is exactly $1/2$. Hence, the non-negligible advantage ε of A results solely from the case that (g_1, u_1, g_2, u_2) has the Diffie–Hellman property.

In order to decide whether (g_1, u_1, g_2, u_2) has the Diffie–Hellman property, S repeatedly randomly generates private key elements (x, y, z) and runs the attack with A. If A correctly determines which of the messages m_b was encrypted by S in significantly more than half of the iterations, then S can be almost certain that (g_1, u_1, g_2, u_2) has the Diffie–Hellman property. Otherwise, S is almost certain that it does not. $\square$

Exercises

1. Let $n \in \mathbb{N}$. We consider the *affine cipher* modulo n. It is a symmetric encryption scheme. A key (a, b) consists of a unit $a \in \mathbb{Z}_n^*$ and an element $b \in \mathbb{Z}_n$. A message $m \in \mathbb{Z}_n$ is encrypted as $a \cdot m + b$.
   Is the affine cipher perfectly secret if we randomly (and uniformly) choose a key for each message m to be encrypted?

2. Let E be an encryption algorithm which encrypts plaintexts $m \in M$ as ciphertexts $c \in C$, and let K denote the secret key used to decrypt ciphertexts.
   Show that an adversary's uncertainty about the secret key is at least as great as her uncertainty about the plaintext: $\mathrm{H}(K|C) \geq \mathrm{H}(M|C)$.

3. ElGamal encryption (Section 3.4.1) is probabilistic. Is it ciphertext-indistinguishable?

4. Consider Definition 9.14 of ciphertext-indistinguishable encryptions. Show that an encryption algorithm $E(i, m)$ is ciphertext-indistinguishable if

and only if for every probabilistic polynomial distinguishing algorithm $A(i, m_0, m_1, c)$ and every probabilistic polynomial sampling algorithm S, which on input $i \in I$ yields $S(i) = \{m_0, m_1\} \subset M_i$, and every positive polynomial $P \in \mathbb{Z}[X]$, there is a $k_0 \in \mathbb{N}$ such that for all $k \geq k_0$,

$$\mathrm{prob}(A(i, m_0, m_1, c) = m_0 : i \leftarrow K(1^k), \{m_0, m_1\} \leftarrow S(i), c \leftarrow E(m_0))$$
$$- \mathrm{prob}(A(i, m_0, m_1, c) = m_0 : i \leftarrow K(1^k), \{m_0, m_1\} \leftarrow S(i), c \leftarrow E(m_1))$$
$$\leq \frac{1}{P(k)}.$$

5.. Let $I_k := \{(n, e) \mid n = pq,\ p, q \text{ distinct primes}, |p| = |q| = k, e \in \mathbb{Z}^*_{\varphi(n)}\}$ and $(n, e) \xleftarrow{u} I_k$ be a randomly chosen RSA key. Encrypt messages $m \in \{0, 1\}^r$, where $r \leq \log_2(|n|)$, in the following way. Pad m with leading random bits to get a padded plaintext $\overline{m}$ with $|\overline{m}| = |n|$. If $\overline{m} > n$, then repeat the padding of m until $\overline{m} < n$. Then encrypt m as $E(m) := c := \overline{m}^e \bmod n$.
Prove that this scheme is ciphertext-indistinguishable.
Hint: use Exercise 7 in Chapter 8.

6. Let $I = (I_k)_{k \in \mathbb{N}}$ be a key set with security parameter, and let $f = (f_i : D_i \longrightarrow D_i)_{i \in I}$ be a family of one-way trapdoor permutations with hard-core predicate $B = (B_i : D_i \longrightarrow \{0, 1\})_{i \in I}$ and key generator K. Consider the following probabilistic public-key encryption scheme ([GolMic84]; [GolBel01]): Let Q be a polynomial and $n := Q(k)$. A bit string $m := m_1 \ldots m_n$ is encrypted as a concatenation $c_1 \| \ldots \| c_n$, where $c_j := f_i(x_j)$ and x_j is a randomly selected element of D_i with $B_i(x_j) = m_j$.
Describe the decryption procedure and show that the encryption scheme is ciphertext-indistinguishable.
Hints: Apply Exercise 4. Given a pair $\{m_0, m_1\}$ of plaintexts, construct a sequence of messages $\tilde{m}_1 := m_0, \tilde{m}_2, \ldots, \tilde{m}_n := m_1$ such that $\tilde{m}_{j+1}$ differs from $\tilde{m}_j$ in at most one bit. Then consider the sequence of distributions $c \leftarrow E(i, \tilde{m}_j)$ (also see the proof of Proposition 8.4).

7. We consider the Goldwasser–Micali probabilistic encryption scheme ([GolMic84]). Let $I_k := \{n \mid n = pq, p, q \text{ distinct primes}, |p| = |q| = k\}$ and $I := (I_k)_{k \in \mathbb{N}}$. As his public key, each user Bob randomly chooses an $n \xleftarrow{u} I_k$ (by first randomly choosing the secret primes p and q) and a quadratic non-residue $z \xleftarrow{u} QNR_n$ with Jacobi symbol $\left(\frac{z}{n}\right) = 1$ (he can do this easily, since he knows the primes p and q; see Appendix A.7). A bit string $m = m_1 \ldots m_l$ is encrypted as a concatenation $c_1 \| \ldots \| c_l$, where $c_j = x_j^2$ if $m_j = 1$, and $c_j = z x_j^2$ if $m_j = 0$, with a randomly chosen $x_j \xleftarrow{u} \mathbb{Z}_n^*$. In other words: A 1 bit is encrypted by a random quadratic residue, a 0 bit by a random non-residue.
Describe the decryption procedure and show that the encryption scheme

is ciphertext-indistinguishable, provided the quadratic residuosity assumption (Definition 6.11) is true.

Hint: The proof is similar to the proof of Exercise 6. Use Exercise 9 in Chapter 6.

8. The decisional composite residuosity assumption says that it is infeasible to distinguish n-th residues and non-residues in $\mathbb{Z}_{n^2}^*$ (see Section A.9). More precisely:

Let $I_k := \{n \in I \mid n = pq, p, q \text{ distinct primes}, |p| = |q| = k\}$ for $k \in \mathbb{N}$. Then, for every probabilistic polynomial algorithm $A(n, x)$ and every positive polynomial $Q(X) \in \mathbb{Z}[X]$, there exists a $k_0 \in \mathbb{N}$ such that

$$|\mathrm{prob}(A(n, x^n) = 1 : n \xleftarrow{u} I_k, x \xleftarrow{u} \mathbb{Z}_{n^2}^*) -$$

$$\mathrm{prob}(A(n, x) = 1 : n \xleftarrow{u} I_k, x \xleftarrow{u} \mathbb{Z}_{n^2}^*)| \le \frac{1}{Q(k)}$$

for $k \ge k_0$.

Show that Paillier encryption (see Section 3.6.2) is ciphertext-indistinguishable if the decisional composite residuosity assumption holds.

10. Unconditional Security of Cryptosystems

The security of many currently used cryptosystems, in particular that of all public-key cryptosystems, is based on the hardness of an underlying computational problem, such as factoring integers or computing discrete logarithms. Security proofs for these systems show that the ability of an adversary to perform a successful attack contradicts the assumed difficulty of the computational problem. Security proofs of this type were presented in Chapter 9. For example, we proved that public-key one-time pads induced by one-way permutations with a hard-core predicate are ciphertext-indistinguishable. The security of the encryption scheme is reduced to the one-way feature of function families, such as the RSA or modular squaring families, and the one-way feature of these families is, in turn, based on the assumed hardness of inverting modular exponentiation or factoring a large integer (see Chapter 6). The security proof is conditional, and there is some risk that in the future, the underlying condition will turn out to be false.

On the other hand, Shannon's information-theoretic model of security provides unconditional security. The perfect secrecy of Vernam's one-time pad (Section 9.1) is not dependent on the hardness of a computational problem or limits on the computational power of an adversary.

Although perfect secrecy is not reachable in most practical situations, there have been various promising attempts to design practical cryptosystems whose security is not based on assumptions and which provably come close to perfect information-theoretic security.

An important approach is quantum cryptography, introduced by Bennett and Brassard. Here, two parties agree on a secret key by transmitting polarized photons over a fiber-optic channel. The secrecy of the key is based on the uncertainty principle of quantum mechanics ([BraCre96]).

In other approaches, the unconditional security of a cryptosystem is based on the fact that communication channels are noisy (and hence, an eavesdropper never gets all the information), or on the limited storage capacity of an adversary (see, e.g., [Maurer99] for an overview of information-theoretic cryptography).

In this chapter, we give an introduction to unconditionally secure cryptosystems and the techniques applied in them. Systems whose security is based on the limited storage capacity of an adversary or on the unavoidable noise in a communication channel are discussed in Sections 10.1 and 10.2. The classical methods for authentication, namely digital signatures (such as RSA and DSA) and classical message authentication codes (such as HMAC), rely on computational-hardness assumptions and cannot provide unconditional security. In Section 10.3, we explain how unconditionally secure message authentication codes can be constructed from almost universal classes of hash functions.

One of the essential techniques is privacy amplification, also called entropy smoothing. Almost all of the Rényi entropy of a bit string can be transformed into a random bit string. Privacy amplification enables the communication partners to derive an unconditionally secure shared secret key from information that they have exchanged through a public channel. In Section 10.4, we discuss the basic properties of Rényi entropy, give a proof of the Privacy Amplification Theorem and explain the extraction of secure keys.

Finally, in Section 10.5, we give a detailed introduction to quantum cryptography. We describe the famous BB84 protocol of Bennett and Brassard for quantum key distribution and prove its security, assuming intercept-and-resend attacks.

10.1 The Bounded Storage Model

We will give a short introduction to encryption schemes designed by Maurer et al., whose unconditional security is guaranteed by a limit on the total amount of storage capacity available to an adversary. Most of the encryption schemes studied in this approach are similar to the one proposed in [Maurer92], which we will describe now.

Alice wants to transmit messages $m \in M := \{0,1\}^n$ to Bob. She uses a one-time pad for encryption, i.e., she XORs the message m bitwise with a one-time key k. As usual, we have a probability distribution on M which, together with the probabilistic choice of the keys, yields a probability distribution on the set C of ciphertexts. Without loss of generality, we assume that $\mathrm{prob}(m) > 0$ for all $m \in M$, and $\mathrm{prob}(c) > 0$ for all $c \in C$. The eavesdropper Eve is a passive attacker: by observing the ciphertext $c \in C \subseteq \{0,1\}^n$, she tries to get information about the plaintext.

Alice and Bob extract the key k for the encryption of a message m from a publicly accessible "table of random bits". Security is achieved if Eve has access only to some part of the table. This requires some clever realization of the public table of random bits. A possibly realistic scenario (the "satellite scenario") is that the random bits are broadcast by some radio source (e.g., a satellite or a natural deep-space radio source) with a sufficiently high data rate (see the example below), and Eve can store only parts of the resulting table.

So, we assume that there is a public source broadcasting truly (uniformly distributed) random bits to Alice, Bob and Eve at a high speed. The communication channels are assumed to be error free.

Alice and Bob select their key bits from the bit stream over some time period T, according to some private strategy not known to Eve. The ciphertext is transmitted later, after the end of T. Due to her limited storage resources, Eve can store only a small part of the bits broadcast during T.

To extract the one-time key k for a message $m \in \{0,1\}^n$, Alice and Bob synchronize on the source and listen to the broadcast bits over the time period T. Let R (called the randomizer) be the random bits transmitted during T, and let $(r_{i,j} \mid 1 \leq i \leq l, 0 \leq j \leq t-1)$ be these bits arranged as elements of a matrix with l rows and t columns. Thus, R contains $|R| = lt$ bits. Typically, l is small (about 50) and t is huge, even when compared with the length n of the plaintext message m.

Alice and Bob have agreed on a (short) private key $(s_1, \ldots, s_l)$ in advance, with s_i randomly chosen in $\{0, \ldots, t-1\}$ (with respect to the uniform distribution), and take as the key $k := k_0 \ldots k_{n-1}$, with

$$k_j := r_{1,(s_1+j) \bmod t} \oplus r_{2,(s_2+j) \bmod t} \cdots \oplus r_{l,(s_l+j) \bmod t}.$$

In other words, Alice and Bob select from each row i a bit string b_i of length n, starting at the randomly chosen "seed" s_i (jumping back to the beginning if the end of the row is reached), and then get their key k by XORing these strings b_i.

The attacker Eve also listens to the random source during T and stores some of the bits, hoping that this will help her to extract information when the encrypted message is transmitted after the end of T. Due to her limited storage space, she stores only q of the bits $r_{i,j}$. The bits are selected by some probabilistic algorithm.[1] For each of these q bits, she knows the value and possibly the position in the random bit stream R. Eve's knowledge about the randomizer bits is summarized by the random variable S. One may also consider S as a probabilistic algorithm, returning q positions and bit values. As usual in the study of a one-time pad encryption scheme, Eve may know the distribution on M. We assume that she has no further a priori knowledge about the messages $m \in M$ actually sent to Bob by Alice. The following theorem is proven in [Maurer92].

Theorem 10.1. *There exists an event $\mathcal{E}$ such that for all probabilistic strategies S for storing q bits of the randomizer R,*

$$I(M; CS \mid \mathcal{E}) = 0 \text{ and } \mathrm{prob}(\mathcal{E}) \geq 1 - n\delta^l,$$

where $\delta := q/|R|$ is the fraction of randomizer bits stored by Eve.

---

[1] Note that there are no restrictions on the computing power of Eve.

Proof. We sketch the basic idea and refer the interested reader to [Maurer92]. Let $(s_1, \ldots, s_l)$ be the private key of Alice and Bob, and $k := k_0 \ldots k_{n-1}$ be the key extracted from R by Alice and Bob. Then bit k_j is derived from R by XORing the bits $r_{1,(s_1+j) \bmod t}, r_{2,(s_2+j) \bmod t}, \ldots, r_{l,(s_l+j) \bmod t}$.

If Eve's storage strategy has missed only one of these bits, then the resulting bit k_j appears truly random to her, despite her knowledge S of the randomizer. The probability of the event $\mathcal{F}$ that, with her strategy, Eve stores all the bits $r_{1,(s_1+j) \bmod t}, r_{2,(s_2+j) \bmod t}, \ldots, r_{l,(s_l+j) \bmod t}$, for at least one j, $0 \leq j \leq n-1$, is very small – it turns out to be $\leq n\delta^l$.

The "security event" $\mathcal{E}$ is defined as the complement of $\mathcal{F}$. If $\mathcal{E}$ occurs, then, from Eve's point of view, the key extracted from R by Alice and Bob is truly random, and we have the situation of Vernam's one-time pad, with a mutual information which equals 0 (see Theorem 9.5). $\qquad\square$

Remarks:

1. The mutual information in Theorem 10.1 is conditional on an event $\mathcal{E}$. This means that all the entropies involved are computed with conditional probabilities assuming $\mathcal{E}$ (see the final remark in Appendix B.5).
2. In [Maurer92], a stronger version is proven. This also includes the case where Eve has some a priori knowledge V of the plaintext messages, where V is jointly distributed with M. Then the mutual information $I(M, CS \mid V, \mathcal{E})$ between M and CS, conditional on V and assuming $\mathcal{E}$, is 0. Conditioning over V means that the mutual information does not include the amount of information about M resulting from the knowledge of V (see Proposition B.43).
3. The adversary Eve cannot gain an advantage from learning the secret key $(s_1, \ldots, s_l)$ after the broadcast of the randomizer R. Therefore, Theorem 10.3 is a statement on *everlasting* security.
4. The model of attack applied here is somewhat restricted. In the first phase, while listening to the random source, the eavesdropper Eve does not exploit her full computing power; she simply stores some of the transmitted bits and does not use the bit stream as input for computations at that time. The general model of attack, where Eve may compute and store arbitrary bits of information about the randomizer, is considered below.

Example. This example is derived from a similar example in [CachMau97]. A satellite broadcasting random bits at a rate of 16 Gbit/s is used for one day to provide a randomizer table R with about $1.5 \cdot 10^{15}$ bits. Let R be arranged in $l := 100$ rows and $t := 1.5 \cdot 10^{13}$ columns. Let the plaintexts be 6 MiB in size, i.e., $n \approx 5 \cdot 10^7$ bits. Alice and Bob have to agree on a private key $(s_1, \ldots, s_l)$ of $100 \cdot \log_2(1.5 \cdot 10^{13}) \approx 4380$ bits. The storage capacity of the adversary Eve is assumed to be 100 TB, which equals about $8.8 \cdot 10^{14}$ bits. Then $\delta \approx 0.587$ and

$$\text{prob(not } \mathcal{E}) \leq 5 \cdot 10^7 \cdot 0.587^{100} \approx 3.7 \cdot 10^{-16} < 10^{-15}.$$

Thus, the probability that Eve gets any additional information about the plaintext by observing the ciphertext and applying an optimal storage strategy is less than 10^{-15}.

Theorem 10.1 may also be interpreted in terms of distinguishing algorithms (see Proposition 9.10). We denote by E the probabilistic encryption algorithm which encrypts m as $c := m \oplus k$, with k randomly chosen as above.

Theorem 10.2. *For every probabilistic storage strategy S storing a fraction δ of all randomizer bits, and every probabilistic distinguishing algorithm $A(m_0, m_1, c, s)$ and all $m_0, m_1 \in M$, with $m_0 \neq m_1$,*

$$\mathrm{prob}(A(m_0, m_1, c, s) = m : m \xleftarrow{u} \{m_0, m_1\}, c \leftarrow E(m), s \leftarrow S) \leq \frac{1}{2} + n\delta^l.$$

Remark. This statement is equivalent to

$$\begin{aligned} | \, \mathrm{prob}(A(m_0, m_1, c, s) &= m_0 : c \leftarrow E(m_0), s \leftarrow S) \\ &- \mathrm{prob}(A(m_0, m_1, c, s) = m_0 : c \leftarrow E(m_1), s \leftarrow S) \, | \leq n\delta^l, \end{aligned}$$

as the same computation as in the proof of Proposition 9.10 shows.

Proof. From Theorem 10.1 and Proposition B.39, we conclude that

$$\mathrm{prob}(c, s \, | \, m_0, \mathcal{E}) = \mathrm{prob}(c, s \, | \, m_1, \mathcal{E})$$

for all $m_0, m_1 \in M$, $c \in C$ and $s \in S$. Computing with probabilities conditional on $\mathcal{E}$, we get

$$\begin{aligned} \mathrm{prob}(A(m_0, m_1, c, s) &= m_0 \, | \, \mathcal{E} : c \leftarrow E(m_0), s \leftarrow S) \\ &= \sum_{c, s} \mathrm{prob}(c, s \, | \, m_0, \mathcal{E}) \cdot \mathrm{prob}(A(m_0, m_1, c, s) = m_0 \, | \, \mathcal{E}) \\ &= \sum_{c, s} \mathrm{prob}(c, s \, | \, m_1, \mathcal{E}) \cdot \mathrm{prob}(A(m_0, m_1, c, s) = m_0 \, | \, \mathcal{E}) \\ &= \mathrm{prob}(A(m_0, m_1, c, s) = m_0 \, | \, \mathcal{E} : c \leftarrow E(m_1), s \leftarrow S). \end{aligned}$$

Using Lemma B.4 we get

$$\begin{aligned} | \, \mathrm{prob}(A(m_0, m_1, c, s) &= m_0 : c \leftarrow E(m_0), s \leftarrow S) \\ &- \mathrm{prob}(A(m_0, m_1, c, s) = m_0 : c \leftarrow E(m_1), s \leftarrow S) \, | \\ \leq \mathrm{prob}(\text{not } \mathcal{E}) &\leq n\delta^l, \end{aligned}$$

as desired. $\qquad\qquad\qquad\qquad\qquad\qquad\qquad\qquad\qquad\qquad\qquad\qquad\qquad\quad \square$

As we observed above, the model of attack just used is somewhat restricted, because the eavesdropper Eve does not use the bit stream as input for computations in the first phase, while listening to the random source. Security proofs for the general model of attack, where Eve may use her (unlimited) computing power at any time – without any restrictions – were given only recently in a series of papers ([AumRab99]; [AumDinRab02]; [DinRab02]; [DziMau02]; [Lu02]; [Vadhan03]).

The results obtained by Aumann and Rabin ([AumRab99]) were still restricted. The *randomness efficiency*, i.e., the ratio $\delta = q/|R|$ of Eve's storage capacity to the size of the randomizer, was very small.

Major progress on the bounded storage model was then achieved by Aumann, Ding and Rabin ([AumDinRab02]; [Ding01]; [DinRab02]), who proved the security of schemes with a randomness efficiency of about 0.2.

Strong security results for the general model of attack were presented by Dziembowski and Maurer in [DziMau02] (an extended version is presented in [DziMau04a]). These authors proved that keys k whose length n is much longer than the length of the initial private key can be securely generated with a randomness efficiency which may be arbitrarily close to 1. A randomness efficiency of 0.1 is possible for reasonable parameter sizes which appear possible in practice (for example, with the attacker Eve's storage capacity ≈ 125 TB, a derived key k of 1 GB for the one-time pad, an initial private key < 1 KB and a statistical distance $< 2^{-29}$ between the distribution of the derived key k and the uniform distribution, from Eve's point of view).

The schemes investigated in the papers cited are all very similar to the scheme of [Maurer92] which we have explained here. Of course, security proofs for the general model of attack require more sophisticated methods of probability and information theory.

In all of the bounded-storage-model schemes that we have referred to above one assumes that Alice and Bob share an initial secret key s, usually without considering how such a key s is obtained by Alice and Bob. A natural way would be to exchange the initial key by using a public-key key agreement protocol, for example the Diffie–Hellman protocol (see Section 4.1.2). At first glance, this approach may appear useless, since the information-theoretic security against a computationally unbounded adversary Eve is lost – Eve could break the public-key protocol with her unlimited resources. However, if Eve is an attacker who gains her infinite computing power (and then the initial secret key) only after the broadcast of the randomizer, then the security of the scheme might be preserved (see [DziMau04b] for a detailed discussion).

In [CachMau97], another approach is discussed. In a variant of that scheme, the key k for the one-time pad is generated within the bounded storage model, and Alice and Bob need not share an initial secret key s. The general model of attack is applied in [CachMau97] – the adversary Eve may use her unlimited computing power at any time.[2]

---

[2] Unfortunately, either the schemes arising are impractical or the attacker Eve's probability of success is non-negligible; see below.

Some of the techniques used there are basic.[3] To illustrate these techniques, we give a short overview of some parts of [CachMau97].

As before, there is some source of truly random bits. Alice, Bob and Eve receive these random bits over perfect channels without any errors. We are looking for bit sequences of length n to serve as keys in a one-time pad for encrypting messages $m \in \{0,1\}^n$. The random bit source generates N bits $R := (r_1, \ldots, r_N)$. The storage capacity q of Eve is smaller than N, so she is not able to store the whole randomizer. In contrast to the preceding model, she not only stores q bits of the randomizer, but also executes some probabilistic algorithm U while listening to the random source, to compute q bits of information from R (and store them in her memory). As before, we denote by $\delta := q/N$ the fraction of Eve's storage capacity with respect to the total number of randomizer bits.

In the first phase, called *advantage distillation*, Alice and Bob extract sufficiently many, say l, bits $S := (s_1, \ldots, s_l)$ from R, at randomly chosen positions $P := (p_1, \ldots, p_l)$:

$$s_1 := r_{p_1}, s_2 := r_{p_2}, \ldots, s_l := r_{p_l}.$$

It is necessary that the positions are chosen pairwise independently. The positions $p_1, \ldots, p_l$ are kept secret until the broadcast of the randomizer is finished.

Alice and Bob can select the bits in two ways.

1. **Private-key scenario:** Alice and Bob agree on the positions $p_1, \ldots, p_l$ in advance and share these positions as an initial secret key.
2. **Key agreement solely by public discussion:** Independently, Alice and Bob each select and store w bits of the randomizer R. The positions of the selected bits $t_1, \ldots, t_w$ and $u_1, \ldots, u_w$ are randomly selected, pairwise independent and uniformly distributed. When the broadcast of the randomizer is finished, Alice and Bob exchange the chosen positions over a public channel. Let $\{p_1, \ldots, p_l\} = \{t_1, \ldots, t_w\} \cap \{u_1, \ldots, u_w\}$. Then, Alice and Bob share the l randomizer bits at the positions $p_1, \ldots, p_l$. It is easy to see that the expected number l of common positions is w^2/N (see, for example, Corollary B.25). Hence, on average, they have to select and store $\sqrt{lN}$ randomizer bits to obtain l common bits.

Since Eve can store at most q bits, her information about S is incomplete. For example, it can be proven that Eve knows at most a fraction δ of the l bits in S (in the information-theoretic sense). Thus, Alice and Bob have distilled an advantage. Let e be the integer part of Eve's uncertainty $H(S \mid$ Eve's knowledge$)$ about S. Then Eve lacks approximately e bits of information about S.

---

[3] They are also applied, for example, in the noisy channel model, which we discuss in Section 10.2.

In the second phase, Alice and Bob apply a powerful technique, called *privacy amplification* or *entropy smoothing*, to extract f bits from S in such a way that Eve has almost no information about the resulting string $\tilde{S}$. Here, f is given by the Rényi entropy of order 2 (see below). Since this entropy is less than or equal to the Shannon entropy, we have $f \leq e$. Eve's uncertainty about $\tilde{S}$ is close to f, so from Eve's point of view, $\tilde{S}$ appears almost truly random. Thus, it can serve Alice and Bob as a provably secure key k in a one-time pad.

Privacy amplification is accomplished by randomly selecting a member from a so-called universal class of hash functions (see below). Alice randomly selects an element h from such a universal class $\mathcal{H}$ (with respect to the uniform distribution) and sends h to Bob via a public channel. Thus, Eve may even know $\mathcal{H}$ and h. Alice and Bob both apply h to S in order to obtain their key $k := h(S)$ for the one-time pad.

Rényi entropy and privacy amplification are described in detail in Section 10.4.

Let H and K be the random variables describing the probabilistic choice of the function h and the probabilistic choice of the key k. The following theorem is proven in [CachMau97].

Theorem 10.3. *Given a fixed storage capacity q of Eve and $\epsilon_1, \epsilon_2 > 0$, there exists a security event $\mathcal{E}$ such that*

$$\mathrm{prob}(\mathcal{E}) \geq 1 - \epsilon_1 \ and \ \mathrm{I}(K; H \,|\, U = u, P = p, \mathcal{E}) \leq \epsilon_2,$$

and hence in particular

$$\mathrm{I}(K; UHP \,|\, \mathcal{E}) \leq \epsilon_2 \ and \ \mathrm{I}(K; UH \,|\, \mathcal{E}) \leq \epsilon_2,$$

provided the size N of the randomizer R and the number l of elements selected from R by S are sufficiently large.

Remarks:

1. Explicit formulae are derived in [CachMau97] which connect the bounds ϵ_1, ϵ_2, the size N of the randomizer, Eve's storage capacity q, the number l of chosen positions and the number f of derived key bits.

2. The third inequality follows from the second by Proposition B.42, and the fourth inequality from the third by Proposition B.43 (observe also the final remark in Appendix B.5).

3. $\mathrm{I}(K; H \,|\, U = u, P = p, \mathcal{E}) \leq \epsilon_2$ means the following. Assume Eve has the specific knowledge $U = u$ about the randomizer and has learned the positions P of the bits selected from R by Alice and Bob after the broadcast of the randomizer R. Then the average amount of information (measured in bits in the information-theoretic sense) that Eve can derive about the key k from learning the hash function h is less than ϵ_2, provided the security event $\mathcal{E}$ occurs. Thus, in the private-key scenario, the bound

ϵ_2 also holds if the secret key shared by Alice and Bob (i.e., the positions of the selected randomizer bits) is later compromised. The security is everlasting.

As we mentioned above, a key step in the proof of Theorem 10.3 is privacy amplification to transform almost all the entropy of a bit string into a random bit string. For this purpose, it is not sufficient to work with the classical Shannon entropy as defined in Appendix B.5. Instead, it is necessary to use more general information measures: the *Rényi entropies* of order α ($0 \leq \alpha \leq \infty$, see [Rényi61]; [Rényi70]). Here, in particular, the Rényi entropy of order 2 – also called the *collision entropy* – is needed.

Definition 10.4. Let S be a random variable with values in the finite set $\mathcal{S}$. The *collision probability* $\mathrm{prob}_c(S)$ of S is defined as

$$\mathrm{prob}_c(S) := \sum_{s \in \mathcal{S}} \mathrm{prob}(S = s)^2.$$

The *collision entropy* or *Rényi entropy* (of order 2) of S is

$$H_2(S) := -\log_2(\mathrm{prob}_c(S)) = -\log_2\left(\sum_{s \in \mathcal{S}} \mathrm{prob}(S = s)^2\right).$$

$\mathrm{prob}_c(S)$ is the probability that two independent executions of S yield the same result. $H_2(S)$ measures the uncertainty that two independent executions of the random experiment S yield the same result.

The mathematical foundation of privacy amplification is the *Smoothing Entropy Theorem* (Theorem 10.12). This states that almost all the collision entropy of a random variable S may be converted into uniform random bits by selecting a function h randomly from a universal class of hash functions and applying h to S (for details and a proof, see Section 10.4 below). Universal classes of hash functions were introduced by Carter and Wegman ([CarWeg79]; [WegCar81]).

Definition 10.5. A set $\mathcal{H}$ of functions $h : X \longrightarrow Y$ is called a *universal class of hash functions* if, for all distinct $x, x' \in X$,

$$\mathrm{prob}(h(x) = h(x') : h \xleftarrow{u} \mathcal{H}) = \frac{1}{|Y|}.$$

$\mathcal{H}$ is called a *strongly universal class of hash functions* if for all distinct $x, x' \in X$ and all (not necessarily distinct) $y, y' \in Y$,

$$\mathrm{prob}(h(x) = y, h(x') = y' : h \xleftarrow{u} \mathcal{H}) = \frac{1}{|Y|^2}.$$

Remark. If $\mathcal{H}$ is a strongly universal class of hash functions and $x, x' \in X, y \in Y, x \neq x'$, then

$$\text{prob}(h(x) = y : h \xleftarrow{u} \mathcal{H}) = \sum_{y' \in Y} \text{prob}(h(x) = y, h(x') = y' : h \xleftarrow{u} \mathcal{H})$$

$$= \sum_{y' \in Y} \frac{1}{|Y|^2} = \frac{|Y|}{|Y|^2} = \frac{1}{|Y|}$$

and

$$\text{prob}(h(x) = h(x') : h \xleftarrow{u} \mathcal{H}) = \sum_{y' \in Y} \text{prob}(h(x) = y', h(x') = y' : h \xleftarrow{u} \mathcal{H})$$

$$= \sum_{y' \in Y} \frac{1}{|Y|^2} = \frac{|Y|}{|Y|^2} = \frac{1}{|Y|}.$$

In particular, a strongly universal class is a universal class. (Strongly) universal classes of hash functions behave like completely random functions with respect to collisions (or value pairs).

Example. A straightforward computation shows that the set of linear mappings $\{0,1\}^l \longrightarrow \{0,1\}^f$ is a strongly universal class of hash functions. There are smaller classes ([Stinson94]). For example, the set

$$\mathcal{G}_{l,f} := \{h_{a_0, a_1} : \mathbb{F}_{2^l} \longrightarrow \mathbb{F}_{2^f}, \ x \longmapsto \text{msb}_f(a_0 \cdot x + a_1) \mid a_0, a_1 \in \mathbb{F}_{2^l}\}$$

is a strongly universal class ($l \geq f$), and the set

$$\mathcal{H}_{l,f} := \{h_a : \mathbb{F}_{2^l} \longrightarrow \mathbb{F}_{2^f}, \ x \longmapsto \text{msb}_f(a \cdot x) \mid a \in \mathbb{F}_{2^l}\}$$

is a universal class. Here, we consider $\{0,1\}^m$ as equipped with the structure $\mathbb{F}_{2^m}$ of the Galois field with 2^m elements (see Appendix A.5), and msb_f denotes the f most-significant bits. See Exercise 1.

Remark. In the key generation scheme discussed, Alice and Bob select w (or l) bits at pairwise independent random positions from the N bits broadcast by the random source. They have to store the positions of these bits. At first glance, $w \cdot \log_2(N)$ bits are necessary to describe w positions. Since w is large, a huge number of bits have to be stored and transferred between Alice and Bob. Strongly universal classes of hash functions also provide a solution to this problem.

Assume $N = 2^m$, and consider $\{0,1\}^m$ as $\mathbb{F}_{2^m}$. Alice and Bob may work with the strongly universal class

$$\mathcal{H} = \{h : \mathbb{F}_{2^m} \longrightarrow \mathbb{F}_{2^m}, \ x \longmapsto a_0 \cdot x + a_1 \mid a_0, a_1 \in \mathbb{F}_{2^m}\}$$

of hash functions. They fix pairwise different elements $x_1, \ldots, x_w \in \mathbb{F}_{2^m}$ in advance. $\mathcal{H}$ and the x_i may be known to Eve. Now, to select w positions randomly – uniformly distributed and pairwise independent – Alice or Bob randomly chooses some h from $\mathcal{H}$ (with respect to the uniform distribution) and applies h to the x_i. This yields w uniformly distributed and pairwise independent positions

$$h(x_1), h(x_2), \ldots, h(x_w).$$

Thus, the random choice of the w positions reduces to the random choice of an element h in $\mathcal{H}$, and this requires the random choice of $2m = 2\log_2(N)$ bits.

Example. Assume that Alice and Bob do not share an initial private key and the key is derived solely by a public discussion. Using the explicit formulae, we get the following example for the Cachin–Maurer scheme (see [CachMau97] for more details). A satellite broadcasting random bits at a rate of 40 Gbit/s is used for $2 \cdot 10^5$ seconds (about 2 days) to provide a randomizer R with about $N = 8.6 \cdot 10^{15}$ bits. The storage capacity of the adversary Eve is assumed to be $1/2$ PB, which equals about $4.5 \cdot 10^{15}$ bits. To get $l = 1.3 \cdot 10^7$ common positions and common random bits, Alice and Bob each have to select and store $w = \sqrt{lN} = 3.3 \cdot 10^{11}$ bits (or about 39 GB) from R. By privacy amplification, they get a key k of about 61 KB and Eve knows no more than 10^{-20} bits of k, provided that the security event $\mathcal{E}$ occurs. The probability of $\mathcal{E}$ is $\geq 1 - \varepsilon_1$ with $\varepsilon_1 = 10^{-3}$. Since l is of the order of $1/\varepsilon_1^2$, the probability that the security event $\mathcal{E}$ does not occur cannot be reduced to significantly smaller values without increasing the storage requirements for Alice to unreasonably high values. To choose the w positions of the randomizer bits which they store, Alice and Bob each randomly select a strongly universal hash function (see the preceding remark). So, to exchange these positions, Alice and Bob have to transmit a strongly universal hash function in each direction, which requires $2\log_2(N) \approx 106$ bits. For privacy amplification, either Alice or Bob chooses the random universal hash function and communicates it, which takes about l bits ≈ 1.5 MiB. The large size of the hash functions may be substantially reduced by using "almost universal" hash functions. Almost universal classes of hash functions are discussed in detail in Section 10.3 below.

Remarks:

1. Alice and Bob need – as the example demonstrates – a very large capacity to store the positions and the values of the randomizer bits, and the size of this storage increases rapidly if the probability of the security event $\mathcal{E}$ is required to be significantly closer to 1 than 10^{-3}, which is certainly not negligible. To restrict the adversary Eve's probability of success to negligible values would require unreasonably high storage capacities of Alice and Bob. This is also true for the private-key scenario.

2. If the key k is derived solely by a public discussion, then both Alice and Bob need storage on the order of $\sqrt{N}$, which is also on the order of $\sqrt{q}$ (recall that N is the size of the randomizer and q is the storage size of the

attacker). It was shown in [DziMau04b] that these storage requirements cannot be reduced. The Cachin–Maurer scheme is essentially optimal in terms of the ratio between the storage capacity of Alice and Bob and the storage capacity of the adversary Eve. The practicality of schemes in the bounded storage model which do not rely on a shared initial secret key is therefore highly questionable.

10.2 The Noisy Channel Model

An introduction to the noisy channel model was given, for example, in the survey article [Wolf98]. As before, we use the "satellite scenario". Random bits are broadcast by some radio source. Alice and Bob receive these bits and generate a key from them by a public discussion. The eavesdropper, Eve, also receives the random bits and can listen to the communication channel between Alice and Bob. Again we assume that Eve is a passive adversary. There are other models including an active adversary (see, e.g., [Maurer97]; [MauWol97]). Though all communications are public, Eve gains hardly any information about the key. Thus, the generated key appears almost random to Eve and can be used in a provably secure one-time pad. The secrecy of the key is based on the fact that no information channel is error-free. The system also works in the case where Eve receives the random bits via a much better channel than Alice and Bob.

The key agreement works in three phases. As in Section 10.1, it starts with advantage distillation and ends with privacy amplification. There is an additional intermediate phase called *information reconciliation*.

During advantage distillation, Alice chooses a truly random key k, for example from the radio source. Before transmitting it to Bob, she uses a random bit string r to mask k and to make the transmission to Bob highly reliable (by applying a suitable error detection code randomized by r). The random bit string r is taken from the radio source and is commonly available to all participants. If sufficient redundancy and randomness are built in, which means that the random bit string r is sufficiently long, the error probability of the adversary is higher than the error probability of the legitimate recipient Bob. In this way, Alice and Bob gain an advantage over Eve.

When phase 1 is finished, the random string k held by Alice may still differ from the string k' received by Bob. Now, Alice and Bob start information reconciliation and interactively modify k and k' such that at the end, the probability that $k \neq k'$ is negligibly small. This must be performed without leaking too much information to the adversary Eve. Alice and Bob may, for example, try to detect differing positions in the string by comparing the parity bits of randomly chosen substrings. In Section 10.5.5 below, we give a detailed description of the important information reconciliation protocol Cascade.

After phase 2, the same random string k of size l is available to Alice and Bob with a very high probability, and they have an advantage over Eve. Eve's information about k, measured by Rényi entropies, is incomplete. By applying the privacy amplification techniques sketched in Section 10.1, Alice and Bob obtain their desired key.

10.3 Unconditionally Secure Message Authentication

Information-theoretically secure message authentication codes can be calculated from keyed universal families of hash functions. The technique is symmetric – the communication partners have to agree on a shared secret key.

10.3.1 Almost Universal Classes of Hash Functions

Almost universal classes generalize the notion of universal classes of hash functions (Definition 10.5). Our definition follows [AtiSti96, Section 3].

Definition 10.6. Let X, Y be finite sets and let $\mathcal{H}$ be a set of functions $h : X \longrightarrow Y$. Let $\varepsilon > 0$.

1. $\mathcal{H}$ is called an ε-*universal class of hash functions* if, for all distinct $x_1, x_2 \in X$,
$$\mathrm{prob}(h(x_1) = h(x_2) : h \xleftarrow{u} \mathcal{H}) \leq \varepsilon.$$

2. $\mathcal{H}$ is called an ε-*almost strongly universal class of hash functions* if:
   a. for all $x \in X$ and all $y \in Y$,
   $$\mathrm{prob}(h(x) = y : h \xleftarrow{u} \mathcal{H}) = \frac{1}{|Y|} \text{ and}$$

   b. for all distinct $x_1, x_2 \in X$ and all (not necessarily distinct) $y_1, y_2 \in Y$,
   $$\mathrm{prob}(h(x_1) = y_1, h(x_2) = y_2 : h \xleftarrow{u} \mathcal{H}) \leq \frac{\varepsilon}{|Y|}.$$

Remarks:

1. A hash family is universal (Definition 10.5) if and only if it is $1/|Y|$-universal.
2. A hash family is strongly universal (Definition 10.5) if and only if it is $1/|Y|$-almost strongly universal.
3. An ε-almost strongly universal class of hash functions is ε-universal.

The following construction of an ε-almost strongly universal class $\tilde{\mathcal{H}}$ of hash functions $h : \{0,1\}^r \longrightarrow \{0,1\}^n$ was introduced in [WegCar81, Section 3, p. 272].

Let $s := n + \lceil \log_2(\log_2(r)) \rceil$ and let $\mathcal{G}$ be a strongly universal class of hash functions $h : \{0,1\}^{2s} \longrightarrow \{0,1\}^s$ (take, for example, the class $\mathcal{G}_{2s,s}$ defined on page 330). The elements of $\tilde{\mathcal{H}}$ are tuples $H := (h_1, h_2, \ldots, h_k)$ of functions from $\mathcal{G}$. The length k of the tuples depends on the parameters r, n (see below). The hash value $H(m)$ of $m \in \{0,1\}^r$ is computed iteratively. We start with $x := m$ and $i := 1$ and repeat until x has length s:

1. We subdivide x into subblocks of length $2s$: $x = x_1 \| x_2 \ldots \| x_l$. If necessary, we pad out the last block x_l with zeros to get a block of length $2s$.
2. $x := h_i(x_1) \| h_i(x_2) \| \ldots \| h_i(x_l), \quad i := i + 1$.

The hash value $H(m)$ consists of the least-significant bits of the final value x:

$$H(m) := \mathrm{Lsb}_n(x).$$

The key needed to specify H is the concatenation of the keys needed to specify the functions $h_1, h_2, \ldots, h_k$. In each iteration, we roughly halve the length of the message. Thus, we need about $\log_2(r/s) = (\log_2(r) - \log_2(s)) \approx (\log_2(r) - \log_2(n))$ functions.

Example. The size of the keys which are needed to identify a function in an ε-almost strongly universal class of hash functions is substantially smaller than the size of the keys which are needed in a truly strongly universal class. Consider, for example, the strongly universal classes of hash functions

$$\mathcal{G}_{l,f} := \{h_{a_0,a_1} : \mathbb{F}_{2^l} \longrightarrow \mathbb{F}_{2^f}, \; x \longmapsto \mathrm{msb}_f(a_0 \cdot x + a_1) \mid a_0, a_1 \in \mathbb{F}_{2^l}\},$$

which we defined on page 330. Functions are identified by the two l-bit elements a_0, a_1. If we use the strongly universal class $\mathcal{G}_{r,n}$ for hashing from $\{0,1\}^r$ to $\{0,1\}^n$, then we need $2 \cdot r$ key bits to describe a function. If we use the ε-almost universal class $\tilde{\mathcal{H}}$, with $\mathcal{G}_{2s,s}$ as the basic building block, then $\log_2(r/s) \cdot 2 \cdot 2s < 4s \log_2(r)$ key bits are sufficient.

Theorem 10.7 (*Wegman and Carter*). $\tilde{\mathcal{H}}$ *is an* $1/2^{n-1}$-*almost strongly universal class of hash functions.*

Proof. Choose $H = (h_1, h_2, \ldots, h_k) \in \tilde{\mathcal{H}}$ at random. Let $m \in \{0,1\}^r$. Compute $H(m)$ and let v be the intermediate value x before the last iteration. The function h_k, which is applied in the last iteration, is chosen randomly from a strongly universal class. Hence, all elements in $\{0,1\}^s$ and then all elements in $\{0,1\}^n$ have the same probability to be the value $h_k(v)$, which is $H(m)$. We see that the first requirement for ε-almost strongly universal hash families is fulfilled.

Now, let $m_1, m_2 \in \{0,1\}^r, m_1 \neq m_2$, and compute $H(m_1)$ and $H(m_2)$ in parallel. Let v_1 and v_2 be the intermediate values x before the last iteration in the computation of $H(m_1)$ and $H(m_2)$. In each iteration, the probability that two distinct strings result in identical output strings is $1/2^s$, because $\mathcal{H}$ is a universal class. Since we iterate about $(\log_2(r) - \log_2(s))$ times, the chance that $v_1 = v_2$ is $\leq \log_2(r) \cdot 1/2^s \leq \log_2(r)/2^n 2^{\log_2(\log_2(r))} = 1/2^n$.

Now, let $t_1, t_2 \in \{0,1\}^n$. Since $\mathcal{G}$ is a strongly universal class, we get, if $v_1 \neq v_2$,

$$\text{prob}(h_k(v_1) = y_1, h_k(v_2) = y_2 \mid v_1 \neq v_2 : h_k \overset{u}{\leftarrow} \mathcal{G}) = \frac{1}{2^{2s}}$$

for all $y_1, y_2 \in \{0,1\}^s$. Hence,

$$\text{prob}(H(m_1) = t_1, H(m_2) = t_2 \mid v_1 \neq v_2 : H \overset{u}{\leftarrow} \tilde{\mathcal{H}}) = \frac{1}{2^{2n}}.$$

If $v_1 = v_2$, then

$$\text{prob}(h_k(v_1) = h_k(v_2) = y \mid v_1 = v_2 : h_k \overset{u}{\leftarrow} \mathcal{G}) = \frac{1}{2^s}$$

for all $y \in \{0,1\}^s$, hence

$$\text{prob}(H(m_1) = t_1, H(m_2) = t_2 \mid v_1 = v_2 : H \overset{u}{\leftarrow} \tilde{\mathcal{H}}) = \begin{cases} 1/2^n & \text{if } t_1 = t_2, \\ 0 & \text{otherwise.} \end{cases}$$

Summarizing, we obtain

$$\begin{aligned}
&\text{prob}(H(m_1) = t_1, H(m_2) = t_2 : H \overset{u}{\leftarrow} \tilde{\mathcal{H}}) \\
&= \text{prob}(v_1 \neq v_2) \cdot \text{prob}(H(m_1) = t_1, H(m_2) = t_2 \mid v_1 \neq v_2 : H \overset{u}{\leftarrow} \tilde{\mathcal{H}}) \\
&\quad + \text{prob}(v_1 = v_2) \cdot \text{prob}(H(m_1) = t_1, H(m_2) = t_2 \mid v_1 = v_2 : H \overset{u}{\leftarrow} \tilde{\mathcal{H}}) \\
&\leq \text{prob}(v_1 \neq v_2) \cdot \frac{1}{2^{2n}} + \text{prob}(v_1 = v_2) \cdot \frac{1}{2^n} \\
&\leq 1 \cdot \frac{1}{2^{2n}} + \frac{1}{2^n} \cdot \frac{1}{2^n} = 2 \cdot \frac{1}{2^n} \cdot \frac{1}{2^n} = \frac{1}{2^{n-1}} \cdot \frac{1}{2^n}.
\end{aligned}$$

$\square$

10.3.2 Message Authentication with Universal Hash Families

Wegman and Carter ([WegCar81]) proposed authentication codes that are computed by using ε-almost strongly universal hash families. Let $\mathcal{H}$ be an ε-almost strongly universal class of hash functions $h : \{0,1\}^r \longrightarrow \{0,1\}^n$. To authenticate the origin and integrity of a message $m \in \{0,1\}^r$ from Alice to Bob, Alice and Bob agree on a randomly chosen function $h \in \mathcal{H}$ as their shared secret key. Alice supplements her message m with the authentication code $a = h(m)$ and sends (m, a) to Bob. Bob checks the code by comparing $h(m)$ with a, and if both values match he can be sure that the message originates from Alice and was not modified during transmission. The key h is a one-time key; it is only used for one message.

This authentication scheme is unconditionally secure. We need not restrict the computing power of an adversary Eve. Assume that Alice and Bob have agreed on a random shared secret key $h \in \mathcal{H}$. Even if Eve intercepts a valid message (m, a) authenticated by using h, the probability of Eve finding the correct authentication code for a message $m' \neq m$ is $\leq \varepsilon$.

Proposition 10.8. *Let Eve be an adversary with unlimited resources. Assume that Eve knows a valid authenticated message (m, a), where $a = h(m)$ is computed with the randomly chosen secret key $h \in \mathcal{H}$. Then the probability that Eve is able to find the valid authentication code $h(m')$ for some message $m' \in \{0,1\}^r, m' \neq m$, is $\leq \varepsilon$.*

Proof. Let $(m', a'), m' \neq m$, be an authenticated message which is forged by Eve. The only information which Eve has available is that $h(m) = a$, where h is the randomly chosen secret key. Therefore, in computing the probability that Eve's forgery is valid, we have to consider the conditional probability given $h(m) = a$. The probability that a' is the valid authentication code is

$$\text{prob}(h(m') = a' \,|\, h(m) = a : h \xleftarrow{u} \mathcal{H})$$
$$= \frac{\text{prob}(h(m') = a', h(m) = a : h \xleftarrow{u} \mathcal{H})}{\text{prob}(h(m) = a : h \xleftarrow{u} \mathcal{H})} \leq \frac{\varepsilon/2^n}{1/2^n} = \varepsilon.$$

$\square$

10.3.3 Authenticating Multiple Messages

Wegman and Carter extended their scheme so that it could authenticate a sequence of messages with one key ([WegCar81, Section 4]). Let $\mathcal{H}$ be an ε-almost strongly universal class of hash functions

$$h : \{0,1\}^r \longrightarrow \{0,1\}^n.$$

To authenticate a sequence $m_1, m_2, \ldots, m_l \in \{0,1\}^r$ of messages, Alice and Bob agree on a randomly selected secret key consisting of a function $h \in \mathcal{H}$ and elements $b_1, b_2, \ldots, b_l \in \{0,1\}^n$.

The messages must contain a unique message number. Let i be the message number of m_i. Then the authentication code a_i of m_i is defined as

$$a_i := h(m_i) \oplus b_i.$$

Proposition 10.9. *Let Eve be an adversary with unlimited resources. Let $(h, b_1, b_2, \ldots, b_l), h \in \mathcal{H}, b_i \in \{0,1\}^n, 1 \leq i \leq l$, be a randomly chosen key. Assume that Eve knows the valid authentication codes a_i for messages $m_i, 1 \leq i \leq l$. Then the probability that Eve is able to find the valid authentication code for some message $m' \in \{0,1\}^r, m' \notin \{m_1, \ldots, m_l\}$, is $\leq \varepsilon$.*

Proof. Let $(m', a'), m' \notin \{m_1, \ldots, m_l\}$, be an authenticated message which is forged by Eve. The only information which Eve has available is that $h(m_i) = a_i \oplus b_i$, whereas she does not know h and the b_i. a' is a valid authentication code only if $a' = h(m') \oplus b_j$ for some $j, 1 \leq j \leq l$, say $j = 1$. In computing the probability that Eve's forgery is valid, we have to consider the conditional probability given $h(m_i) = a_i \oplus b_i, 1 \leq i \leq l$. The probability that a' is the valid authentication code is

$$\text{prob}(h(m') = a' \oplus b_1 \,|\, h(m_i) = a_i \oplus b_i, 1 \leq i \leq l:$$
$$h \xleftarrow{u} \mathcal{H}, b_i \xleftarrow{u} \{0,1\}^n, 1 \leq i \leq l)$$
$$= \text{prob}(h(m') = a' \oplus b_1 \,|\, h(m_1) = a_1 \oplus b_1 : h \xleftarrow{u} \mathcal{H}, b_1 \xleftarrow{u} \{0,1\}^n)$$
$$= \sum_{y \in \{0,1\}^n} \text{prob}(b_1 = y) \cdot \text{prob}(h(m') = a' \oplus y \,|\, h(m_1) = a_1 \oplus y : h \xleftarrow{u} \mathcal{H})$$
$$= \frac{1}{2^n} \sum_{y \in \{0,1\}^n} \frac{\text{prob}(h(m') = a' \oplus y, h(m_1) = a_1 \oplus y : h \xleftarrow{u} \mathcal{H})}{\text{prob}(h(m_1) = a_1 \oplus y : h \xleftarrow{u} \mathcal{H})}$$
$$\leq \frac{1}{2^n} \sum_{y \in \{0,1\}^n} \frac{\varepsilon \cdot 1/2^n}{1/2^n} = \frac{1}{2^n} \cdot 2^n \cdot \varepsilon = \varepsilon.$$

The first equality holds because the choice of $b_i, 2 \leq i \leq l$, is independent of the choice of h and b_1. $\qquad\qquad\square$

Remarks. Let $\mathcal{H} = \tilde{\mathcal{H}}$ be the almost universal class of hash functions constructed from a universal class $\mathcal{G}$ of hash functions (see page 333).

1. $\mathcal{H}$ is $1/2^{n-1}$-almost universal (Theorem 10.7). An adversary's chance of successfully forging an authentication code is $\leq 1/2^{n-1}$, which is negligibly small for sufficiently large n.
2. Assume that $\mathcal{G} = \mathcal{G}_{2s,s}$ is used. Then the length of a key is less than

$$4s \log_2(r) + l \cdot n \approx 4(n + \log_2(\log_2(r))) \log_2(r) + l \cdot n$$

$$= (4 \log_2(r) + l)n + 4 \log_2(\log_2(r)) \log_2(r),$$

which is about $(4 \cdot 23 + l)n + 416$ for $r = 1 \text{ MiB} = 2^{20} \text{ B} = 2^{23}$ bits. To authenticate $l = 2^{10}$ 1 MiB messages with a security level of $1/2^{127}$, which means $n = 128$, we need a key of length 17.5 KiB.
3. There are various designs for unconditionally secure authentication codes. Many are similar to the approach of Wegman and Carter. An overview and comparison of such codes can be found in [AtiSti96, Section 3.1].
4. There are schemes for the authentication of multiple messages which do not require that each message has a counter attached to it (see [AtiSti96]).

10.4 Collision Entropy and Privacy Amplification

In Section 10.1, we introduced the Rényi entropy of order 2, also called the collision entropy, of a random variable X (Definition 10.4). As always, we can assume in our setting that the range of X is a finite set $\mathcal{X}$.

The Rényi entropy $H_2(X)$ measures the uncertainty that two independent executions X_1 and X_2 of X yield the same result:

$$H_2(X) := -\log_2(\text{prob}_c(X)),$$

where $\mathrm{prob}_c(X)$ is the collision probability of X:

$$
\begin{aligned}
\mathrm{prob}_c(X) &:= \mathrm{prob}(X_1 = X_2) \\
&= \sum_{x \in \mathcal{X}} \mathrm{prob}(X_1 = x, X_2 = x) \\
&= \sum_{x \in \mathcal{X}} \mathrm{prob}(X_1 = x) \cdot \mathrm{prob}(X_2 = x) \\
&= \sum_{x \in \mathcal{X}} \mathrm{prob}(X = x)^2.
\end{aligned}
$$

In this section, we review some basic properties of the Rényi entropy, give a proof of the Privacy Amplification Theorem and explain the extraction of provably secure keys.

10.4.1 Rényi Entropy

If X and Y are jointly distributed random variables with values in $\mathcal{X}$ and $\mathcal{Y}$, respectively, then we can consider the conditional Rényi entropy. We denote the random variable X conditional on $y \in \mathcal{Y}$ by $X \mid Y = y$, and its probability distribution by $\mathrm{prob}_{X \mid Y = y}$:

$$
\mathrm{prob}_{X \mid Y = y}(x) = \mathrm{prob}(X = x \mid Y = y) = \frac{\mathrm{prob}(X = x, Y = y)}{\mathrm{prob}(Y = y)} \text{ for } x \in \mathcal{X}.
$$

The conditional Rényi entropies are

$$
H_2(X \mid Y = y) := -\log_2 \left(\sum_{x \in \mathcal{X}} \mathrm{prob}(X = x \mid Y = y)^2 \right),
$$

$$
H_2(X \mid Y) := \sum_{y \in \mathcal{Y}} \mathrm{prob}(Y = y) H_2(X \mid Y = y).
$$

These definitions are completely analogous to the definitions for the Shannon entropy (Definition B.37).

Proposition 10.10. *Let X, Y be jointly distributed random variables.*

1. $H_2(X) \leq H_2(XY)$.
2. If X and Y are independent, then

$$
H_2(XY) = H_2(X) + H_2(Y).
$$

Proof.

1. The statement follows from the inequality

$$\text{prob}(X = x)^2 = \left(\sum_y \text{prob}(X = x, Y = y) \right)^2 \geq \sum_y \text{prob}(X = x, Y = y)^2.$$

2.

$$\begin{aligned}
\text{prob}_c(XY) &= \sum_{x,y} \text{prob}(X = x, Y = y)^2 \\
&= \sum_{x,y} \text{prob}(X = x)^2 \cdot \text{prob}(Y = y)^2 \\
&= \left(\sum_x \text{prob}(X = x)^2 \right) \cdot \left(\sum_y \text{prob}(Y = y)^2 \right) \\
&= \text{prob}_c(X) \cdot \text{prob}_c(Y).
\end{aligned}$$

$\square$

Remark. In Proposition B.38, we list some basic properties of the Shannon entropy. For the Rényi entropy, most of these properties do not hold in general. We give some examples:

1. Whereas $H(Y) + H(X \mid Y) = H(XY) \geq H(X)$ by Proposition B.38 (items 2 and 4), we may have $H_2(Y) + H_2(X \mid Y) < H_2(XY)$ and even $H_2(Y) + H_2(X \mid Y) < H_2(X)$. See Exercise 2 for an example.
2. Whereas $H(X) + H(Y) \geq H(XY)$ by Proposition B.38 (item 3), we may have $H_2(X) + H_2(Y) < H_2(XY)$. For an example, see [Rényi70, Section IX.6, Theorem 4].
3. The conditional entropy $H(X \mid Y)$ measures the uncertainty about X knowing Y. Since $H(X) - H(X \mid Y) = I(X; Y) \geq 0$ (Proposition B.38), we have $H(X \mid Y) \leq H(X)$. This means that acquiring knowledge reduces the uncertainty. In contrast to the Shannon entropy, additional knowledge does not always reduce the Rényi entropy, it may even increase it. We may have $H_2(X \mid Y) > H_2(X)$. In this case, Y is *spoiling knowledge*. See Exercise 3 for an example.

The Shannon entropy is an upper bound for the Rényi entropy. The two entropies coincide for uniform distributions.

Proposition 10.11. *Let X be a random variable with values in $\mathcal{X}$. Then*

$$H_2(X) \leq H(X),$$

with equality if and only if the probability distribution of X is the uniform distribution over $\mathcal{X}$ or a subset of $\mathcal{X}$. If X and Y are jointly distributed random variables with values in $\mathcal{X}$ and $\mathcal{Y}$, respectively, then

$$H_2(X|Y) \leq H(X|Y),$$

with equality if and only if, for all $y \in \mathcal{Y}$ with $\mathrm{prob}(Y = y) > 0$, the conditional distribution $\mathrm{prob}_{X|Y=y}$ is the uniform distribution over $\mathcal{X}$ or a subset of $\mathcal{X}$.

Proof. Consider the random variable $Z : \mathcal{X} \longrightarrow \mathbb{R}$, $x \longmapsto \mathrm{prob}(X = x)$. Then we may consider the entropies as expected values:

$$H(X) = E(-\log_2(Z)) \quad \text{and} \quad H_2(X) = -\log_2(E(Z)).$$

Since $-\log_2$ is a strictly convex function, the first statement follows from Jensen's inequality (Proposition B.22). The second statement then follows from the definitions of $H(X|Y)$ and $H_2(X|Y)$:

$$H(X|Y) = \sum_{y \in \mathcal{Y}} \mathrm{prob}(Y = y) H(X|Y = y),$$

$$H_2(X|Y) = \sum_{y \in \mathcal{Y}} \mathrm{prob}(Y = y) H_2(X|Y = y).$$

$\square$

10.4.2 Privacy Amplification

Privacy amplification, also called *entropy smoothing*, converts almost all the Rényi entropy of a bit string into a uniformly distributed random bit string. We can achieve this by applying a universal hash function. The following privacy amplification theorem was stated and proven by Bennett, Brassard, Crépeau and Maurer in 1995 ([BenBraCréMau95]). Earlier work on this subject may be found, for example, in [ImpLevLub89] and [BenBraRob88].

Theorem 10.12 *(Privacy Amplification Theorem). Let X be a random variable with values in the finite set $\mathcal{X}$ and Rényi entropy $H_2(X)$. Let $\mathcal{H}$ be a universal class of hash functions $\mathcal{X} \longrightarrow \{0,1\}^r$. Let G be the random variable describing the random choice of a function $g \in \mathcal{H}$ (with respect to the uniform distribution), and let $Q := G(X)$ be the random variable which applies a randomly and uniformly chosen function $g \in \mathcal{H}$ to the output of X. Then*

$$H(Q|G) \geq H_2(Q|G) \geq r - \log_2\left(1 + 2^{r-H_2(X)}\right) \geq r - \frac{2^{r-H_2(X)}}{\ln(2)}.$$

Proof. The first inequality follows from Proposition 10.11. For the second inequality, we start by writing

$$H_2(Q|G) = \sum_{g \in \mathcal{H}} \text{prob}(g) \cdot H_2(Q|G = g)$$

$$= \sum_{g \in \mathcal{H}} -\text{prob}(g) \cdot \log_2(\text{prob}_c(Q|G = g))$$

$$\geq -\log_2 \left(\sum_{g \in \mathcal{H}} \text{prob}(g) \cdot \text{prob}_c(Q|G = g) \right).$$

The last step in the above computation follows from Jensen's inequality (Proposition B.21).

$$\sum_{g \in \mathcal{H}} \text{prob}(g) \cdot \text{prob}_c(Q|G = g)$$

$$= \sum_{g \in \mathcal{H}} \text{prob}(g) \cdot \text{prob}(g(x_1) = g(x_2) : x_1 \leftarrow X, x_2 \leftarrow X)$$

$$= \text{prob}(g(x_1) = g(x_2) : x_1 \leftarrow X, x_2 \leftarrow X, g \overset{u}{\leftarrow} \mathcal{H})$$

$$= \text{prob}(x_1 = x_2 : x_1 \leftarrow X, x_2 \leftarrow X)$$
$$+ \text{prob}(x_1 \neq x_2 : x_1 \leftarrow X, x_2 \leftarrow X)$$
$$\qquad \cdot \text{prob}(g(x_1) = g(x_2)|x_1 \neq x_2 : x_1 \leftarrow X, x_2 \leftarrow X, g \overset{u}{\leftarrow} \mathcal{H})$$

$$= \text{prob}_c(X) + (1 - \text{prob}_c(X)) \cdot 2^{-r} \quad \text{(since } \mathcal{H} \text{ is a universal class)}$$
$$= 2^{-H_2(X)} + (1 - \text{prob}_c(X)) \cdot 2^{-r} \quad \text{(by the definition of } H_2(X))$$
$$\leq 2^{-H_2(X)} + 2^{-r} = 2^{-r} \left(2^{r - H_2(X)} + 1 \right).$$

Taking $-\log_2$ on both sides, we obtain the second inequality. The third inequality holds because $1 + x \leq e^x$, and hence $\ln(1 + x) \leq x$ for all $x \in \mathbb{R}$. $\square$

Remark. There is an alternative method for privacy amplification based on *extractors* ([NisZuc96], [MauWol00a]).

10.4.3 Extraction of a Secret Key

Privacy amplification is an essential building block in protocols that establish a provably secure shared key, such as the bounded storage and noisy channel protocols (Sections 10.1 and 10.2) and quantum key distribution (Section 10.5 below).

In these protocols, Alice and Bob first exchange a preliminary key A consisting of n randomly chosen bits.[4] An adversary Eve may be able to get some knowledge about A. Let V be the random variable that subsumes Eve's knowledge, and let a, v be the particular values of the preliminary key and Eve's knowledge $A = a, V = v$. Eve's knowledge about A is incomplete and,

---

[4] In quantum key distribution, A is called the *raw key*.

therefore, her conditional (Shannon and Rényi) entropy is positive. Assume that Alice and Bob are able to compute a lower bound l for Eve's conditional Rényi entropy, $l \leq H_2(A|V = v)$.

Alice and Bob then randomly select a function $g \in \mathcal{H}$ from a universal class $\mathcal{H}$ of hash functions $\{0,1\}^n \longrightarrow \{0,1\}^r$, where $r \in \mathbb{N}, r \leq l$. They publicly exchange this function, so Eve learns g. Alice and Bob apply g to their preliminary shared key a and obtain the final shared key $k = g(a)$. The generation of k is described by the random variable $K := G(A)$, where G is the random variable describing the choice of g.

The key k is provably secure – Eve's uncertainty about K is almost maximal, which means that the probability distribution of K is almost uniform in Eve's view. More precisely, Theorem 10.12 applied to the random variable A conditional on $V = v$, which has the probability distribution $\text{prob}_{A|V=v}$, implies the following result.

Corollary 10.13.

$$H(K|G, V = v) \geq r - \log_2\left(1 + 2^{r-l}\right) \geq r - \frac{2^{-(l-r)}}{\ln(2)}.$$

Remarks:

1. Eve learns the function g, so we have to consider the entropy conditional on $G = g$.
2. Eve's information about K, $H(K) - H(K|G, V = v)$, is $\leq 2^{-(l-r)}/\ln(2)$, which is exponentially small in the security parameter $s := l - r$.
3. In the entropy $H(K|G, V = v)$, the average is taken over all functions $g \in \mathcal{H}$. Therefore, there may exist functions g for which Eve obtains a significant amount of information about k. But the probability that such a g is selected by Alice and Bob is negligibly small.

Example. In contrast to the Rényi entropy, the Shannon entropy cannot always be transformed into a secret key. Consider the following example from [BenBraCréMau95, Section III]. Assume that Eve obtains, with probability p, the complete bit string A that Alice sends, and a totally random string in all other cases, i.e., $V = A$ with probability p and $V \xleftarrow{u} \{0,1\}^n$ with probability $1 - p$. Let $\mathcal{E}$ denote the event "Eve obtains the complete bit string". Eve does not know which case occurs. We have

$$\text{prob}(A = x | V = v, \mathcal{E}) = \begin{cases} 1 & \text{if } x = v, \\ 0 & \text{if } x \neq v, \end{cases}$$

and $\text{prob}(A = x | V = v, \overline{\mathcal{E}}) = 2^{-n}$.

Then, Eve's entropy $H(A|V = v)$ is slightly more than $(1 - p)n$, i.e., Eve knows slightly less than a fraction p of the bits of A (see Exercise 5). But it is not possible to transform Eve's uncertainty into a secret key. Namely, let $K = g(A)$ be the resulting key, extracted by applying $g \in \mathcal{H}$. We have

$$\text{prob}(K = z \,|\, V = v, G = g, \mathcal{E}) = \begin{cases} 1 & \text{if } z = g(v), \\ 0 & \text{if } z \neq g(v), \end{cases}$$

and Eve's uncertainty $H(K \,|\, V = v, G = g)$ about K is slightly more than $(1 - p)r$ (see Exercise 5). This means that Eve has almost $p \cdot r$ bits of information about K, regardless of which function g is chosen.

Eve's Rényi entropy is small:

$$\text{prob}_c(A \,|\, V = v)$$
$$= \sum_x \text{prob}(A = x \,|\, V = v)^2$$
$$= \sum_x \left(\text{prob}\,(\mathcal{E})\,\text{prob}\,(A = x \,|\, V = v, \mathcal{E}) + \text{prob}\,(\overline{\mathcal{E}})\,\text{prob}\,(A = x \,|\, V = v, \overline{\mathcal{E}})\right)^2$$
$$= p^2 + p(1-p)2^{-(n-1)} + 2^n (1-p)^2 2^{-2n}$$
$$= p^2 + p(1-p)2^{-(n-1)} + (1-p)^2 2^{-n}$$
$$> p^2,$$

hence $H_2(A \,|\, V = v) < -2\log_2(p)$.

10.5 Quantum Key Distribution

Quantum key distribution, often simply called *quantum cryptography*, enables Alice and Bob to establish an unconditionally secure shared secret key. This key may then be used for protecting messages by using methods from symmetric cryptography such as a one-time pad or message authentication codes. The unconditional security of quantum cryptography is a consequence of the probabilistic nature of quantum systems: It is impossible to distinguish non-orthogonal states with certainty, and measurements disturb the system. In a typical implementation, the quantum behavior of photons is used. Alice sends quantum bits encoded as polarized photons over an optical channel to Bob. They get a preliminary shared key. The bits are encoded by non-orthogonal quantum states. Hence, they cannot be read without introducing detectable disturbances. Alice and Bob cannot prevent eavesdropping, but they can detect it. It is even possible to calculate the amount of information that has been intercepted. By applying privacy amplification, the remaining, non-intercepted bits of information can be extracted from the preliminary key in order to get a secure key.

Quantum key distribution works as a "key expander". Alice and Bob need an initial shared secret key, which is necessary for mutual authentication. This "bootstrap key" can then be expanded to a secure key of arbitrary length.

Comprehensive surveys of quantum cryptography have been given in, for example, [Assche06], [DusLütHen06]. The present section is simply an introduction. We will describe Bennett and Brassard's famous BB84 protocol ([BenBra84]) and give a security proof, assuming intercept-and-resend attacks. In Section 10.5.4, intercept-and-resend attacks are analyzed comprehensively in detail. Section 10.5.1 summarizes some essential basics of quantum mechanics, such as Hilbert spaces and quantum measurements. Section 10.5.5 covers information reconciliation. We outline the protocol Cascade and study the effects of information reconciliation on the adversary's Rényi entropy. Section 10.5 closes with an example (Section 10.5.6) and an overview of general attacks and security proofs (Section 10.5.7).

10.5.1 Quantum Bits and Quantum Measurements

Postulate I of quantum mechanics says that the states of an isolated physical system can be modeled by unit vectors in a Hilbert space (see, for example, [CohDiuLal77, Chapter III] and [NieChu00, Chapter 2.2]).

Hilbert Spaces. We consider the n-dimensional vector space $\mathcal{H} := \mathbb{C}^n$ over the field $\mathbb{C}$ of complex numbers. In quantum mechanics, the *bra–ket notation* or *Dirac notation*[5] is a standard notation for vectors in $\mathcal{H}$. Let $v = (x_1, x_2, \ldots, x_n) \in \mathcal{H}$. The *ket vector* $|v\rangle$ is the column vector v and the *bra vector* $\langle v|$ is the complex conjugate of the row vector v:

$$\langle v| := (\overline{x_1}, \overline{x_2}, \ldots, \overline{x_n}), \quad |v\rangle := \begin{pmatrix} x_1 \\ \vdots \\ x_n \end{pmatrix}.$$

We have a *Hermitian*[6] *inner product* $\langle \cdot | \cdot \rangle$ on $\mathcal{H}$: For vectors $v = (x_1, x_2, \ldots, x_n)$, $w = (y_1, y_2, \ldots, y_n) \in \mathcal{H}$, the inner product of v and w is defined as

$$\langle v|w\rangle := \langle v| \cdot |w\rangle = (\overline{x_1}, \overline{x_2}, \ldots, \overline{x_n}) \cdot \begin{pmatrix} y_1 \\ \vdots \\ y_n \end{pmatrix} = \overline{x_1}y_1 + \ldots \overline{x_n}y_n.$$

The inner product is the analogue of the scalar product on the Euclidean space $\mathbb{R}^n$, and on $\mathbb{R}^n \subset \mathbb{C}^n$ it coincides with the usual scalar product. We summarize some properties and definitions:

1. $\langle v|w\rangle = \overline{\langle w|v\rangle}$.
2. The inner product is linear in the second argument and antilinear in the first:

$$\langle v| \sum_i \lambda_i w_i\rangle = \sum_i \lambda_i \langle v|w_i\rangle, \quad \langle \sum_i \lambda_i v_i|w\rangle = \sum_i \overline{\lambda_i} \langle v_i|w\rangle.$$

---

[5] This notation was introduced by Paul Dirac (1902–1984), an English theoretical physicist.
[6] Charles Hermite (1822–1901) was a French mathematician.

3. $\langle v|v \rangle \geq 0$, with equality if and only if $|v\rangle = 0$.
4. The *length* (or *norm*) $||v||$ of v is $||v|| := \sqrt{\langle v|v\rangle}$. Vectors of length 1 are called *unit vectors*.
5. v and w are *orthogonal* if and only if $\langle v|w\rangle = 0$.
6. A set $\mathcal{V}$ of vectors in $\mathcal{H}$ is *orthonormal* if and only if the elements of $\mathcal{V}$ are unit vectors and pairwise orthogonal.
7. A *unitary operator* on $\mathcal{H}$ is a $\mathbb{C}$-linear invertible map $\mathcal{H} \longrightarrow \mathcal{H}$ which preserves the inner product and hence the length of vectors. A unitary operator is described by a unitary matrix A, which means that A is invertible and $A^{-1} = A^* := \overline{A}^t$.

The vector space $\mathcal{H}$ together with the inner product is called a *Hilbert space*.[7] $\mathcal{H}$ is a simple example of a Hilbert space because it is finite-dimensional. Hilbert spaces can have infinite dimension.

Quantum Bits. The states of a quantum bit are represented by unit vectors in the two-dimensional Hilbert space $\mathcal{H} := \mathbb{C}^2$, which we call the *Hilbert space of a quantum bit*. A classical bit has two possible states – it is either 0 or 1. Its state is an element of $\{0,1\}$. A *quantum bit*, or *qubit* for short, can also take on these two states 0 and 1, and we represent these two states by the standard basis vectors $|H\rangle := (1,0)$ and $|V\rangle := (0,1)$ of $\mathcal{H}$.[8] The difference between a qubit and a classical bit, which allows only the states 0 and 1, is that a qubit can also take on states

$$\alpha \cdot |H\rangle + \beta \cdot |V\rangle, \ \alpha, \beta \in \mathbb{C}, |\alpha|^2 + |\beta|^2 = 1,$$

between $|H\rangle$ and $|V\rangle$. Such states are called *superpositions*. The states $|H\rangle$ and $|V\rangle$ are orthogonal. They constitute the orthonormal standard basis of $\mathcal{H}$.

It is not necessary to represent the classical bits 0 and 1 by $|H\rangle$ and $|V\rangle$. Any orthonormal basis of our Hilbert space $\mathcal{H}$ can be used for this purpose. We characterize the orthonormal bases of $\mathcal{H}$ in the following proposition.

Proposition 10.14. *Let $\mathcal{H} = \mathbb{C}^2$ be the Hilbert space of a quantum bit and (v, w) be an orthonormal basis of $\mathcal{H}$. Let $U := (|v\rangle\ |w\rangle)$ be the matrix with column vectors $|v\rangle, |w\rangle$. Then there exist elements $\eta, \alpha, \beta, \theta \in [-\pi, \pi[$ such that*

$$U = e^{i\eta/2}\, U', \quad U' = \begin{pmatrix} e^{i\alpha}\cos(\theta) & e^{i\beta}\sin(\theta) \\ -e^{-i\beta}\sin(\theta) & e^{-i\alpha}\cos(\theta) \end{pmatrix}.$$

Conversely, each such matrix U yields an orthonormal basis of $\mathcal{H}$.

---

[7] David Hilbert (1862–1943) was a German mathematician. He is recognized as one of the most outstanding mathematicians of the nineteenth and early twentieth centuries.

[8] H for "horizontal" and V for "vertical".

Proof. The fact that the columns of U are orthonormal means that $\overline{U}^t \cdot U = \begin{pmatrix} 1 & 0 \\ 0 & 1 \end{pmatrix}$, hence $|\det(U)| = 1$ and $U \cdot \overline{U}^t = \begin{pmatrix} 1 & 0 \\ 0 & 1 \end{pmatrix}$. Transposing, we get

$\overline{U} \cdot U^t = \begin{pmatrix} 1 & 0 \\ 0 & 1 \end{pmatrix}$, which means that the rows of U are also orthonormal.

Let $e^{i\eta} = \det(U), \eta \in [-\pi, \pi[$. Then $U' := e^{-i\eta/2} U$ has determinant 1. Let $U' = \begin{pmatrix} a & b \\ c & d \end{pmatrix}$. Then $|a|^2 + |c|^2 = |a|^2 + |b|^2 = |b|^2 + |d|^2 = |c|^2 + |d|^2 = 1$,

and hence there is some $\theta \in [-\pi, \pi[$ such that $|b|^2 = |c|^2 = \sin^2(\theta)$ and $|a|^2 = |d|^2 = \cos^2(\theta)$. Let

$$a = e^{i\alpha} \cos(\theta), b = e^{i\beta} \sin(\theta), c = -e^{i\gamma} \sin(\theta), d = e^{i\delta} \cos(\theta)$$

be polar forms of a, b, c, d, where $\alpha, \beta, \gamma, \delta \in [-\pi, \pi[$. From $\overline{a}b + \overline{c}d = 0$, we get $\cos(\theta)\sin(\theta)(e^{i(\beta-\alpha)} - e^{i(\delta-\gamma)}) = 0$. Assume for a moment that $\theta \neq k \cdot \pi/2, k \in \mathbb{Z}$. Then $\beta - \alpha = \delta - \gamma$, i.e., $\alpha + \delta = \beta + \gamma$. From $\det(U') = ad - cb = 1$, we get $1 = \cos^2(\theta)e^{i(\alpha+\delta)} + \sin^2(\theta)e^{i(\beta+\gamma)} = e^{i(\alpha+\delta)}(\cos^2(\theta) + \sin^2(\theta)) = e^{i(\alpha+\delta)}$, and hence $\alpha = -\delta$ and $\gamma = -\beta$. If $\theta = k \cdot \pi/2, k \in \mathbb{Z}$, then either

$\sin(\theta) = 0, \cos(\theta) = \pm 1, e^{i(\alpha+\delta)} = 1$ and hence $U' = \pm \begin{pmatrix} e^{i\alpha} & 0 \\ 0 & e^{-i\alpha} \end{pmatrix}$, or

$\sin(\theta) = \pm 1, \cos(\theta) = 0, e^{i(\beta+\gamma)} = 1$ and hence $U' = \pm \begin{pmatrix} 0 & e^{i\beta} \\ -e^{-i\beta} & 0 \end{pmatrix}$. $\square$

Remark. The group consisting of the unitary matrices is called the *unitary group* U(2). The subgroup consisting of the unitary matrices with determinant 1 is called the *special unitary group* SU(2). U is a *unitary matrix*, i.e., $U \in$ U(2), and U' is unitary with $\det(U') = 1$, i.e., $U' \in$ SU(2). Conversely, all members of U(2) and SU(2) are of this form.

Some examples of orthonormal bases that can be used for the representation of the classical bits 0 and 1 are the following:

1. The standard basis $(|H\rangle, |V\rangle)$ is called the *rectilinear basis*.
2. $\eta = \alpha = \beta = 0, \theta = \pi/4$ yield the *diagonal basis*

$$|-\rangle := \frac{\sqrt{2}}{2}(|H\rangle - |V\rangle), \quad |+\rangle := \frac{\sqrt{2}}{2}(|H\rangle + |V\rangle).$$

The diagonal basis is obtained from the rectilinear basis by applying a clockwise rotation through an angle $\pi/4$.

3. $\eta = \pi/2, \alpha = \beta = -\pi/4, \theta = \pi/4$ yield the *circular basis*

$$|R\rangle := \frac{\sqrt{2}}{2}(|H\rangle - i \cdot |V\rangle), \quad |L\rangle := \frac{\sqrt{2}}{2}(|H\rangle + i \cdot |V\rangle).$$

4. $\eta = \alpha = \beta = 0, \theta \in [0, 2\pi[$ yield the basis

$$|v_\theta\rangle := \cos(\theta) \cdot |H\rangle - \sin(\theta) \cdot |V\rangle, \quad |w_\theta\rangle := \sin(\theta) \cdot |H\rangle + \cos(\theta) \cdot |V\rangle.$$

This basis is obtained by rotating the rectilinear basis clockwise through an angle θ.

5. Halfway between the bases $(|H\rangle, |V\rangle)$ and $(|+\rangle, |-\rangle)$, we obtain the *Breid-bart basis*

$$|v_{\pi/8}\rangle := \cos(\pi/8) \cdot |H\rangle - \sin(\pi/8) \cdot |V\rangle,$$

$$|w_{\pi/8}\rangle := \sin(\pi/8) \cdot |H\rangle + \cos(\pi/8) \cdot |V\rangle.$$

Physical Encoding of Quantum Bits. As already suggested in the classical BB84 quantum cryptography protocol of Bennett and Brassard ([BenBra84]), quantum bits can be physically encoded in the polarization of photons. We can use linearly polarized photons with perpendicular polarization axes as realizations of $|H\rangle$ and $|V\rangle$.[9] The bases $(|v_\theta\rangle, |w_\theta\rangle)$ are all the bases that represent *linearly polarized photons*. Compared with the rectilinear basis, their polarization axes are rotated through θ. In particular, the angle between the polarization axes of $|-\rangle$ and $|H\rangle$ (or $|+\rangle$ and $|V\rangle$) is $\pi/4$. The angle between the polarization axes of the Breidbart basis and the axes of the rectilinear basis $(|H\rangle, |V\rangle)$ and the diagonal basis $(|-\rangle, |+\rangle)$ is $\pm\pi/8$. The circular basis $(|R\rangle, |L\rangle)$ is associated with (right and left) *circularly polarized photons*. General orthonormal bases are represented by *elliptically polarized photons*.

Measurements. Measurement in quantum systems is substantially different from measurement in classical physics. For any measurement, there is an underlying orthonormal basis of the associated Hilbert space, which we call the *measurement basis*, and the measurement can distinguish with certainty only among the orthogonal state vectors of the measurement basis. If the system is in a state which is a linear combination of basis vectors, then the measurement causes a random state transition. The resulting state corresponds to one of the vectors of the measurement basis. It is this key feature of quantum systems that enables us to detect any eavesdropping, which necessarily includes some measurement.

Let $\mathcal{H} = \mathbb{C}^2$ be the Hilbert space of a quantum bit, and let Bob measure a quantum bit that is in a state $|\psi\rangle \in \mathcal{H}, \|\psi\| = 1$. Assume that Bob measures in the orthonormal measurement basis $(|v\rangle, |w\rangle)$. If polarized photons are used in the implementation, this means that Bob's device is able to distinguish between these two orthogonal polarization states.[10] Postulate II of quantum

---

[9] Polarization modulators or several laser diodes each emitting one of the prescribed polarization directions can be used; see, for example, [Assche06, Section 10.2] and [DusLütHen06, Section 4].

[10] His measurement device typically includes a polarizing beam splitter (a Wollaston prism, for example), which is able to split an incoming stream of photons into two orthogonally polarized beams (see, for example, [Assche06, Section 10.2] and [DusLütHen06, Section 4]).

mechanics says that with probability $|\langle \psi | v \rangle|^2$ Bob measures $|v\rangle$, and with probability $|\langle \psi | w \rangle|^2$ he measures $|w\rangle$. After the measurement, the quantum bit is in the state $|v\rangle$ or $|w\rangle$, respectively.

Remark. The measurement statistics are not changed if the state vectors $|\psi\rangle, |v\rangle, |w\rangle$ are multiplied by *phase factors* $e^{i\eta}$, because these phase factors have absolute value 1. The probabilities of the measurement results remain the same if we use $e^{i\rho}|\psi\rangle, e^{i\sigma}|v\rangle, e^{i\tau}|w\rangle$ instead of $|\psi\rangle, |v\rangle, |w\rangle$.

Each orthonormal basis $(|v\rangle, |w\rangle)$ of $\mathcal{H}$ can be written as

$$|v\rangle = e^{i\eta/2} \begin{pmatrix} e^{i\alpha}\cos(\theta) \\ -e^{-i\beta}\sin(\theta) \end{pmatrix}, \quad |w\rangle = e^{i\eta/2} \begin{pmatrix} e^{i\beta}\sin(\theta) \\ e^{-i\alpha}\cos(\theta) \end{pmatrix},$$

where $\eta, \alpha, \beta, \theta \in [-\pi, \pi[$ (Proposition 10.14). Without affecting the measurement results, we can multiply $|v\rangle$ by $e^{-i(\alpha+\eta/2)}$ and $|w\rangle$ by $e^{-i(\beta+\eta/2)}$ and assume that

$$|v\rangle = \begin{pmatrix} \cos(\theta) \\ -e^{i\gamma}\sin(\theta) \end{pmatrix} \text{ and } |w\rangle = \begin{pmatrix} \sin(\theta) \\ e^{i\gamma}\cos(\theta) \end{pmatrix},$$

with $\gamma = -(\alpha + \beta)$. We denote this basis by $\mathcal{B}_{\gamma,\theta}$:

$$\mathcal{B}_{\gamma,\theta} := \begin{pmatrix} \cos(\theta) & \sin(\theta) \\ -e^{i\gamma}\sin(\theta) & e^{i\gamma}\cos(\theta) \end{pmatrix}.$$

Remark. Quantum measurements can be used to generate truly random bits. Measuring the state $\sqrt{2}/2\,(|H\rangle + |V\rangle)$ in the rectilinear basis $(|H\rangle, |V\rangle)$ produces a random bit. This principle is implemented in random generators from the company ID Quantique ([ID Quantique]).

Sending a Bit Through a Quantum Channel. Alice wants to send a bit to Bob. For this purpose, Alice chooses an orthonormal basis $(|v_A\rangle, |w_A\rangle)$ of $\mathcal{H}$ as her encoding basis, and Bob chooses an orthonormal basis $(|v_B\rangle, |w_B\rangle)$ of $\mathcal{H}$ as his measurement basis. Alice encodes 0 by $|v_A\rangle$ and 1 by $|w_A\rangle$, and Bob decodes $|v_B\rangle$ to 0 and $|w_B\rangle$ to 1. To send a bit $x \in \{0,1\}$ to Bob, Alice encodes x as $|v_A\rangle$ or $|w_A\rangle$ and sends this quantum state to Bob.[11] Bob measures the incoming quantum state and obtains either $|v_B\rangle$ or $|w_B\rangle$, which he decodes to 0 or 1. This communication channel is called the *quantum channel*.

As observed in the preceding section on measurements, we can assume without loss of generality that $(|v_A\rangle, |w_A\rangle) = \mathcal{B}_{\gamma,\theta}$ and $(|v_B\rangle, |w_B\rangle) = \mathcal{B}_{\gamma',\theta'}$.

The communication channel from Alice to Bob is a *binary symmetric channel*. Such a channel is *binary* – the sender transmits a bit $x \in \{0,1\}$ and

---

[11] In the typical implementation with photons, the associated photon is sent through the open air or an optical fiber.

the receiver receives a bit y – and it is *symmetric* – the probability ε that a transmission error occurs, which means that Bob receives the complementary bit $y = \overline{x}$, is the same for $x = 0$ and $x = 1$. In information theory, the sender and receiver for a communication channel are modeled as random variables X and Y. The variable X gives the message that is sent, and Y gives the message that is received. In a binary symmetric channel, we have

$$\text{prob}(Y = 1 \mid X = 0) = \text{prob}(Y = 0 \mid X = 1) = \varepsilon,$$
$$\text{prob}(Y = 0 \mid X = 0) = \text{prob}(Y = 1 \mid X = 1) = 1 - \varepsilon.$$

We now analyze the error probability of the binary symmetric quantum channel from Alice to Bob. We assume that only errors that result from the measurement's randomness occur. In reality, the channel is not physically perfect and there are transmission errors that are not caused by the measurement. These errors increase the error rate.

Lemma 10.15. *The probabilities* $\text{prob}(Y = y \mid X = x)$ *that Bob receives bit* y *if Alice sends bit* x *are*

$$\text{prob}(Y = 0 \mid X = 0) = \text{prob}(Y = 1 \mid X = 1)$$
$$= \cos^2(\theta - \theta') - \frac{1}{2}(1 - \cos(\gamma' - \gamma)) \sin(2\theta) \sin(2\theta'),$$
$$\text{prob}(Y = 1 \mid X = 0) = \text{prob}(Y = 0 \mid X = 1)$$
$$= \sin^2(\theta - \theta') + \frac{1}{2}(1 - \cos(\gamma' - \gamma)) \sin(2\theta) \sin(2\theta').$$

Proof.

$$\text{prob}(Y = 0 \mid X = 0) = |\langle v_A | v_B \rangle|^2$$
$$= \text{prob}(Y = 1 \mid X = 1) = |\langle w_A | w_B \rangle|^2$$
$$= \left(\cos(\theta) \cos(\theta') + e^{+i(\gamma' - \gamma)} \sin(\theta) \sin(\theta') \right)$$
$$\cdot \left(\cos(\theta) \cos(\theta') + e^{-i(\gamma' - \gamma)} \sin(\theta) \sin(\theta') \right)$$
$$= \cos^2(\theta) \cos^2(\theta') + \sin^2(\theta) \sin^2(\theta')$$
$$+ 2 \cos(\gamma' - \gamma) \cos(\theta) \cos(\theta') \sin(\theta) \sin(\theta')$$
$$= (\cos(\theta) \cos(\theta') + \sin(\theta) \sin(\theta'))^2$$
$$- 2(1 - \cos(\gamma' - \gamma)) \cos(\theta) \cos(\theta') \sin(\theta) \sin(\theta')$$
$$= \cos^2(\theta - \theta') - \frac{1}{2}(1 - \cos(\gamma' - \gamma)) \sin(2\theta) \sin(2\theta').$$
$$\text{prob}(Y = 1 \mid X = 0) = |\langle v_A | w_B \rangle|^2$$
$$= \text{prob}(Y = 0 \mid X = 1) = |\langle w_A | v_B \rangle|^2$$
$$= 1 - \text{prob}(Y = 0 \mid X = 0)$$
$$= \sin^2(\theta - \theta') + \frac{1}{2}(1 - \cos(\gamma' - \gamma)) \sin(2\theta) \sin(2\theta').$$

$\square$

Definition 10.16. The orthonormal bases $(|v_A\rangle, |w_A\rangle)$ and $(|v_B\rangle, |w_B\rangle)$ are called *conjugate* if and only if

$$\text{prob}(Y = 0 \mid X = 0) = \text{prob}(Y = 1 \mid X = 0)$$
$$= \text{prob}(Y = 1 \mid X = 1) = \text{prob}(Y = 0 \mid X = 1) = \frac{1}{2}.$$

Examples:

1. Let Bob's measurement basis be the same as Alice's encoding basis. Bob's device is then able to distinguish deterministically between the two orthogonal polarization states used by Alice. Thus, Bob measures with certainty the bit that Alice has sent.
2. If Alice and Bob's bases are conjugate, then, instead of measuring, Bob could equally well toss a coin.
3. The rectilinear basis $\mathcal{B}_{0,0} = (|H\rangle, |V\rangle)$ and the basis $\mathcal{B}_{\gamma,\theta}$ are conjugate if and only if $\theta = (2k + 1)\pi/4, k \in \mathbb{Z}$.
4. The rectilinear basis $\mathcal{B}_{0,0} = (|H\rangle, |V\rangle)$ and the diagonal basis $\mathcal{B}_{0,\pi/4} = (|+\rangle, |-\rangle)$ are conjugate.
5. The rectilinear basis and the circular basis $\mathcal{B}_{\pi/2,\pi/4} = (|R\rangle, |L\rangle)$ are conjugate.
6. The circular basis and the diagonal basis are conjugate.
7. Let Alice encode by using one of the conjugate bases $(|H\rangle, |V\rangle)$ and $\mathcal{B}_{\gamma,\pi/4}$. The Breidbart basis $\mathcal{B}_{\gamma,\pi/8}$ (see page 347) is halfway between these bases. If Bob measures in this Breidbart basis, then he obtains with probability $\cos^2(\pi/8) = 0.854$ the bit x which Alice has sent, and with probability 0.146 the complementary bit $\bar{x}$, independent of Alice's choice of the encoding basis.

10.5.2 The BB84 Protocol

Quantum cryptography dates back to 1984. Bennett and Brassard had an idea for how to establish a shared secret key with the help of quantum mechanics. In their famous paper [BenBra84], they published the *BB84 protocol* for quantum key distribution (QKD for short). Other protocols have been invented since then. But the BB84 protocol is still an outstanding one. It is the protocol most often analyzed and implemented, and there are commercial implementations (see, for example, [ID Quantique] and [MagiQ]). The consecutive phases of key establishment – from quantum encoding to privacy amplification – and the methods applied are typical of QKD protocols. We therefore take BB84 as our example in this introduction.

Typically, photons are used to implement the quantum bits in the BB84 protocol.[12] The photons can be sent through the open air or an optical fiber.

---

[12] Producing single photons is a technological challenge. Alternatively, single-photon states can be approximated by *weak coherent states* with a low average number of photons. See, for example, [Assche06, Chapter 10] or [DusLütHen06, Section 5] for an overview of the technology.

A bit can be encoded in the polarization of a photon, as described above, or in the phase of a photon (which we do not discuss here). Polarization encoding is preferred for open-air communication, and phase encoding for fiber communication (see, e.g., [Assche06, Chapter 10] and [DusLütHen06, Section 4]).

In addition to the quantum channel for transmitting quantum bits, Alice and Bob need an auxiliary classical communication channel to exchange synchronization information. This communication channel is public and can be eavesdropped on – Alice and Bob lead a "public discussion". An adversary, Eve, with access to the channel could intentionally modify messages or even act as a (wo)man-in-the-middle to impersonate Alice or Bob. Therefore, the auxiliary channel must be authenticated. The origin of messages must be authenticated and the integrity of the messages protected. If we apply digital signatures (Sections 3.3.4 and 3.4.3) or classical message authentication codes (MACs) (Section 2.2.3) for this purpose, we cannot achieve unconditional security – both methods rely on computational-hardness assumptions. Therefore, unconditionally secure MACs are applied (see Section 10.3 above). We have to assume that Alice and Bob share an initial secret key k for authentication, a "bootstrap key". Using k, Alice and Bob can compute unconditionally secure authentication codes for the messages, which they exchange through the auxiliary channel during the first round of their QKD protocol. As soon as they have established a shared secret key K in a round of QKD, they can use a part of K to authenticate the messages in the next round. Quantum key distribution works as an "expander": some initial shared secret key is needed, and then this key can be expanded arbitrarily.

Except for the transmission of the quantum bits, Alice and Bob communicate through the public authenticated auxiliary channel. An eavesdropper can learn all of the information exchanged.

Beforehand, Alice and Bob agree on a pair of conjugate orthonormal bases $\mathcal{B}_1 = (|v_1\rangle, |w_1\rangle)$ and $\mathcal{B}_2 = (|v_2\rangle, |w_2\rangle)$ of the space $\mathcal{H}$ of quantum bits. They consider $|v_1\rangle$ and $|v_2\rangle$ as encodings of bit 0 and $|w_1\rangle$ and $|w_2\rangle$ as encodings of bit 1. They could agree, for example, on $\mathcal{B}_1 = (|H\rangle, |V\rangle)$, the rectilinear basis, and $\mathcal{B}_2 = (|+\rangle, |-\rangle)$, the diagonal basis, or they could take, as Bennett and Brassard did in their classic paper [BenBra84], the rectilinear and the circular basis.

Alice and Bob establish a shared secret key by executing the following steps:

1. Generation, encoding and transmission of qubits.
2. Sifting.
3. Estimation of the error rate.
4. Information reconciliation.
5. Privacy amplification.

Generation, Encoding and Transmission of Qubits. Alice generates a sequence of random bits $b_1, b_2, \ldots$. For each bit $b_i, i = 1, 2, \ldots$, she randomly selects one of the two bases, encodes b_i in this basis as a quantum bit and sends the resulting polarized photon to Bob. For $b_i = 0$, she sends $|v_1\rangle$ if basis $\mathcal{B}_1$ is selected, and $|v_2\rangle$ if basis $\mathcal{B}_2$ is selected. For $b_i = 1$, she sends $|w_1\rangle$ or $|w_2\rangle$.

For each bit, Bob chooses one of the two bases, randomly and independently from Alice. He measures the quantum bit that he receives in this basis and decodes it to bit b_i'.

If Bob and Alice have both chosen the same basis for b_i, then Bob's measurement is deterministic. Bob obtains, with certainty, the bit that Alice has sent: $b_i' = b_i$, provided there was no transmission error. If Bob and Alice have chosen different bases for b_i, then Bob's measurement result is random. With probability $1/2$, Bob obtains the complementary bit $\overline{b_i}$: $\mathrm{prob}(b_i' = b_i) = \mathrm{prob}(b_i' = \overline{b_i}) = 1/2$. He could equally well toss a coin.

Sifting. Alice and Bob exchange their choices of encoding and measurement bases through the public auxiliary channel. Of course, they do not send any information about the generated bits b_i or the results of the measurements. They discard all the key bits for which Alice's encoding basis is different from Bob's measurement basis. The remaining key bits constitute the *sifted key*. It is expected that half of the key bits will be discarded.

Estimation of the Error Rate. Alice and Bob compare a fraction of the sifted key through the public auxiliary channel and estimate the error rate (see Section 10.5.3 below for details). The compared bits become known to an eavesdropper and are also discarded.

Definition 10.17. The remaining key bits are called the *raw key*.

Alice and Bob need such an estimate of the error rate for the initialization of the information reconciliation procedure (Section 10.5.5 below). Later, during information reconciliation, they then learn the exact number of errors that occurred. Eavesdropping causes errors, and Alice and Bob can infer from the error rate that an eavesdropper is active. Of course, errors can also occur for many other reasons, and in a real physical system such errors always occur. Alice and Bob cannot distinguish them from errors caused by eavesdropping.[13] Therefore, if Alice and Bob were to abort and restart the protocol as soon as they detected errors and suspected eavesdropping, they would probably never end up with a shared secret key. Instead, Alice and Bob proceed with the following two steps, provided that the error rate is not too high. They apply sophisticated methods from information theory to distill a secret shared key from the raw key.

Information Reconciliation. Due to transmission errors, the raw keys of Alice and Bob still differ. They now iteratively reveal parity bits of subsets of their key elements. In this way, Bob can detect and correct transmission

---

[13] For example, there are dark counts in photon detectors, where a photon is counted even though there is no incoming photon.

errors. At the end of this phase, Alice and Bob share the same raw key, with overwhelmingly high probability. A detailed discussion of reconciliation will be given in Section 10.5.5.

Privacy Amplification. After the previous step, Alice and Bob share a raw key a of binary length n. Alice selects the bits of a randomly. Therefore, the selection of a is described by a uniformly distributed random variable A with values in $\{0,1\}^n$. Unfortunately, an eavesdropper Eve has some knowledge about A. Let V be the random variable that captures Eve's information about A. Eve obtains this information by eavesdropping on the quantum communication from Alice to Bob and by listening to the public auxiliary channel. Thus, $V = (E, S, P)$, where E is the key string that Eve obtains by intercepting and measuring quantum bits on their way from Alice to Bob,[14] S is the information about the encoding bases used by Alice and P consists of the parity bits exchanged for the reconciliation of the raw key.

As shown in Section 10.4, the Rényi entropy of the adversary Eve can be converted into a provably secure shared secret key by applying the powerful technique of *privacy amplification*. As a result of her eavesdropping, Eve has reduced her Rényi entropy about A from the initial value $H_2(A) = H(A) = n$.[15]

Let a, v be the particular values of the raw key and Eve's knowledge $A = a, V = v$. Eve's conditional Rényi entropy about A is $H_2(A | V = v)$. In Sections 10.5.4 and 10.5.5, we will see how Alice and Bob are able to figure out a lower bound for Eve's conditional Rényi entropy $H_2(A | V = v)$.

As described in Section 10.4, Alice and Bob select parameters $r, l \in \mathbb{N}$ with $r \leq l \leq H_2(A | V = v)$, apply a randomly selected function $g \in \mathcal{H}$ to the shared raw key a and obtain the final shared key $k = g(a)$, where $\mathcal{H}$ is a universal class of hash functions $\{0,1\}^n \longrightarrow \{0,1\}^r$. The generation of k is described by the random variable $K := G(A)$, where G is the random variable describing the choice of g.

Corollary 10.13 says that Eve's uncertainty about K is almost maximal, which means that the probability distribution of K is almost uniform in Eve's view:

$$H(K | G, V = v) \geq r - \log_2\left(1 + 2^{r-l}\right) \geq r - \frac{2^{-(l-r)}}{\ln(2)}.$$

10.5.3 Estimation of the Error Rate

Alice and Bob estimate the error rate ε by comparing m randomly selected bits from the sifted key. They calculate the empirical error rate $\overline{\varepsilon} = {}^r/_m$, where r is the number of bits which were not transmitted correctly to Bob. For the

---

[14] To simplify the discussion, we restrict our considerations to intercept-and-resend attacks; see Sections 10.5.4 and 10.5.7 below.

[15] Recall that the Rényi entropy equals the Shannon entropy for uniformly distributed random variables (Proposition 10.11).

comparison, the selected bits are sent through the public auxiliary channel. They are then known to Eve and must be discarded. Typical error rates ε are between 0.01 and 0.3. To get an idea of an adequate size for m, we assume that single bit errors occur independently and that the error probability is ε for all bits. Then the number of erroneous bits is distributed binomially with probability ε. The binomial distribution can be approximated by a Gaussian normal distribution with variance $\sigma = m\varepsilon(1 - \varepsilon)$ and expected value $m\varepsilon$, provided the size m of the sample is sufficiently large, say ≥ 0.2 KiB. We apply the theory of confidence intervals from statistics. If

$$z_{1-\alpha/2}\sqrt{\frac{\overline{\varepsilon}(1 - \overline{\varepsilon})}{m}} \leq \beta,$$

where z_θ denotes the θ-quantile of the standard normal distribution, or, equivalently,

$$m \geq \frac{z_{1-\alpha/2}^2 \overline{\varepsilon}(1 - \overline{\varepsilon})}{\beta^2},$$

then, with probability $1 - \alpha$, the true ε deviates from $\overline{\varepsilon}$ by at most β (see, for example, [DevBer12, Section 8.2, Eq. 8.11]). The condition is fulfilled for $\alpha = 0.05, \beta = 0.01, \overline{\varepsilon} \leq 0.12$ and $m \geq 0.5$ KiB, for example.

10.5.4 Intercept-and-Resend Attacks

To understand the effect of eavesdropping, we consider the *intercept-and-resend attack*, a simple and very intuitive attack. The adversary Eve intercepts photons on their way from Alice to Bob and measures them. According to her measurement result, she prepares a new photon and sends it to Bob. The intercept-and-resend attack belongs to the class of *individual attacks*: Eve considers each quantum bit separately.

There are other types of attacks, for example individual attacks where Eve clones the quantum bit. The cloning is necessarily imperfect, because cloning an unknown quantum state perfectly is impossible. Other classes of attacks are *collective attacks* and *coherent attacks* (see Section 10.5.7).

We now study the intercept-and-resend attack in detail. First, we consider the transmission of a single key bit. Since Eve measures each quantum bit separately and independently, it is then easy to extend our results to the transmission of a key string of arbitrary length.

Recall that Alice and Bob have agreed beforehand on two conjugate bases that they will use for the encoding of the bits. By applying a unitary coordinate transformation, we may assume that one of the bases is the rectilinear basis $(|H\rangle, |V\rangle)$. The rectilinear basis and the basis $\mathcal{B}_{\alpha,\varphi}$ are conjugate if and only if $\varphi = (2k + 1)\pi/4, k \in \mathbb{Z}$ (see the remarks after Definition 10.16 above). If $\mathcal{B}_{\alpha,\varphi} = (|v_A\rangle, |w_A\rangle)$, then $\mathcal{B}_{\alpha,\varphi+\pi/2} = (-|w_A\rangle, |v_A\rangle)$. Therefore, we may assume without loss of generality that the other basis is $\mathcal{B}_{\alpha,\pi/4}$. We call this basis the *conjugate basis*.

Transmission of a Single Key Bit. We have to study the following scenario ([BenBesBraSalSmo92]):

1. Alice generates a random bit b. Then she randomly selects an encoding basis, either the rectilinear or the conjugate one. She encodes b in this basis and sends the resulting quantum bit through the quantum channel.
2. Eve intercepts the quantum bit and measures it by using her measurement basis $\mathcal{B}_{\gamma,\theta}$. She resends the quantum bit by encoding her measurement result in the basis $\mathcal{B}_{\gamma',\theta'}$ and sending it to Bob.[16] Eve's re-encoding basis is not necessarily the same as her measurement basis.
3. Bob measures the incoming quantum bit in the same basis that Alice used for the encoding. He obtains a bit b'.

We have to consider only bits where Alice and Bob use the same basis, since all other key bits are discarded in the sifting phase. Eve's measurement result is probabilistic. Eve measures the correct value of Alice's bit b only with some probability. We may therefore model eavesdropping as a communication channel from Alice to Eve, where errors occur with some probability. Eve's measurement changes the state of the quantum bit and in this way causes errors in the quantum channel from Alice to Bob.

In the following, we analyze the error probabilities of both channels. As before, let A, B, E be the binary random variables which give the key bit that Alice generates, the bit that Bob receives and the bit that Eve measures, respectively. Let S be the random variable that captures Alice's choice of the encoding basis. From Lemma 10.15 we obtain the following result immediately.

Lemma 10.18. *Assume that the channel from Alice to Eve is physically perfect, i.e., all errors that occur result from the measurement's randomness. Then, the error probability of the quantum communication channel from Alice to Eve depends on Alice's choice of the encoding basis:*

$$\text{prob}(E = \overline{b} \,|\, A = b, S = \text{rectilinear}) = \sin^2(\theta),$$

$$\text{prob}(E = \overline{b} \,|\, A = b, S = \text{conjugate}) = \sin^2\left(\theta - \frac{\pi}{4}\right)$$
$$+ \frac{1}{2}(1 - \cos(\gamma - \alpha))\sin(2\theta).$$

Here, $b \in \{0, 1\}$ denotes a bit and $\overline{b}$ its complementary bit $1 - b$.

If Eve measures in the Breidbart basis $\mathcal{B}_{\alpha,\pi/8}$, the error rate is independent of the encoding basis:

$$\text{prob}(E = \overline{b} \,|\, A = b, S = \text{rectilinear}) = \text{prob}(E = \overline{b} \,|\, A = b, S = \text{conjugate})$$

---

[16] We do not assume in our discussion that Eve uses the same bases for all bits, she may change her measurement and re-encoding bases for every bit.

$$= \operatorname{prob}(E = \bar{b} \mid A = b) = \sin^2\left(\frac{\pi}{8}\right) = 0.146.$$

Remark. The error probability for $S = $ conjugate can also be expressed as

$$\frac{1}{2}(1 - \cos(\gamma - \alpha)\sin(2\theta)),$$

because

$$\sin^2\left(\theta - \frac{\pi}{4}\right) + \frac{1}{2}\left(1 - \cos\left(\gamma - \alpha\right)\right)\sin\left(2\theta\right) - \frac{1}{2}\left(1 - \cos\left(\gamma - \alpha\right)\sin\left(2\theta\right)\right)$$

$$= \frac{1}{2}\left(2\sin^2\left(\theta - \frac{\pi}{4}\right) - 1 + \sin\left(2\theta\right)\right)$$

$$= \frac{1}{2}\left(\sin^2\left(\theta - \frac{\pi}{4}\right) - \cos^2\left(\theta - \frac{\pi}{4}\right) + \sin\left(2\theta\right)\right)$$

$$= \frac{1}{2}\left(-\cos\left(2\theta - \frac{\pi}{2}\right) + \sin\left(2\theta\right)\right) = 0.$$

We now analyze the error rate of the channel from Alice to Bob.

Proposition 10.19. *Assume that the channel from Alice to Bob is physically perfect, i.e., all errors that occur are caused by Eve's measurement and re-encoding. Then, the error probability $\varepsilon = \operatorname{prob}(B = \bar{b} \mid A = b)$[17] of the quantum communication channel from Alice to Bob is bounded as follows:*

$$\frac{1}{4} \leq \varepsilon \leq \frac{3}{4}.$$

If Eve uses the same basis $B_{\gamma,\theta}$ for measurement and re-encoding and $\gamma = \alpha$ or $\gamma = \alpha + \pi$, then $\varepsilon = 1/4$.

For a proof, see Exercise 7.

Remark. Bob can turn an error probability $\varepsilon \geq 1/2$ into an error probability $1 - \varepsilon$, which is $\leq 1/2$. He needs only to switch the role of 0 and 1 during decoding. Therefore, we can summarize Proposition 10.19 as follows: Bob's error rate is between $1/4$ and $1/2$.

By her measurement, Eve gains information about A, which reduces her uncertainty and her Rényi entropy for A. We are interested in the amount of this reduction. By listening to the public auxiliary channel, Eve learns during the sifting phase which encoding basis Alice selected. Therefore, E and S together describe the complete knowledge of Eve about A, before information reconciliation starts.

Alice generates the key bit uniformly. Hence, A is a uniformly distributed random bit, and the Shannon and Rényi entropies of A are maximal: $H(A) = H_2(A) = 1$ (Propositions B.36 and 10.11).

The communication channels from Alice to Eve and from Alice to Bob are binary symmetric channels (see page 348). We define the *binary entropy functions* h and h_2.

---

[17] The probability is also taken over Alice's random choice of the encoding basis.

Definition 10.20. Let $0 \leq \varepsilon \leq 1$. Then

$$h(\varepsilon) := -\varepsilon \log_2(\varepsilon) - (1-\varepsilon) \log_2(1-\varepsilon) \text{ for } 0 < \varepsilon < 1, \quad h(0) := h(1) := 0,$$

and

$$h_2(\varepsilon) := -\log_2(\varepsilon^2 + (1-\varepsilon)^2) \text{ for } 0 < \varepsilon < 1, \quad h_2(0) := h_2(1) := 0.$$

We will apply the following general result on the entropies of binary symmetric channels.

Proposition 10.21. *Consider a binary symmetric channel with sender X and receiver Y and an error probability ε. Let $x, y \in \{0,1\}$.*

1. *If X is uniformly distributed, then Y is also uniformly distributed and $\mathrm{prob}(X = x \,|\, Y = y) = \mathrm{prob}(Y = y \,|\, X = x)$.*
2. *$\mathrm{H}(Y|X) = \mathrm{H}(Y|X = x) = h(\varepsilon)$.*
3. *If X is uniformly distributed, then $\mathrm{H}(X|Y = y) = \mathrm{H}(Y|X = x) = h(\varepsilon)$ and $\mathrm{H}(X|Y) = \mathrm{H}(Y|X) = h(\varepsilon)$.*
4. *$\mathrm{H}_2(Y|X) = \mathrm{H}_2(Y|X = x) = h_2(\varepsilon)$.*
5. *If X is uniformly distributed, then $\mathrm{H}_2(X|Y = y) = \mathrm{H}_2(Y|X = x) = h_2(\varepsilon)$ and $\mathrm{H}_2(X|Y) = \mathrm{H}_2(Y|X) = h_2(\varepsilon)$.*
6. *We have $h(\varepsilon) = h(1-\varepsilon)$ and $h_2(\varepsilon) = h_2(1-\varepsilon)$. The two entropies $h(\varepsilon)$ and $h_2(\varepsilon)$ are strictly monotonically increasing for $0 \leq \varepsilon \leq 1/2$ and strictly monotonically decreasing for $1/2 \leq \varepsilon \leq 1$. The minimum value is 0, $h(0) = h_2(0) = h(1) = h_2(1) = 0$, and the maximum value is 1, $h(1/2) = h_2(1/2) = 1$.*

Proof. The statements can be proven by straightforward computations. See Exercise 4. □

Proposition 10.22. *Let Eve measure in the basis $\mathcal{B}_{\gamma,\theta}$. Assume that the channel from Alice to Eve is physically perfect, i.e., the only errors that occur result from the measurement's randomness. Let e be the value measured by Eve. Then Eve's conditional entropies are*

$$\mathrm{H}(A \,|\, E = e, S = \text{rectilinear}) = h(\sin^2(\theta)),$$
$$\mathrm{H}(A \,|\, E = e, S = \text{conjugate}) \geq h(\sin^2(\theta - \pi/4)),$$
$$\mathrm{H}_2(A \,|\, E = e, S = \text{rectilinear}) = h_2(\sin^2(\theta)),$$
$$\mathrm{H}_2(A \,|\, E = e, S = \text{conjugate}) \geq h_2(\sin^2(\theta - \pi/4)).$$

The inequalities are equalities if $\gamma = \alpha$ or $\gamma = \alpha + \pi$.

Recall that α is the phase parameter of the conjugate basis $\mathcal{B}_{\alpha,\pi/4}$.

Proof. We apply Proposition 10.21 to A and E conditional on Alice's choice $S = s$ of the encoding basis, i.e., $X = A\,|\,(S = s)$ and $Y = E\,|\,(S = s)$. Since Alice chooses the key bit and the encoding basis independently, the distribution of $A\,|\,(S = s)$ is the same as the distribution of A: $A = A\,|\,(S = s)$. Since A is uniformly distributed, the conditional entropies $H(A\,|\,E = e, S = s)$ and $H_2(A\,|\,E = e, S = s)$ are equal to $h(\varepsilon)$ and $h_2(\varepsilon)$, respectively, where ε is the error rate of the channel from Alice to Eve (Proposition 10.21, items 3 and 5). If Alice uses the rectilinear basis ($S = $ rectilinear), the error probability is $\sin^2(\theta)$ (Lemma 10.18) and the rectilinear case follows. Now, assume that Alice uses the conjugate basis ($S = $ conjugate). Then, the error probability is $\varepsilon = \varepsilon(\delta, \theta) = \sin^2(\theta - \pi/4) + \frac{1}{2}(1 - \cos(\delta))\sin(2\theta)$, with $\delta := \gamma - \alpha$ (Lemma 10.18). Let $H(\delta, \theta) := h(\varepsilon(\delta, \theta))$ and $H_2(\delta, \theta) := h_2(\varepsilon(\delta, \theta))$. To determine the minimum values, we consider the derivatives

$$\frac{\partial \varepsilon}{\partial \delta} = \frac{1}{2}\sin(\delta)\sin(2\theta),$$

$$\frac{\partial H}{\partial \delta} = \frac{1}{2}(-\log_2(\varepsilon) + \log_2(1 - \varepsilon))\sin(\delta)\sin(2\theta),$$

$$\frac{\partial H_2}{\partial \delta} = -\frac{1}{\ln(2)} \cdot \frac{2\varepsilon - 1}{\varepsilon^2 + (1 - \varepsilon)^2}\sin(\delta)\sin(2\theta).$$

Each of the derivatives $\partial H/\partial \delta$ and $\partial H_2/\partial \delta$ is 0 if and only if

(a) $\theta = k \cdot \pi/2, k \in \mathbb{Z}$, or
(b) $\varepsilon = 1/2$, or
(c) $\delta = k\pi, k \in \mathbb{Z}$.

In cases (a) and (b), the error probability is $\varepsilon = 1/2$, and hence $H = 1$ and $H_2 = 1$ attain their maximum value 1. In case (c), the error rate is $\varepsilon_1 = \sin^2(\theta - \pi/4)$ if k is even and $\varepsilon_2 = \sin^2(\theta - \pi/4) + \sin(2\theta)$ if k is odd. We have

$$(1 - \varepsilon_1) - \varepsilon_2 = \cos^2(\theta - \pi/4) - \sin^2(\theta - \pi/4) - \sin(2\theta)$$

$$= \cos(2\theta - \pi/2) - \sin(2\theta) = 0,$$

and hence $h(\varepsilon_1) = h(1 - \varepsilon_1) = h(\varepsilon_2)$ and $h_2(\varepsilon_1) = h_2(1 - \varepsilon_1) = h_2(\varepsilon_2)$. We see that in case (c), H and H_2 attain the values $h(\sin^2(\theta - \pi/4))$ and $h_2(\sin^2(\theta - \pi/4))$, respectively. This value is the minimum value – there are no more zeros of the derivatives left. Note that H and H_2 are 2π-periodic smooth functions of δ with values between 0 and 1. Therefore there is a minimum. $\qquad\square$

Remarks:

1. Eve knows her measurement result e and, after the sifting, the encoding base s which Alice has used. Therefore, we have to consider the entropies conditional on the concrete values e and s.

2. In reality, the channel is not physically perfect and there are errors which are not caused by the measurement. These errors increase the error rate and hence also Eve's conditional entropies. Proposition 10.22 gives lower bounds for Eve's entropies in reality also.

3. If Eve uses the Breidbart basis ($\gamma = \alpha, \theta = \pi/8$) , then her entropies are independent of Alice's choice of the encoding basis:

$$H(A|E = e, S = s) = H(A|E = e) = h(\sin^2(\pi/8)) = 0.601,$$

$$H_2(A|E = e, S = s) = H_2(A|E = e) = h_2(\sin^2(\pi/8)) = 0.415$$

(see Lemma 10.18). If Eve's measurement basis mb is the rectilinear or the conjugate basis ($\gamma = \theta = 0$ or $\gamma = \alpha, \theta = \pi/4$), then she measures the correct value of A with certainty if Alice used the same basis, and a completely random value if Alice used the other basis, i.e.,

$$H(A|E = e, S = mb) = H_2(A|E = e, S = mb) = 0,$$

$$H(A|E = e, S \neq mb) = H_2(A|E = e, S \neq mb) = 1.$$

4. Eve is interested in using the measurement basis which leads to the maximum reduction in her uncertainties. At the time when she selects the measurement basis, she does not know the encoding basis, nor the result of her measurement. Therefore, she can only consider the expected values

$$H(A|E, S) \geq \frac{1}{2}\left(h\left(\sin^2(\theta)\right) + h\left(\sin^2\left(\theta - \frac{\pi}{4}\right)\right)\right),$$

$$H_2(A|E, S) \geq \frac{1}{2}\left(h_2\left(\sin^2(\theta)\right) + h_2\left(\sin^2\left(\theta - \frac{\pi}{4}\right)\right)\right).$$

For each θ, her entropies are minimum if $\gamma = \alpha$. Then equality holds. Assume $\gamma = \alpha$. Eve's expected Rényi entropy $H_2(A|E, S)$ is strictly decreasing between the maximum of 0.5 at $\theta = 0$ and the minimum of 0.415 at $\theta = \pi/8$ (Breidbart basis). In contrast, her expected Shannon entropy is strictly increasing between its minimum of 0.5 at $\theta = 0$ and its maximum of 0.601 at $\theta = \pi/8$. See Lemma 10.23 below.

Except when Eve measures in the rectilinear or the conjugate basis ($\theta = 0$ or $\theta = \pi/4$), her reduction in Rényi entropy exceeds her reduction in Shannon entropy, i.e., her gain of Rényi information is larger than her gain of Shannon information. If she measures in the rectilinear ·or the conjugate basis, the gains are equal.

Lemma 10.23. *Let $f, g : \mathbb{R} \longrightarrow \mathbb{R}$ be defined by*

$$f(\theta) := \frac{1}{2}\left(h\left(\sin^2(\theta)\right) + h\left(\sin^2\left(\theta - \frac{\pi}{4}\right)\right)\right) \ and$$

$$g(\theta) := \frac{1}{2}\left(h_2\left(\sin^2(\theta)\right) + h_2\left(\sin^2\left(\theta - \frac{\pi}{4}\right)\right)\right).$$

1. f, g are periodic with period $\pi/4$.
2. f is strictly increasing between 0 and $\pi/8$ and strictly decreasing between $\pi/8$ and $\pi/4$. The minimum value is $f(0) = \frac{1}{2}h(0.5) = 0.5$ and the maximum value is $f(\pi/8) = h(\sin^2(\pi/8)) = 0.601$.
3. g is strictly decreasing between 0 and $\pi/8$ and strictly increasing between $\pi/8$ and $\pi/4$. The maximum value is $g(0) = \frac{1}{2}h_2(0.5) = 0.5$ and the minimum value is $g(\pi/8) = h_2(\sin^2(\pi/8)) = 0.415$.

Proof. The first statement is obvious. The other statements can be verified by computing the derivatives of f and g. □

Transmission of a Key String. Up to now, we have studied the transmission of one bit. Since each key bit is handled separately and independently by all participants if we assume an intercept-and-resend attack, we can easily extend our results to arbitrary key lengths. Let A and B now be the raw keys of Alice and Bob, after the estimation of the error rate and before information reconciliation starts. Let E and S be the random variables that capture the bits measured by Eve and Alice's choices of the encoding basis. Let n be the length of the raw key.

Proposition 10.24. *Let Eve execute an intercept-and-resend attack. Assume that Eve intercepts $m \leq n$ bits of the raw key. Let e be the bit values measured by Eve and let s be Alice's choices of the encoding basis. Let $p \in [0,1]$. Then, with probability $1 - p$,*

$$\mathrm{H}(A \,|\, E = e, S = s) \geq n - m \cdot 0.5 - \frac{\sqrt{m}}{2} \cdot z_{1-p/2},$$

$$\mathrm{H}_2(A \,|\, E = e, S = s) \geq n - m \cdot 0.585 - \frac{\sqrt{m}}{2} \cdot z_{1-p/2}.$$

Here, z_β denotes the β-quantile of the standard Gaussian normal distribution.

Proof. Assume that Alice has encoded k of the bits which Eve measures by using the rectilinear basis and $l := m - k$ bits by using the conjugate basis. Alice chooses the encoding basis randomly. Therefore, the number k is binomially distributed with variance $\sigma^2 = m/4$ and expected value $\mu = m/2$. Since m is large, we can approximate the binomial distribution by the Gaussian normal distribution and find that, with probability $1 - p$,

$$\left| k - \frac{m}{2} \right| \leq \sigma \cdot z_{1-p/2} = \frac{\sqrt{m}}{2} \cdot z_{1-p/2}.$$

The entropies are the sums of the entropies for a single bit (Propositions B.39 and 10.10). We note that $l - m/2 = -(k - m/2)$ and $0 \leq |h(\sin^2(\alpha)) - h(\sin^2(\alpha - \pi/4))| \leq 1$. By applying Proposition 10.22 and Lemma 10.23, we obtain the result that, with probability $1 - p$,

$$H(A \mid E = e, S = s)$$

$$\geq k \cdot h\left(\sin^2(\theta)\right) + l \cdot h\left(\sin^2\left(\theta - \frac{\pi}{4}\right)\right) + (n - m)$$

$$= \frac{m}{2}\left(h\left(\sin^2(\theta)\right) + h\left(\sin^2\left(\theta - \frac{\pi}{4}\right)\right)\right)$$

$$\qquad + \left(k - \frac{m}{2}\right) h\left(\sin^2(\theta)\right) + \left(l - \frac{m}{2}\right) h\left(\sin^2\left(\theta - \frac{\pi}{4}\right)\right) + (n - m)$$

$$= \frac{m}{2}\left(h\left(\sin^2(\theta)\right) + h\left(\sin^2\left(\theta - \frac{\pi}{4}\right)\right)\right)$$

$$\qquad + \left(k - \frac{m}{2}\right)\left(h\left(\sin^2(\theta)\right) - h\left(\sin^2\left(\theta - \frac{\pi}{4}\right)\right)\right) + (n - m)$$

$$\geq \frac{m}{2}\left(h\left(\sin^2(\theta)\right) + h\left(\sin^2\left(\theta - \frac{\pi}{4}\right)\right)\right) - \left|k - \frac{m}{2}\right| + (n - m)$$

$$\geq \frac{m}{2}\left(h\left(\sin^2(\theta)\right) + h\left(\sin^2\left(\theta - \frac{\pi}{4}\right)\right)\right) - \frac{\sqrt{m}}{2} \cdot z_{1-p/2} + (n - m)$$

$$\geq m \cdot 0.5 - \frac{\sqrt{m}}{2} \cdot z_{1-p/2} + (n - m) = n - m \cdot 0.5 - \frac{\sqrt{m}}{2} \cdot z_{1-p/2}$$

and

$$H_2(A \mid E = e, S = s)$$

$$\geq k \cdot h_2\left(\sin^2(\theta)\right) + l \cdot h_2\left(\sin^2\left(\theta - \frac{\pi}{4}\right)\right) + (n - m)$$

$$\geq \frac{m}{2}\left(h_2\left(\sin^2(\theta)\right) + h_2\left(\sin^2\left(\theta - \frac{\pi}{4}\right)\right)\right) - \frac{\sqrt{m}}{2} \cdot z_{1-p/2} + (n - m)$$

$$\geq m \cdot 0.415 - \frac{\sqrt{m}}{2} \cdot z_{1-p/2} + (n - m) = n - m \cdot 0.585 - \frac{\sqrt{m}}{2} \cdot z_{1-p/2}.$$

$$\square$$

Eavesdropping necessarily causes errors and therefore does not go unnoticed. We now generalize the single-bit error rate that Eve causes in the channel from Alice to Bob to arbitrary key lengths.

Proposition 10.25. *Let $n \geq 1$ be the length of Alice and Bob's raw key. Let Eve execute an intercept-and-resend attack. Assume that Eve intercepts a fraction $\eta, 0 \leq \eta \leq 1$, of all transmitted bits. Then the expected number of erroneous bits in Bob's raw key B is $\geq n \cdot \eta/4$.*

Proposition 10.25 is an immediate consequence of Proposition 10.19.

From the empirical error rate, Alice and Bob can estimate the number of intercepted bits.

Corollary 10.26. *Let $n \geq 1$ be the length of Alice and Bob's raw key, and let Eve execute an intercept-and-resend attack. Assume that the number of erroneous bits in Bob's raw key B is r. Let $\bar{\varepsilon} := r/n$ be the empirical error rate. Then, with probability $1 - p$, $0 < p < 1$, the fraction η of bits that Eve intercepts is*

$$\eta \leq 4\bar{\varepsilon} + 2 \cdot \frac{z_{1-p}}{\sqrt{n}},$$

where z_{1-p} denotes the $(1-p)$-quantile of the standard Gaussian normal distribution.

Proof. We may assume that all errors are caused by Eve's measurements and that $\eta > 0$. Let X_i be the binary random variable with $X_i = 1$ if bit i is erroneous, and $X_i = 0$ otherwise. Then $X := \sum_{i=1}^{n} X_i$ counts the number of errors. The central limit theorem of probability theory (see, for example, [Billingsley12, Theorem 27.3]) says that X/n is normally distributed for large n.[18] The variance σ^2 of X/n is

$$\sigma^2 = \frac{1}{n^2} \sum_{i=1}^{n} \sigma_i^2,$$

where $\sigma_i^2 = \varepsilon_i(1 - \varepsilon_i)$ is the variance at position i, with ε_i being the error rate at position i. By Proposition 10.19, $3/16 \leq \sigma_i^2 \leq 1/4$ if Eve measures at position i, and $\sigma_i^2 = 0$ otherwise. Hence $\sigma^2 \leq \eta/4n$, i.e., $\sigma \leq \sqrt{\eta}/2\sqrt{n}$. With probability $1 - p$,

$$\varepsilon - \bar{\varepsilon} \leq \sigma \cdot z_{1-p},$$

where ε is the expected value of $\bar{\varepsilon}$. The expected value ε is $\geq \eta/4$ by Proposition 10.25; hence, with probability $1 - p$,

$$\eta \leq 4 \cdot \varepsilon = 4 \cdot \bar{\varepsilon} + 4 \cdot (\varepsilon - \bar{\varepsilon}) \leq 4 \cdot \bar{\varepsilon} + 4\sigma z_{1-p} \leq 4 \cdot \bar{\varepsilon} + 2 \cdot \frac{z_{1-p}}{\sqrt{n}}.$$

$\square$

10.5.5 Information Reconciliation

The quantum channel between Alice and Bob is not physically perfect. Noise is introduced into the messages, and the eavesdropping causes errors. Therefore, the raw key A that Alice sends and the raw key B that Bob receives are not identical. By exchanging messages, Alice and Bob want to reconcile their keys: Bob is able to recover the raw key A from his version B and the additional information P that he obtains from Alice.

Information reconciliation can be either one-way or interactive. Alice and Bob communicate through the public auxiliary classical channel. All information that they exchange can become known to an adversary Eve.

As an example, we sketch the interactive protocol *Cascade*, which is the preferred de facto standard for QKD implementations. A comprehensive introduction to reconciliation protocols can be found, for example, in [Assche06, Chapter 8].

---

[18] We do not assume that the X_i are equally distributed. The central limit theorem is applicable because the Lyapunov condition is fulfilled, which can easily be checked.

The Reconciliation Protocol Cascade. Cascade ([BraSal93]) is an enhancement of the error correction protocol which was originally designed by Bennett et al. ([BenBesBraSalSmo92]). Errors in physical transmission media are typically not uniformly distributed. For example, they often exhibit burst structures, i.e., a sequence of consecutive errors is more likely to occur. A popular method to randomize the location of errors is that Alice and Bob first agree on a random permutation and re-order their bit strings. This strategy is also applied in Cascade.

Cascade is an interactive protocol and works in several passes. Typically, there are four passes. When the protocol is finished, the raw keys A of Alice and B of Bob are identical, with overwhelming probability.

In each pass, Alice and Bob agree on a random permutation and re-order their bit strings. Then, they subdivide their strings into blocks of equal length w, Alice sends the parity bits of these subblocks to Bob, and Bob compares these with the parity bits of his subblocks. Since the block length is increased and the bit positions are randomly permuted, the subdivision changes from pass to pass. When Bob's parity of a subblock differs from Alice's parity, at least one erroneous bit is contained in Bob's block. The block is then bisected. Alice sends the parity bit of one half, Bob compares the parity bits of the halves, the half with different parity is bisected again and so on. By this binary search, Alice and Bob identify an erroneous bit and correct it – Bob switches his bit. Alice and Bob keep track of the parity bits of all blocks investigated, including the blocks that are generated by the bisections. A block is listed either in the set $\mathcal{B}_{eq}$ of blocks with matching parity or in the set $\mathcal{B}_{diff}$ of blocks with different parity. Until $\mathcal{B}_{diff}$ is empty, Alice and Bob take a block of minimum length from $\mathcal{B}_{diff}$ and correct an erroneous bit by applying the bisection procedure. The sets $\mathcal{B}_{eq}$ and $\mathcal{B}_{diff}$ are updated – blocks which contain the switched bit are moved to the other set, and the newly generated blocks are added. The next pass is started after all known divergent parities have been corrected, i.e., when $\mathcal{B}_{diff}$ becomes empty. If a bit b is switched in the second or a later pass, then matching-parity blocks from previous passes containing b become non-matching, which immediately leads to the correction of at least one more erroneous bit, b'.

The subblock size w depends on the iteration. The initial block size is chosen such that about one error can be expected in each block. Therefore, the error rate ε of the channel must be estimated beforehand (see Section 10.5.2 above). In the original Cascade ([BraSal93]), the block size w_1 in the first pass was $w_1 \approx 0.73/\varepsilon$ (see [Assche06, Section 8.3.2]) and $w_i = 2w_{i-1}$.

If only a small number of errors are left, the block parity disclosure procedure becomes ineffective. A lot of parity bits are disclosed and only a few errors can be detected. At this point, Alice and Bob switch to the BICONF strategy ([BraSal93]), in which they repeatedly compare the parity of a randomly selected subset of their entire bit strings. Upon finding a parity mismatch, they correct an error by applying a bisective search. Each comparison indicates a difference between the strings with probability $1/2$ (see Corollary 10.28 below). Hence, if the parities match in l consecutive comparisons, then Alice and Bob's strings are identical with probability $1 - 2^{-l}$, which is overwhelmingly high for, say, $l \geq 50$.

Remarks:

1. A crucial parameter for the efficiency of Cascade is the initial block size. The ideal case is that exactly one error is contained in each block. Then all errors would be removed in the first pass. Procedures for adapting the block size to the error rates actually observed were developed in [RasKol10, Sections 4.3 and 4.5]. Instead of estimating the error rate, the initial block size can also be derived by using knowledge about error distributions in finished protocol runs (see [RasKol10, Sections 4.3 and 4.4]).

2. The number of parity bits revealed by Cascade is roughly $(1.1+\varepsilon)\cdot h(\varepsilon)\cdot n$, where n is the length of the raw key (see [Assche06, Section 8.3.2]). Of course, the exact number of parity bits exchanged is known at the end of information reconciliation.

3. Alice and Bob identify all transmission errors. Therefore, they learn the exact number of errors and the exact empirical error rate $\bar{\varepsilon}$.

4. When speaking of the error rate, we tacitly assume that the quantum channel from Alice to Bob is (the n-th extension of) a binary symmetric channel with a constant error probability. This is an adequate model in most cases. Namely, we assume an individual attack, which means that errors at the various bit positions occur independently. In Cascade, the bits are re-ordered randomly, which means that the positions of the intercepted bits are random. However, if Eve changes her strategy and changes her bases for measuring and re-encoding, the error probability may vary. If this happens, the security of the QKD protocol is not compromised: we are not dependent on estimates of the error rate and estimates of the number of parity bits exchanged. After information reconciliation, the exact numbers of parity bits and errors are known.

5. If the error probability $\varepsilon \geq 0.25$, then Cascade discloses more than n bits and hence potentially the complete raw key A.

6. Cascade was developed further in [Nguyen02] and [RasKol10].

7. There are other binary interactive reconciliation protocols, which combine the ideas of [BenBesBraSalSmo92] and Cascade and the use of error-correcting codes (see, e.g., [FurYam01], [BuLaToNiDoPe03]).

8. In one-way reconciliation (see [Assche06] for an introduction), Alice supplements her bit string with error-correcting codes before sending it to Bob, and Bob is then able to correct the errors.

Lemma 10.27. *Let M be a finite non-empty set. Then exactly one half of the subsets of M contain an even number of elements and exactly one half of the subsets of M contain an odd number of elements.*

Proof. We use induction on $m := |M|$. If $m = 1$, then there is one subset with 0 elements and one subset with 1 element. Assume that the assertion is true for m, and let $|M| = m + 1$. Let $x \in M$. The map $N \mapsto N \cup \{x\}$ is a bijection between the subsets of $M \setminus \{x\}$ and the subsets of M which contain x. Sets with even and odd numbers of elements are mapped to sets with odd and even numbers of elements, respectively. $\qquad\square$

Corollary 10.28. *Let* $a = a_1 a_2 \ldots a_n, b = b_1 b_2 \ldots b_n \in \{0,1\}^n, a \neq b,$ *be distinct bit strings of length* n. *Then the parities* $\sum_{i \in J} a_i \bmod 2$ *of* a *and* $\sum_{i \in J} b_i \bmod 2$ *of* b *disagree with probability* $1/2$ *if* J *is a randomly chosen subset of* $\{1, 2, \ldots, n\}$.

Proof. Let $I := \{i \in \{1, 2, \ldots, n\} \mid a_i \neq b_i\}$. By Lemma 10.27, the probability that $I \cap J$ contains an even number of elements is $1/2$. $\qquad\square$

Reduction of Rényi Entropy. As before, let A and B be Alice and Bob's versions of the raw key, and let P be the additional information exchanged over the public channel. In Cascade, P consists of the parity bits that Alice sends to Bob. A, B and P are random variables. Since Alice chooses the key bits uniformly at random, A is uniformly distributed with values in $\{0,1\}^n$, where n denotes the length of the raw key.

There is a lower bound on the amount of information that is needed for the reconciliation.

Proposition 10.29. *A reconciliation protocol reveals at least* $H(A \mid B)$ *bits of information about the key* A.

Proof. Using P, Bob is able to fully reconstruct A from B. This implies that $H(A \mid PB) = 0$. Then

$$H(P) \geq H(P \mid B) \geq H(P \mid B) - H(P \mid AB) = I(P; A \mid B)$$

$$= I(A; P \mid B) = H(A \mid B) - H(A \mid PB) = H(A \mid B).$$

$\qquad\square$

Corollary 10.30. *Assume that the channel from Alice to Bob is a binary symmetric channel with error probability* ε. *Then, a reconciliation protocol reveals at least* $n \cdot h(\varepsilon)$ *bits of information about the raw key* A.

Remark. We see that the number of bits disclosed in Cascade is close to the theoretical lower bound.

The error-correcting information P is sent through the public auxiliary channel and becomes known to Eve. It reduces Eve's Rényi entropy about the raw key A. To estimate the amount of reduction, we can apply the following upper bound, which was proven by Cachin in his dissertation ([Cachin97, Theorem 5.2]).

Theorem 10.31 *(Reduction of Rényi entropy). Let X and U be jointly distributed random variables with values in $\mathcal{X}$ and $\mathcal{U}$, respectively. Let $\kappa > 0$ be a security parameter. Then, with probability $\geq 1 - 2^{-\kappa}$, U takes on a value u such that*

$$H_2(X) - H_2(X \mid U = u) \leq \log_2(|\mathcal{U}|) + 2\kappa + 2.$$

Proof. Our proof follows the proof at [Cachin97, Theorem 4.17].

$$H_2(XU) = -\log_2 \left(\sum_{x \in \mathcal{X}, u \in \mathcal{U}} \mathrm{prob}(X = x, U = u)^2 \right)$$

$$= -\log_2 \left(\sum_{u \in \mathcal{U}} \mathrm{prob}(U = u)^2 \sum_{x \in \mathcal{X}} \mathrm{prob}(X = x \mid U = u)^2 \right)$$

$$= -\log_2 \left(\sum_{u \in \mathcal{U}} \mathrm{prob}(U = u) \cdot 2^{\log_2(\mathrm{prob}(U=u)) - H_2(X \mid U=u)} \right).$$

Let $\beta(u) := H_2(X \mid U = u)$ and $\mathrm{prob}_U(u) := \mathrm{prob}(U = u)$, and consider the random variables $\mathrm{prob}_U(U)$ and $\beta(U)$. Exponentiating both sides of the equation, we get

$$2^{-H_2(XU)} = E \left(2^{\log_2(\mathrm{prob}_U(U)) - \beta(U)} \right)$$

and then, for every $r > 0$,

$$2^{-r} = E \left(2^{\log_2(\mathrm{prob}_U(U)) - \beta(U) + H_2(XU) - r} \right).$$

From the inequality $\mathrm{prob}(Z \geq r) \leq E(2^{Z-r})$ (Corollary B.19), we get

$$\mathrm{prob} \left(\log_2(\mathrm{prob}_U(U)) - \beta(U) + H_2(XU) \geq r \right) \leq 2^{-r}.$$

This means that with probability $\geq 1 - 2^{-r}$, U takes on a value u such that

$$H_2(X \mid U = u) \geq H_2(XU) + \log_2(\mathrm{prob}(U = u)) - r. \qquad (1)$$

For every $t > 0$, the probability that U takes on a value u such that

$$\mathrm{prob}(U = u) < \frac{2^{-t}}{|\mathcal{U}|}$$

is $\sum_{u:\mathrm{prob}(U=u)<2^{-t}/|\mathcal{U}|} \mathrm{prob}(U = u) < 2^{-t}$. Hence, with probability $\geq 1 - 2^{-t}$, U takes on a value u such that

$$\mathrm{prob}(U = u) \geq \frac{2^{-t}}{|\mathcal{U}|}, \quad \text{i.e., } \log_2(\mathrm{prob}(U = u)) \geq -t - \log_2(|\mathcal{U}|). \qquad (2)$$

Then both of the inequalities (1) and (2) are valid with probability $\geq 1 - 2^{-r} - 2^{-t}$. Summarizing, we find that with probability $\geq 1 - 2^{-r} - 2^{-t}$, U takes on a value u such that

$$H_2(X\,|\,U = u) \geq H_2(XU) - \log_2(|\mathcal{U}|) - r - t \geq H_2(X) - \log_2(|\mathcal{U}|) - r - t.$$

Here, we recall that $H_2(X) \leq H_2(XU)$ (Proposition 10.10). Taking $r = t = \kappa + 1$, we see that with probability $\geq 1 - 2^{-\kappa}$,

$$H_2(X) - H_2(X\,|\,U = u) \leq \log_2(|\mathcal{U}|) + 2\kappa + 2.$$

$\square$

We apply Theorem 10.31 to the following random variables X and U. X is Alice's raw key A conditional on Eve's knowledge which she has obtained from her measurements and from listening to the public auxiliary channel before information reconciliation starts, i.e., X is A conditional on $E = e$ and $S = s$. $U = P$ consists of the parity bits which are sent from Alice to Bob through the public channel and hence are accessible to Eve. Let m be the number of parity bits, and let p be the concrete parity values.

Corollary 10.32. *Let $\kappa > 0$ be a security parameter. Then*

$$H_2(A\,|\,E = e, S = s) - H_2(A\,|\,E = e, S = s, P = p) \leq m + 2\kappa + 2,$$

with probability $\geq 1 - 2^{-\kappa}$.

10.5.6 Exchanging a Secure Key – An Example

Assuming that the adversary Eve executes an intercept-and-resend attack, we can derive the following example from the theory developed in this chapter.

Let Alice generate and transmit a random 20 KiB bit string through the quantum channel to Bob. In the sifting phase, it is expected that half of the bits will be discarded, and the sifted key will therefore have a length of about 10 KiB.

Before starting information reconciliation, Alice and Bob estimate the error rate ε of their quantum channel. To be able to specify the quality of their estimate, they assume an upper bound on the empirical error rate, say 12%. Later, after information reconciliation, Alice and Bob can validate their assumption – they then know the exact number of errors. For their estimate, Alice and Bob exchange randomly chosen bits. By exchanging 0.5 KiB, for example, Alice and Bob can estimate ε very exactly, with a probability of 95%: see the example in Section 10.5.3. The exchanged bits are known to Eve and are discarded. The remaining raw key has a length n of 9.5 KiB.

Alice and Bob reconcile their strings interactively by applying Cascade. During reconciliation, they learn the exact number of errors. Assume that the exact empirical error rate turns out to be $\bar{\varepsilon} = 10\%$. Then Eve has intercepted, with probability $1 - 2^{-30}$, at most a fraction $\eta = 4 \cdot \bar{\varepsilon} + 2 \cdot z_{1-2^{-30}}/\sqrt{n} = 0.45$ of the bits (Corollary 10.26). From Proposition 10.24 we conclude that, with probability $p = 1 - 2^{-30}$, Eve's conditional Rényi entropy after the eavesdropping and before the reconciliation is at least

$$H_2(A \mid E = e, S = s) \geq n - \eta n \cdot 0.585 - \frac{\sqrt{\eta n}}{2} \cdot z_{1-p/2}$$

$$= (1 - 0.45 \cdot 0.585) \cdot n - \frac{\sqrt{0.45 n}}{2} \cdot z_{1-p/2}$$

$$= 0.737 \cdot n - \frac{\sqrt{0.45 n}}{2} \cdot z_{1-p/2}$$

$$= 0.737 \cdot n - \frac{0.671 \sqrt{n}}{2} \cdot z_{1-2^{-31}}$$

$$= 7.00 \text{ KiB} - 0.07 \text{ KiB} = 6.93 \text{ KiB}.$$

In reality, Alice and Bob know the exact number of parity bits that they exchanged during information reconciliation. In our example, we use the estimate given in Remark 2 on page 364. The number of parity bits exchanged through the public channel is roughly $(1.1+\varepsilon) \cdot h(\varepsilon) \cdot n = (1.1+0.1) \cdot h(0.1) \cdot n = 0.563 \cdot n = 0.563 \cdot 9.5 \text{ KiB} \approx 5.35 \text{ KiB}$.

The parity bits $P = p$ that Alice and Bob exchange for information reconciliation reduce Eve's Rényi entropy. With probability $1 - 2^{-30}$, Eve's remaining entropy can be bounded as follows (Corollary 10.32, taking $\kappa = 30$):

$$H_2(A \mid V = v) = H_2(A \mid E = e, S = s, P = p) \geq 6.93 - 5.35 - 0.01 \geq 1.57 \text{ KiB}.$$

Here, as above, we subsume Eve's information $v = (e, s, p)$ into the random variable $V = (E, S, P)$.

Corollary 10.13 says that, by applying a randomly chosen element from a universal class of hash functions, Alice and Bob can generate a shared 1.56 KiB key k with $H(A \mid G, V = v) \geq 1.56 \text{ KiB} - 2^{-81}/\ln(2) \approx 1.56 \text{ KiB}$, which means that, from Eve's point of view, k is a randomly and uniformly chosen secret.

10.5.7 General Attacks and Security Proofs

In the preceding sections, we gave a detailed security analysis of the intercept-and-resend attack: Eve intercepts the photons, performs measurements on them and resends them to Bob. The theory of quantum mechanical measurement says that any eavesdropping involves an interaction between a probe and the signals: Eve prepares a system and lets it interact with the quantum system that carries the information. Eve obtains information by measuring the probe.

There are three basic types of eavesdropping attacks: individual, collective and coherent attacks.

Individual Attacks. In an individual attack, Eve lets each signal interact with a separate, independent probe. Eve performs her measurements on each probe separately. The intercept-and-resend attack is such an individual attack. There are other individual attacks, in which Eve clones the quantum bit. The cloning is necessarily imperfect – perfectly cloning an unknown quantum state is impossible. See, for example, [Assche06, Section 10.3.2] or [Schauer10, Section 5.2.2].

Collective Attacks. As in an individual attack, Eve lets each signal interact with a separate probe. But now she is able to perform measurements which operate on all probes coherently.

Coherent Attacks. Coherent attacks, also called joint attacks, are the most general attacks. It is assumed that Eve has access to all signals at the same time. The sequence of signals constitutes a single quantum state and Eve is able to perform measurements by using a single probe. She is able to introduce correlations between consecutive signals.

Security proofs for this most general class of attacks are typically based on an equivalence of the quantum key distribution protocol and its secret-key distillation to an *entanglement purification protocol*. An example is Shor and Preskill's security proof for the BB84 protocol ([ShoPre00]). Other security proofs for the BB84 protocol have been given, for example, by Mayers ([Mayers96]) and Biham et al. ([BiBoBoMoRo06]). An alternative approach to general security proofs is to estimate the size of Eve's quantum memory and apply the concept of selectable knowledge developed by König, Maurer and Renner ([KoeMauRen04], [KoeMauRen05]).

The general security proof for the BB84 protocol relies on the fact that the protocol can be translated into an entanglement purification protocol if the universal class of hash functions $\{h_a : \mathbb{F}_{2^l} \longrightarrow \mathbb{F}_{2^f}, \ x \longmapsto \mathrm{msb}_f(a \cdot x) \ | \ a \in \mathbb{F}_{2^l}\}$ is used for privacy amplification and if information reconciliation is one-way and based on specific quantum codes. If Cascade is used, problems arise from the bisection which reveals values of single bits. A solution to this problem proposed by Lo ([Lo03]) requires that Alice and Bob encrypt the reconciliation messages (see [Assche06, Section 12.2.4] for a discussion).

König, Renner, Bariska and Maurer discovered a problem with quantum key distribution security proofs ([KoeRenBarMau07]). Renner was able to overcome this problem by introducing a stronger definition of secure keys and giving a generic security proof that applies to known protocols such as BB84 ([Renner05]; [RenKoe05]). Security proofs based on entanglement purification can be adapted to meet the stronger requirements on keys, as shown in [KoeRenBarMau07].

A short overview of security proofs for the most general class of attacks – coherent attacks (also called joint attacks) – was given in [DusLütHen06, Section 8.4]. A comprehensive exposition of security proofs, in particular proofs that rely on an equivalence to entanglement purification protocols, can be found in [Assche06, Chapter 12].

We conclude this section with some final remarks:

1. The security proofs for QKD do not rely on an unproven assumption that a computational problem, such as the factoring of integers, is intractable in practice. The resources of an attacker are not limited. Therefore, we speak of unconditional security. This does not mean that QKD is perfectly secure. The results proven are probabilistic. They state that security features hold with some probability, provided the underlying assumptions are fulfilled.

2. Of course, a valid security proof for a cryptographic scheme does not guarantee that a practical implementation is secure. It is always questionable to what extent the underlying assumptions are fulfilled in an implementation. Security risks result from vulnerabilities in the platform and components that are used in the implementation.

   There are examples of successful "quantum hacking". Researchers have been able to implement perfect intercept-and-resend attacks against practical QKD systems, including commercial ones ([Makarov09]; [Merali09]; [LyWiWiElSkMa10]), where they intercepted the exchanged key without leaving any trace of their eavesdropping. In these attacks, a weakness in the detectors is exploited. By shining a continous laser at Bob's detector, Eve blinds the detector. The detector no longer functions as a "quantum detector" that distinguishes between orthogonal quantum states. But it still works as a "classical detector" for light pulses. This enables Eve to intercept the quantum bits from Alice and resend them as a classical signal to Bob. The communication line from Alice to Bob is split into a quantum channel from Alice to Eve and a classical optical channel from Eve to Bob. Bob obtains exactly the same bits that Eve measures.

3. It is not only the practical security of QKD that has been challenged. Hirota and Yuen initiated a debate about the validity of security proofs for QKD ([Yuen12]; [MiKiGrWePeEuSc13]). In a recent paper, Renner refuted their critique ([Renner12]).

Exercises

1. Let $l \geq f$. Prove that
   a. $\mathcal{H}_{l,f} := \{h_a : \mathbb{F}_{2^l} \longrightarrow \mathbb{F}_{2^f}, \ x \longmapsto \mathrm{msb}_f(a \cdot x) \mid a \in \mathbb{F}_{2^l}\}$ is a universal class of hash functions.
   b. $\mathcal{G}_{l,f} := \{h_{a_0,a_1} : \mathbb{F}_{2^l} \longrightarrow \mathbb{F}_{2^f}, \ x \longmapsto \mathrm{msb}_f(a_0 \cdot x + a_1) \mid a_0, a_1 \in \mathbb{F}_{2^l}\}$ is a strongly universal class of hash functions.

   Here we consider $\{0,1\}^m$ as equipped with the structure $\mathbb{F}_{2^m}$ of the Galois field with 2^m elements. As before, msb_f denotes the f most-significant bits.

2. The following example is taken from [Cachin97, Example 5.1]. Let X be a random variable with values in $\mathcal{X} := \{a_1, \ldots, a_{10}, b_1, \ldots, b_{10}\}$ and let

$\text{prob}(X = a_i) = 0.01$ and $\text{prob}(X = b_i) = 0.09$ for $1 \leq i \leq 10$. Let $f : \mathcal{X} \longrightarrow \{0, 1\}$ be the map

$$x \mapsto \begin{cases} 0 & \text{if } x \in \{a_1, \ldots, a_9, b_{10}\}, \\ 1 & \text{if } x \in \{a_{10}, b_1, \ldots, b_9\}, \end{cases}$$

and let $Y := f(X)$ Compute $H_2(X), H_2(X \mid Y)$ and $H_2(Y)$, and show that $H_2(Y) + H_2(X \mid Y) < H_2(X)$.

3. In contrast to the Shannon entropy, additional knowledge does not always reduce the Rényi entropy; it may even increase it. The following example was given in [BenBraCréMau95, Section VI].

Let $x_0 \in \{0, 1\}^n$ and let X be the random variable with values in $\{0, 1\}^n$ and distribution

$$\text{prob}(X = x) := \begin{cases} 2^{-n/4} & \text{if } x = x_0, \\ \dfrac{1 - 2^{-n/4}}{2^n - 1} & \text{if } x \neq x_0. \end{cases}$$

Let Y be the random variable

$$Y := \begin{cases} 0 & \text{if } X = x_0, \\ 1 & \text{otherwise.} \end{cases}$$

Show that

$$H_2(X \mid Y) > H_2(X).$$

4. Give a proof of Proposition 10.21.

5. Let X be a random variable with values in $\{x_1, x_2, \ldots, x_m\}$. Let $\mathcal{E}$ be an event with $\text{prob}(\mathcal{E}) = p$ and assume that X is constant if $\mathcal{E}$ occurs, i.e.,

$$\text{prob}(X = x_i \mid \mathcal{E}) = \begin{cases} 1 & \text{if } i = i_0, \\ 0 & \text{if } i \neq i_0. \end{cases}$$

Show that

$$H(X) \leq (1 - p) \log_2(m) + h(p),$$

where $h(p) := -p \log_2(p) - (1 - p) \log_2(1 - p)$.[19]
Now, assume additionally that $\text{prob}(X = x_i \mid \overline{\mathcal{E}}) = 1/m, 1 \leq i \leq m$, i.e., X is uniformly distributed in the case of $\overline{\mathcal{E}}$, and $m \gg 1$. Show that $H(X)$ is then slightly more than $(1 - p) \log_2(m)$.

6. Derive the following addition theorem for sines and cosines: For $x, y \in \mathbb{R}$,

$$\cos^2(x) \cos^2(y) + \sin^2(x) \sin^2(y) = \frac{1}{2}(1 + \cos(2x) \cos(2y)).$$

---

[19] Note that $0 \leq h(p) \leq 1$ (Proposition 10.21).

7. In this exercise, we develop a proof of Proposition 10.19.

We denote by $(|v\rangle, |w\rangle) := (|H\rangle, |V\rangle)$ and $(|v\rangle', |w\rangle') := \mathcal{B}_{\alpha, \pi/4}$ the encoding bases used by Alice and Bob. Let $(|v_E\rangle, |w_E\rangle) := \mathcal{B}_{\gamma, \theta}$ be Eve's measurement basis and $(|v_E'\rangle, |w_E'\rangle) := \mathcal{B}_{\gamma', \theta'}$ be Eve's re-encoding basis. Let A, B be the random variables that capture the quantum states that Alice sends and Bob measures, respectively.

a. Show that the error rate ε is

$$
\begin{aligned}
\varepsilon &= \varepsilon(\theta, \theta', \delta, \delta') \\
&= \frac{1}{2}(\mathrm{prob}(B = |w\rangle \,|\, A = |v\rangle)) + \mathrm{prob}(B = |w\rangle' \,|\, A = |v\rangle')) \\
&= \frac{1}{4}(2 - \cos(2\theta)\cos(2\theta') - \sin(2\theta)\sin(2\theta')\cos(\delta)\cos(\delta')),
\end{aligned}
$$

where $\delta := \gamma - \alpha$ and $\delta' := \gamma' - \alpha$.

Hint: Use Lemma 10.18 and the subsequent remark, and the addition theorem obtained in Exercise 6.

b. Determine the zeros of the partial derivatives $\partial \varepsilon / \partial \delta$ and $\partial \varepsilon / \partial \delta'$ of ε with respect to δ and δ'. Then, for θ, θ' fixed, find the minimum and maximum values of $\varepsilon(\theta, \theta', \delta, \delta')$.

11. Provably Secure Digital Signatures

In previous sections, we discussed signature schemes (Full-Domain-Hash RSA signatures and PSS in Section 3.3.5; the Fiat–Shamir signature scheme in Section 4.2.5) that include a hash function h and whose security can be proven in the random oracle model. It is assumed that the hash function h is a random oracle, i.e., it behaves like a perfectly random function (see Sections 2.2.4 and 3.3.5). Perfectly random means that for all messages m, each of the k bits of the hash value $h(m)$ is determined by tossing a coin, or, equivalently, that the map $h : X \longrightarrow Y$ is randomly chosen from the set $\mathcal{F}(X, Y)$ of all functions from X to Y. In general, $\mathcal{F}(X, Y)$ is tremendously large. For example, if $X = \{0, 1\}^n$ and $Y = \{0, 1\}^k$, then $|\mathcal{F}(X, Y)| = 2^{k2^n}$. Thus, it is obvious that perfectly random oracles cannot be implemented.

Moreover, examples of cryptographic schemes were constructed that are provably secure in the random oracle model, but are insecure in any real-world implementation, where the random oracle is replaced by a real hash function. Although these examples are contrived, doubts on the random oracle model arose (see the remark on page 312 in Section 9.5).

Therefore, it is desirable to have signature schemes whose security can be proven solely under standard assumptions (like the RSA or the discrete logarithm assumption). Examples of such signature schemes are given in this chapter.

11.1 Attacks and Levels of Security

Digital signature schemes are public-key cryptosystems and are based on the one-way feature of a number-theoretical function. A *digital signature scheme*, for example the basic RSA signature scheme (Section 3.3.4) or ElGamal's signature scheme (Section 3.4.2), consists of the following:

1. A key generation algorithm K, which on input 1^k (k being the security parameter) produces a pair (pk, sk) consisting of a public key pk and a secret (private) key sk.
2. A signing algorithm $S(sk, m)$, which given the secret key sk of user Alice and a message m to be signed, generates Alice's signature σ for m.

3. A verification algorithm $V(pk, m, \sigma)$, which given Alice's public key pk, a message m and a signature σ, checks whether σ is a valid signature of Alice for m. Valid means that σ might be output by $S(sk, m)$, where sk is Alice's secret key.

Of course, all algorithms must be polynomial. The key generation algorithm K is always a probabilistic algorithm. In many cases, the signing algorithm is also probabilistic (see, e.g., ElGamal or PSS). The verification algorithm might be probabilistic, but in practice it usually is deterministic.

As with encryption schemes, there are different types of attacks on signature schemes. We may distinguish between (see [GolMicRiv88]):

1. *Key-only attack.* The adversary Eve only knows the public key of the signer Alice.
2. *Known-signature attack.* Eve knows the public key of Alice and has seen message-signature pairs produced by Alice.
3. *Chosen-message attack.* Eve may choose a list $(m_1, \ldots, m_t)$ of messages and ask Alice to sign these messages.
4. *Adaptively-chosen-message attack.* Eve can adaptively choose messages to be signed by Alice. She can choose some messages and gets the corresponding signatures. Then she can do cryptanalysis and, depending on the outcome of her analysis, she can choose the next message to be signed, and so on.

Adversary Eve's level of success may be described in increasing order as (see [GolMicRiv88]):

1. *Existential forgery.* Eve is able to forge the signature of at least one message, not necessarily the one of her choice.
2. *Selective forgery.* Eve succeeds in forging the signature of some messages of her choice.
3. *Universal forgery.* Although unable to find Alice's secret key, Eve is able to forge the signature of any message.
4. *Retrieval of secret keys.* Eve finds out Alice's secret key.

As we have seen before, signatures in the basic RSA, ElGamal and DSA schemes, without first applying a suitable hash function, can be easily existentially forged using a key-only attack (see Section 3). In the basic Rabin scheme, secret keys may be retrieved by a chosen-message attack (see Section 3.5.1). We may define the level of security of a signature scheme by the level of success of an adversary performing a certain type of attack. Different levels of security may be required in different applications.

In this chapter, we are interested in signature schemes which provide the maximum level of security. The adversary Eve cannot succeed in an existential forgery with a significant probability, even if she is able to perform an adaptively-chosen-message attack. As usual, the adversary is modeled as a probabilistic polynomial algorithm.

Definition 11.1. Let $\mathcal{D}$ be a digital signature scheme, with key generation algorithm K, signing algorithm S and verification algorithm V. An *existential forger* F for $\mathcal{D}$ is a probabilistic polynomial algorithm F that on input of a public key pk outputs a message-signature pair $(m, \sigma) := F(pk)$. F is successful on pk if σ is a valid signature of m with respect to pk, i.e., $V(pk, F(pk)) = $ accept. F performs an adaptively-chosen-message attack if, while computing $F(pk)$, F can repeatedly generate a message $\tilde{m}$ and then is supplied with a valid signature $\tilde{\sigma}$ for $\tilde{m}$.

Remarks. Let F be an existential forger performing an adaptively-chosen-message attack:

1. Let (pk, sk) be a key of security parameter k. Since the running time of $F(pk)$ is bounded by a polynomial in k (note that pk is generated in polynomial time from 1^k), the number of messages for which F requests a signature is bounded by $T(k)$, where T is a polynomial.
2. The definition leaves open who supplies F with the valid signatures. If F is used in an attack against the legitimate signer, then the signatures $\tilde{\sigma}$ are supplied by the signing algorithm S, $S(sk, \tilde{m}) = \tilde{\sigma}$, where sk is the private key associated with pk. In a typical security proof, the signatures are supplied by a "simulated signer", who is able to generate valid signatures without knowing the trapdoor information that is necessary to derive sk from pk. This sounds mysterious and impossible. Actually, for some time it was believed that a security proof for a signature scheme is not possible, because it would necessarily yield an algorithm for inverting the underlying one-way function. However, the security proof given by Goldwasser, Micali and Rivest for their GMR scheme (discussed in Section 11.3) proved the contrary. See [GolMicRiv88], Section 4: The paradox of proving signature schemes secure. The key idea for solving the paradox is that the simulated signer constructs signatures for keys whose form is a very specific one, whereas their probability distribution is the same as the distribution of the original keys (see, e.g., the proof of Theorem 11.12).
3. The signatures $\sigma_i, 1 \leq i \leq T(k)$, supplied to F are, besides pk, inputs to F. The messages $m_i, 1 \leq i \leq T(k)$, for which F requests signatures, are outputs of F. Let M_i be the random variable describing the i-th message m_i. Since F adaptively chooses the messages, message m_i may depend on the messages m_j and the signatures σ_j supplied to F for $m_j, 1 \leq j < i$. Thus M_i may be considered as a probabilistic algorithm with inputs pk and $(m_j, \sigma_j)_{1 \leq j < i}$.
   The *probability of success* of F for security parameter k is then computed as[1]

---

[1] Unless otherwise stated, we always mean F's probability of success when the signatures for the adaptively chosen messages are supplied by the legitimate signer S.

$$\mathrm{prob}(V(pk, F(pk, (\sigma_i)_{1 \leq i \leq T(k)})) = \mathrm{accept} : (pk, sk) \leftarrow K(1^k),$$
$$m_i \leftarrow M_i(pk, (m_j, \sigma_j)_{1 \leq j < i}), \sigma_i \leftarrow S(sk, m_i), 1 \leq i \leq T(k)).$$

Definition 11.2. A digital signature scheme *is secure against adaptively-chosen-message attacks* if and only if for every existential forger F performing an adaptively-chosen-message attack and every positive polynomial P, there is a $k_0 \in \mathbb{N}$ such that for all security parameters $k \geq k_0$, the probability of success of F is $\leq 1/P(k)$.

Remark. Fail-stop signature schemes provide an additional security feature. If a forger – even if he has unlimited computing power and can do an exponential amount of work – succeeds in generating a valid signature, then the legitimate signer Alice can prove with a high probability that the signature is forged. In particular, Alice can detect forgeries and then stop using the signing mechanism ("fail then stop"). The signature scheme is based on the assumed hardness of a computational problem, and the proof of forgery is performed by showing that this underlying assumption has been compromised. Fail-stop signature schemes were introduced by Waidner and Pfitzmann ([WaiPfi89]). We do not discuss fail-stop signatures here (see, e.g., [Pfitzmann96]; [MenOorVan96]; [Stinson95]; [BarPfi97]).

11.2 Claw-Free Pairs and Collision-Resistant Hash Functions

In many digital signature schemes, the message to be signed is first hashed with a collision-resistant hash function. Provably collision-resistant hash functions can be constructed from claw-free pairs of trapdoor permutations. In Section 11.3 we will discuss the GMR signature scheme introduced by Goldwasser, Micali and Rivest. It was the first signature scheme that was provably secure against adaptively-chosen-message attacks (without depending on the random oracle model), and it is based on claw-free pairs.

Definition 11.3. Let $f_0 : D \longrightarrow D$ and $f_1 : D \longrightarrow D$ be permutations of the same domain D. A pair (x, y) is called a *claw* of f_0 and f_1 if $f_0(x) = f_1(y)$.

Let $I = (I_k)_{k \in \mathbb{N}}$ be a key set with security parameter k. We consider families

$$f_0 = (f_{0,i} : D_i \longrightarrow D_i)_{i \in I}, f_1 = (f_{1,i} : D_i \longrightarrow D_i)_{i \in I}$$

of one-way permutations with common key generator K that are defined on the same domains.

Definition 11.4. (f_0, f_1) is called a *claw-free pair of one-way permutations* if it is infeasible to compute claws; i.e., for every probabilistic polynomial algorithm A which on input i outputs distinct elements $x, y \in D_i$, and for every positive polynomial P, there is a $k_0 \in \mathbb{N}$ such that for all $k \geq k_0$

$$\text{prob}(f_{0,i}(x) = f_{1,i}(y) : i \leftarrow K(1^k), \{x, y\} \leftarrow A(i)) \leq \frac{1}{P(k)}.$$

Claw-free pairs of one-way permutations exist, if, for example, factoring is hard.

Proposition 11.5. *Let* $I := \{n \mid n = pq,\ p, q \ primes, p \equiv 3 \bmod 8, q \equiv 7 \bmod 8\}$. *If the factoring assumption (Definition 6.9) holds, then*

$$\text{CQ} := (f_n, g_n : \text{QR}_n \longrightarrow \text{QR}_n)_{n \in I},$$

where $f_n(x) := x^2$ *and* $g_n(x) := 4x^2$, *is a claw-free pair of one-way permutations (with trapdoor).*

Proof. Let $n = pq \in I$. Since both primes, p and q, are congruent 3 modulo 4, f_n is a permutation of QR_n (Proposition A.72). Four is a square and it is a unit in $\mathbb{Z}_n$. Thus, g_n is also a permutation of QR_n. Now, let $x, y \in \text{QR}_n$ with $x^2 = 4y^2 \bmod n$ be a claw. From $n \equiv 1 \bmod 4$ and $n \equiv -3 \bmod 8$, we conclude $\left(\frac{-1}{n}\right) = 1$ and $\left(\frac{2}{n}\right) = -1$ (Theorem A.63). We get $\left(\frac{\pm 2y}{n}\right) = \left(\frac{\pm 1}{n}\right) \cdot \left(\frac{2}{n}\right) \cdot \left(\frac{y}{n}\right) = 1 \cdot (-1) \cdot 1 = -1$, whereas $\left(\frac{x}{n}\right) = 1$ since x is a square. Thus $x \neq \pm 2y$, and the Euclidean algorithm yields a factorization of n. Namely, $0 = x^2 - 4y^2 = (x - 2y)(x + 2y) = 0$ and thus $\gcd(x^2 - 4y^2, n)$ is a non-trivial divisor of n. We see that an algorithm which, with some probability, finds claws of CQ yields an algorithm factoring n with the same probability, which is a contradiction to the factoring assumption.   □

Claw-free pairs of one-way permutations can be used to construct collision-resistant hash functions.

Definition 11.6. Let $I = (I_k)_{k \in \mathbb{N}}$ be a key set with security parameter k, and let K be a probabilistic polynomial sampling algorithm for I, which on input 1^k outputs a key $i \in I_k$. Let $k(i)$ be the security parameter of i (i.e., $k(i) = k$ for $i \in I_k$), and $g : \mathbb{N} \longrightarrow \mathbb{N}$ be a polynomial function. A family

$$\mathcal{H} = \left(h_i : \{0,1\}^* \longrightarrow \{0,1\}^{g(k(i))}\right)_{i \in I}$$

of hash functions is called a family of *collision-resistant* (or *collision-free*) hash functions (or a *collision-resistant hash function* for short) with key generator K, if:

1. The hash values $h_i(x)$ can be computed by a polynomial algorithm H with inputs $i \in I$ and $x \in \{0,1\}^*$.
2. It is computationally infeasible to find a collision; i.e., for every probabilistic polynomial algorithm A which on input $i \in I$ outputs messages $m_0, m_1 \in \{0,1\}^*, m_0 \neq m_1$, and for every positive polynomial P, there is a $k_0 \in \mathbb{N}$ such that

$$\text{prob}(h_i(m_0) = h_i(m_1) : i \leftarrow K(1^k), \{m_0, m_1\} \leftarrow A(i)) \leq \frac{1}{P(k)},$$

   for all $k \geq k_0$.

Remark. Collision-resistant hash functions are one way (see Exercise 2).

Let (f_0, f_1) be a pair of one-way permutations, as above. For every $m \in \{0,1\}^*$, we may derive a family $f_m = (f_{m,i} : D_i \longrightarrow D_i)_{i \in I}$ as follows. For $m := m_1 \ldots m_l \in \{0,1\}^l$ and $x \in D_i$, let

$$f_{m,i}(x) := f_{m_1,i}(f_{m_2,i}(\ldots f_{m_l,i}(x) \ldots)).$$

If $m \in \{0,1\}^*$ is the concatenation $m := m_1 \| m_2$ of strings m_1 and m_2, then obviously $f_{m,i}(x) = f_{m_1,i}(f_{m_2,i}(x))$.

This family may now be used to construct a family $\mathcal{H} = (h_j)_{j \in J}$ of hash functions. Let $J_k := \{(i,x) \mid i \in I_k, x \in D_i\}$ and $J := \bigcup_{k \in \mathbb{N}} J_k$. We define

$$F_j(m) := f_{m,i}(x) \in D_i \text{ for } j = (i,x) \in J \text{ and } m \in \{0,1\}^*.$$

Our goal is a collision-resistant family of hash functions $\mathcal{H}$. To achieve this, we have to modify our construction a little and first encode our messages m in a prefix-free way. Let $[m]$ denote a prefix-free (binary) encoding of the messages m in $\{0,1\}^*$. Prefix-free means that no encoded $[m]$ appears as a prefix[2] in the encoding $[m']$ of an $m' \neq m$. For example, we might encode each 1 by 1, each 0 by 00 and terminate all encoded messages by 01.[3] We define

$$h_j(m) := F_j([m]), \text{ for } m \in \{0,1\}^* \text{ and } j \in J.$$

We will prove that $\mathcal{H}$ is a collision-resistant family of hash functions if the pair (f_0, f_1) is claw-free. Thus, we obtain the following proposition.

Proposition 11.7. *If claw-free pairs of one-way permutations exist, then collision-resistant hash functions also exist.*

Proof. Let $\mathcal{H}$ be the family of hash functions constructed above. Assume that $\mathcal{H}$ is not collision-resistant. This means that there is a positive polynomial P and a probabilistic polynomial algorithm A which on input $j = (i,x) \in J_k$ finds a collision $\{m, m'\}$ of h_j with non-negligible probability $\geq 1/P(k)$ (where P is a positive polynomial) for infinitely many k. Collision means that $f_{[m],i}(x) = f_{[m'],i}(x)$. Let $[m] = m_1 \ldots m_r$ and $[m'] = m_1' \ldots m_{r'}'$ ($m_j, m_{\tilde{j}}' \in \{0,1\}$), and let l be the smallest index u with $m_u \neq m_u'$. Such an index l exists, since $[m]$ is not a prefix of $[m']$, nor vice versa. We have $f_{m_l \ldots m_r,i}(x) = f_{m_l' \ldots m_{r'}',i}(x)$, since $f_{0,i}$ and $f_{1,i}$ are injective. Then $(f_{m_{l+1} \ldots m_r,i}(x), f_{m_{l+1}' \ldots m_{r'}',i}(x))$ is a claw of (f_0, f_1). The binary lengths r and r' of m and m' are bounded by a polynomial in k, since m and m' are computed by the polynomial algorithm A. Thus, the claw of (f_0, f_1) can be computed from the collision $\{m, m'\}$ in polynomial time. Hence, we can compute claws with non-negligible probability $\geq 1/P(k)$, for infinitely many k, which is a contradiction. □

---

[2] A string s is called a prefix of a string t if $t = s \| s'$ is the concatenation of s and another string s'.

[3] Efficient prefix-free encodings exist such that $[m]$ has almost the same length as m (see, e.g., [BerPer85]).

Remarks:

1. The constructed hash functions are rather inefficient. Thus, in practice, custom-designed hash functions such as SHA are used, whose collision resistance cannot be proven rigorously (see Section 2.2).
2. Larger sets of pairwise claw-free one-way permutations may be used in the construction instead of one pair, for example sets with 2^s elements. Then s bits of the messages m are processed in one step. There are larger sets of pairwise claw-free one-way permutations that are based on the assumptions that factoring and the computation of discrete logarithms are infeasible (see [Damgård87]).
3. Another method of constructing provably collision-resistant hash functions is given in Exercise 5 in Chapter 3. It is based on the assumed infeasibility of computing discrete logarithms.

11.3 Authentication-Tree-Based Signatures

Again, we consider a claw-free pair (f_0, f_1) of one-way permutations, as above. In addition, we assume that f_0 and f_1 have trapdoors. Such a claw-free pair of trapdoor permutations and the induced functions f_m, as defined above, may be used to generate probabilistic signatures. Namely, Alice randomly chooses some $i \in I$ (with a sufficiently large security parameter k) and some $x \in D_i$, and publishes (i, x) as her public key. Then Alice, by using her trapdoor information, computes her signature $\sigma(i, x, m)$ for a message $m \in \{0, 1\}^*$ as

$$\sigma(i, x, m) := f_{[m],i}^{-1}(x),$$

where $[m]$ denotes some (fixed) prefix-free encoding of m. Bob can verify Alice's signature σ by comparing $f_{[m],i}(\sigma)$ with x. Since $f_{[m],i}$ is one way and $m \mapsto f_{[m],i}(\sigma)$ is collision resistant, as we have just seen in the proof of Proposition 11.7, only Alice can compute the signature for m, and no one can use one signature σ for two different messages m and m'. Unfortunately, this scheme is a *one-time signature scheme*. This means that only one message can be signed by Alice with her key (i, x); otherwise the security is not guaranteed.[4] If two messages $m \neq m'$ were signed with the same reference value x, then a claw of f_0 and f_1 can be easily computed from $\sigma(i, x, m)$ and $\sigma(i, x, m')$ (see Exercise 5)[5], and this can be a severe security risk. If we use, for example, the claw-free pair of Proposition 11.5, then Alice's secret key (the factors of the modulus n) can be easily retrieved from the computed claw.

---

[4] More examples of one-time signature schemes may be found, e.g., in [MenOorVan96] and [Stinson95].

[5] It is not a contradiction to the assumed claw-freeness of the pair that the claw can be computed, because here the adversary is additionally supplied with two signatures which can only be computed by use of the secret trapdoor information.

In the *GMR signature scheme* ([GolMicRiv88]), Goldwasser, Micali and Rivest overcome this difficulty by using a new random reference value for each message m. Of course, it is not possible to publish all these reference values as a public key in advance, so the reference values are attached to the signatures. Then it is necessary to authenticate these reference values, and this is accomplished by a second claw-free pair of trapdoor permutations. The GMR scheme is based on two claw-free pairs

$$(f_0, f_1) = (f_{0,i}, f_{1,i} : D_i \longrightarrow D_i)_{i \in I}, (g_0, g_1) = (g_{0,j}, g_{1,j} : E_j \longrightarrow E_j)_{j \in J}$$

of trapdoor permutations, defined over $I = (I_k)_{k \in \mathbb{N}}$ and $J = (J_k)_{k \in \mathbb{N}}$. Each user Alice runs a key-generation algorithm $K(1^k)$ to randomly choose an $i \in I_k$ and an $j \in J_k$ (and generate the associated trapdoor information). Moreover, Alice generates a binary "authentication tree" of depth d. The nodes of this tree are randomly chosen elements in D_i. Then Alice publishes (i, j, r) as her public key, where r is the root of the authentication tree. The authentication tree has 2^d leaves $v_l, 1 \leq l \leq 2^d$. Alice can now sign up to 2^d messages. To sign the l-th message m, she takes the l-th leaf v_l and takes as the first part $\sigma_1(m)$ of the signature the previously defined probabilistic signature $f_{[m],i}^{-1}(v_l)$ with respect to the reference value v_l.[6] The second part $\sigma_2(m)$ of the signature authenticates v_l. It contains the elements $x_0 := r, x_1, \ldots, x_d := v_l$ on the path from the root r to the leaf v_l in the authentication tree and authentication values for each node $x_m, 1 \leq m \leq d$. The authentication value of x_m contains the parent x_{m-1} and both of its children c_0 and c_1 (one of them is x_m) and the signature $g_{[c_0 \| c_1], j}^{-1}(x_{m-1})$ of the concatenated children with respect to the reference value x_{m-1}, computed by the second claw-free pair. The children of a node are authenticated jointly. To verify Alice's signatures, Bob has to climb up the tree from the leaf in the obvious way. If he finally computes the correct root r, he accepts the signature.

Theorem 11.8. *The GMR signature scheme is secure against adaptively-chosen-message attacks.*

Proof. See [GolMicRiv88].   □

Remarks:

1. The full authentication tree and the authentication values for its nodes could be constructed in advance and stored. However, it is more efficient to develop it dynamically, as it is needed for signatures, and to store only the necessary information about its current state.

2. The size of a GMR signature is of order $O(kd)$ if the inputs of the claw-free pairs are of order $O(k)$ (as in the pair given in Proposition 11.5). In practice, this size is considerable if the desired number n of signatures and hence $d = \log_2(n)$ increases. For example, think only of 10000

---

[6] Here we simplify a little and omit the "bridge items".

signatures with a security parameter $k = 1024$. The size of the signatures and the number of computations of f – necessary to generate and to verify signatures – could be substantially reduced if authentication trees with much larger branching degrees could be used instead of the binary one, thus reducing the distance from a leaf to the root. Such signature schemes have been developed, for example, by Dwork and Naor ([DwoNao94]) and Cramer and Damgård ([CraDam96]). They are quite efficient and provably provide security against adaptively-chosen-message attacks if the RSA assumption 6.7 holds. For example, taking a 1024-bit RSA modulus, a branching degree of 1000 and a tree of depth 3 in the Cramer–Damgård scheme, Alice could sign up to 10^9 messages, with the size of each signature being less than 4000 bits.

11.4 A State-Free Signature Scheme

The signing algorithm in GMR or in other authentication-tree-based signature schemes is not state free: the signing algorithm has to store the current state of the authentication tree which depends on the already generated signatures, and the next signature depends on this state. In this section, we will describe a provably secure and quite efficient state-free digital signature scheme introduced by Cramer and Shoup ([CraSho00]). The scheme is secure against adaptively-chosen-message attacks, provided the so-called *strong* RSA assumption (see below) holds. Another state-free signature scheme based on the strong RSA assumption, has been, for example, introduced in [GenHalRab99]. The security proof, which we will give below, shows the typical features of such a proof. It runs with a contradiction: a successful forging algorithm is used to construct an attacker A who successfully inverts the underlying one-way function. The main problem is that in a chosen-message attack, the forger F is only successful if he can request signatures from the legitimate signer. Now the legitimate signer, who uses his secret key, cannot be called during the execution of F, because A is only allowed to use publicly accessible information. Thus, a major problem is to substitute the legitimate signer by a simulation.

The moduli in the Cramer–Shoup signature scheme are defined with special types of primes.

Definition 11.9. A prime p is called a *Sophie Germain prime* if $2p + 1$ is also a prime.[7]

Remark. In the Cramer–Shoup signature scheme we have to assume that sufficiently many Sophie Germain primes exist. Otherwise there is no guarantee that keys can be generated in polynomial time. The security proof

---

[7] Sophie Germain (1776–1831) proved the first case of Fermat's Last Theorem for prime exponents p, for which $2p + 1$ is also prime ([Osen74]).

given below also relies on this assumption (see Lemma 11.11). There must be a positive polynomial P such that the number of k-bit Sophie Germain primes p is $\geq 2^k/P(k)$. Today, there is no rigorous mathematical proof for this. It is not even known whether there are infinitely many Sophie Germain primes. On the other hand, it is conjectured and there are heuristic arguments and numerical evidence that the number of k-bit Sophie Germain primes is asymptotically equal to $c \cdot 2^k/k^2$, where c is an explicitly computable constant ([Koblitz88]; [BatHor62]; [BatHor65]). Thus, there is convincing evidence for the existence of sufficiently many Sophie Germain primes.

The Strong RSA Assumption. The security of the Cramer–Shoup signature scheme is based on the following *strong RSA assumption* introduced in [BarPfi97].

Let $I := \{n \in \mathbb{N} \mid n = pq,\ p \neq q \text{ prime numbers },\ |p| = |q|\}$ be the set of RSA moduli and $I_k := \{n \in I \mid n = pq,\ |p| = |q| = k\}$.

Definition 11.10 (*strong RSA assumption*). For every positive polynomial Q and every probabilistic polynomial algorithm A which on inputs $n \in I$ and $y \in \mathbb{Z}_n^*$ outputs an exponent $e > 1$ and an $x \in \mathbb{Z}_n^*$, there exists a $k_0 \in \mathbb{N}$ such that

$$\text{prob}(x^e = y : n \xleftarrow{u} I_k, y \xleftarrow{u} \mathbb{Z}_n^*, (e, x) \leftarrow A(n, y)) \leq \frac{1}{Q(k)}$$

for $k \geq k_0$.

Remark. The strong RSA assumption implies the classical RSA assumption (Definition 6.7). In the classical RSA assumption, the attacking algorithm has to find an e-th root for $y \in \mathbb{Z}_n^*$, for a given e. Here the exponent is not given. The adversary is successful if, given some $y \in \mathbb{Z}_n^*$, she can find an exponent $e > 1$ such that she is able to extract the e-th root x of y. Today, the only known method for breaking either assumption is to solve the factorization problem.

Let
$$I_{\text{SG}} := \{n \in I \mid n = pq, p = 2\tilde{p} + 1, q = 2\tilde{q} + 1, \tilde{p}, \tilde{q} \text{ Sophie Germain primes}\}$$
and $I_{\text{SG},k} := I_{\text{SG}} \cap I_k$.

Lemma 11.11. *Assume that there is a positive polynomial P such that the number of k-bit Sophie Germain primes is $\geq 2^k/P(k)$. Then the strong RSA assumption implies that for every positive polynomial Q and every probabilistic polynomial algorithm A which on inputs $n \in I_{\text{SG}}$ and $y \in \mathbb{Z}_n^*$ outputs an exponent $e > 1$ and an $x \in \mathbb{Z}_n^*$, there exists a $k_0 \in \mathbb{N}$ such that*

$$\text{prob}(x^e = y : n \xleftarrow{u} I_{\text{SG},k}, y \xleftarrow{u} \mathbb{Z}_n^*, (e, x) \leftarrow A(n, y)) \leq \frac{1}{Q(k)}$$

for $k \geq k_0$.

Proof. The distribution $n \xleftarrow{u} I_{SG,k}$ is polynomially bounded by the distribution $n \xleftarrow{u} I_k$ if the existence of sufficiently many Sophie Germain primes is assumed. Thus, we may replace $n \xleftarrow{u} I_k$ by $n \xleftarrow{u} I_{SG,k}$ in the strong RSA assumption (Proposition B.34). $\qquad\Box$

The Cramer–Shoup Signature Scheme. In the key generation and in the signing procedure, a probabilistic polynomial algorithm *GenPrime*(1^k) with the following properties is used:

1. On input 1^k, *GenPrime* outputs a k-bit prime.
2. If *GenPrime*(1^k) is executed $R(k)$ times (R a positive polynomial), then the probability that any two of the generated primes are equal is negligibly small; i.e., for every positive polynomial P there is a $k_0 \in \mathbb{N}$ such that for all $k \geq k_0$

$$\mathrm{prob}(e_{j_1} = e_{j_2} \text{ for some } j_1 \neq j_2 : e_j \leftarrow GenPrime(1^k), 1 \leq j \leq R(k))$$

$$\leq \frac{1}{P(k)}.$$

Such algorithms exist. For example, an algorithm that randomly and uniformly chooses primes of binary length k satisfies the requirements. Namely, the probability that $e_{j_1} = e_{j_2}$ (j_1 and j_2 fixed, $j_1 \neq j_2$) is about $k/2^k$ by the Prime Number Theorem (Theorem A.82), and there are $\binom{R(k)}{2} < R(k)^2/2$ subsets $\{j_1, j_2\}$ of $\{1, \ldots, R(k)\}$. There are suitable implementations of *GenPrime* which are much more efficient than the uniform sampling algorithm (see [CraSho00]).

Let $N \in \mathbb{N}, N > 1$ be a constant. To set up a Cramer–Shoup signature scheme, we choose two security parameters k and l, with $k^{1/N} < l+1 < k-1$. Then we choose a collision-resistant hash function $h : \{0,1\}^* \longrightarrow \{0,1\}^l$. More precisely, by using $\mathcal{H}$'s key generator, we randomly select a hash function from $\mathcal{H}_l$, where $\mathcal{H}$ is a collision-resistant family of hash functions and $\mathcal{H}_l$ is the subset of functions with security parameter l (without loss of generality, we assume that the functions in $\mathcal{H}_l$ map to $\{0,1\}^l$). We proved in Section 11.2 that such collision-resistant families exist if the RSA assumption and, as a consequence, the factoring assumption hold. The output of h is considered as a number in $\{0, \ldots, 2^l - 1\}$. All users of the scheme generate their signatures by using the hash function h.[8]

Given k, l and h, each user Alice generates her public and secret key.

Key Generation.

1. Alice randomly chooses a modulus $n \xleftarrow{u} I_{SG,k}$, i.e., she randomly and uniformly chooses Sophie Germain primes $\tilde{p}$ and $\tilde{q}$ of length $k-1$ and sets $n := pq$, $p := 2\tilde{p} + 1$ and $q := 2\tilde{q} + 1$.

---

[8] In practice we may use, for example, $k = 512, l = 160$ and $h =$ SHA-1, which is believed to be collision resistant.

2. She chooses $g \overset{u}{\leftarrow} \mathrm{QR}_n$ and $x \overset{u}{\leftarrow} \mathrm{QR}_n$ at random and generates an $(l+1)$-bit prime $\tilde{e} := GenPrime(1^{l+1})$.
3. $(n, g, x, \tilde{e})$ is the public key; (p, q) is the secret key.

Remark. Using Sophie Germain primes ensures that the order $\frac{p-1}{2} \cdot \frac{q-1}{2} = \tilde{p} \cdot \tilde{q}$ of QR_n is a product of distinct primes. Thus it is a cyclic group; it is the cyclic subgroup of order $\tilde{p}\tilde{q}$ of $\mathbb{Z}_n^*$.

In the following, all computations are done in $\mathbb{Z}_n^*$ unless otherwise stated and, as usual, we identify $\mathbb{Z}_n = \{0, \dots, n-1\}$.

Signing. It is possible to sign arbitrary messages $m \in \{0,1\}^*$. To sign m, Alice generates an $(l+1)$-bit prime $e := GenPrime(1^{l+1})$ and randomly chooses $\tilde{y} \overset{u}{\leftarrow} \mathrm{QR}_n$. She computes

$$\tilde{x} := \tilde{y}^{\tilde{e}} \cdot g^{-h(m)},$$

$$y := \left(x \cdot g^{h(\tilde{x})} \right)^{e^{-1}},$$

where e^{-1} is the inverse of e in $\mathbb{Z}_{\varphi(n)}^*$ (the powers are computed in $\mathbb{Z}_n^*$, which is of order $\varphi(n)$). The signature σ of m is $(e, y, \tilde{y})$.

Remarks:

1. Taking the e^{-1}-th power in the computation of y means computing the e-th root in $\mathbb{Z}_n^*$. Alice needs her secret key for this computation. Since $|e| = l+1 < k-1 = |\tilde{p}| = |\tilde{q}|$, the prime e does not divide $\varphi(n) = 4\tilde{p}\tilde{q}$. Hence, Alice can easily compute the inverse e^{-1} of e in $\mathbb{Z}_{\varphi(n)}^*$, by using her secret $\tilde{p}$ and $\tilde{q}$ (and the extended Euclidean algorithm, see Proposition A.17).
2. Signing is a probabilistic algorithm, because the prime e is generated probabilistically and a random quadratic residue $\tilde{y}$ is chosen. After these choices, the computation of the signature is deterministic. Therefore, we can describe the signature σ of m as the value of a mathematical function sign:

$$\sigma = \mathrm{sign}(h, n, g, x, \tilde{e}, e, \tilde{y}, m).$$

To compute the function sign by an algorithm, Alice has to use her knowledge about the prime factors of n.

Verification. Recipient Bob verifies a signature $\sigma = (e, y, \tilde{y})$ of Alice for message m as follows:

1. First, he checks whether e is an odd $(l+1)$-bit number that is not divisible by $\tilde{e}$.
2. Then he computes

$$\tilde{x} := \tilde{y}^{\tilde{e}} \cdot g^{-h(m)}$$

and checks whether

$$x = y^e \cdot g^{-h(\tilde{x})}.$$

He accepts if both checks are affirmative; otherwise he rejects.

Remarks:

1. Note that the verification algorithm does not verify that e is a prime; it only checks whether e is odd. A primality test would considerably decrease the efficiency of verification, and the security of the scheme does not require it (as the security proof shows).
2. If Alice generates a signature $(e, y, \tilde{y})$ with $e = \tilde{e}$, then this signature is not accepted by the verification procedure. However, since both e and $\tilde{e}$ are generated by *GenPrime*, this happens only with negligible probability, and Alice could simply generate a new prime in this case.

Theorem 11.12. *If the strong RSA assumption holds, "many" Sophie Germain primes exist and $\mathcal{H}$ is collision resistant, then the Cramer–Shoup signature scheme is secure against adaptively-chosen-message attacks.*

Remark. There is a variant of the signature scheme which does not require the collision resistance of the hash function (see [CraSho00]). The family $\mathcal{H}$ is only assumed to be a *universal one-way family of hash functions* ([NaoYun89]; [BelRog97]). The universal one-way property is weaker than full collision resistance: if an adversary Eve first chooses a message m and then a random key i is chosen, it should be infeasible for Eve to find $m' \neq m$ with $h_i(m) = h_i(m')$. Note that the size of the key can grow with the length of m.

In the proof of the theorem we need the following technical lemma.

Lemma 11.13. *There is a deterministic polynomial algorithm that for all k, given $n \in I_{SG,k}$, an odd natural number e with $|e| < k - 1$, a number f and elements $u, v \in \mathbb{Z}_n^*$ with $u^e = v^f$ as inputs, computes the r-th root $v^{r^{-1}} \in \mathbb{Z}_n^*$ of v for $r := e/d$ and $d := \gcd(e, f)$.*

Proof. e and hence r and d are prime to $\varphi(n)$, since $\varphi(n) = 4\tilde{p}\tilde{q}$, with Sophie Germain primes $\tilde{p}$ and $\tilde{q}$ of binary length $k - 1$, and e is an odd number with $|e| < k - 1$. Thus, the inverse elements r^{-1} of r and d^{-1} of d in $\mathbb{Z}_{\varphi(n)}^*$ exist. Let $s := f/d$. Since r is prime to s, the extended Euclidean algorithm (Algorithm A.6) computes integers m and m', with $sm + rm' = 1$. We have

$$u^r = \left(u^e\right)^{d^{-1}} = \left(v^f\right)^{d^{-1}} = v^s$$

and

$$\left(u^m \cdot v^{m'}\right)^r = (v^s)^m \cdot (v^{m'})^r = v^{sm+rm'} = v.$$

By setting

$$v^{r^{-1}} = u^m \cdot v^{m'}$$

we obtain the r-th root of v. $\qquad\qquad\square$

Proof (of Theorem 11.12). The proof runs by contradiction.

Let $Forger(h, n, g, x, \tilde{e})$ be a probabilistic polynomial forging algorithm which adaptively requests the signatures for t messages, where $t = R(k)$ for some polynomial R, and then produces a valid forgery with non-negligible probability for infinitely many security parameters (k, l). By non-negligible, we mean that the probability is $> 1/Q(k)$ for some positive polynomial Q.

We will define an attacking algorithm A that on inputs $n \in I_{SG}$ and $z \in \mathbb{Z}_n^*$ successfully computes an r-th root modulo n of z, without knowing the prime factors of n (contradicting the strong RSA assumption, by Lemma 11.11).

On inputs $n \in I_{SG,k}$ and $z \in \mathbb{Z}_n^*$, A works as follows:

1. Randomly and uniformly select the second security parameter l and randomly choose a hash function $h \in \mathcal{H}_l$ (by using $\mathcal{H}$'s key generator).
2. In a clever way, generate the missing elements g, x and $\tilde{e}$ of a public key $(n, g, x, \tilde{e})$.
3. Interact with *Forger* to obtain a forged signature (m, σ) for the public key $(n, g, x, \tilde{e})$.

   Since the prime factors of n are not known in this setting, *Forger* cannot get the signatures he requests from the original signing algorithm. Instead, he obtains them from A. Since g, x and $\tilde{e}$ were chosen in a clever way, A is able to supply *Forger* with valid signatures without knowing the prime factors of n.
4. By use of the forged signature (m, σ), compute an r-th root modulo n of z for some $r > 1$.

A simulates the legitimate signer in step 3. We therefore also say that *Forger* runs against a simulated signer. Simulating the signer is the core of the proof. To ensure that *Forger* yields a valid signature with a non-negligible probability, the probabilistic setting where *Forger* operates must be identical (or at least very close) to the setting where *Forger* runs against the legitimate signer. This means, in particular, that the keys generated in step 2 must be distributed as are the keys in the original signature scheme, and the signatures supplied to *Forger* in step 3 must be distributed as if they were generated by the legitimate signer.

We denote by $m_i, 1 \leq i \leq t$, the messages for which signatures are requested by *Forger*, and by $\sigma_i = (e_i, y_i, \tilde{y}_i)$ the corresponding signatures supplied to *Forger*. Let (m, σ) be the output of *Forger*, i.e., m is a message $\neq m_i, 1 \leq i \leq t$, and $\sigma = (e, y, \tilde{y})$ is the forged signature of m. Let $\tilde{x}_i := \tilde{y}_i^{\tilde{e}} \cdot g^{-h(m_i)}$ and $\tilde{x} := \tilde{y}^{\tilde{e}} \cdot g^{-h(m)}$.

We distinguish three (overlapping) types of forgery:

1. Type 1. For some i, $1 \leq i \leq t$, e_i divides e and $\tilde{x} = \tilde{x}_i$.
2. Type 2. For some i, $1 \leq i \leq t$, e_i divides e and $\tilde{x} \neq \tilde{x}_i$.
3. Type 3. For all i, $1 \leq i \leq t$, e_i does not divide e.

Here, note that the number e in the forged signature can be non-prime (the verification procedure does not test whether e is a prime). The numbers e_i are primes (see below).

We may define a forging algorithm $Forger_1$ which yields the output of $Forger$ if it is of type 1, and otherwise returns some message and signature not satisfying the verification condition. Analogously, we define $Forger_2$ and $Forger_3$. Then the valid forgeries of $Forger_1$ are of type 1, those of $Forger_2$ are of type 2 and those of $Forger_3$ of type 3. If $Forger$ succeeds with non-negligible probability, then at least one of the three "single-type forgers" succeeds with non-negligible probability. Replacing $Forger$ by this algorithm, we may assume from now on that $Forger$ generates valid forgeries of one type.

Case 1: Forger is of type 1.
To generate the public key in step 2, A proceeds as follows:

1. Generate $(l+1)$-bit primes $e_i := GenPrime(1^{l+1})$, $1 \leq i \leq t$. Set

$$g := z^{2\prod_{1 \leq i \leq t} e_i}.$$

2. Randomly choose $w \xleftarrow{u} \mathbb{Z}_n^*$ and set

$$x := w^{2\prod_{1 \leq i \leq t} e_i}.$$

3. Set $\tilde{e} := GenPrime(1^{l+1})$.

To generate the signature for the i-th message m_i, A randomly chooses $\tilde{y}_i \xleftarrow{u} QR_n$ and computes

$$\tilde{x}_i := \tilde{y}_i^{\tilde{e}} \cdot g^{-h(m_i)} \text{ and } y_i := \left(x \cdot g^{h(\tilde{x}_i)}\right)^{e_i^{-1}}.$$

Though A does not know the prime factors of n, she can easily compute the e_i-th root to get y_i, because she knows the e_i-th root of x and g by construction.

$Forger$ then outputs a forged signature $\sigma = (e, y, \tilde{y})$ for a message $m \notin \{m_1, \ldots, m_t\}$. If the forged signature does not pass the verification procedure, then the forger did not produce a valid signature. In this case, A returns a random exponent and a random element in $\mathbb{Z}_n^*$, and stops. Otherwise, $Forger$ yields a valid type-1 forgery. Thus, for some j, $1 \leq j \leq t$, we have $e_j \mid e$, $\tilde{x} = \tilde{x}_j$ and

$$\tilde{y}^{\tilde{e}} = \tilde{x} \cdot g^{h(m)}, \quad \tilde{y}_j^{\tilde{e}} = \tilde{x} \cdot g^{h(m_j)}.$$

Now A has to compute an r-th root of z. If $h(m) \neq h(m_j)$, which happens almost with certainty, since $\mathcal{H}$ is collision resistant, we get, by dividing the two equations, the equation

$$\tilde{z}^{\tilde{e}} = g^a = z^{2a\prod_{1 \leq i \leq t} e_i},$$

with $0 < a < 2^l$. Here recall that h yields l-bit numbers and these are $< 2^l$. Then $a := h(m) - h(m_j)$ if $h(m) - h(m_j) > 0$; otherwise $a := h(m_j) - h(m)$. Since $\tilde{e}$ is an $(l+1)$-bit prime and thus $\geq 2^l$, $\tilde{e}$ does not divide a. Moreover, $\tilde{e}$ and the e_i were chosen by $GenPrime$. Thus, with a high probability (stated more precisely below), $\tilde{e} \neq e_i, 1 \leq i \leq t$. In this case, A can compute $z^{\tilde{e}^{-1}}$ using Lemma 11.13 (note that $\tilde{e}$ is an $(l+1)$-bit prime and $l+1 < k-1$). A returns the exponent $\tilde{e}$ and the $\tilde{e}$-th root $z^{\tilde{e}^{-1}}$.

We still have to compute the probability of success of A. Interacting with the legitimate signer, $Forger$ is assumed to be successful. This means that there exists a positive polynomial P, an infinite subset $\mathcal{K} \subseteq \mathbb{N}$ and an $l(k)$ for each $k \in \mathcal{K}$ such that $Forger$ produces a valid signature with a probability $\geq 1/P(k)$ for all the pairs $(k, l(k)), k \in \mathcal{K}$, of security parameters. To state this more precisely, let $M_i(h, n, g, x, \tilde{e}, (m_j, \sigma_j)_{1 \leq j < i})$ be the random variable describing the choice of m_i by $Forger$ $(1 \leq i \leq t)$. The signatures σ_i are additional inputs to $Forger$. We have for $k \in \mathcal{K}$ and $l := l(k)$:

$$p_{\text{success}, Forger}(k, l) = \text{prob}(Verify(h, n, g, x, \tilde{e}, m, \sigma) = \text{accept} :$$
$$h \leftarrow \mathcal{H}_l, n \stackrel{u}{\leftarrow} I_{\text{SG},k},$$
$$g \stackrel{u}{\leftarrow} \text{QR}_n, x \stackrel{u}{\leftarrow} \text{QR}_n, \tilde{e} \leftarrow GenPrime(1^{l+1}),$$
$$e_i \leftarrow GenPrime(1^{l+1}), \tilde{y}_i \stackrel{u}{\leftarrow} \text{QR}_n,$$
$$m_i \leftarrow M_i(h, n, g, x, \tilde{e}, (m_j, \sigma_j)_{1 \leq j < i}),$$
$$\sigma_i = \text{sign}(h, n, g, x, \tilde{e}, e_i, \tilde{y}_i, m_i), 1 \leq i \leq t,$$
$$(m, \sigma) \leftarrow Forger(h, n, g, x, \tilde{e}, (\sigma_i)_{1 \leq i \leq t}))$$
$$\geq \frac{1}{P(k)}.$$

Here recall that the signature σ_i of m_i can be derived deterministically from $h, n, g, x, \tilde{e}, e_i, \tilde{y}_i$ and m_i as the value of a mathematical function sign (see p. 384).

Let $\psi(z, (e_i)_{1 \leq i \leq t}) := g = z^{2 \prod_{1 \leq i \leq t} e_i}$ and $\chi(w, (e_i)_{1 \leq i \leq t}) := x = w^{2 \prod_{1 \leq i \leq t} e_i}$ be the specific g and x constructed by A. $A(n, z)$ succeeds in computing the root $z^{\tilde{e}^{-1}}$ if the $Forger$ produces a valid signature, if $h(m) \neq h(m_i)$ and if $\tilde{e} \neq e_i, 1 \leq i \leq t$. All other steps in A are deterministic. Therefore, we may compute the probability of success of A (for security parameter k) as follows:

$$\text{prob}(v^r = z : n \stackrel{u}{\leftarrow} I_{\text{SG},k}, z \stackrel{u}{\leftarrow} \mathbb{Z}_n^*, (v, r) \leftarrow A(n, z))$$
$$\geq \text{prob}(Verify(h, n, \psi(z, (e_i)_i), \chi(w, (e_i)_i), \tilde{e}, m, \sigma) = \text{accept},$$
$$h(m) \neq h(m_i), \tilde{e} \neq e_i, 1 \leq i \leq t :$$
$$h \leftarrow \mathcal{H}_l, n \stackrel{u}{\leftarrow} I_{\text{SG},k}, z \stackrel{u}{\leftarrow} \mathbb{Z}_n^*, w \stackrel{u}{\leftarrow} \mathbb{Z}_n^*, \tilde{e} \leftarrow GenPrime(1^{l+1}),$$
$$e_i \leftarrow GenPrime(1^{l+1}), \tilde{y}_i \stackrel{u}{\leftarrow} \text{QR}_n,$$
$$m_i \leftarrow M_i(h, n, \psi(z, (e_i)_i), \chi(w, (e_i)_i), \tilde{e}, (m_j, \sigma_j)_{1 \leq j < i}),$$

$$\sigma_i = \text{sign}(h, n, \psi(z, (e_i)_i), \chi(w, (e_i)_i), \tilde{e}, e_i, \tilde{y}_i, m_i), 1 \leq i \leq t,$$
$$(m, \sigma) \leftarrow Forger(h, n, \psi(z, (e_i)_i), \chi(w, (e_i)_i), \tilde{e}, (\sigma_i)_{1 \leq i \leq t}))$$

$$=: p_1$$

Let Q be a positive polynomial. *GenPrime*, when called polynomially times, generates the same prime more than once only with negligible probability, and h is randomly chosen from a collision-resistant family $\mathcal{H}$ of hash functions. Thus, there is some k_0 such that both the probability that $\tilde{e} \neq e_i$ for $1 \leq i \leq t$ and the probability that $h(m) \neq h(m_i)$ for $1 \leq i \leq t$ are $\geq 1 - 1/Q(k)$, for $k \geq k_0$.[9] Hence, we get for $k \geq k_0$ that

$$p_1 \geq \text{prob}(Verify(h, n, \psi(z, (e_i)_i), \chi(w, (e_i)_i, \tilde{e}, m, \sigma) = \text{accept} :$$
$$h \leftarrow \mathcal{H}_l, n \xleftarrow{u} I_{\text{SG},k}, z \xleftarrow{u} \mathbb{Z}_n^*, w \xleftarrow{u} \mathbb{Z}_n^*, \tilde{e} \leftarrow GenPrime(1^{l+1}),$$
$$e_i \leftarrow GenPrime(1^{l+1}), \tilde{y}_i \xleftarrow{u} \text{QR}_n,$$
$$m_i \leftarrow M_i(h, n, \psi(z, (e_i)_i), \chi(w, (e_i)_i), \tilde{e}, (m_j, \sigma_j)_{1 \leq j < i}),$$
$$\sigma_i = \text{sign}(h, n, \psi(z, (e_i)_i), \chi(w, (e_i)_i), \tilde{e}, e_i, \tilde{y}_i, m_i), 1 \leq i \leq t,$$
$$(m, \sigma) \leftarrow Forger(h, n, \psi(z, (e_i)_i), \chi(w, (e_i)_i), \tilde{e}, (\sigma_i)_{1 \leq i \leq t}))$$
$$\cdot \left(1 - \frac{1}{Q(k)}\right) \cdot \left(1 - \frac{1}{Q(k)}\right)$$

$$=: p_2$$

The first factor in p_2 is the probability that *Forger* successfully yields a valid signature when interacting with the simulated signer. This probability is equal to *Forger*'s probability of success, $p_{\text{success},Forger}(k, l)$, when he interacts with the legitimate signer. Namely, $\psi(z, (e_i)_i)$ and $\chi(w, (e_i)_i)$ are uniformly distributed quadratic residues, independent of the distribution of the e_i, since $z \xleftarrow{u} \mathbb{Z}_n^*$ and $w \xleftarrow{u} \mathbb{Z}_n^*$.[10] Thus, we may replace $\psi(z, (e_i)_i), z \xleftarrow{u} \mathbb{Z}_n^*$ and $\chi(w, (e_i)_i), w \xleftarrow{u} \mathbb{Z}_n^*$ by $g \xleftarrow{u} \text{QR}_n$ and $x \xleftarrow{u} \text{QR}_n$. We get

$$p_2 = p_{\text{success},Forger}(k, l) \cdot \left(1 - \frac{1}{Q(k)}\right)^2.$$

For $k \in \mathcal{K}$ and $l = l(k)$, we have $p_{\text{success},Forger}(k, l) \geq 1/P(k)$. The probability that A chooses $l(k)$ in her first step is $\geq 1/k$, and we finally obtain that

$$\text{prob}(v^r = z : n \xleftarrow{u} I_{\text{SG},k}, z \xleftarrow{u} \mathbb{Z}_n^*, (v, r) \leftarrow A(n, z)) \geq \frac{1}{P(k)} \cdot \frac{1}{k} \cdot \left(1 - \frac{1}{Q(k)}\right)^2,$$

for the infinitely many $k \in \mathcal{K}$. This contradicts the strong RSA assumption. The proof of Theorem 11.12 is finished in the case where the forger is of type 1. The other cases are proven below. $\square$

---

[9] Note that the messages m and m_i are generated by a probabilistic polynomial algorithm, namely *Forger*.

[10] Here note that $\prod_i e_i$ and $\tilde{e}$ are prime to $\varphi(n) = 4\tilde{p}\tilde{q}$, because $\tilde{e}$ and the e_i are $(l+1)$-bit primes and $l + 1 < k - 1 = |\tilde{p}| = |\tilde{q}|$.

Remark. Before we study the next cases, let us have a look back at the proof just finished. The core of the proof is to simulate the legitimate signer and to supply valid signatures to the forger, without knowing the prime factors of the modulus n. We managed to do this by a clever choice of the second parts of the public keys. Here, the key point is that given a fixed first part of the public key, i.e., given a modulus n, the joint distribution of the second part $(g, x, \tilde{e})$ of the public key and the generated signatures is the same in the legitimate signer and in the simulation by A. This fact is often referred to as "the simulated signer perfectly simulates the legitimate signer".

Proof (of cases 2 and 3).

Case 2: Forger is of type 2.
We may assume that the j with $e_j \mid e$ and $\tilde{x} \neq \tilde{x}_j$ is fixed. Namely, we may guess the correct j, i.e., we iterate over the polynomially many cases j and assume j fixed in each case (see Chapter 7, proof of Theorem 7.7, for an analogous argument).

To generate the missing elements g, x and $\tilde{e}$ of the public key, A proceeds as follows:

1. Generate $(l+1)$–bit primes $e_i := GenPrime(1^{l+1})$, $1 \le i \le t$. Choose a further prime $\tilde{e} := GenPrime(1^{l+1})$. Set

$$g := z^{2\tilde{e} \prod_{i \neq j} e_i}.$$

2. Randomly choose $w \stackrel{u}{\leftarrow} \mathbb{Z}_n^*$ and $u \stackrel{u}{\leftarrow} \mathbb{Z}_n^*$. Set

$$y_j := w^{2 \prod_{i \neq j} e_i} \text{ and } \tilde{x}_j := u^{2\tilde{e}}.$$

3. Let $x := y_j^{e_j} \cdot g^{-h(\tilde{x}_j)}$.

Then g and x are uniformly distributed quadratic residues, since z, w and u are uniformly distributed (and the exponents $\tilde{e}$ and e_i are prime to $\varphi(n)$, see the footnote on p. 389).

To generate the signature $(e_i, y_i, \tilde{y}_i)$ for the i-th message m_i, requested by *Forger*, A proceeds as follows:

1. If $i \neq j$, then A randomly chooses $\tilde{y}_i \stackrel{u}{\leftarrow} QR_n$ and computes

$$\tilde{x}_i := \tilde{y}_i^{\tilde{e}} \cdot g^{-h(m_i)} \text{ and } y_i := \left(x \cdot g^{h(\tilde{x}_i)} \right)^{e_i^{-1}}.$$

   A can compute the e_i-th root, because the e_i-th roots of g and x are known to her by construction.

2. If $i = j$, then the value of y_j has already been computed above. Moreover, A can compute the correct value of $\tilde{y}_j = (\tilde{x}_j g^{h(m_j)})^{\tilde{e}^{-1}}$, because she knows the $\tilde{e}$-th root of $\tilde{x}_j$ and g. Note that $\tilde{y}_j$ is uniformly distributed, as required, since $\tilde{x}_j$ is uniformly distributed.

It is obvious from the construction that A generates signatures which satisfy the verification condition. *Forger* outputs a forged signature $\sigma = (e, y, \tilde{y})$ for a message $m \notin \{m_1, \ldots, m_t\}$. If the forged signature does not pass the verification procedure, A returns a random exponent and a random element in $\mathbb{Z}_n^*$, and stops. Otherwise, *Forger* yields a valid type-2 forgery such that e_j divides e, i.e., $e = e_j \cdot f$ and $\tilde{y}^{\tilde{e}} \cdot g^{-h(m)} = \tilde{x} \neq \tilde{x}_j$. We have

$$y^e = \left(y^f\right)^{e_j} = x \cdot g^{h(\tilde{x})} \text{ and } y_j{}^{e_j} = x \cdot g^{h(\tilde{x}_j)}.$$

Now A has to compute an r-th root of z. If $h(\tilde{x}) \neq h(\tilde{x}_j)$, which happens almost with certainty, since $\mathcal{H}$ is collision resistant, we get, by dividing the two equations, the equation

$$\tilde{z}^{e_j} = g^a = z^{2a\tilde{e}\prod_{i \neq j} e_i},$$

with $0 < a < 2^l$. Since all $(l+1)$-bit primes are chosen by $GenPrime(1^{l+1})$, the probability that e_j is equal to $\tilde{e}$ or equal to an $e_i, i \neq j$, is negligibly small. If e_j is different from $\tilde{e}$ and the $e_i, i \neq j$, then we can compute $z^{e_j^{-1}}$ by Lemma 11.13. In this case, A returns the exponent e_j and the e_j-th root $z^{e_j^{-1}}$.

$A(n, z)$ succeeds in computing the root $z^{e_j^{-1}}$ if the *Forger* produces a valid signature, if $h(\tilde{x}) \neq h(\tilde{x}_j)$ and if e_j is different from $\tilde{e}$ and the $e_i, i \neq j$. As in case 1, it follows from the construction that A perfectly simulates the choice of the key together with the generation of signatures supplied to *Forger*. Thus, as we have seen in case 1, the *Forger*'s probability of success if he interacts with the simulated signer is the same as if he interacted with the legitimate signer. Computing the probabilities in a completely analogous way as in case 1, we derive that

$$\text{prob}(v^r = z : n \xleftarrow{u} I_{SG,k}, z \xleftarrow{u} \mathbb{Z}_n^*, (v, r) \leftarrow A(n, z)) \geq \frac{1}{S(k)},$$

for some positive polynomial S and infinitely many k. This contradicts the strong RSA assumption (Lemma 11.11).

Case 3: Forger is of type 3.
To complete the public key by g, x and $\tilde{e}$, A proceeds as follows:

1. Generate $(l+1)$-bit primes $\tilde{e}$ and $e_i, 1 \leq i \leq t$, by applying $GenPrime(1^{l+1})$. Set

$$g := z^{2\tilde{e}\prod_i e_i}.$$

2. Choose $a \xleftarrow{u} \{1, \ldots, n^2\}$ and set $x := g^a$.

A can easily generate valid signatures $(e_i, y_i, \tilde{y}_i)$ for messages $m_i, 1 \leq i \leq t$, requested by *Forger*. Namely, A chooses $\tilde{y}_i \xleftarrow{u} QR_n$ and computes $\tilde{x}_i = \tilde{y}_i^{e_i} \cdot g^{-h(m_i)}$ and $y_i = (x \cdot g^{h(\tilde{x}_i)})^{e_i^{-1}}$. The latter computation works, because due to the construction, the e_i-th roots of g and x can be immediately derived for $1 \leq i \leq t$.

Then *Forger* outputs a forged signature $\sigma = (e, y, \tilde{y})$ for a message $m \notin \{m_1, \ldots, m_t\}$. The signature is of type 3, i.e., e_i does not divide $e, 1 \leq i \leq t$. If the signature is valid, we get the equation

$$y^e = x \cdot g^{h(\tilde{x})} = z^f, \quad \text{where } f = 2\tilde{e} \prod_i e_i \cdot (a + h(\tilde{x})),$$

with $\tilde{x} = \tilde{y}^{\tilde{e}} \cdot g^{-h(m)}$.

To compute a root of z, let $d = \gcd(e, f)$. Observe that e may be a non-prime, because the verification algorithm only tests whether e is odd. If $d < e$, i.e., e does not divide f, then $r := e/d > 1$, and we can compute the r-th root $z^{r^{-1}}$ by Lemma 11.13.

Thus, A succeeds in computing a root if *Forger* yields a valid forgery and if e does not divide f. To compute the probability that *Forger* yields a valid forgery, we have to consider the distribution of the keys (g, x), generated by A in step 2.

By the definition of n, QR_n is a cyclic group of order $\tilde{p}\tilde{q}$ with distinct Sophie Germain primes $\tilde{p}$ and $\tilde{q}$.

Note that $\tilde{e}$ and the e_i are $(l + 1)$-bit primes, and $l + 1 < k - 1$. Thus, $\tilde{e}$ and none of the e_i are equal to $\tilde{p}$ or $\tilde{q}$, which are $(k - 1)$-bit primes. Hence, $\tilde{e} \prod_i e_i$ is prime to $\varphi(n) = 4\tilde{p}\tilde{q}$. We conclude that g is uniformly distributed in QR_n, since z is uniformly chosen from $\mathbb{Z}_n^*$.

Let $a = b\tilde{p}\tilde{q} + c, 0 \leq c < \tilde{p}\tilde{q}$ (division with remainder). Now $2^{k-2} < \tilde{p}, \tilde{q} < 2^{k-1}$, $n^2 \approx 2^{4k}$, and $a \xleftarrow{u} \{1, \ldots, n^2\}$ is uniformly chosen. This implies that the probability of a remainder $c, 0 \leq c < \tilde{p}\tilde{q}$, differs from $1/\tilde{p}\tilde{q}$ by at most $1/n^2 \approx 1/2^{4k}$. This means that the distribution of c is polynomially close to the uniform distribution. This in turn implies that the conditional distribution of x, assuming that g is a generator of QR_n, is polynomially close to the uniform distribution on QR_n. However, $\mathrm{QR}_n \cong \mathbb{Z}_{\tilde{p}\tilde{q}} \cong \mathbb{Z}_{\tilde{p}} \times \mathbb{Z}_{\tilde{q}}$ and therefore $(\tilde{p}-1)(\tilde{q}-1)$ of the $\tilde{p}\tilde{q}$ elements in QR_n are generators. Thus, the probability that g is a generator of QR_n is $\geq 1 - 1/2^{k-3}$, which is exponentially close to 1. Summarizing, we get that the distribution of x is polynomially close to the uniform distribution on QR_n.

We see that A almost perfectly simulates the legitimate signer. The distributions of the keys and the signatures supplied to *Forger* are polynomially close to the distributions when *Forger* interacts with the legitimate signer. By Lemmas B.29 and B.32, we conclude that the probability that *Forger* produces a valid signature, if he interacts with the simulated signer in A, cannot be polynomially distinguished from his probability of success when interacting with the legitimate signer. Thus, *Forger* produces in step 3 of A a valid signature with probability $\geq 1/Q(k)$ for some positive polynomial Q and infinitely many k.

We still have to study the conditional probability that e does not divide f, assuming that *Forger* produces a valid signature. It is sufficient to prove that this probability is non-negligible. We will show that it is $> 1/2$. If we could prove this estimate assuming $h, n, g, x, \tilde{e}$ and the forged signed message (m, σ) fixed, for every $h \in \mathcal{H}_l, n \in I_{SG,k}$, every g, x and $\tilde{e}$ possibly generated by A, and every valid (m, σ) possibly output by *Forger*, then we are done (take the sum over all $h, n, g, x, \tilde{e}, m$ and σ). Therefore, we now assume that h, n, g, x and $\tilde{e}$, and m and σ are fixed. This implies that c and $\tilde{x}$ are also fixed.

Let s be a prime dividing e. Then $s > 2$ and $s \neq \tilde{e}$, because otherwise the verification condition was not satisfied. Moreover, $s \neq e_i$, since the forgery is of type 3. Thus, it suffices to prove that s does not divide $a + h(\tilde{x})$ with probability $\geq 1/2$, assuming $h, n, g, x, \tilde{e}, m$ and σ fixed. Let $a = b\tilde{p}\tilde{q} + c$, as above. $a + h(\tilde{x}) = b\tilde{p}\tilde{q} + c + h(\tilde{x}) = L(b)$, with L a linear function (note that c and $\tilde{x}$ are fixed). The probability that s divides $a + h(\tilde{x})$ is the same as the probability that $L(b) \equiv 0 \bmod s$. Now the conditional distribution of b, assuming c fixed, is also polynomially close to the uniform distribution on $\{0, \ldots, \lfloor n^2/\tilde{p}\tilde{q} \rfloor\}$. Thus, the distribution of $b \bmod s$ is polynomially close to the uniform distribution. s does not divide $\tilde{p}\tilde{q}$, because $|s| \leq l+1 < k-1$ and $|\tilde{p}| = |\tilde{q}| = k-1$. Thus, $L(b) \equiv 0 \bmod s$ is a non-vanishing linear equation over $\mathbb{Z}_s$. Hence, the probability that $L(b) = 0 \bmod s$ is very close to $1/s$. This means that s and hence e do not divide f, with a probability $\geq 1 - 1/(s-1) \geq 1/2$ (recall that $s > 2$).

Now the proof of case 3 is finished, and the proof of Theorem 11.12 is complete.   □

Exercises

1. Consider the construction of collision-resistant hash functions in Section 11.2. Explain how the prefix-free encoding of the messages can be avoided by applying Merkle's meta method (Section 2.2.2).

2. Let $I = (I_k)_{k \in \mathbb{N}}$ be a key set with security parameter k, and let $\mathcal{H} = \left(h_i : \{0,1\}^* \longrightarrow \{0,1\}^{g(k(i))} \right)_{i \in I}$ be a family of collision-resistant hash functions. Let $l_i \geq g(k(i)) + k(i)$ for all $i \in I$, and let $\{0,1\}^{\leq l_i} := \{m \in \{0,1\}^* \mid 1 \leq |m| \leq l_i\}$ be the bit strings of length $\leq l_i$. Show that the family

$$\left(h_i : \{0,1\}^{\leq l_i} \longrightarrow \{0,1\}^{g(k(i))} \right)_{i \in I}$$

is a family of one-way functions (with respect to $\mathcal{H}$'s key generator).

3. The RSA and ElGamal signature schemes are introduced in Chapter 3. There, various attacks against the basic schemes (no hash function is applied to the messages) are discussed. Classify these attacks and their levels of success (according to Section 11.1).

4. The following signature scheme was suggested by Ong, Schnorr and Shamir ([OngSchSha84]). Alice chooses two random large distinct primes p and q, and a random $x \in \mathbb{Z}_n^*$, where $n := pq$. Then she computes $y := -x^{-2} \in \mathbb{Z}_n^*$ and publishes (n, y) as her public key. Her secret is x (she does not need to know the prime factors of n). To sign a message $m \in \mathbb{Z}_n$, she randomly selects an $r \in \mathbb{Z}_n^*$ and calculates (in $\mathbb{Z}_n$)

$$s_1 := 2^{-1} \left(r^{-1}m + r \right) \text{ and } s_2 := 2^{-1}x \left(r^{-1}m - r \right).$$

(s_1, s_2) is the signature of m. To verify a signature (s_1, s_2), Bob checks that $m = s_1^2 + ys_2^2$ (in $\mathbb{Z}_n$):

   a. Prove that retrieving the secret key by a key-only attack is equivalent to the factoring of n.
   b. The scheme is existentially forgeable by a key-only attack.
   c. The probability that a randomly chosen pair (s_1, s_2) is a signature for a given message m is negligibly small.
   d. State the problem, which an adversary has to solve, of forging signatures for messages of his choice by a key-only attack.
   (In fact, Pollard has broken the scheme by an efficient algorithm for this problem, see [PolSch87].)

5. Let $I = (I_k)_{k \in \mathbb{N}}$ be a key set with security parameter k, and let $f_0 = (f_{0,i} : D_i \longrightarrow D_i)_{i \in I}, f_1 = (f_{1,i} : D_i \longrightarrow D_i)_{i \in I}$ be a claw-free pair of trapdoor permutations with key generator K. We consider the following signature scheme. As her public key, Alice randomly chooses an index $i \in I_k$ – by computing $K(1^k)$ – and a reference value $x \overset{u}{\leftarrow} D_i$. Her private key is the trapdoor information of $f_{0,i}$ and $f_{1,i}$. We encode the messages $m \in \{0,1\}^*$ in a prefix-free way (see Section 11.2), and denote the encoded m by $[m]$. Then Alice's signature of a message m is $\sigma(i, x, m) := f_{[m],i}^{-1}(x)$, where $f_{[m],i}$ is defined as in Section 11.2. Bob can verify Alice's signature σ by comparing $f_{[m],i}(\sigma)$ with x. Study the security of the scheme. More precisely, show

   a. A claw of f_0 and f_1 can be computed from $\sigma(i, x, m)$ and $\sigma(i, x, m')$ if the messages m and m' are distinct.
   b. The scheme is secure against existential forgery by a key-only attack.
   c. Assume that the message space has polynomial cardinality. More precisely, let $c \in \mathbb{N}$ and assume that only messages $m \in \{0,1\}^{c \lfloor \log_2(k) \rfloor}$ are signed by the scheme. Then the scheme is secure against adaptively-chosen-message attacks if used as a *one-time signature scheme* (i.e., Alice signs at most one message with her key).
   d. Which problem do you face in the security proof in c if the scheme is used to sign arbitrary messages in $\{0,1\}^*$?

6. We consider the same setting as in Exercise 5. We assume that the generation of the messages $m \in \{0,1\}^*$ to be signed can be uniformly modeled for all users by a probabilistic polynomial algorithm $M(i)$. In particular,

this means that the messages Alice wants to sign do not depend on her reference value x.

Let (i, x) be the public key of Alice, and let m_j be the j-th message to be signed by Alice. The signature $\sigma(i, x, m_j)$ of m_j is defined as $\sigma(i, x, m_j) := (s_j, [m_1]\| \ldots \|[m_{j-1}])$, with $s_j := f^{-1}_{[m_j], i}(s_{j-1})$. Here $[m_1]\| \ldots \|[m_{j-1}]$ denotes the concatenation of the prefix-free encoded messages, and $s_0 := x$ is Alice's randomly chosen reference value (see [GolBel01]):

a. Show by an example that in order to prevent forging by known signature attacks, the verification procedure has to check whether the bit string $\hat{m}$ in a signature $(s, \hat{m})$ is well formed with respect to the prefix-free encoding.

Give the complete verification condition.

b. Prove that no one can existentially forge a signature by a known-signature attack.

A. Algebra and Number Theory

Public-key cryptosystems are based on modular arithmetic. In this section, we summarize the concepts and results from algebra and number theory which are necessary for an understanding of cryptographic methods. Textbooks on number theory and modular arithmetic include [HarWri79], [IreRos82], [Rose94], [Forster96] and [Rosen00]. This section is also intended to establish notation. We assume that the reader is familiar with elementary notions of algebra such as groups, rings and fields.

A.1 The Integers

$\mathbb{Z}$ denotes the ring of integers; $\mathbb{N} = \{z \in \mathbb{Z} \mid z > 0\}$ denotes the subset of natural numbers.

We first introduce the notion of divisors and the fundamental Euclidean algorithm which computes the greatest common divisor of two numbers.

Definition A.1. Let $a, b \in \mathbb{Z}$:

1. *a divides b* if there is some $c \in \mathbb{Z}$ with $b = ac$. b is called a *multiple* of a. We write $a \mid b$ for "a divides b".
2. $d \in \mathbb{N}$ is called the *greatest common divisor* of a and b if:
   a. d divides a and d divides b.
   b. If $d' \in \mathbb{Z}$ divides both a and b, then d' divides d.
   The greatest common divisor is denoted by $\gcd(a, b)$.
3. If $\gcd(a, b) = 1$, then a is called *relatively prime to b*, or *prime to b* for short.
4. $m \in \mathbb{N}$ is called the *least common multiple* of a and b if:
   a. m is a multiple of a and of b.
   b. If $m' \in \mathbb{Z}$ is a multiple of a and b, then m divides m'.
   The least common multiple is denoted by $\mathrm{lcm}(a, b)$.

Proposition A.2. *Let* $a, b \in \mathbb{Z}$. *Then* $\gcd(a, b) \cdot \mathrm{lcm}(a, b) = |a \cdot b|$.

Proof. We may assume that $a, b \geq 0$. If $a = 0$ or $b = 0$, then $\operatorname{lcm}(a, b) = 0$ and the equation holds. Let $a, b > 0$ and $m := \operatorname{lcm}(a, b)$. Since ab is a multiple of a and b, m divides ab, i.e., $ab = md$. We now show that $d = {ab}/{m} = \gcd(a, b)$. We have $a = {m}/{b}d$ and $b = {m}/{a}d$, which means that d divides a and b. Let d' be a divisor of a and b. Then $a{b}/{d'} = {a}/{d'}b$ is a multiple of a and b. Hence m divides ${a}/{d'}b$. It follows that d' divides $d = {ab}/{m}$. □

Theorem A.3 *(Division with remainder). Let $z, a \in \mathbb{Z}, a \neq 0$. Then there are unique numbers $q, r \in \mathbb{Z}$ such that $z = q \cdot a + r$ and $0 \leq r < |a|$.*

Proof. In the first step, we prove that such q and r exist. If $a > 0$ and $z \geq 0$, we may apply induction on z. For $0 \leq z < a$ we obviously have $z = 0 \cdot a + z$. If $z \geq a$, then, by induction, $z - a = q \cdot a + r$ for some q and $r, 0 \leq r < a$, and hence $z = (q + 1) \cdot a + r$. If $z < 0$ and $a > 0$, then we have just shown the existence of an equation $-z = q \cdot a + r, 0 \leq r < a$. Then $z = -q \cdot a$ if $r = 0$, and $z = -q \cdot a - r = -q \cdot a - a + (a - r) = -(q + 1) \cdot a + (a - r)$ and $0 < a - r < a$. If $a < 0$, then $-a > 0$. Hence $z = q \cdot (-a) + r = -q \cdot a + r$, with $0 \leq r < |a|$.

To prove uniqueness, consider $z = q_1 \cdot a + r_1 = q_2 \cdot a + r_2$. Then $0 = (q_1 - q_2) \cdot a + (r_1 - r_2)$. Hence a divides $(r_1 - r_2)$. Since $|r_1 - r_2| < |a|$, this implies $r_1 = r_2$, and then also $q_1 = q_2$. □

Remark. r is called the *remainder* of z modulo a. We write $z \bmod a$ for r. The number q is the *(integer) quotient* of z and a. We write $z \operatorname{div} a$ for q.

The Euclidean Algorithm. Let $a, b \in \mathbb{Z}, a > b > 0$. The greatest common divisor $\gcd(a, b)$ can be computed by an iterated division with remainder. Let $r_0 := a, r_1 := b$ and

$$
\begin{aligned}
r_0 &= q_1 r_1 + r_2, & 0 < r_2 < r_1, \\
r_1 &= q_2 r_2 + r_3, & 0 < r_3 < r_2, \\
&\vdots \\
r_{k-1} &= q_k r_k + r_{k+1}, & 0 < r_{k+1} < r_k, \\
&\vdots \\
r_{n-2} &= q_{n-1} r_{n-1} + r_n, & 0 < r_n < r_{n-1}, \\
r_{n-1} &= q_n r_n + r_{n+1}, & 0 = r_{n+1}.
\end{aligned}
$$

By construction, $r_1 > r_2 > \ldots$. Therefore, the remainder becomes 0 after a finite number of steps. The last remainder $\neq 0$ is the greatest common divisor, as is shown in the next proposition.

Proposition A.4.

1. $r_n = \gcd(a, b)$.
2. *There are numbers $d, e \in \mathbb{Z}$ with $\gcd(a, b) = da + eb$.*

Proof. 1. From the equations considered in reverse order, we conclude that r_n divides r_k, $k = n-1, n-2 \ldots$. In particular, r_n divides $r_1 = b$ and $r_0 = a$. Now let t be a divisor of $a = r_0$ and $b = r_1$. Then $t \mid r_k, k = 2, 3, \ldots$, and hence $t \mid r_n$. Thus, r_n is the greatest common divisor.

2. Iteratively substituting r_{k+1} by $r_{k-1} - q_k r_k$, we get

$$
\begin{aligned}
r_n &= r_{n-2} - q_{n-1} \cdot r_{n-1} \\
&= r_{n-2} - q_{n-1} \cdot (r_{n-3} - q_{n-2} \cdot r_{n-2}) \\
&= (1 + q_{n-1} q_{n-2}) \cdot r_{n-2} - q_{n-1} \cdot r_{n-3} \\
&\quad \vdots \\
&= da + eb,
\end{aligned}
$$

with integers d and e. □

We have shown that the following algorithm, called *Euclid's algorithm*, outputs the greatest common divisor. $\mathrm{abs}(a)$ denotes the absolute value of a.

Algorithm A.5.
```
int gcd(int a, b)
1   while b ≠ 0 do
2       r ← a mod b
3       a ← b
4       b ← r
5   return abs(a)
```

We now extend the algorithm such that not only $\gcd(a, b)$ but also the coefficients d and e of the linear combination $\gcd(a, b) = da + eb$ are computed. For this purpose, we write the recursion

$$r_{k-1} = q_k r_k + r_{k+1}$$

using matrices

$$\begin{pmatrix} r_k \\ r_{k+1} \end{pmatrix} = Q_k \begin{pmatrix} r_{k-1} \\ r_k \end{pmatrix}, \text{ where } Q_k = \begin{pmatrix} 0 & 1 \\ 1 & -q_k \end{pmatrix}, k = 1, \ldots, n.$$

Multiplying the matrices, we get

$$\begin{pmatrix} r_n \\ r_{n+1} \end{pmatrix} = Q_n \cdot Q_{n-1} \cdot \ldots \cdot Q_1 \begin{pmatrix} r_0 \\ r_1 \end{pmatrix}.$$

The first component of this equation yields the desired linear combination for $r_n = \gcd(a, b)$. Therefore, we have to compute $Q_n \cdot Q_{n-1} \cdot \ldots \cdot Q_1$. This is accomplished by iteratively computing the matrices

$$\Lambda_0 = \begin{pmatrix} 1 & 0 \\ 0 & 1 \end{pmatrix}, \quad \Lambda_k = \begin{pmatrix} 0 & 1 \\ 1 & -q_k \end{pmatrix} \Lambda_{k-1}, \quad k = 1, \ldots, n,$$

to finally get $\Lambda_n = Q_n \cdot Q_{n-1} \cdot \ldots \cdot Q_1$. In this way, we have derived the following algorithm, called the *extended Euclidean algorithm*. On inputs a and b it outputs the greatest common divisor and the coefficients d and e of the linear combination $\gcd(a, b) = da + eb$.

Algorithm A.6.

```
    int array gcdCoef (int a, b)
 1      λ11 ← 1, λ22 ← 1, λ12 ← 0, λ21 ← 0
 2      while b ≠ 0 do
 3          q ← a div b
 4          r ← a mod b
 5          a ← b
 6          b ← r
 7          t21 ← λ21; t22 ← λ22
 8          λ21 ← λ11 − q · λ21
 9          λ22 ← λ12 − q · λ22
10          λ11 ← t21
11          λ12 ← t22
12      return (abs(a), λ11, λ12)
```

We analyze the running time of the Euclidean algorithm. Here we meet the Fibonacci numbers.

Definition A.7. The *Fibonacci numbers* f_n are recursively defined by

$$f_0 := 0, \quad f_1 := 1,$$
$$f_n := f_{n-1} + f_{n-2}, \text{ for } n \geq 2.$$

Remark. The Fibonacci numbers can be non-recursively computed using the formula

$$f_n = \frac{1}{\sqrt{5}}(g^n - \tilde{g}^n),$$

where g and $\tilde{g}$ are the solutions of the equation $x^2 = x + 1$:

$$g := \frac{1}{2}\left(1 + \sqrt{5}\right) \text{ and } \tilde{g} := 1 - g = -\frac{1}{g} = \frac{1}{2}\left(1 - \sqrt{5}\right).$$

See, for example, [Forster96].

Definition A.8. The number g is called the *Golden Ratio*.[1]

Lemma A.9. *For $n \geq 2$, $f_n \geq g^{n-2}$. In particular, the Fibonacci numbers grow exponentially fast.*

---

[1] It is the proportion of length to width which the Greeks found most beautiful.

Proof. The statement is clear for $n = 2$. By induction on n, assuming that the statement holds for $\leq n$, we get

$$f_{n+1} = f_n + f_{n-1} \geq g^{n-2} + g^{n-3} = g^{n-3}(1 + g) = g^{n-3}g^2 = g^{n-1}.$$

$\square$

Proposition A.10. *Let $a, b \in \mathbb{Z}$, $a > b > 0$. Assume that computing $\gcd(a, b)$ by the Euclidean algorithm takes n iterations (i.e., using n divisions with remainder). Then $a \geq f_{n+1}$ and $b \geq f_n$.*

Proof. Let $r_0 := a, r_1 := b$ and consider

$$
\begin{aligned}
r_0 &= q_1 r_1 + r_2, & f_{n+1} &= f_n + f_{n-1}, \\
r_1 &= q_2 r_2 + r_3, & f_n &= f_{n-1} + f_{n-2}, \\
&\;\;\vdots & \text{and} \qquad &\;\;\vdots \\
r_{n-2} &= q_{n-1} r_{n-1} + r_n, & f_3 &= f_2 + f_1, \\
r_{n-1} &= q_n r_n, & f_2 &= f_1.
\end{aligned}
$$

By induction, starting with $i = n$ and descending, we show that $r_i \geq f_{n+1-i}$. For $i = n$, we have $r_n \geq f_1 = 1$. Now assume the inequality proven for $\geq i$. Then

$$r_{i-1} = q_i r_i + r_{i+1} \geq r_i + r_{i+1} \geq f_{n+1-i} + f_{n+1-(i+1)} = f_{n+1-(i-1)}.$$

Hence $a = r_0 \geq f_{n+1}$ and $b = r_1 \geq f_n$. $\square$

Notation. As is common use, we denote by $\lfloor x \rfloor$ the greatest integer less than or equal to x (the *"floor"* of x), and by $\lceil x \rceil$ the smallest integer greater than or equal to x (the *"ceiling"* of x).

Corollary A.11. *Let $a, b \in \mathbb{Z}$. Then the Euclidean algorithm computes $\gcd(a, b)$ in at most $\lfloor \log_g(a) \rfloor + 1$ iterations.*

Proof. Let n be the number of iterations. From $a \geq f_{n+1} \geq g^{n-1}$ (Lemma A.9) we conclude $n - 1 \leq \lfloor \log_g(a) \rfloor$. $\square$

The Binary Encoding of Numbers. Studying algorithms with numbers as inputs and outputs, we need *binary encodings* of numbers (and residues, see below). We always assume that integers $n \geq 0$ are encoded in the standard way as unsigned integers:

the sequence $z_{k-1} z_{k-2} \ldots z_1 z_0$ of bits $z_i \in \{0, 1\}, 0 \leq i \leq k-1$, is the encoding of

$$n = z_0 + z_1 \cdot 2^1 + \ldots + z_{k-2} \cdot 2^{k-2} + z_{k-1} \cdot 2^{k-1} = \sum_{i=0}^{k-1} z_i \cdot 2^i.$$

If the leading digit z_{k-1} is not zero (i.e., $z_{k-1} = 1$), we call n a *k-bit integer*, and k is called the *binary length* of n. The binary length of $n \in \mathbb{N}$ is usually denoted by $|n|$. Of course, we only use this notation if it cannot be confused with the absolute value. The binary length of $n \in \mathbb{N}$ is $\lfloor \log_2(n) \rfloor + 1$. The numbers of binary length k are the numbers $n \in \mathbb{N}$ with $2^{k-1} \leq n \leq 2^k - 1$.

The Big-O Notation. To state estimates, the *big-O* notation is useful. Suppose $f(k)$ and $g(k)$ are functions of the positive integers k which take positive (not necessarily integer) values. We say that $f(k) = O(g(k))$ if there is a constant C such that $f(k) \leq C \cdot g(k)$ for all sufficiently large k. For example, $2k^2 + k + 1 = O(k^2)$ because $2k^2 + k + 1 \leq 4k^2$ for all $k \geq 1$. In our examples, the constant C is always "small", and we use the big-O notation for convenience. We do not want to state a precise value of C.

Remark. Applying the classical grade school methods, we see that adding and subtracting two k-bit numbers requires $O(k)$ binary operations. Multiplication and division with remainder can be done with $O(k^2)$ binary operations (see [Knuth98] for a more detailed discussion of time estimates for doing arithmetic). Thus, the greatest common divisor of two k-bit numbers can be computed by the Euclidean algorithm with $O(k^3)$ binary operations.

Next we will show that every natural number can be uniquely decomposed into prime numbers.

Definition A.12. Let $p \in \mathbb{N}, p \geq 2$. p is called a *prime* (or a *prime number*) if 1 and p are the only positive divisors of p. A number $n \in \mathbb{N}$ which is not a prime is called a *composite*.

Remark. If p is a prime and $p \mid ab$, $a, b \in \mathbb{Z}$, then either $p \mid a$ or $p \mid b$.

Proof. Assume that p does not divide a and does not divide b. Then there are $d_1, d_2, e_1, e_2 \in \mathbb{Z}$, with $1 = d_1 p + e_1 a, 1 = d_2 p + e_2 b$ (Proposition A.4). Then $1 = d_1 d_2 p^2 + d_1 e_2 bp + e_1 a d_2 p + e_1 e_2 ab$. If p divided ab, then p would divide 1, which is impossible. Thus, p does not divide ab. □

Theorem A.13 *(Fundamental Theorem of Arithmetic). Let $n \in \mathbb{N}, n \geq 2$. There are pairwise distinct primes $p_1, \ldots, p_r$ and exponents $e_1, \ldots, e_r \in \mathbb{N}, e_i \geq 1, i = 1, \ldots, r$, such that*

$$n = \prod_{i=1}^{r} p_i^{e_i}.$$

The primes $p_1, \ldots, p_r$ and exponents $e_1, \ldots, e_r$ are unique.

Proof. By induction on n we obtain the existence of such a decomposition. $n = 2$ is a prime. Now assume that the existence is proven for numbers $\leq n$. Either $n + 1$ is a prime or $n + 1 = l \cdot m$, with $l, m < n + 1$. By assumption, there are decompositions of l and m and hence also for $n + 1$.
In order to prove uniqueness, we assume that there are two different decompositions of n. Dividing both decompositions by all common primes, we get

(not necessarily distinct) primes $p_1, \ldots, p_s$ and $q_1, \ldots, q_t$, with $\{p_1, \ldots, p_s\} \cap \{q_1, \ldots, q_t\} = \emptyset$ and $p_1 \cdot \ldots \cdot p_s = q_1 \cdot \ldots \cdot q_t$. Since $p_1 \mid q_1 \cdot \ldots \cdot q_t$, we conclude from the preceding remark that there is an $i, 1 \leq i \leq t$, with $p_1 \mid q_i$. This is a contradiction. $\qquad\square$

A.2 Residues

In public-key cryptography, we usually have to compute with remainders modulo n. This means that the computations take place in the residue class ring $\mathbb{Z}_n$.

Definition A.14. Let n be a positive integer.

1. $a, b \in \mathbb{Z}$ are *congruent modulo n*, written as

$$a \equiv b \bmod n,$$

   if n divides $a - b$. This means that a and b have the same remainder when divided by n: $a \bmod n = b \bmod n$.
2. Let $a \in \mathbb{Z}$. $[a] := \{x \in \mathbb{Z} \mid x \equiv a \bmod n\}$ is called the *residue class* of a modulo m.
3. $\mathbb{Z}_n := \{[a] \mid a \in \mathbb{Z}\}$ is the set of residue classes modulo n.

Remark. As is easily seen, "congruent modulo n" is a symmetric, reflexive and transitive relation, i.e., it is an equivalence relation. The residue classes are the equivalence classes. A residue class $[a]$ is completely determined by one of its members. If $a' \in [a]$, then $[a] = [a']$. An element $x \in [a]$ is called a *representative* of $[a]$. Division with remainder by n yields the remainders $0, \ldots, n - 1$. Therefore, there are n residue classes in $\mathbb{Z}_n$:

$$\mathbb{Z}_n = \{[0], \ldots, [n - 1]\}.$$

The integers $0, \ldots, n - 1$ are called the *natural representatives*. The natural representative of $[x] \in \mathbb{Z}_n$ is just the remainder ($x \bmod n$) of x modulo n (see division with remainder, Theorem A.3). If, in the given context, no confusion is possible, we sometimes identify the residue classes with their natural representatives.

Since we will study algorithms whose inputs and outputs are residue classes, we need *binary encodings* of the residue classes. The binary encoding of $[x] \in \mathbb{Z}_n$ is the binary encoding of the natural representative $x \bmod n$ as an unsigned integer (see our remark on the binary encoding of non-negative integers in Section A.1).

Definition A.15. By defining addition and multiplication as

$$[a] + [b] = [a + b] \text{ and } [a] \cdot [b] = [a \cdot b],$$

$\mathbb{Z}_n$ becomes a commutative ring, with unit element $[1]$. It is called the *residue class ring* modulo n.

Remark. The sum $[a] + [b]$ and the product $[a] \cdot [b]$ do not depend on the choice of the representatives by which they are computed, as straightforward computations show. For example, let $a' \in [a]$ and $b' \in [b]$. Then $n \mid a' - a$ and $n \mid b' - b$. Hence $n \mid a' + b' - (a + b)$, and therefore $[a + b] = [a' + b']$.

Doing multiplications in a ring, we are interested in those elements which have a multiplicative inverse. They are called the units.

Definition A.16. Let R be a commutative ring with unit element e. An element $x \in R$ is called a *unit* if there is an element $y \in R$ with $x \cdot y = e$. We call y a *multiplicative inverse* of x. The subset of units is denoted by R^*.

Remark. The multiplicative inverse of a unit x is uniquely determined, and we denote it by x^{-1}. The set of units R^* is a subgroup of R with respect to multiplication.

Example. In $\mathbb{Z}$, elements a and b satisfy $a \cdot b = 1$ if and only if both a and b are equal to 1, or both are equal to -1. Thus, 1 and -1 are the only units in $\mathbb{Z}$. The residue class rings $\mathbb{Z}_n$ contain many more units, as the subsequent considerations show. For example, if p is a prime then every residue class in $\mathbb{Z}_p$ different from $[0]$ is a unit. An element $[x] \in \mathbb{Z}_n$ in a residue class ring is a unit if there is a residue class $[y] \in \mathbb{Z}_n$ with $[x] \cdot [y] = [1]$, i.e., n divides $x \cdot y - 1$.

Proposition A.17. *An element $[x] \in \mathbb{Z}_n$ is a unit if and only if $\gcd(x, n) = 1$. The multiplicative inverse $[x]^{-1}$ of a unit $[x]$ can be computed using the extended Euclidean algorithm.*

Proof. If $\gcd(x, n) = 1$, then there is an equation $xb + nc = 1$ in $\mathbb{Z}$, and the coefficients $b, c \in \mathbb{Z}$ can be computed using the extended Euclidean algorithm A.6. The residue class $[b]$ is an inverse of $[x]$. Conversely, if $[x]$ is a unit, then there are $y, k \in \mathbb{Z}$ with $x \cdot y = 1 + k \cdot n$. This implies $\gcd(x, n) = 1$. $\square$

Corollary A.18. *Let p be a prime. Then every $[x] \neq [0]$ in $\mathbb{Z}_p$ is a unit. Thus, $\mathbb{Z}_p$ is a field.*

Definition A.19. The subgroup

$$\mathbb{Z}_n^* := \{x \in \mathbb{Z}_n \mid x \text{ is a unit in } \mathbb{Z}_n\}$$

of units in $\mathbb{Z}_n$ is called the *prime residue class group modulo n*.

Definition A.20. Let M be a finite set. The number of elements in M is called the *cardinality* or *order* of M. It is denoted by $|M|$.

We introduce the Euler phi function, which gives the number of units modulo n.

Definition A.21.
$$\varphi : \mathbb{N} \longrightarrow \mathbb{N}, \ n \longmapsto |\mathbb{Z}_n^*|$$
is called the *Euler phi function* or the *Euler totient function*.

Proposition A.22 *(Euler).*

$$\sum_{d \mid n} \varphi(d) = n.$$

Proof. If d is a divisor of n, let $Z_d := \{x \mid 1 \leq x \leq n, \gcd(x,n) = d\}$. Each $k \in \{1, \ldots, n\}$ belongs to exactly one Z_d. Thus $n = \sum_{d \mid n} |Z_d|$. Since $x \mapsto {}^x/d$ is a bijective map from Z_d to $\mathbb{Z}_{n/d}^*$, we have $|Z_d| = \varphi(n/d)$, and hence $n = \sum_{d \mid n} \varphi(n/d) = \sum_{d \mid n} \varphi(d)$. $\square$

Corollary A.23. *Let p be a prime and $k \in \mathbb{N}$. Then $\varphi(p^k) = p^{k-1}(p-1)$.*

Proof. By Euler's result, $\varphi(1) + \varphi(p) + \ldots + \varphi(p^k) = p^k$ and $\varphi(1) + \varphi(p) + \ldots + \varphi(p^{k-1}) = p^{k-1}$. Subtracting both equations yields $\varphi(p^k) = p^k - p^{k-1} = p^{k-1}(p-1)$. $\square$

Remarks:

1. By using the Chinese Remainder Theorem below (Section A.3), we will also get a formula for $\varphi(n)$ if n is not a power of a prime (Corollary A.31).
2. At some points in the book we need a lower bound for the fraction $\varphi(n)/n$ of units in $\mathbb{Z}_n$. In [RosSch62] it is proven that

$$\varphi(n) > \frac{n}{e^\gamma \log(\log(n)) + \frac{2.6}{\log(\log(n))}}, \ \text{with Euler's constant } \gamma = 0.5772 \ldots .$$

This inequality implies, for example, that

$$\varphi(n) > \frac{n}{6 \log(\log(n))} \ \text{for } n \geq 1.3 \cdot 10^6,$$

as a straightforward computation shows.

The RSA cryptosystem is based on old results by Fermat and Euler.[2] These results are special cases of the following proposition.

Proposition A.24. *Let G be a finite group and e be the unit element of G. Then $x^{|G|} = e$ for all $x \in G$.*

---

[2] Pierre de Fermat (1601–1665) and Leonhard Euler (1707–1783).

Proof. Since we apply this result only to Abelian groups, we assume in our proof that the group G is Abelian. A proof for the general case may be found in most introductory textbooks on algebra.

The map $\mu_x : G \longrightarrow G$, $g \longmapsto xg$, multiplying group elements by x, is a bijective map (multiplying by x^{-1} is the inverse map). Hence,

$$\prod_{g \in G} g = \prod_{g \in G} xg = x^{|G|} \prod_{g \in G} g,$$

and this implies $x^{|G|} = e$.   □

As a first corollary of Proposition A.24, we get Fermat's Little Theorem.

Proposition A.25 *(Fermat). Let p be a prime and $a \in \mathbb{Z}$ be a number that is prime to p (i.e., p does not divide a). Then*

$$a^{p-1} \equiv 1 \bmod p.$$

Proof. The residue class $[a]$ of a modulo p is a unit, because a is prime to p (Proposition A.17). Since $|\mathbb{Z}_p^*| = p - 1$ (Corollary A.18), we have $[a]^{p-1} = 1$ by Proposition A.24.   □

Remark. Fermat stated a famous conjecture known as Fermat's Last Theorem. It says that the equation $x^n + y^n = z^n$ has no solutions with non-zero integers x, y and z, for $n \geq 3$. For more than 300 years, Fermat's conjecture was one of the outstanding challenges of mathematics. It was finally proven in 1995 by Andrew Wiles.

Euler generalized Fermat's Little Theorem.

Proposition A.26 *(Euler). Let $n \in \mathbb{N}$ and let $a \in \mathbb{Z}$ be a number that is prime to n. Then*
$$a^{\varphi(n)} \equiv 1 \bmod n.$$

Proof. It follows from Proposition A.24, in the same way as Proposition A.25. The residue class $[a]$ of a modulo n is a unit and $|\mathbb{Z}_n^*| = \varphi(n)$.   □

Fast Modular Exponentiation. In cryptography, we often have to compute a power x^e or a modular power $x^e \bmod n$. This can be done efficiently by the fast exponentiation algorithm, also called the square-and-multiply algorithm. The idea is that if the exponent e is a power of 2, say $e = 2^k$, then we can exponentiate by successively squaring:

$$x^e = x^{2^k} = ((((\ldots (x^2)^2)^2 \ldots)^2)^2)^2.$$

In this way we compute x^e by k squarings. For example, $x^{16} = (((x^2)^2)^2)^2$.

If the exponent is not a power of 2, then we use its binary representation. Assume that e is a k-bit number, $2^{k-1} \leq e < 2^k$. Then

$$e = 2^{k-1}e_{k-1} + 2^{k-2}e_{k-2} + \ldots + 2^1 e_1 + 2^0 e_0, \quad \text{(with } e_{k-1} = 1)$$
$$= (2^{k-2}e_{k-1} + 2^{k-3}e_{k-2} + \ldots + e_1) \cdot 2 + e_0$$
$$= (\ldots((2e_{k-1} + e_{k-2}) \cdot 2 + e_{k-3}) \cdot 2 + \ldots + e_1) \cdot 2 + e_0.$$

Hence,

$$x^e = x^{(\ldots((2e_{k-1}+e_{k-2})\cdot 2+e_{k-3})\cdot 2+\ldots+e_1)\cdot 2+e_0} =$$
$$= (x^{(\ldots((2e_{k-1}+e_{k-2}))\cdot 2+e_{k-3})\cdot 2+\ldots+e_1})^2 \cdot x^{e_0} =$$
$$= (\ldots(((x^2 \cdot x^{e_{k-2}})^2 \cdot x^{e_{k-3}})^2 \cdot \ldots)^2 \cdot x^{e_1})^2 \cdot x^{e_0}.$$

We see that x^e can be computed in $k-1$ steps, with each step consisting of squaring the intermediate result and, if the corresponding binary digit e_i of e is 1, an additional multiplication by x. If we want to compute the modular power $x^e \bmod n$, then we take the remainder modulo n after each squaring and multiplication:

$$x^e \bmod n =$$

$$(\ldots(((x^2 \cdot x^{e_{k-2}} \bmod n)^2 \cdot x^{e_{k-3}} \bmod n)^2 \cdot \ldots)^2 \cdot x^{e_1} \bmod n)^2 \cdot x^{e_0} \bmod n.$$

We obtain the following square-and-multiply algorithm for fast modular exponentiation.

Algorithm A.27.
  int *ModPower*(int x, n, bitString $e_{k-1} \ldots e_0$)
  1   $y \leftarrow x$;
  2   for $i \leftarrow k-2$ downto 0 do
  3       $y \leftarrow y^2 \cdot x^{e_i} \bmod n$
  4   return y

In particular, we get

Proposition A.28. *Let* $l = \lfloor \log_2 e \rfloor$. *The computation of* $x^e \bmod n$ *can be done by* l *squarings,* l *multiplications and* l *divisions.*

Proof. The binary length k of e is $\lfloor \log_2(e) \rfloor + 1$. $\square$

A.3 The Chinese Remainder Theorem

The Chinese Remainder Theorem provides a method of solving systems of congruences. The solutions can be found using an easy and efficient algorithm.

Theorem A.29. *Let* $n_1, \ldots, n_r \in \mathbb{N}$ *be pairwise relatively prime numbers, i.e.,* $\gcd(n_i, n_j) = 1$ *for* $i \neq j$. *Let* $b_1, b_2, \ldots, b_r$ *be arbitrary integers. Then there is an integer* b *such that*

$$b \equiv b_i \bmod n_i, \quad i = 1, \ldots, r.$$

Furthermore, the remainder $b \bmod n$ *is unique, where* $n = n_1 \cdot \ldots \cdot n_r$.

The statement means that there is a one-to-one correspondence between the residue classes modulo n and tuples of residue classes modulo $n_1, \ldots, n_r$. This one-to-one correspondence preserves the additive and multiplicative structure. Therefore, we have the following ring-theoretic formulation of Theorem A.29.

Theorem A.30 *(Chinese Remainder Theorem). Let $n_1, \ldots, n_r \in \mathbb{N}$ be pairwise relatively prime numbers, i.e., $\gcd(n_i, n_j) = 1$, for $i \neq j$. Let $n = n_1 \cdot \ldots \cdot n_r$. Then the map*

$$\psi : \mathbb{Z}_n \longrightarrow \mathbb{Z}_{n_1} \times \ldots \times \mathbb{Z}_{n_r}, \quad [x] \longmapsto ([x \bmod n_1], \ldots, [x \bmod n_r])$$

is an isomorphism of rings.

Remark. Before we give a proof, we review the notion of an *"isomorphism"*. It means that ψ is a homomorphism and bijective. "Homomorphism" means that ψ preserves the additive and multiplicative structure. More precisely, a map $f : R \longrightarrow R'$ between rings with unit elements e and e' is called a *(ring) homomorphism* if

$$f(e) = e' \text{ and } f(a + b) = f(a) + f(b), f(a \cdot b) = f(a) \cdot f(b) \text{ for all } a, b \in R.$$

If f is a bijective homomorphism, then, automatically, the inverse map $g = f^{-1}$ is also a homomorphism. Namely, let $a', b' \in R'$. Then $a' = f(a)$ and $b' = f(b)$, and $g(a' \cdot b') = g(f(a) \cdot f(b)) = g(f(a \cdot b)) = a \cdot b = g(a') \cdot g(b')$ (analogously for $+$ instead of $\cdot$).

Being an isomorphism, as ψ is, is an extremely nice feature. It means, in particular, that a is a unit in R if and only if $f(a)$ is a unit R' (to see this, compute $e' = f(e) = f(a \cdot a^{-1}) = f(a) \cdot f(a^{-1})$, hence $f(a^{-1})$ is an inverse of $f(a)$). And the "same" equations hold in domain and range. For example, we have $a^2 = b$ in R if and only if $f(a)^2 = f(b)$ (note that $f(a)^2 = f(a^2)$). Thus, b is a square if and only if $f(b)$ is a square (we will use this example in Section A.8).

Isomorphism means that the domain and range may be considered to be the same for all questions concerning addition and multiplication.

Proof (of Theorem A.30). Since each n_i divides n, the map is well defined, and it obviously is a ring homomorphism. The domain and range of the map have the same cardinality (i.e., they contain the same number of elements). Thus, it suffices to prove that ψ is surjective.

Let $t_i := n/n_i = \prod_{k \neq i} n_k$. Then $t_i \equiv 0 \bmod n_k$ for all $k \neq i$, and $\gcd(t_i, n_i) = 1$. Hence, there is a $d_i \in \mathbb{Z}$ with $d_i \cdot t_i \equiv 1 \bmod n_i$ (Proposition A.17). Setting $u_i := d_i \cdot t_i$, we have

$$u_i \equiv 0 \bmod n_k, \text{ for all } k \neq i, \text{ and } u_i \equiv 1 \bmod n_i.$$

This means that the element $(0, \ldots, 0, 1, 0, \ldots, 0)$ (the i-th component is 1, all other components are 0) is in the image of ψ. If $([x_1], \ldots, [x_r]) \in \mathbb{Z}_{n_1} \times \ldots \times \mathbb{Z}_{n_r}$ is an arbitrary element, then $\psi(\sum_{i=1}^{r} x_i \cdot u_i) = ([x_1], \ldots, [x_r])$. $\square$

Remarks:

1. Actually, the proof describes an efficient algorithm for computing a number b, with $b \equiv b_i \bmod n_i$, $i = 1, \ldots, r$ (recall our first formulation of the Chinese Remainder Theorem in Theorem A.29). In a preprocessing step, the inverse elements $[d_i] = [t_i]^{-1}$ are computed modulo n_i using the extended Euclidean algorithm (Proposition A.17). Then b can be computed as $b = \sum_{i=1}^{r} b_i \cdot d_i \cdot t_i$, for any given integers b_i, $1 \le i \le r$.

   We mainly apply the Chinese Remainder Theorem with $r = 2$ (for example, in the RSA cryptosystem). Here we simply compute coefficients d and e with $1 = d \cdot n_1 + e \cdot n_2$ (using the extended Euclidean algorithm A.6), and then $b = d \cdot n_1 \cdot b_2 + e \cdot n_2 \cdot b_1$.

2. The Chinese Remainder Theorem can be used to make arithmetic computations modulo n easier and (much) more efficient. We map the operands to $\mathbb{Z}_{n_1} \times \ldots \times \mathbb{Z}_{n_r}$ by ψ and do our computation there. $\mathbb{Z}_{n_1} \times \ldots \times \mathbb{Z}_{n_r}$ is a direct product of rings. Addition and multiplication are done componentwise, i.e., we perform the computation modulo n_i, for $i = 1, \ldots, r$. Here we work with (much) smaller numbers.[3] Finally, we map the result back to $\mathbb{Z}_n$ by ψ^{-1} (which is easily done, as we have seen in the preceding remark).

As a corollary of the Chinese Remainder Theorem, we get a formula for Euler's phi function for composite inputs.

Corollary A.31. *Let $n \in \mathbb{N}$ and $n = p_1^{e_1} \cdot \ldots \cdot p_r^{e_r}$ be the decomposition of n into primes (as stated in Theorem A.13). Then:*

1. *$\mathbb{Z}_n$ is isomorphic to $\mathbb{Z}_{p_1^{e_1}} \times \ldots \times \mathbb{Z}_{p_r^{e_r}}$.*
2. *$\mathbb{Z}_n^*$ is isomorphic to $\mathbb{Z}_{p_1^{e_1}}^* \times \ldots \times \mathbb{Z}_{p_r^{e_r}}^*$.*

   In particular, we have for Euler's phi function that

$$\varphi(n) = n \prod_{i=1}^{r} \left(1 - \frac{1}{p_i}\right).$$

Proof. The ring isomorphism of Theorem A.30 induces, in particular, an isomorphism on the units. Hence,

$$\varphi(n) = \varphi(p_1^{e_1}) \cdot \ldots \cdot \varphi(p_r^{e_r}),$$

and the formula follows from Corollary A.23. □

A.4 Primitive Roots and the Discrete Logarithm

Definition A.32. Let G be a finite group and let e be the unit element of G. Let $x \in G$. The smallest $n \in \mathbb{N}$ with $x^n = e$ is called the *order of x*. We write this as $\mathrm{ord}(x)$.

---

[3] For example, if $n = pq$ (as in an RSA scheme) with 512-bit numbers p and q, then we compute with 512-bit numbers instead of with 1024-bit numbers.

Remark. There are exponents $n \in \mathbb{N}$, with $x^n = e$. Namely, since G is finite, there are exponents m and $m', m < m'$, with $x^m = x^{m'}$. Then $m' - m > 0$ and $x^{m'-m} = e$.

Lemma A.33. *Let G be a finite group and $x \in G$. Let $n \in \mathbb{N}$ with $x^n = e$. Then $\mathrm{ord}(x)$ divides n.*

Proof. Let $n = q \cdot \mathrm{ord}(x) + r, 0 \le r < \mathrm{ord}(x)$ (division with remainder). Then $x^r = e$. Since $0 \le r < \mathrm{ord}(x)$, this implies $r = 0$. $\square$

Corollary A.34. *Let G be a finite group and $x \in G$. Then $\mathrm{ord}(x)$ divides the order $|G|$ of G.*

Proof. By Proposition A.24, $x^{|G|} = e$. $\square$

Lemma A.35. *Let G be a finite group and $x \in G$. Let $l \in \mathbb{Z}$ and $d = \gcd(l, \mathrm{ord}(x))$. Then $\mathrm{ord}(x^l) = \mathrm{ord}(x)/d$.*

Proof. Let $r = \mathrm{ord}(x^l)$. From $(x^l)^{\mathrm{ord}(x)/d} = (x^{\mathrm{ord}(x)})^{l/d} = e$ we conclude $r \le \mathrm{ord}(x)/d$. Choose numbers a and b with $d = a \cdot l + b \cdot \mathrm{ord}(x)$ (Proposition A.4). From $x^{r \cdot d} = x^{r \cdot a \cdot l + r \cdot b \cdot \mathrm{ord}(x)} = x^{l \cdot r \cdot a} = e$, we derive $\mathrm{ord}(x) \le r \cdot d$. $\square$

Definition A.36. Let G be a finite group. G is called *cyclic* if there is an $x \in G$ which generates G, i.e., $G = \{x, x^2, x^3, \ldots, x^{\mathrm{ord}(x)-1}, x^{\mathrm{ord}(x)} = e\}$. Such an element x is called a *generator* of G.

Theorem A.37. *Let p be a prime. Then $\mathbb{Z}_p^*$ is cyclic, and the number of generators is $\varphi(p-1)$.*

Proof. For $1 \le d \le p - 1$, let $S_d = \{x \in \mathbb{Z}_p^* \mid \mathrm{ord}(x) = d\}$ be the units of order d. If $S_d \ne \emptyset$, let $a \in S_d$. The equation $X^d - 1$ has at most d solutions in $\mathbb{Z}_p$, since $\mathbb{Z}_p$ is a field (Corollary A.18). Hence, the solutions of $X^d - 1$ are just the elements of $A := \{a, a^2, \ldots, a^d\}$. Each $x \in S_d$ is a solution of $X^d - 1$, and therefore $S_d \subset A$. Using Lemma A.35 we derive that $S_d = \{a^c \mid 1 \le c < d, \gcd(c, d) = 1\}$. In particular, we conclude that $|S_d| = \varphi(d)$ if $S_d \ne \emptyset$ (and an $a \in S_d$ exists).
By Fermat's Little Theorem (Proposition A.25), $\mathbb{Z}_p^*$ is the disjoint union of the sets $S_d, d \mid p - 1$. Hence $|\mathbb{Z}_p^*| = p - 1 = \sum_{d \mid p-1} |S_d|$. On the other hand, $p - 1 = \sum_{d \mid p-1} \varphi(d)$ (Proposition A.22), and we see that $|S_d| = \varphi(d)$ must hold for all divisors d of $p - 1$. In particular, $|S_{p-1}| = \varphi(p-1)$. This means that there are $\varphi(p-1)$ generators of $\mathbb{Z}_p^*$. $\square$

Definition A.38. Let p be a prime. A generator g of the cyclic group $\mathbb{Z}_p^*$ is called a *primitive root* of $\mathbb{Z}_p^*$ or a *primitive root modulo p*.

Remark. It can be proven that $\mathbb{Z}_n^*$ is cyclic if and only if n is one of the following numbers: $1, 2, 4, p^k$ or $2p^k$; p a prime, $p \ge 3$, $k \ge 1$.

Proposition A.39. *Let p be a prime. Then $x \in \mathbb{Z}_p^*$ is a primitive root if and only if $x^{(p-1)/q} \neq [1]$ for every prime q which divides $p - 1$.*

Proof. An element x is a primitive root if and only if x has order $p-1$. Since $\operatorname{ord}(x)$ divides $p - 1$ (Corollary A.34), either $x^{(p-1)/q} = [1]$ for some prime divisor q of $p - 1$ or $\operatorname{ord}(x) = p - 1$.   □

We may use Proposition A.39 to generate a primitive root for those primes p for which we know (or can efficiently compute) the prime factors of $p - 1$.

Algorithm A.40.
    int *PrimitiveRoot*(prime p)
    1   Randomly choose an integer g, with $0 < g < p-1$
    2   if $g^{(p-1)\ \text{div}\ q} \not\equiv 1 \bmod p$, for all primes q dividing $p - 1$
    3   then return g
    4   else go to 1

Since $\varphi(p-1) > (p-1)/6\log(\log(p-1))$ (see Section A.2), we expect to find a primitive element after $O(\log(\log(p)))$ iterations (see Lemma B.12).

No efficient algorithm is known for the computation of primitive roots for arbitrary primes. The problem is to compute the prime factors of $p-1$, which we need in Algorithm A.40. Often there are primitive roots which are small.

Algorithm A.40 is used, for example, in the key-generation procedure of the ElGamal cryptosystem (see Section 3.4.1). There the primes p are chosen in such a way that the prime factors of $p - 1$ can be derived efficiently.

Lemma A.41. *Let p be a prime and let q be a prime that divides $p-1$. Then the set*

$$G_q = \{x \in \mathbb{Z}_p^* \mid \operatorname{ord}(x) = q \text{ or } x = [1]\},$$

which consists of the unit element $[1]$ and the elements of order q, is a subgroup of $\mathbb{Z}_p^*$. G_q is a cyclic group, and every element $x \in \mathbb{Z}_p^*$ of order q, i.e., every element $x \in G_q$, $x \neq [1]$, is a generator. G_q is generated, for example, by $g^{(p-1)/q}$, where g is a primitive root modulo p. G_q is the only subgroup of G of order q.

Proof. Let $x, y \in G_q$. Then $(xy)^q = x^q y^q = [1]$, and therefore $\operatorname{ord}(xy)$ divides q. Since q is a prime, we conclude that $\operatorname{ord}(xy)$ is 1 or q. Thus $xy \in G_q$, and G_q is a subgroup of $\mathbb{Z}_p^*$. Let $h \in \mathbb{Z}_p^*$ be an element of order q, for example, $h := g^{p-1/q}$, where g is a primitive root modulo p. Then $\{h^0, h^1, h^2, \ldots h^{q-1}\} \subseteq G_q$. The elements of G_q are solutions of the equation $X^q - 1$ in $\mathbb{Z}_p$. This equation has at most q solutions in $\mathbb{Z}_p$, since $\mathbb{Z}_p$ is a field (Corollary A.18). Therefore $\{h^0, h^1, h^2, \ldots h^{q-1}\} = G_q$, and h is a generator of G_q. If H is any subgroup of order q and $z \in H$, $z \neq [1]$, then $\operatorname{ord}(z)$ divides q, and hence $\operatorname{ord}(z) = q$, because q is a prime. Thus $z \in G_q$, and we conclude that $H = G_q$.   □

Computing Modulo a Prime. The security of many cryptographic schemes is based on the discrete logarithm assumption, which says that $x \mapsto g^x \bmod p$ is a one-way function. Here p is a large prime and the base element g is

1. either a primitive root modulo p, i.e., a generator of $\mathbb{Z}_p^*$, or
2. it is an element of order q in $\mathbb{Z}_p^*$, i.e., a generator of the subgroup G_q of order q, and q is a (large) prime that divides $p - 1$.

Examples of such schemes which we discuss in this book are ElGamal encryption and digital signatures, the digital signature standard DSS (see Section 3.4), commitment schemes (see Section 4.3.2), electronic elections (see Section 4.5) and digital cash (see Section 4.8).

When setting up such schemes, generators g of $\mathbb{Z}_p^*$ or G_q have to be selected. This can be difficult or even infeasible in the first case, because we must know the prime factors of $p-1$ in order to test whether a given element g is a primitive root (see Algorithm A.40 above). On the other hand, it is easy to find a generator g of G_q. We simply take a random element $h \in \mathbb{Z}_p^*$ and set $g := h^{(p-1)/q}$. The order of g divides q, because $g^q = h^{p-1} = [1]$. Since q is a prime, we conclude that $\mathrm{ord}(g) = 1$ or $\mathrm{ord}(g) = q$. Therefore, if $g \neq [1]$, then $\mathrm{ord}(g) = q$ and g is a generator of G_q.

To implement cryptographic operations, we have to compute in $\mathbb{Z}_p^*$ or in the subgroup G_q. The following rules simplify these computations.

1. Let $x \in \mathbb{Z}_p^*$. Then $x^k = x^{k'}$, if $k \equiv k' \bmod (p - 1)$.
   In particular, $x^k = x^{k \bmod (p-1)}$, i.e., exponents can be reduced by modulo $(p - 1)$, and $x^{-k} = x^{p-1-k}$.
2. Let $x \in \mathbb{Z}_p^*$ be an element of order q, i.e., $x \in G_q$. Then $x^k = x^{k'}$, if $k \equiv k' \bmod q$.
   In particular, $x^k = x^{k \bmod q}$, i.e., exponents can be reduced by modulo q, and $x^{-k} = x^{q-k}$.

The rules state that the exponents are added and multiplied modulo $(p - 1)$ or modulo q. The rules hold, because $x^{p-1} = [1]$ for $x \in \mathbb{Z}_p^*$ (Proposition A.25) and $x^q = [1]$ for $x \in G_q$, which implies that

$$x^{k+l\cdot(p-1)} = x^k x^{l\cdot(p-1)} = x^k \left(x^{p-1}\right)^l = x^k [1]^l = x^k \text{ for } x \in \mathbb{Z}_p^*$$

$$\text{and } x^{k+l\cdot q} = x^k x^{l\cdot q} = x^k \left(x^q\right)^l = x^k [1]^l = x^k \text{ for } x \in G_q.$$

These rules can be very useful in computations. For example, let $x \in \mathbb{Z}_p^*$ and $k \in \{0, 1, \ldots, p - 2\}$. Then you can compute the inverse x^{-k} of x^k by raising x to the $(p - 1 - k)$-th power, $x^{-k} = x^{p-1-k}$, without explicitly computing an inverse by using, for example, the Euclidean algorithm. Note that $(p - 1 - k)$ is a positive exponent. Powers of x are efficiently computed by the fast exponentiation algorithm (Algorithm A.27).

In many cases it is also possible to compute the k-th root of elements in $\mathbb{Z}_p^*$.

1. Let $x \in \mathbb{Z}_p^*$ and $k \in \mathbb{N}$ with $\gcd(k, p-1) = 1$, i.e., k is a unit modulo $p-1$. Let k^{-1} be the inverse of k modulo $p - 1$, i.e., $k \cdot k^{-1} \equiv 1 \bmod (p - 1)$. Then $\left(x^{k^{-1}}\right)^k = x$, i.e., $x^{k^{-1}}$ is a k-th root of x in $\mathbb{Z}_p^*$.

2. Let $x \in \mathbb{Z}_p^*$ be an element of order q, i.e., $x \in G_q$, and $k \in \mathbb{N}$ with $1 \leq k < q$. Let k^{-1} be the inverse of k modulo q, i.e., $k \cdot k^{-1} \equiv 1 \bmod q$. Then $\left(x^{k^{-1}}\right)^k = x$, i.e., $x^{k^{-1}}$ is a k-th root of x in $\mathbb{Z}_p^*$.

It is common practice to denote the k-th root $x^{k^{-1}}$ by $x^{1/k}$.

You can apply analogous rules of computation to elements g^k in any finite group G. Proposition A.24, which says that $g^{|G|}$ is the unit element, implies that exponents k are added and multiplied modulo the order $|G|$ of G.

A.5 Polynomials and Finite Fields

A *finite field* is a field with a finite number of elements. In Section A.2, we met examples of finite fields: the residue class ring $\mathbb{Z}_n$ is a field, if and only if n is a prime. The fields $\mathbb{Z}_p$, p a prime number, are called the finite *prime fields*, and they are also denoted by $\mathbb{F}_p$. Finite fields are extensions of these prime fields. Field extensions are constructed by using polynomials. So we first study the ring of polynomials with coefficients in a field k.

A.5.1 The Ring of Polynomials

Let $k[X]$ be the ring of polynomials in one variable X over a (not necessarily finite) field k. The elements of $k[X]$ are the polynomials

$$F(X) = a_0 + a_1 X + a_2 X^2 + \ldots a_d X^d = \sum_{i=0}^{d} a_i X^i,$$

with coefficients $a_i \in k, 0 \leq i \leq d$.

If we assume that $a_d \neq 0$, then the *leading term* $a_d X^d$ really appears in the polynomial, and we call d the *degree* of F, $\deg(F)$ for short. The polynomials of degree 0 are just the elements of k.

The polynomials in $k[X]$ are added and multiplied as usual:

1. We add two polynomials $F = \sum_{i=0}^{d} a_i X^i$ and $G = \sum_{i=0}^{e} b_i X^i$, assume $d \leq e$, by adding the coefficients (set $a_i = 0$ for $d < i \leq e$):

$$F + G = \sum_{i=0}^{e} (a_i + b_i) X^i.$$

2. The product of two polynomials $F = \sum_{i=0}^{d} a_i X^i$ and $G = \sum_{i=0}^{e} b_i X^i$ is

$$F \cdot G = \sum_{i=0}^{d+e} \left(\sum_{k=0}^{i} a_k b_{i-k} \right) X^i.$$

With this addition and multiplication, $k[X]$ becomes a commutative ring with unit element. The unit element of $k[X]$ is the unit element 1 of k. The ring $k[X]$ has no zero divisors, i.e., if F and G are non-zero polynomials, then the product $F \cdot G$ is also non-zero.

The algebraic properties of the ring $k[X]$ of polynomials are analogous to the algebraic properties of the ring of integers.

Analogously to Definition A.1, we define for polynomials F and G what it means that F divides G and the *greatest common divisor* of F and G. The greatest common divisor is unique up to a factor $c \in k, c \neq 0$, i.e., if A is a greatest common divisor of F and G, then $c \cdot A$ is also a greatest common divisor, for $c \in k^* = k \setminus \{0\}$.

A polynomial F is *(relatively) prime to* G if the only common divisors of F and G are the units k^* of k.

Division with remainder works as with the integers. The difference is that the "size" of a polynomial is measured by using the degree, whereas the absolute value was used for an integer.

Theorem A.42 *(Division with remainder).* Let $F, G \in k[X], G \neq 0$. Then there are unique polynomials $Q, R \in k[X]$ such that $F = Q \cdot G + R$ and $0 \leq \deg(R) < \deg(G)$.

Proof. The proof runs exactly in the same way as the proof of Theorem A.3: replace the absolute value with the degree. □

R is called the *remainder* of F modulo G. We write $F \bmod G$ for R. The polynomial Q is the *quotient* of F and G. We write $F \operatorname{div} G$ for Q.

You can compute a greatest common divisor of polynomials F and G by using the *Euclidean algorithm*, and the *extended Euclidean algorithm* yields the coefficients $C, D \in k[X]$ of a linear combination

$$A = C \cdot F + D \cdot G,$$

with A a greatest common divisor of F and G.

If you have obtained such a linear combination for one greatest common divisor, then you immediately get a linear combination for any other greatest common divisor by multiplying with a unit from k^*.

In particular, if F is prime to G, then the extended Euclidean algorithm computes a linear combination

$$1 = C \cdot F + D \cdot G.$$

We also have the analogue of prime numbers.

Definition A.43. Let $P \in k[X], P \notin k$. P is called *irreducible* (or a *prime*) if the only divisors of P are the elements $c \in k^*$ and $c \cdot P, c \in k^*$, or, equivalently, if whenever one can write $P = F \cdot G$ with $F, G \in k[X]$, then $F \in k^*$ or $G \in k^*$. A polynomial $Q \in k[X]$ which is not irreducible is called *reducible* or a *composite*.

As the ring $\mathbb{Z}$ of integers, the ring $k[X]$ of polynomials is factorial, i.e., every element has a unique decomposition into irreducible elements.

Theorem A.44. *Let $F \in k[X], F \neq 0$, be a non-zero polynomial. There are pairwise distinct irreducible polynomials $P_1, \ldots, P_r, r \geq 0$, exponents $e_1, \ldots, e_r \in \mathbb{N}, e_i \geq 1, i = 1, \ldots, r$, and a unit $u \in k^*$ such that*

$$F = u \prod_{i=1}^{r} P_i^{e_i}.$$

This factorization is unique in the following sense: if

$$F = v \prod_{i=1}^{s} Q_i^{f_i}$$

is another factorization of F, then we have $r = s$, and after a permutation of the indices i we have $Q_i = u_i P_i$, with $u_i \in k^*$, and $e_i = f_i$ for $1 \leq i \leq r$.

Proof. The proof runs in the same way as the proof of the Fundamental Theorem of Arithmetic (Theorem A.13). □

A.5.2 Residue Class Rings

As in the ring of integers, we can consider residue classes in $k[X]$ and residue class rings.

Definition A.45. Let $P \in k[X]$ be a polynomial of degree ≥ 1:

1. $F, G \in k[X]$ are *congruent modulo* P, written as

$$F \equiv G \bmod P,$$

if P divides $F - G$. This means that F and G have the same remainder when divided by P, i.e., $F \bmod P = G \bmod P$.
2. Let $F \in k[X]$. $[F] := \{G \in k[X] \mid G \equiv F \bmod P\}$ is called the *residue class* of F modulo P.

As before, "congruent modulo" is an equivalence relation, the equivalence classes are the residue classes, and the set of residue classes

$$k[X]/Pk[X] := \{[F] \mid F \in k[X]\}$$

is a ring. Residue classes are added and multiplied by adding and multiplying a representative:

$$[F] + [G] := [F + G], \quad [F] \cdot [G] := [F \cdot G].$$

We also have a natural representative of $[F]$, the remainder $F \bmod P$ of F modulo P: $[F] = [F \bmod P]$. As remainders modulo P, we get all the polynomials which have a degree $< \deg(P)$. Therefore, we have a one-to-one correspondence between $k[X]/Pk[X]$ and the set of residues $\{F \in k[X] \mid \deg(F) < \deg(P)\}$. We often identify both sets:

$$k[X]/Pk[X] = \{F \in k[X] \mid \deg(F) < \deg(P)\}.$$

Two residues F and G are added or multiplied by first adding or multiplying them as polynomials and then taking the residue modulo P. Since the sum of two residues F and G has a degree $< \deg(P)$, it is a residue, and we do not have to reduce. After a multiplication, we have, in general, to take the remainder.

Addition : $(F, G) \longmapsto F + G$,   Multiplication : $(F, G) \longmapsto F \cdot G \bmod P$.

Let $n := \deg(P)$ be the degree of P. The residue class ring $k[X]/Pk[X]$ is an n-dimensional vector space over k. A basis of this vector space is given by the elements $[1], [X], [X^2], \ldots, [X^{n-1}]$. If k is a finite field with q elements, then $k[X]/Pk[X]$ consists of q^n elements.

Example. Let $k = \mathbb{F}_2 = \mathbb{Z}_2 = \{0, 1\}$ be the field with two elements 0 and 1 consisting of the residues modulo 2, and $P := X^8 + X^4 + X^3 + X + 1 \in k[X]$. The elements of $k[X]/Pk[X]$ may be identified with the binary polynomials $b_7 X^7 + b_6 X^6 + \ldots + b_1 X + b_0, b_i \in \{0, 1\}, 0 \leq i \leq 7$, of degree ≤ 7. The ring $k[X]/Pk[X]$ contains $2^8 = 256$ elements. We have, for example,

$$
\begin{aligned}
(X^6 + X^3 &+ X^2 + 1) \cdot (X^5 + X^2 + 1) \\
&= X^{11} + X^7 + X^6 + X^4 + X^3 + 1 \\
&= X^3 \cdot (X^8 + X^4 + X^3 + X + 1) + 1 \\
&\equiv 1 \bmod (X^8 + X^4 + X^3 + X + 1).
\end{aligned}
$$

Thus, $X^6 + X^3 + X^2 + 1$ is a unit in $k[X]/Pk[X]$, and its inverse is $X^5 + X^2 + 1$.

We may characterize units as in the integer case.

Proposition A.46. *An element $[F] \in k[X]/Pk[X]$ is a unit if and only if F is prime to P. The multiplicative inverse $[F]^{-1}$ of a unit $[F]$ can be computed using the extended Euclidean algorithm.*

Proof. The proof is the same as the proof in the integer case (see Proposition A.17). Recall that the inverse may be calculated as follows: if F is prime to P, then the extended Euclidean algorithm produces a linear combination

$$C \cdot F + D \cdot P = 1, \text{ with polynomials } C, D \in k[X].$$

We see that $C \cdot F \equiv 1 \bmod P$. Hence, $[C]$ is the inverse $[F]^{-1}$. $\qquad\square$

If the polynomial P is irreducible, then all residues modulo P, i.e., all polynomials with a degree $< \deg(P)$, are prime to P. So we get the same corollary as in the integer case.

Corollary A.47. *Let P be irreducible. Then every $[F] \neq [0]$ in $k[X]/Pk[X]$ is a unit. Thus, $k[X]/Pk[X]$ is a field.*

Remarks:

1. Let P be an irreducible polynomial of degree n. The field k is a subset of the larger field $k[X]/Pk[X]$. We therefore call $k[X]/Pk[X]$ an *extension field* of k of degree n.
2. If P is reducible, then $P = F \cdot G$, with polynomials F, G of degree $< \deg(P)$. Then $[F] \neq [0]$ and $[G] \neq [0]$, but $[F] \cdot [G] = [P] = [0]$. $[F]$ and $[G]$ are "zero divisors". They have no inverse, and we see that $k[X]/Pk[X]$ is not a field.

A.5.3 Finite Fields

Now, let $k = \mathbb{Z}_p = \mathbb{F}_p$ be the prime field of residues modulo p, $p \in \mathbb{Z}$ a prime number, and let $P \in \mathbb{F}_p[X]$ be an irreducible polynomial of degree n. Then $k[X]/Pk[X] = \mathbb{F}_p[X]/P\mathbb{F}_p[X]$ is an extension field of $\mathbb{F}_p$. It is an n-dimensional vector space over $\mathbb{F}_p$, and it contains p^n elements.

In general, there is more than one irreducible polynomial of degree n over $\mathbb{F}_p$. Therefore there are more finite fields with p^n elements. For example, if $Q \in \mathbb{F}_p[X]$ is another irreducible polynomial of degree n, $Q \neq cP$ for all $c \in k$, then $\mathbb{F}_p[X]/Q\mathbb{F}_p[X]$ is a field with p^n elements, different from $k[X]/Pk[X]$. But one can show that all the finite fields with p^n elements are isomorphic to each other in a very natural way. As the mathematicians state it, up to canonical isomorphism, there is only one finite field with p^n elements. It is denoted by $\mathbb{F}_{p^n}$ or by $\mathrm{GF}(p^n)$.[4]

If you need a concrete representation of $\mathbb{F}_{p^n}$, then you choose an irreducible polynomial $P \in \mathbb{F}_p[X]$ of degree n, and you have $\mathbb{F}_{p^n} = \mathbb{F}_p[X]/P\mathbb{F}_p[X]$. But there are different representations, reflecting your degrees of freedom when choosing the irreducible polynomial.

---

[4] Finite fields are also called *Galois fields*, in honor of the French mathematician Évariste Galois (1811–1832).

One can also prove that in every finite field k, the number $|k|$ of elements in k must be a power p^n of a prime number p. Therefore, the fields $\mathbb{F}_{p^n}$ are all the finite fields that exist.

In cryptography, finite fields play an important role in many places. For example, the classical ElGamal cryptosystems are based on the discrete logarithm problem in a finite prime field (see Section 3.4), the elliptic curves used in cryptography are defined over finite fields, and the basic encryption operations of the Advanced Encryption Standard (AES) are algebraic operations in the field $\mathbb{F}_{2^8}$ with 2^8 elements. The AES is discussed in this book (see Section 2.1.4). This motivates the following closer look at the fields $\mathbb{F}_{2^n}$.

We identify $\mathbb{F}_2 = \mathbb{Z}_2 = \{0, 1\}$. Let $P = X^n + a_{n-1}X^{n-1} + \ldots + a_1 X + a_0$, $a_i \in \{0, 1\}$, $0 \le i \le n - 1$ be a binary irreducible polynomial of degree n. Then $\mathbb{F}_{2^n} = \mathbb{F}_p[X]/P\mathbb{F}_p[X]$, and we may consider the binary polynomials $A = b_{n-1}X^{n-1} + b_{n-2}X^{n-2} + \ldots + b_1 X + b_0$ of degree $\le n - 1$ ($b_i \in \{0, 1\}$, $0 \le i \le n - 1$) as the elements of $\mathbb{F}_{2^n}$. Adding two of these polynomials in $\mathbb{F}_{2^n}$ means to add them as polynomials, and multiplying them means to first multiply them as polynomials and then take the remainder modulo P.

Now we can represent the polynomial A by the n-dimensional vector $b_{n-1}b_{n-2}\ldots b_1 b_0$ of its coefficients. In this way, we get a binary representation of the elements of $\mathbb{F}_{2^n}$; the elements of $\mathbb{F}_{2^n}$ are just the bit strings of length n. To add two of these elements means to add them as binary vectors, i.e., you add them bitwise modulo 2, which is the same as bitwise XORing:

$$b_{n-1}b_{n-2}\ldots b_1 b_0 \quad + \quad c_{n-1}c_{n-2}\ldots c_1 c_0$$

$$= (b_{n-1} \oplus c_{n-1})(b_{n-2} \oplus c_{n-2})\ldots(b_1 \oplus c_1)(b_0 \oplus c_0).$$

To multiply two elements is more complicated: you have to convert the bit strings to polynomials, multiply them as polynomials, reduce modulo P and take the coefficients of the remainder. The 0-element of $\mathbb{F}_{2^n}$ is $00\ldots00$ and the 1-element is $00\ldots001$.

In the Advanced Encryption Standard (AES), encryption depends on algebraic operations in the finite field $\mathbb{F}_{2^8}$. The irreducible binary polynomial $P := X^8 + X^4 + X^3 + X + 1$ is taken to represent $\mathbb{F}_{2^8}$ as $\mathbb{F}_2[X]/P\mathbb{F}_2[X]$ (we already used this polynomial in an example above). Then the elements of $\mathbb{F}_{2^8}$ are just strings of 8 bits. In this way, a byte is an element of $\mathbb{F}_{2^8}$ and vice versa. One of the core operations of AES is the so-called S-Box. The AES S-Box maps a byte x to its inverse x^{-1} in $\mathbb{F}_{2^8}$ and then modifies the result by an $\mathbb{F}_2$-affine transformation (see Section 2.1.4). We conclude this section with examples for adding, multiplying and inverting bytes in $\mathbb{F}_{2^8}$.

$$01001101 + 00100101 = 01101000,$$
$$10111101 \cdot 01101001 = 11111100,$$
$$01001101 \cdot 00100101 = 00000001,$$
$$01001101^{-1} = 00100101.$$

As is common practice, we sometimes represent a byte and hence an element of $\mathbb{F}_{2^8}$ by two hexadecimal digits. Then the examples read as follows:

$$4D + 25 = 68, \ BD \cdot 69 = FC, \ 4D \cdot 25 = 01, \ 4D^{-1} = 25.$$

A.6 Solving Quadratic Equations in Binary Fields

The elementary method of solving a quadratic equation $X^2 + aX + b = 0$, by completing the square, includes a division by 2. Thus, this method does not work in binary fields $\mathbb{F}_{2^n}$. Here, instead, the trace and half-trace functions can be used to solve quadratic equations (see [CohFre06, Section II.11.2.6]).

Throughout this section, we consider the finite binary field $\mathbb{F}_{2^n}$ with 2^n elements, $n \in \mathbb{N}$ (see Section A.5.3 for the basics of finite fields).

Definition A.48. Let $x \in \mathbb{F}_{2^n}$. The *trace* $\mathrm{Tr}(x)$ of x is

$$\mathrm{Tr}(x) = \sum_{j=0}^{n-1} x^{2^j} = x + x^2 + x^{2^2} + \ldots + x^{2^{n-1}}.$$

In a binary field, $(a + b)^2 = a^2 + b^2$, and from Corollary A.34 we conclude that $x^{2^n} = x \cdot x^{2^n - 1} = x \cdot 1 = x$ for $x \in \mathbb{F}_{2^n}, x \neq 0$. Applying this observation, we get the following result.

Lemma A.49. *Let $x \in \mathbb{F}_{2^n}$. Then $\mathrm{Tr}(x)^2 = \mathrm{Tr}(x^2) = \mathrm{Tr}(x)$.*

Proof. The statement follows from

$$\left(\sum_{j=0}^{n-1} x^{2^j} \right)^2 = \sum_{j=0}^{n-1} \left(x^{2^j} \right)^2 = \sum_{j=0}^{n-1} \left(x^2 \right)^{2^j} = \sum_{j=0}^{n-1} x^{2^{j+1}} = \sum_{j=0}^{n-1} x^{2^j}.$$

$\square$

The solutions of $X^2 = X$ are 0 and 1. We see that $\mathrm{Tr}(x) = 0$ or $\mathrm{Tr}(x) = 1$, i.e., $\mathrm{Tr}(x) \in \mathbb{F}_2 \subseteq \mathbb{F}_{2^n}$.

Definition A.50. The map

$$\mathrm{Tr} : \mathbb{F}_{2^n} \longrightarrow \mathbb{F}_2, \ x \longmapsto \mathrm{Tr}(x)$$

is called the *trace map*.

Obviously, the trace map is linear over $\mathbb{F}_2$, which means that $\mathrm{Tr}(x + y) = \mathrm{Tr}(x) + \mathrm{Tr}(y)$.

Proposition A.51. *The number of elements x in $\mathbb{F}_{2^n}$ with $\mathrm{Tr}(x) = 1$ equals the number of elements with $\mathrm{Tr}(x) = 0$.*

Proof. To see that not all elements have trace 0, we consider the polynomial $f(X) = X + X^2 + X^{2^2} + \ldots + X^{2^{n-1}}$. The degree of $f(X)$ is 2^{n-1}. Thus, the number of roots is $\leq 2^{n-1}$, and not all of the 2^n elements of $\mathbb{F}_{2^n}$ can be roots of $f(X)$. Thus, there exists an element d with trace 1. The translation by d

$$\tau_d : \{x \in \mathbb{F}_{2^n} \mid \mathrm{Tr}(x) = 1\} \longrightarrow \{x \in \mathbb{F}_{2^n} \mid \mathrm{Tr}(x) = 0\}, \ x \longmapsto x + d,$$

is bijective, thereby proving our proposition. $\qquad\square$

We are now ready to solve quadratic equations

$$X^2 + aX + b = 0.$$

In $\mathbb{F}_{2^n}$, every element x is a square: $\left(x^{2^{n-1}}\right)^2 = x^{2^n} = x$. If $a = 0$, then $b^{2^{n-1}}$ is a solution (of multiplicity 2) of $X^2 = b$.

If $a \neq 0$, we substitute $X = aX$ and get the equation $X^2 + X + ba^{-2} = 0$. If x is a solution to this equation, then ax is a solution to the original equation. Thus, we only have to care about equations

$$X^2 + X + b = 0.$$

If x is a solution to this equation, then $x + 1$ is the second solution and

$$0 = \mathrm{Tr}(x^2 + x + b) = \mathrm{Tr}(x^2) + \mathrm{Tr}(x) + \mathrm{Tr}(b) = 2\,\mathrm{Tr}(x) + \mathrm{Tr}(b) = \mathrm{Tr}(b).$$

The condition $\mathrm{Tr}(b) = 0$ is also sufficient for the existence of a solution. In order to see this, we distinguish two cases.

If n is odd, the *half-trace* $\tau(b)$ is a solution. This is defined by

$$\tau(x) := \sum_{j=0}^{(n-1)/2} x^{2^{2j}} = x + x^{2^2} + x^{2^4} + \ldots + x^{2^{n-1}}.$$

The equality

$$\tau(x)^2 + \tau(x) = \sum_{j=0}^{(n-1)/2} x^{2^{2j+1}} + \sum_{j=0}^{(n-1)/2} x^{2^{2j}} = x + \mathrm{Tr}(x)$$

shows that $\mathrm{Tr}(x) = x + \tau(x) + \tau(x)^2$. Hence, $\tau(b)^2 + \tau(b) + b = 0$ if $\mathrm{Tr}(b) = 0$.

If n is even, let $d \in \mathbb{F}_{2^n}$ with $\mathrm{Tr}(d) = 1$, and let

$$x := \sum_{i=0}^{n-2} \left(\sum_{j=i+1}^{n-1} d^{2^j} \right) b^{2^i}.$$

Then

$$x^2 + x = \sum_{i=1}^{n-1}\left(\sum_{j=i+1}^{n} d^{2^j}\right) b^{2^i} + \sum_{i=0}^{n-2}\left(\sum_{j=i+1}^{n-1} d^{2^j}\right) b^{2^i}$$

$$= \sum_{i=1}^{n-1}\left(\sum_{j=i+1}^{n-1} d^{2^j}\right) b^{2^i} + d\sum_{i=1}^{n-1} b^{2^i} + \sum_{i=0}^{n-2}\left(\sum_{j=i+1}^{n-1} d^{2^j}\right) b^{2^i}$$

$$= d\sum_{i=1}^{n-1} b^{2^i} + \left(\sum_{j=1}^{n-1} d^{2^j}\right) b$$

$$= d(\mathrm{Tr}(b) + b) + (\mathrm{Tr}(d) + d)b$$

$$= d \cdot \mathrm{Tr}(b) + b.$$

We see that x is a solution to the equation if $\mathrm{Tr}(b) = 0$. An element $d \in \mathbb{F}_{2^n}$ with $\mathrm{Tr}(d) = 1$ can easily be generated by repeatedly choosing a random element until one with trace 1 is found. We expect to be successful after two trials.

Summarizing, we get the following proposition.

Proposition A.52. *Let* $b \in \mathbb{F}_{2^n}$ *with* $\mathrm{Tr}(b) = 0$.

1. *If n is odd, then $\tau(b)$ and $\tau(b) + 1$ are the roots of $X^2 + X + b = 0$.*
2. *If n is even and $d \in \mathbb{F}_{2^n}$ with $\mathrm{Tr}(d) = 1$ and*

$$x = \sum_{i=0}^{n-2}\left(\sum_{j=i+1}^{n-1} d^{2^j}\right) b^{2^i},$$

then x and $x + 1$ are the solutions of $X^2 + X + b = 0$.

A.7 Quadratic Residues

We now study the question as to which of the residues modulo n are squares.

Definition A.53. *Let* $n \in \mathbb{N}$ *and* $x \in \mathbb{Z}$. *We say that x is a* quadratic residue modulo n *if there is an element* $y \in \mathbb{Z}$ *with* $x \equiv y^2 \bmod n$. *Otherwise, x is called a* quadratic non-residue modulo n.

Examples:

1. The numbers $0, 1, 4, 5, 6$ and 9 are the quadratic residues modulo 10.
2. The numbers $0, 1, 3, 4, 5$ and 9 are the quadratic residues modulo 11.

Remark. The property of being a quadratic residue depends only on the residue class $[x] \in \mathbb{Z}_n$ of x modulo n. An integer x is a quadratic residue modulo n if and only if its residue class $[x]$ is a square in the residue class ring $\mathbb{Z}_n$ (i.e., if and only if there is some $[y] \in \mathbb{Z}_n$ with $[x] = [y]^2$). The residue class $[x]$ is often also called a quadratic residue.

In most cases we are only interested in the quadratic residues x which are units modulo n (i.e., x and n are relatively prime, see Proposition A.17).

Definition A.54. The subgroup of $\mathbb{Z}_n^*$ that consists of the residue classes represented by a quadratic residue is denoted by QR_n:

$$\mathrm{QR}_n = \{[x] \in \mathbb{Z}_n^* \mid \text{There is a } [y] \in \mathbb{Z}_n^* \text{ with } [x] = [y]^2\}.$$

It is called the *subgroup of quadratic residues* or the *subgroup of squares*. The complement of QR_n is denoted by $\mathrm{QNR}_n := \mathbb{Z}_n^* \setminus \mathrm{QR}_n$. It is called the *subset of quadratic non-residues*.

We give a criterion for determining the quadratic residues modulo a prime.

Lemma A.55. *Let p be a prime > 2 and $g \in \mathbb{Z}_p^*$ be a primitive root of $\mathbb{Z}_p^*$. Let $x \in \mathbb{Z}_p^*$. Then $x \in \mathrm{QR}_p$ if and only if $x = g^t$ for some even number $t, 0 \leq t \leq p - 2$.*

Proof. Recall that $\mathbb{Z}_p^*$ is a cyclic group generated by g (Theorem A.37). If $x \in \mathrm{QR}_p$, then $x = y^2$, and $y = g^s$ for some s. Then $x = g^{2s} = g^t$, with $t := 2s \bmod (p-1)$ (the order of g is $p-1$) and $0 \leq t \leq p-2$. Since $p-1$ is even, t is also even.

Conversely, if $x = g^t$, and t is even, then $x = (g^{t/2})^2$, which means that $x \in \mathrm{QR}_p$. $\qquad\square$

Proposition A.56. *Let p be a prime > 2. Exactly half of the elements of $\mathbb{Z}_p^*$ are squares, i.e., $|\mathrm{QR}_p| = (p-1)/2$.*

Proof. Since half of the integers x with $0 \leq x \leq p-2$ are even, the proposition follows from the preceding lemma. $\qquad\square$

Definition A.57. Let p be a prime > 2, and let $x \in \mathbb{Z}$ be prime to p.

$$\left(\frac{x}{p}\right) := \begin{cases} +1 & \text{if } [x] \in \mathrm{QR}_p, \\ -1 & \text{if } [x] \notin \mathrm{QR}_p, \end{cases}$$

is called the *Legendre symbol* of $x \bmod p$. For $x \in \mathbb{Z}$ with $p|x$, we set $\left(\frac{x}{p}\right) := 0$.

Proposition A.58 *(Euler's criterion). Let p be a prime > 2, and let $x \in \mathbb{Z}$. Then*

$$\left(\frac{x}{p}\right) \equiv x^{(p-1)/2} \bmod p.$$

Proof. If p divides x, then both sides are congruent 0 modulo p. Suppose p does not divide x. Let $[g] \in \mathbb{Z}_p^*$ be a primitive element.

We first observe that $g^{(p-1)/2} \equiv -1 \bmod p$. Namely, $[g]^{(p-1)/2}$ is a solution of the equation $X^2 - 1$ over the field $\mathbb{Z}_p^*$. Hence, $g^{(p-1)/2} \equiv \pm 1 \bmod p$. However, $g^{(p-1)/2} \bmod p \neq 1$, because the order of $[g]$ is $p-1$.

Let $[x] = [g]^t, 0 \leq t \leq p-2$. By Lemma A.55, $[x] \in \mathrm{QR}_p$ if and only if t is even. On the other hand, $x^{(p-1)/2} \equiv g^{t(p-1)/2} \equiv \pm 1 \bmod p$, and it is $\equiv 1 \bmod p$ if and only if t is even. This completes the proof. $\qquad\square$

Remarks:

1. The Legendre symbol is multiplicative in x:

$$\left(\frac{xy}{p}\right) = \left(\frac{x}{p}\right) \cdot \left(\frac{y}{p}\right).$$

This immediately follows, for example, from Euler's criterion. It means that $[xy] \in QR_p$ if and only if either both $[x], [y] \in QR_p$ or both $[x], [y] \notin QR_p$.

2. The Legendre symbol $\left(\frac{x}{p}\right)$ depends only on $x \bmod p$, and the map

$$\mathbb{Z}_p^* \longrightarrow \{1, -1\}, \ x \longmapsto \left(\frac{x}{p}\right)$$

is a homomorphism of groups.

We do not give proofs of the following two important results. Proofs may be found, for example, in [HarWri79], [Rosen00], [Koblitz94] and [Forster96].

Theorem A.59. *Let p be a prime > 2. Then:*

1. $\left(\frac{-1}{p}\right) = (-1)^{(p-1)/2} = \begin{cases} +1 & \text{if } p \equiv 1 \bmod 4, \\ -1 & \text{if } p \equiv 3 \bmod 4. \end{cases}$

2. $\left(\frac{2}{p}\right) = (-1)^{(p^2-1)/8} = \begin{cases} +1 & \text{if } p \equiv \pm 1 \bmod 8, \\ -1 & \text{if } p \equiv \pm 3 \bmod 8. \end{cases}$

Theorem A.60 *(Law of Quadratic Reciprocity). Let p and q be primes > 2, $p \neq q$. Then*

$$\left(\frac{p}{q}\right)\left(\frac{q}{p}\right) = (-1)^{(p-1)(q-1)/4}.$$

We generalize the Legendre symbol for composite numbers.

Definition A.61. *Let $n \in \mathbb{Z}$ be a positive odd number and $n = \prod_{i=1}^{r} p_i^{e_i}$ be the decomposition of n into primes. Let $x \in \mathbb{Z}$.*

$$\left(\frac{x}{n}\right) := \prod_{i=1}^{r} \left(\frac{x}{p_i}\right)^{e_i}$$

is called the Jacobi symbol of $x \bmod n$.

Remarks:

1. The value of $\left(\frac{x}{n}\right)$ only depends on the residue class $[x] \in \mathbb{Z}_n$.
2. If $[x] \in QR_n$, then $[x] \in QR_p$ for all primes p that divide n. Hence, $\left(\frac{x}{n}\right) = 1$. The converse is not true, in general. For example, let $n = pq$ be the product of two primes. Then $\left(\frac{x}{n}\right) = \left(\frac{x}{p}\right) \cdot \left(\frac{x}{q}\right)$ can be 1, whereas both $\left(\frac{x}{p}\right)$ and $\left(\frac{x}{q}\right)$ are -1. This means that $x \bmod p$ (and $x \bmod q$), and hence $x \bmod n$ are not squares.

3. The Jacobi symbol is multiplicative in both arguments:

$$\left(\frac{xy}{n}\right) = \left(\frac{x}{n}\right) \cdot \left(\frac{y}{n}\right) \text{ and } \left(\frac{x}{mn}\right) = \left(\frac{x}{m}\right) \cdot \left(\frac{x}{n}\right).$$

4. The map $\mathbb{Z}_n^* \longrightarrow \{1, -1\}$, $[x] \longmapsto \left(\frac{x}{n}\right)$ is a homomorphism of groups.
5. $J_n^{+1} := \{[x] \in \mathbb{Z}_n^* \mid \left(\frac{x}{n}\right) = 1\}$ is a subgroup of $\mathbb{Z}_n^*$.

Lemma A.62. *Let $n \geq 3$ be an odd integer. If n is a square (in $\mathbb{Z}$), then $\left(\frac{x}{n}\right) = 1$ for all x. Otherwise, half of the elements of $\mathbb{Z}_n^*$ have a Jacobi symbol of 1, i.e., $|J_n^{+1}| = \varphi(n)/2$.*

Proof. If n is a square, then the exponents e_i in the prime factorization of n are all even (notation as above), and the Jacobi symbol is always 1. If n is not a square, then there is an odd e_i, say e_1. By the Chinese Remainder Theorem (Theorem A.30), we find a unit x which is a quadratic non-residue modulo p_1 and a quadratic residue modulo p_i for $i = 2, \dots, r$. Then $\left(\frac{x}{n}\right) = -1$, and mapping $[y]$ to $[y \cdot x]$ yields a one-to-one map from J_n^{+1} to $\mathbb{Z}_n^* \setminus J_n^{+1}$.   □

Theorem A.63. *Let $n \geq 3$ be an odd integer. Then:*

1. $\left(\frac{-1}{n}\right) = (-1)^{(n-1)/2} = \begin{cases} +1 & \text{if } n \equiv 1 \bmod 4, \\ -1 & \text{if } n \equiv 3 \bmod 4. \end{cases}$

2. $\left(\frac{2}{n}\right) = (-1)^{(n^2-1)/8} = \begin{cases} +1 & \text{if } n \equiv \pm 1 \bmod 8, \\ -1 & \text{if } n \equiv \pm 3 \bmod 8. \end{cases}$

Proof. Let $f(n) = (-1)^{(n-1)/2}$ for statement 1 and $f(n) = (-1)^{(n^2-1)/8}$ for statement 2. You can easily check that $f(n_1 n_2) = f(n_1) f(n_2)$ for odd numbers n_1 and n_2 (for statement 2, consider the different cases of $n_1, n_2 \bmod 8$). Thus, both sides of the equations $\left(\frac{-1}{n}\right) = (-1)^{(n-1)/2}$ and $\left(\frac{2}{n}\right) = (-1)^{(n^2-1)/8}$ are multiplicative in n, and the proposition follows from Theorem A.59.   □

Theorem A.64 *(Law of Quadratic Reciprocity). Let $n, m \geq 3$ be odd integers. Then*

$$\left(\frac{m}{n}\right) = (-1)^{(n-1)(m-1)/4} \left(\frac{n}{m}\right).$$

Proof. If m and n have a common factor, then both sides are zero by the definition of the symbols. So we can suppose that m is prime to n. We write $m = p_1 p_2 \dots p_r$ and $n = q_1 q_2 \dots q_s$ as a product of primes. Converting from $\left(\frac{m}{n}\right) = \prod_{i,j} \left(\frac{p_i}{q_j}\right)$ to $\left(\frac{n}{m}\right) = \prod_{i,j} \left(\frac{q_j}{p_i}\right)$, we apply the reciprocity law for the Legendre symbol (Theorem A.60) for each of the factors. We get rs multipliers $\varepsilon_{ij} = (-1)^{(p_i-1)(q_j-1)/4}$. As in the previous proof, we use that $f(n) = (-1)^{(n-1)/2}$ is multiplicative in n and get

$$\prod_{i,j}(-1)^{(p_i-1)(q_j-1)/4} = \prod_{j}\left(\prod_{i}(-1)^{(p_i-1)/2}\right)^{(q_j-1)/2}$$

$$= \prod_{j}\left((-1)^{(m-1)/2}\right)^{(q_j-1)/2} = \left(\prod_{j}(-1)^{(q_j-1)/2}\right)^{(m-1)/2}$$

$$= \left((-1)^{(n-1)/2}\right)^{(m-1)/2} = (-1)^{(n-1)(m-1)/4},$$

as desired.   □

Remark. Computing a Jacobi symbol $\left(\frac{m}{n}\right)$ simply by using the definition requires knowing the prime factors of n. No algorithm is known that can compute the prime factors in polynomial time. However, using the Law of Quadratic Reciprocity (Theorem A.64) and Theorem A.63, we can efficiently compute $\left(\frac{m}{n}\right)$ using the following algorithm, without knowing the factorization of n.

Algorithm A.65.

```
int Jac(int m, n)
 1   m ← m mod n
 2   if m = 0 then return 0
 3   j ← 1; t ← 0
 4   while m is even do
 5        m ← m div 2; t ← t + 1
 6   if n ≡ ±3 mod 8 and t is odd
 7      then j ← −1
 8   if m = 1
 9      then return j
10      else  return j · (−1)^{(m−1)(n−1)/4} · Jac(n, m)
```

An analysis similar to that of the Euclidean algorithm (Algorithm A.5) shows that the algorithm terminates after at most $O(\log_2(n))$ iterations (see Corollary A.11).

Example. We want to determine whether the prime 7331 is a quadratic residue modulo the prime 9859. For this purpose, we have to compute the Legendre symbol $\left(\frac{7331}{9859}\right)$ and could do that using Euler's criterion (Proposition A.58). However, applying Algorithm A.65 is much more efficient:

$$\left(\frac{7331}{9859}\right) = -\left(\frac{9859}{7331}\right) = -\left(\frac{2528}{7331}\right) = -\left(\frac{2}{7331}\right)^5 \cdot \left(\frac{79}{7331}\right)$$

$$= -(-1)^5 \cdot \left(\frac{79}{7331}\right) = \left(\frac{79}{7331}\right) = -\left(\frac{7331}{79}\right) = -\left(\frac{63}{79}\right)$$

$$= \left(\frac{79}{63}\right) = \left(\frac{16}{63}\right) = \left(\frac{2}{63}\right)^4 = 1.$$

Thus, 7331 is a quadratic residue modulo 9859.

A.8 Modular Square Roots

We now discuss how to get the square root of a quadratic residue. Computing square roots modulo n can be a difficult or even an infeasible task if n is a composite number (and, e.g., the Rabin cryptosystem is based on this; see Section 3.5). However, if n is a prime, we can determine square roots using an efficient algorithm.

Proposition A.66. *There is a (probabilistic) polynomial algorithm Sqrt which, given as inputs a prime p and an $a \in \mathrm{QR}_p$, computes a square root $x \in \mathbb{Z}_p^*$ of a: $Sqrt(p, a) = x$ and $x^2 = a$ (in $\mathbb{Z}_p$).*

Remarks:

1. The square roots of $a \in \mathrm{QR}_p$ are the solutions of the equation $X^2 - a = 0$ over $\mathbb{Z}_p$. Hence a has two square roots (for $p > 2$). If x is a square root, then $-x$ is the other root.
2. "Probabilistic" means that random choices[5] are included in the algorithm. Polynomial means that the running time (the number of binary operations) of the algorithm is bounded by a polynomial in the binary length of the inputs. *Sqrt* is a so-called Las Vegas algorithm, i.e., we expect *Sqrt* to return a correct result in polynomial time (for a detailed discussion of the notion of probabilistic polynomial algorithms, see Chapter 5).

Proof. Let $a \in \mathrm{QR}_p$. By Euler's criterion (Proposition A.58), $a^{(p-1)/2} = 1$. Hence $a^{(p+1)/2} = a$. We first consider the (easy) case of $p \equiv 3 \bmod 4$. Since 4 divides $p + 1$, $(p + 1)/4$ is an integer, and $x := a^{(p+1)/4}$ is a square root of a.

Now assume $p \equiv 1 \bmod 4$. The straightforward computation of the square root as in the first case does not work, since $(p + 1)/2$ is not divisible by 2. We choose a quadratic non-residue $b \in \mathrm{QNR}_p$ (here the random choices come into play, see below). By Proposition A.58, $b^{(p-1)/2} = -1$. We have $a^{(p-1)/2} = 1$, and $(p - 1)/2$ is even. Let $(p - 1)/2 = 2^l r$, with r odd and $l \geq 1$. We will compute an exponent s such that $a^r b^{2s} = 1$. Then we are finished. Namely, $a^{r+1} b^{2s} = a$ and $a^{(r+1)/2} b^s$ is a square root of a.

We obtain s in l steps. The intermediate result after step i is a representation $a^{2^{l-i} r} \cdot b^{2s_i} = 1$. We start with $a^{(p-1)/2} \cdot b^0 = a^{(p-1)/2} = 1$ and $s_0 = 0$. Let $y_i = a^{2^{l-i} r} \cdot b^{2s_i}$. In the i-th step we take the square root $y_i' := a^{2^{l-i} r} \cdot b^{s_{i-1}}$ of $y_{i-1} = a^{2^{l-i+1} r} \cdot b^{2s_{i-1}}$. The value of y_i' is either 1 or -1. If $y_i' = 1$, then we

---

[5] In this chapter, all random choices are with respect to the uniform distribution.

take $y_i := y_i'$. If $y_i' = -1$, then we set $y_i := y_i' \cdot b^{(p-1)/2}$. The first time that b appears with an exponent > 0 in the representation is after the first step (if ever), and then b's exponent is $(p-1)/2 = 2^l r$. This implies that s_{i-1} is indeed an even number for $i = 1, \ldots, l$.

Thus, we may compute a square root using the following algorithm.

Algorithm A.67.

   int $Sqrt$(int a, prime p)

  1   if $p \equiv 3 \bmod 4$

  2   then return $a^{(p+1)\,\mathrm{div}\,4} \bmod p$

  3   else

  4   randomly choose $b \in \mathrm{QNR}_p$

  5         $i \leftarrow (p-1)\,\mathrm{div}\,2;\ j \leftarrow 0$

  6   repeat

  7            $i \leftarrow i\,\mathrm{div}\,2;\ j \leftarrow j\,\mathrm{div}\,2$

  8   if $a^i b^j \equiv -1 \bmod p$

  9   then $j \leftarrow j + (p-1)\,\mathrm{div}\,2$

  10   until $i \equiv 1 \bmod 2$

  11   return $a^{(i+1)\,\mathrm{div}\,2}\, b^{j\,\mathrm{div}\,2} \bmod p$

In the algorithm we get a quadratic non-residue by a random choice. For this purpose, we randomly choose an element b of $\mathbb{Z}_p^*$ and test (by Euler's criterion) whether b is a non-residue. Since half of the elements in $\mathbb{Z}_p^*$ are non-residues, we expect (on average) to get a non-residue after 2 random choices (see Lemma B.12).   □

Now let n be a composite number. If n is a product of distinct primes and if we know these primes, we can apply the Chinese Remainder Theorem (Theorem A.30) and reduce the computation of square roots in $\mathbb{Z}_n^*$ to the computation of square roots modulo a prime. There we can apply Algorithm A.67. We discuss this procedure in detail for the RSA and Rabin settings, where $n = pq$, with p and q being distinct primes. The extended Euclidean algorithm yields numbers $d, e \in \mathbb{Z}$ with $1 = dp + eq$. By the Chinese Remainder Theorem, the map

$$\psi_{p,q} : \mathbb{Z}_n \longrightarrow \mathbb{Z}_p \times \mathbb{Z}_q,\ [x] \longmapsto ([x \bmod p], [x \bmod q])$$

is an isomorphism. The inverse map is given by

$$\chi_{p,q} : \mathbb{Z}_p \times \mathbb{Z}_q \longrightarrow \mathbb{Z}_n,\ ([x_1], [x_2]) \longmapsto [dpx_2 + eqx_1].$$

Addition and multiplication in $\mathbb{Z}_p \times \mathbb{Z}_q$ are done component-wise:

$$([x_1], [x_2]) + ([x_1'], [x_2']) = ([x_1 + x_1'], [x_2 + x_2']),$$

$$([x_1], [x_2]) \cdot ([x_1'], [x_2']) = ([x_1 \cdot x_1'], [x_2 \cdot x_2']).$$

For $[x] \in \mathbb{Z}_n$ let $([x_1], [x_2]) := \psi_{p,q}([x])$. We have

$$[x]^2 = [a] \text{ if and only if } [x_1]^2 = [a_1] \text{ and } [x_2]^2 = [a_2]$$

(see the remark after Theorem A.30). Thus, in order to compute the roots of $[a]$, we can compute the square roots of $[a_1]$ and $[a_2]$ and apply the inverse Chinese remainder map $\chi_{p,q}$. In $\mathbb{Z}_p$ and $\mathbb{Z}_q$, we can efficiently compute square roots by using Algorithm A.67.

Recall that $\mathbb{Z}_p$ and $\mathbb{Z}_q$ are fields, and over a field a quadratic equation $X^2 - [a_1] = 0$ has at most 2 solutions. Hence, $[a_1] \in \mathbb{Z}_p$ has at most two square roots. The zero-element $[0]$ has the only square root $[0]$. For $p = 2$, the only non-zero element $[1]$ of $\mathbb{Z}_2$ has the only square root $[1]$. For $p > 2$, a non-zero element $[a_1]$ has either no or two distinct square roots. If $[x_1]$ is a square root of $[a_1]$, then $-[x_1]$ is also a root and for $p > 2$, $[x_1] \neq -[x_1]$.

Combining the square roots of $[a_1]$ in $\mathbb{Z}_p$ with the square roots of $[a_2]$ in $\mathbb{Z}_q$, we get the square roots of $[a]$ in $\mathbb{Z}_n$. We summarize the results in the following proposition.

Proposition A.68. *Let p and q be distinct primes > 2 and $n := pq$. Assume that the prime factors p and q of n are known. Then the square roots of a quadratic residue $[a] \in \mathbb{Z}_n$ can be efficiently computed. Moreover, if $[a] = ([a_1], [a_2])^6$, then:*

1. *$[x] = ([x_1], [x_2])$ is a square root of $[a]$ in $\mathbb{Z}_n$ if and only if $[x_1]$ is a square root of $[a_1]$ in $\mathbb{Z}_p$ and $[x_2]$ is a square root of $[a_2]$ in $\mathbb{Z}_q$.*
2. *If $[x_1]$ and $-[x_1]$ are the square roots of $[a_1]$, and $[x_2]$ and $-[x_2]$ are the square roots of $[a_2]$, then $[u] = ([x_1], [x_2]), [v] = ([x_1], -[x_2]), -[v] = (-[x_1], [x_2])$ and $-[u] = (-[x_1], -[x_2])$ are all the square roots of $[a]$.*
3. *$[0]$ is the only square root of $[0] = ([0], [0])$.*
4. *If $[a_1] \neq [0]$ and $[a_2] \neq [0]$, which means that $[a]$ is a unit in $\mathbb{Z}_n$, then the square roots of $[a]$ given in 2. are pairwise distinct, i.e., $[a]$ has four square roots.*
5. *If $[a_1] = [0]$ (i.e., p divides a) and $[a_2] \neq [0]$ (i.e., q does not divide a), then $[a]$ has only two distinct square roots, $[u]$ and $[v]$.*

Remark. If one of the primes is 2, say $p = 2$, then statements 1–3 of Proposition A.68 are also true. Statements 4 and 5 have to be modified. $[a]$ has only one or two roots, because $[x_1] = -[x_1]$. If $[a_2] = 0$, then there is only one root. If $[a_2] \neq 0$, then the roots $[u]$ and $[v]$ are distinct.

Conversely, the ability to compute square roots modulo n implies the ability to factorize n.

Lemma A.69. *Let $n := pq$, with distinct primes $p, q > 2$. Let $[u]$ and $[v]$ be square roots of $[a] \in \mathrm{QR}_n$ with $[u] \neq \pm[v]$. Then the prime factors of n can be computed from $[u]$ and $[v]$ using the Euclidean algorithm.*

---

[6] We identify $\mathbb{Z}_n$ with $\mathbb{Z}_p \times \mathbb{Z}_q$ via the Chinese remainder isomorphism $\psi_{p,q}$.

Proof. We have $n \mid u^2 - v^2 = (u+v)(u-v)$, but n does not divide $u+v$ and n does not divide $u - v$. Hence, the computation of $\gcd(u+v, n)$ yields one of the prime factors of n. $\square$

Proposition A.70. *Let* $I := \{n \in \mathbb{N} \mid n = pq, \; p, q \text{ distinct primes}\}$. *Then the following statements are equivalent:*

1. *There is a probabilistic polynomial algorithm* A_1 *that on inputs* $n \in I$ *and* $a \in \mathrm{QR}_n$ *returns a square root of* a *in* $\mathbb{Z}_n^*$.
2. *There is a probabilistic polynomial algorithm* A_2 *that on input* $n \in I$ *yields the prime factors of* n.

Proof. Let A_1 be a probabilistic polynomial algorithm that on inputs n and a returns a square root of a modulo n. Then we can find the factors of n in the following way. We randomly select an $x \in \mathbb{Z}_n^*$ and compute $y = A_1(n, x^2)$. Since a has four distinct roots by Proposition A.68, the probability that $x \neq \pm y$ is $1/2$. If $x \neq \pm y$, we easily compute the factors by Lemma A.69. Otherwise we choose a new random x. We expect to be successful after two iterations.

Conversely, if we can compute the prime factors of n using a polynomial algorithm A_2, we can also compute (all the) square roots of arbitrary quadratic residues in polynomial time, as we have seen in Proposition A.68. $\square$

The Chinese Remainder isomorphism can also be used to determine the number of quadratic residues modulo n.

Proposition A.71. *Let* p *and* q *be distinct primes, and* $n := pq$. *Then* $|\mathrm{QR}_n| = (p-1)(q-1)/4$.

Proof. $[a] = ([a_1], [a_2]) \in \mathrm{QR}_n$ if and only if $[a_1] \in \mathrm{QR}_p$ and $[a_2] \in \mathrm{QR}_q$. By Proposition A.56, $|\mathrm{QR}_p| = (p-1)/2$ and $|\mathrm{QR}_q| = (q-1)/2$. $\square$

Proposition A.72. *Let* p *and* q *be distinct primes with* $p, q \equiv 3 \bmod 4$, *and* $n := pq$. *Let* $[a] \in \mathrm{QR}_n$, *and* $[u] = ([x_1], [x_2]), [v] = ([x_1], -[x_2]), -[v] = (-[x_1], [x_2])$ *and* $-[u] = (-[x_1], -[x_2])$ *be the four square roots of* $[a]$ *(see Proposition A.68). Then:*

1. $\left(\frac{u}{n}\right) = -\left(\frac{v}{n}\right)$.
2. *One and only one of the four square roots is in* QR_n.

Proof. 1. We have $u \equiv v \bmod p$ and $u \equiv -v \bmod q$, hence we conclude by Theorem A.59 that

$$\left(\frac{u}{n}\right) = \left(\frac{u}{p}\right)\left(\frac{u}{q}\right) = \left(\frac{v}{p}\right)\left(\frac{v}{q}\right)\left(\frac{-1}{q}\right) = -\left(\frac{v}{p}\right)\left(\frac{v}{q}\right) = -\left(\frac{v}{n}\right).$$

2. By Theorem A.59, $\left(\frac{-1}{p}\right) = \left(\frac{-1}{q}\right) = -1$. Thus, exactly one of the roots $[x_1]$ or $-[x_1]$ is in QR_p, say $[x_1]$, and exactly one of the roots $[x_2]$ or $-[x_2]$ is

in QR_q, say $[x_2]$. Then $[u] = ([x_1], [x_2])$ is the only square root of $[a]$ that is in QR_n.   □

A.9 The Group $\mathbb{Z}^*_{n^2}$

Let $n = pq$, where p and q are distinct primes such that $\gcd(n, \varphi(n)) = 1$. Recall that $\varphi(n) = (p - 1)(q - 1)$ (Corollary A.31). Here, we study the structure of the prime residue class group $\mathbb{Z}^*_{n^2}$. This is the group of the Paillier encryption scheme (Section 3.6.2).

Lemma A.73. *Let $x, y \in \mathbb{N}$, $e \geq 1$.*

1. *Then $(x + ny)^e \equiv x^e + enx^{e-1}y \bmod n^2$. In particular, we have $(x + ny)^n \equiv x^n \bmod n^2$.*
2. *$x^n \equiv 1 \bmod n^2$ if and only if $x \equiv 1 \bmod n$.*

Proof.

1. $(x + ny)^e = \sum_{i=0}^{e} \binom{e}{i} x^{e-i}(ny)^i \equiv x^e + enx^{e-1}y \bmod n^2$.
2. Assume $x^n \equiv 1 \bmod n^2$. Then $x^n \equiv 1 \bmod n$. Since $\gcd(n, \varphi(n)) = 1$, this implies that $x \equiv 1 \bmod n$. Conversely, if $x = 1 + kn$, then $x^n \equiv 1 + kn^2 \equiv 1 \bmod n^2$.

□

Definition A.74. $x \in \mathbb{Z}^*_{n^2}$ *is an n-th residue if and only if $x = y^n$ for some $y \in \mathbb{Z}^*_{n^2}$. The subgroup of $\mathbb{Z}^*_{n^2}$ consisting of the n-th residues is denoted by R_n:*

$$R_n = \{x \in \mathbb{Z}^*_{n^2} \mid x = y^n \text{ for some } y \in \mathbb{Z}^*_{n^2}\}.$$

Proposition A.75. *The map $\varepsilon : \mathbb{Z}^*_n \longrightarrow R_n$, $[x] \longmapsto [x^n]$ is an isomorphism of groups. Thus R_n has $\varphi(n)$ elements.*

Proof. From Lemma A.73, part 1, we conclude that ε is well defined. Obviously, it is a homomorphism. Let $[x] \in \mathbb{Z}^*_n$. Assume that $x^n \equiv 1 \bmod n^2$. Then $x \equiv 1 \bmod n$ (see Lemma A.73, part 2), i.e., ε is injective. Let $y = [y_0 + y_1 n] \in \mathbb{Z}^*_{n^2}, 0 \leq y_0, y_1 \leq n - 1$. Then $\gcd(y_0, n) = 1$ and $y^n = [(y_0 + y_1 n)^n] = [y_0^n]$ (see Lemma A.73, part 1). Thus ε is also surjective. It follows that ε is an isomorphism and $|R_n| = \varphi(n)$.   □

Definition A.76. *A solution of the equation $X^n = [1]$ in $\mathbb{Z}^*_{n^2}$ is called an n-th root of unity. The subgroup of $\mathbb{Z}^*_{n^2}$ that consists of the n-th roots of unity is denoted by W_n:*

$$W_n = \{x \in \mathbb{Z}^*_{n^2} \mid x^n = [1]\}.$$

Proposition A.77. *The subgroup W_n of n-th roots of unity is a cyclic subgroup of $\mathbb{Z}^*_{n^2}$ and contains n elements. An element $w \in W_n$ has order n, i.e., it is a generator of W_n, if and only if $w = [1 + kn], 1 \leq k \leq n - 1, \gcd(k, n) = 1$. In particular, there are $\varphi(n)$ generators of W_n.*

Proof. Let $[x] \in W_n$, $0 \le x < n^2$. Then $x \equiv 1 \bmod n$ (Lemma A.73). Thus $x = 1 + kn$, where $k \in \{0, \ldots, n-1\}$, and we see that $|W_n| \le n$. The elements $[1+n]^k = [1+kn]$, $k = 0, \ldots, n-1$, are in W_n (Lemma A.73) and are pairwise distinct. Hence $|W_n| = n$, W_n is cyclic and $[1+n]$ is a generator. By Lemma A.35, $\mathrm{ord}([1+n]^k) = n$ if and only if $\gcd(k, n) = 1$. Hence, w is a generator of W_n if and only if $w = [1+kn]$, $1 \le k \le n-1$, $\gcd(k, n) = 1$. In particular, there are $\varphi(n)$ generators. $\qquad\square$

Let $g \in \mathbb{Z}_{n^2}^*$ be an element of order n, i.e., let g be a generator of W_n. Then the map
$$\mathrm{Exp}_g : \mathbb{Z}_n \longrightarrow W_n, \ x \longmapsto g^x$$
is an isomorphism of groups. The inverse map is denoted by Log_g:
$$\mathrm{Log}_g : W_n \longrightarrow \mathbb{Z}_n.$$

We define
$$L : W_n \longrightarrow \mathbb{N}, \ [w] \longmapsto \frac{w-1}{n},$$
where w, $0 \le w < n^2$, is the natural representative of $[w]$. Note that $w \equiv 1 \bmod n$ (Lemma A.73). Therefore, $w - 1$ is divisible by n and $(w-1)/n \in \mathbb{N}$.

To simplify the notation in the rest of this section, a residue class and its natural representative are identified. In computations, we always use the natural representative.

Lemma A.73, part 1, tells us that the easily computable function L returns the logarithm of an n-th root of unity $w \in W_n$ with respect to the generator $[1+n]$:
$$\mathrm{Log}_{[1+n]}(w) = [L(w)].$$
More generally, we have the following result.

Proposition A.78. *Let $w \in W_n$. Then*
$$\mathrm{Log}_g(w) = [L(w)] \cdot [L(g)]^{-1}.$$

Proof.
$$g^x = [1+n]^{x\,\mathrm{Log}_{[1+n]}(g)}, \ \text{and hence } \mathrm{Log}_g(w) \cdot \mathrm{Log}_{[1+n]}(g) = \mathrm{Log}_{[1+n]}(w).$$

$\qquad\square$

Proposition A.79. *Let $g \in \mathbb{Z}_{n^2}^*$ be an element of order n, i.e., let g be a generator of W_n. Then*
$$E_g : \mathbb{Z}_n \times \mathbb{Z}_n^* \longrightarrow \mathbb{Z}_{n^2}^*, \ (m, r) \longmapsto g^m r^n$$
is an isomorphism of groups.

Proof. E_g is well defined and a homomorphism of groups. Assume that $g^m r^n = [1]$. Then $g^m = r^{-n}$. Now, $\mathrm{ord}(g^m)$ divides $\mathrm{ord}(g) = n$ and $\mathrm{ord}(r^{-n})$ divides $\varphi(n)$ (Proposition A.75). Since $\gcd(n, \varphi(n)) = 1$ by assumption, we find that $g^m = r^n = [1]$ and then that $m \equiv 0 \bmod n$ and $r \equiv 1 \bmod n$. This shows that E_g is injective. Since $|\mathbb{Z}_n \times \mathbb{Z}_n^*| = n\varphi(n) = \varphi(n^2) = |\mathbb{Z}_{n^2}^*|$, the map E_g is bijective. $\qquad\square$

Proposition A.80. *Let $\lambda := \mathrm{lcm}(p-1, q-1)$ and let $g \in \mathbb{Z}_{n^2}^*$ be an element of order n. Let $w \in \mathbb{Z}_{n^2}^*$ and let $w = g^m r^n$ be the unique representation of w with $m \in \mathbb{Z}_n, r \in \mathbb{Z}_n^*$ (Proposition A.79). Then*

$$m = \left[L\left(w^\lambda\right) \right] \cdot \left[L\left(g^\lambda\right) \right]^{-1}.$$

Proof. Since $p-1$ and $q-1$ divide λ, we have $r^\lambda \equiv 1 \bmod p$ and $r^\lambda \equiv 1 \bmod q$ (Proposition A.25). Then $r^\lambda \equiv 1 \bmod n$ and $r^{n\lambda} \equiv 1 \bmod n^2$ (Lemma A.73). Hence $w^\lambda = g^{m\lambda} r^{n\lambda} = g^{m\lambda} \in W_n$. Since, by assumption, n is prime to $\varphi(n) = (p-1)(q-1)$, we have $\gcd(\lambda, n) = 1$, and therefore g^λ has order n. From Proposition A.78, we conclude that

$$m = \mathrm{Log}_{g^\lambda}\left(w^\lambda\right) = \left[L\left(w^\lambda\right) \right] \cdot \left[L\left(g^\lambda\right) \right]^{-1}.$$

$\qquad\square$

A.10 Primes and Primality Tests

The problem to decide whether a given number is a prime has been studied comprehensively. We will discuss probabilistic methods for quickly recognizing composite numbers. If these tests fail to prove that a number is composite, then we know, at least with very high probability, that this number is a prime.

Theorem A.81 *(Euclid's Theorem). There are infinitely many primes.*

Proof. Assume that there are only a finite number of primes $p_1, \ldots, p_r$. Let $n = 1 + p_1 \cdot \ldots \cdot p_r$. Then p_i does not divide n, $1 \leq i \leq r$. Thus, either n is a prime or it contains a new prime factor different from p_i, $1 \leq i \leq r$. This is a contradiction. $\qquad\square$

There is the following famous result on the distribution of primes. It is called the *Prime Number Theorem* and was proven by Hadamard and de la Vallée Poussin.

Theorem A.82. *Let $\pi(x) = |\{p \text{ prime} \mid p \leq x\}|$. Then, for large x,*

$$\pi(x) \approx \frac{x}{\ln(x)}.$$

A proof can be found, for example, in [HarWri79] or [Newman80].

Corollary A.83. *The frequency of primes among the numbers in the magnitude of x is approximately* $1/\ln(x)$.

Sometimes we are interested in primes which have a given remainder c modulo b.

Theorem A.84 *(Dirichlet's Theorem).* *Let $b, c \in \mathbb{N}$ and $\gcd(b, c) = 1$. Let $\pi_{b,c}(x) = |\{p \text{ prime} \mid p \leq x, p = kb + c, k \in \mathbb{N}\}|$. Then for large x,*

$$\pi_{b,c}(x) \approx \frac{1}{\varphi(b)} \frac{x}{\ln(x)}.$$

Corollary A.85. *Let $b, c \in \mathbb{N}$ and $\gcd(b, c) = 1$. The frequency of primes among the numbers a with $a \bmod b = c$ in the magnitude of x is approximately* $b/\varphi(b)\ln(x)$.

Our goal in this section is to give criteria for prime numbers which can be efficiently checked by an algorithm. We will use the fact that a proper subgroup H of $\mathbb{Z}_n^*$ contains at most $|\mathbb{Z}_n^*|/2 = \varphi(n)/2$ elements. More generally, we prove the following basic result on groups.

Proposition A.86. *Let G be a finite group and $H \subset G$ be a subgroup. Then $|H|$ is a divisor of $|G|$.*

Proof. We consider the following equivalence relation ($\sim$) on G: $g_1 \sim g_2$ if and only if $g_1 \cdot g_2^{-1} \in H$. The equivalence class of an element $g \in G$ is $gH = \{gh \mid h \in H\}$. Thus, all equivalence classes contain the same number, namely $|H|$, of elements. Since G is the disjoint union of the equivalence classes, we have $|G| = |H| \cdot r$, where r is the number of equivalence classes, and we see that $|H|$ divides $|G|$. $\qquad\qquad\square$

Fermat's Little Theorem (Theorem A.25) yields a necessary condition for primes. Let $n \in \mathbb{N}$ be an odd number. If n is a prime and $a \in \mathbb{N}$ with $\gcd(a, n) = 1$, then $a^{n-1} \equiv 1 \bmod n$. If there is an $a \in \mathbb{N}$ with $\gcd(a, n) = 1$ and $a^{n-1} \not\equiv 1 \bmod n$, then n is not a prime.
Unfortunately, the converse is not true: there are composite (i.e., non-prime) numbers n such that $a^{n-1} \equiv 1 \bmod n$, for all $a \in \mathbb{N}$ with $\gcd(a, n) = 1$. The smallest n with this property is $561 = 3 \cdot 11 \cdot 17$.

Definition A.87. *Let $n \in \mathbb{N}, n \geq 3$, be a composite number. We call n a Carmichael number if*

$$a^{n-1} \equiv 1 \bmod n,$$

for all $a \in \mathbb{N}$ with $\gcd(a, n) = 1$.

Proposition A.88. *Let n be a Carmichael number, and let p be a prime that divides n. Then p^2 does not divide n. In other words, the factorization of n does not contain squares.*

Proof. Assume $n = p^k m$, with $k \geq 2$ and p does not divide m. Let $b := 1+pm$. From $b^p = (1+pm)^p = 1+p^2 \cdot \alpha$ we derive that $b^p \equiv 1 \bmod p^2$. Since p does not divide m, we have $b \not\equiv 1 \bmod p^2$ and conclude that b has order p modulo p^2. Now, b is prime to n, because it is prime to p and m, and n is a Carmichael number. Hence $b^{n-1} \equiv 1 \bmod n$, and then, in particular, $b^{n-1} \equiv 1 \bmod p^2$. Thus $p \mid n-1$ (by Lemma A.33), a contradiction to $p \mid n$.   □

Proposition A.89. *Let n be an odd, composite number that does not contain squares. Then n is a Carmichael number if and only if $(p-1) \mid (n-1)$ for all prime divisors p of n.*

Proof. Let $n = p_1 \cdot \ldots \cdot p_r$, with p_i being a prime, $i = 1, \ldots, r$, and $p_i \neq p_j$ for $i \neq j$. n is a Carmichael number if and only if $a^{n-1} \equiv 1 \bmod n$ for all a that are prime to n and, by the Chinese Remainder Theorem this in turn is equivalent to $a^{n-1} \equiv 1 \bmod p_i$, for all a which are not divided by p_i, $i = 1, \ldots, r$. This is the case if and only if $(p_i - 1) \mid (n-1)$, for $i = 1, \ldots, r$. The last equivalence follows from Proposition A.24 and Corollary A.34, since $\mathbb{Z}_{p_i}^*$ is a cyclic group of order $p_i - 1$ (Theorem A.37).   □

Corollary A.90. *Every Carmichael number n contains at least three distinct primes.*

Proof. Assume $n = p_1 \cdot p_2$, with $p_1 < p_2$. Then $n-1 = p_1(p_2-1)+(p_1-1) \equiv (p_1-1) \bmod (p_2-1)$. However, $(p_1-1) \not\equiv 0 \bmod (p_2-1)$, since $0 < p_1 - 1 < p_2 - 1$. Hence, $p_2 - 1$ does not divide $n-1$. This is a contradiction.   □

Though Carmichael numbers are extremely rare (there are only 2163 Carmichael numbers below $25 \cdot 10^9$), the Fermat condition $a^{n-1} \equiv 1 \bmod p$ is not reliable for a primality test.[7] We are looking for other criteria.

Let $n \in \mathbb{N}$ be an odd number. If n is a prime, by Euler's criterion (Proposition A.58), $\left(\frac{a}{n}\right) \equiv a^{(n-1)/2} \bmod n$ for every $a \in \mathbb{N}$ with $\gcd(a,n) = 1$. Here the converse is also true. More precisely:

Proposition A.91. *Let n be an odd and composite number. Let*

$$\overline{E_n} = \left\{ [a] \in \mathbb{Z}_n^* \,\Big|\, \left(\frac{a}{n}\right) \not\equiv a^{(n-1)/2} \bmod n \right\}.$$

Then $|\overline{E_n}| \geq \varphi(n)/2$, i.e., for more than half of the $[a]$, we have $\left(\frac{a}{n}\right) \not\equiv a^{(n-1)/2} \bmod n$.

Proof. Let $E_n := \mathbb{Z}_n^* \setminus \overline{E_n}$ be the complement of $\overline{E_n}$. We have

$$E_n = \{ [a] \in \mathbb{Z}_n^* \mid \left(\frac{a}{n}\right) \equiv a^{(n-1)/2} \bmod n \}$$

$$= \{ [a] \in \mathbb{Z}_n^* \mid a^{(n-1)/2} \cdot \left(\frac{a}{n}\right)^{-1} \equiv 1 \bmod n \}.$$

---

[7] Numbers n satisfying $a^{n-1} \equiv 1 \bmod n$ are called pseudoprimes for the base a.

Since E_n is a subgroup of $\mathbb{Z}_n^*$, we could infer $|E_n| \leq \varphi(n)/2$ if E_n were a proper subset of $\mathbb{Z}_n^*$ (Proposition A.86). Then $|\overline{E_n}| = |\mathbb{Z}_n^*| - |E_n| \geq \varphi(n) - \varphi(n)/2 = \varphi(n)/2$.

Thus it suffices to prove: if $E_n = \mathbb{Z}_n^*$, then n is a prime.

Assume $E_n = \mathbb{Z}_n^*$ and n is not a prime. From $\left(\frac{a}{n}\right) \equiv a^{(n-1)/2} \bmod n$, it follows that $a^{n-1} \equiv 1 \bmod n$. Thus n is a Carmichael number, and hence does not contain squares (Proposition A.88). Let $n = p_1 \cdot \ldots \cdot p_k, k \geq 3$, be the decomposition of n into distinct primes. Let $[v] \in \mathbb{Z}_{p_1}^*$ be a quadratic non-residue, i.e., $\left(\frac{v}{p_1}\right) = -1$. Using the Chinese Remainder Theorem, choose an $[x] \in \mathbb{Z}_n^*$ with $x \equiv v \bmod p_1$ and $x \equiv 1 \bmod n/p_1$. Then $\left(\frac{x}{n}\right) = \left(\frac{x}{p_1}\right) \cdot \ldots \cdot \left(\frac{x}{p_n}\right) = -1$. Since $E_n = \mathbb{Z}_n^*$, $\left(\frac{x}{n}\right) \equiv x^{(n-1)/2} \bmod n$, and hence $x^{(n-1)/2} \equiv -1 \bmod n$, in particular $x^{(n-1)/2} \equiv -1 \bmod p_2$. This is a contradiction. $\square$

The following considerations lead to a necessary condition for primes that does not require the computation of Jacobi symbols. Let $n \in \mathbb{N}$ be an odd number, and let $n - 1 = 2^t m$, with m odd. Suppose that n is a prime. Then $\mathbb{Z}_n$ is a field (Corollary A.18), and hence ± 1 are the only square roots of 1, i.e., the only solutions of $X^2 - 1$, modulo n. Moreover, $a^{n-1} \equiv 1 \bmod n$ for every $a \in \mathbb{N}$ that is prime to n (Theorem A.25). Thus

$a^{n-1} \equiv 1 \bmod n$ and $a^{(n-1)/2} \equiv \pm 1 \bmod n$, and

if $(n-1)/2$ is even and $a^{(n-1)/2} \equiv 1 \bmod n$, then $a^{(n-1)/4} \equiv \pm 1 \bmod n$, and

if $(n-1)/4$ is even and $a^{(n-1)/4} \equiv 1 \bmod n$, then $\ldots$.

We see: if n is a prime, then for every $a \in \mathbb{N}$ with $\gcd(a, n) = 1$, either $a^m \equiv 1 \bmod n$, or there is a $j \in \{0, \ldots, t-1\}$ with $a^{2^j m} \equiv -1 \bmod n$. The converse is also true, i.e., if n is composite, then there exists an $a \in \mathbb{N}$ with $\gcd(a, n) = 1$ such that $a^m \not\equiv 1 \bmod n$ and $a^{2^j m} \not\equiv -1 \bmod n$ for $0 \leq j \leq t-1$. More precisely:

Proposition A.92. *Let $n \in \mathbb{N}$ be a composite odd number. Let $n - 1 = 2^t m$, with m odd. Let*

$$\overline{W_n} = \{[a] \in \mathbb{Z}_n^* \mid a^m \not\equiv 1 \bmod n \text{ and } a^{2^j m} \not\equiv -1 \bmod n \text{ for } 0 \leq j \leq t-1\}.$$

Then $|\overline{W_n}| \geq \varphi(n)/2$.

Proof. Let $W_n := \mathbb{Z}_n^* \setminus \overline{W_n}$ be the complement of $\overline{W_n}$. We will show that W_n is contained in a proper subgroup U of $\mathbb{Z}_n^*$. Then the desired estimate follows by Proposition A.86, as in the proof of Proposition A.91. We distinguish two cases.

Case 1: There is an $[a] \in \mathbb{Z}_n^*$ with $a^{n-1} \not\equiv 1 \bmod n$.

Then $U = \{[a] \in \mathbb{Z}_n^* \mid a^{n-1} \equiv 1 \bmod n\}$ is a proper subgroup of $\mathbb{Z}_n^*$, which contains W_n, and the proof is finished.

Case 2: We have $a^{n-1} \equiv 1 \bmod n$ for all $[a] \in \mathbb{Z}_n^*$.

Then n is a Carmichael number. Hence n does not contain any squares

(Proposition A.88). Let $n = p_1 \cdot \ldots \cdot p_k, k \geq 3$, be the decomposition into distinct primes. We set

$$W_n^i = \{[a] \in \mathbb{Z}_n^* \mid a^{2^i m} \equiv -1 \bmod n\}.$$

W_n^0 is not empty, since $[-1] \in W_n^0$. Let $r = \max\{i \mid W_n^i \neq \emptyset\}$ and

$$U := \{[a] \in \mathbb{Z}_n^* \mid a^{2^r m} \equiv \pm 1 \bmod n\}.$$

U is a subgroup of $\mathbb{Z}_n^*$ and $W_n \subset U$. Let $[a] \in W_n^r$. Using the Chinese Remainder Theorem, we get a $[w] \in \mathbb{Z}_n^*$ with $w \equiv a \bmod p_1$ and $w \equiv 1 \bmod n/p_1$. Then $w^{2^r m} \equiv -1 \bmod p_1$ and $w^{2^r m} \equiv +1 \bmod p_2$, hence $w^{2^r m} \not\equiv \pm 1 \bmod n$. Thus $w \notin U$, and we see that U is indeed a proper subgroup of $\mathbb{Z}_n^*$.   □

Remark. The set $\overline{E_n}$ from Proposition A.91 is a subset of $\overline{W_n}$, and it can even be proven that $|\overline{W_n}| \geq \frac{3}{4}\varphi(n)$ (see, e.g., [Koblitz94]).

Probabilistic Primality Tests. The preceding propositions are the basis of probabilistic algorithms which test whether a given odd number is prime. Proposition A.91 yields the *Solovay–Strassen primality test*, and Proposition A.92 yields the *Miller–Rabin primality test*. The basic procedure is the same in both tests: we define a set W of *witnesses* for the fact that n is composite. Set $W := \overline{E_n}$ (Solovay–Strassen) or $W := \overline{W_n}$ (Miller–Rabin). If we can find a $w \in W$, then $W \neq \emptyset$, and n is a composite number.

To find a witness $w \in W$, we randomly choose (with respect to the uniform distribution) an element $a \in \mathbb{Z}_n^*$ and check whether $a \in W$. Since $|W| \geq \varphi(n)/2$ (Propositions A.91 and A.92), the probability that we get a witness by the random choice, if n is composite, is $\geq 1/2$. By repeating the random choice k times, we can increase the probability of finding a witness if n is composite. The probability is then $\geq 1 - 1/2^k$. If we do not find a witness, n is considered to be a prime.

The tests are probabilistic and the result is not necessarily correct in all cases. However, the error probability is $\leq 1/2^k$ and hence very small, even for moderate values of k.

Remark. The primality test of Miller–Rabin is the better choice:

1. The test condition is easier to compute.
2. A witness for the Solovay–Strassen test is also a witness for the Miller–Rabin test.
3. In the Miller–Rabin test, the probability of obtaining a witness by one random choice is $\geq 3/4$ (we only proved the weaker bound of $1/2$).

Algorithm A.93 implements the Miller–Rabin test (with error probability $\leq 1/4^k$).

Algorithm A.93.

```
boolean MillerRabinTest(int n, k)
 1  if n is even
 2     then return false
 3     m ← (n − 1) div 2; t ← 1
 4  while m is even do
 5         m ← m div 2; t ← t + 1
 6  for i ← 1 to k do
 7         a ← Random(2 . . . n − 2)
 8         u ← a^m mod n
 9         if u ≠ 1
10            then j ← 1
11                 while u ≠ −1 and j < t do
12                        u ← u² mod n; j ← j + 1
13                   if u ≠ −1
14                        then return false
15     return true
```

Remarks:

1. If $\gcd(a, n) > 1$, then a and hence u are non-units modulo n. In particular, u never attains the value 1 or -1, and the algorithm returns the correct value "false". Therefore, it is not necessary to check explicitly that n and a are relatively prime. As 1 and -1 are never witnesses for n being composite, the random function in line 7 returns only values between 2 and $n - 2$.

2. The Miller–Rabin test, a probabilistic polynomial-time algorithm, finds out only with high probability whether a given number is a prime. The long-standing problem of deciding primality deterministically in polynomial time was solved by Agrawal, Kayal and Saxena in [AgrKaySax04]. Their algorithm is called the AKS primality test. It proves with mathematical certainty whether or not a given integer is a prime. The AKS test is primarily of theoretical interest. It is too slow for deciding primality in practical applications.

A.11 Elliptic Curves

We now discuss some elementary properties of elliptic curves. This is necessary to understand the basics of elliptic curve cryptography. In cryptography, we use curves that are defined over a finite field $\mathbb{F}_q$ (see Section A.5.3 above), where either $q = p$ is a large prime p (we then talk of a *prime field*) or $q = 2^l$ is a power of 2 (we then talk of a *binary field*). The binary field $\mathbb{F}_{2^l}$ is an extension of the prime field $\mathbb{F}_2$.

A field k is called *algebraically closed* if every polynomial in $k[X]$ of degree ≥ 1 has a zero in k. Every field k can be embedded into its *algebraic closure* $\overline{k}$. The field $\overline{k}$ is algebraically closed and k is a subfield of $\overline{k}$, i.e., $k \subseteq \overline{k}$. For example, $\mathbb{C}$ is the algebraic closure of $\mathbb{R}$. The algebraic closure of a finite field $\mathbb{F}_q$ is the union of all its finite extension fields (see, e.g., [NieXin09]):

$$\overline{\mathbb{F}}_q = \cup_{n=1}^{\infty} \mathbb{F}_{q^n}.$$

Throughout this section (A.11), let k be a field and $\overline{k}$ be its algebraic closure. As usual, $k^* = k \setminus \{0\}$. The *affine plane* over k is denoted by $\mathbb{A}^2(k)$:

$$\mathbb{A}^2(k) := \{(x, y) \mid x, y \in k\}.$$

A.11.1 Plane Curves

Let $k[X, Y]$ be the polynomial ring in the variables X and Y. The elements of $k[X, Y]$ are the polynomials

$$F(X, Y) = \sum_{i,j \geq 0} a_{i,j} X^i Y^j,$$

with coefficients $a_{i,j} \in k$ (only finitely many $\neq 0$). The *degree* of F is defined by $\deg(F) := \max\{i + j \mid a_{i,j} \neq 0\}$. Polynomials in two variables are added and multiplied in the same way as polynomials in one variable (see Section A.5.1).

Definition A.94. A subset $C \subset \mathbb{A}^2(\overline{k})$ is called a *(plane) affine curve* defined over k if C is the set of zeros of a non-constant polynomial $F(X, Y) \in k[X, Y]$:

$$C = \{(x, y) \in \mathbb{A}^2(\overline{k}) \mid F(x, y) = 0\}.$$

We say that $F = 0$ (or F for short) is a *(defining) equation* for C. An equation of minimal degree defining C is called a *minimal polynomial* of C. The *degree* of C is the degree of a minimal polynomial of C. The subset $C(k)$ of points $(x, y) \in C$ whose coordinates are elements of k,

$$C(k) := \{(x, y) \in C \mid x, y \in k\} = \{(x, y) \in \mathbb{A}^2(k) \mid F(x, y) = 0\},$$

is called the set of *k-rational points* of C.

Example. Figure A.1 shows the $\mathbb{R}$-rational points (usually called real points) $C(\mathbb{R})$ of the curve C defined by $Y^2 = X(X - 1)(X + 1) \in \mathbb{R}[X]$. It is a curve of degree 3, and it is an elliptic curve.

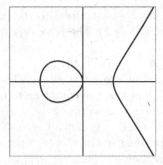

Fig. A.1: The real points of the curve defined by $Y^2 = X(X-1)(X+1)$.

Remark. A curve is expected to be a one-dimensional geometric object, but the k-rational points of a curve defined over k do not necessarily look like a curve. Therefore, a curve is defined as the set of solutions over the algebraic closure $\overline{k}$ of k. Take, for example, the curve C that is defined by $X^2 + Y^2 = 0$ over $\mathbb{R}$. $C(\mathbb{R})$ consists of the single point $(0,0)$. The set of solutions over the algebraic closure $\mathbb{C}$ of $\mathbb{R}$, however, is one-dimensional: $C = C(\mathbb{C})$ is the union of the two lines $X + iY = 0$ and $X - iY = 0$. C is not an elliptic curve. If C is an elliptic curve defined over $\mathbb{R}$, then $C(\mathbb{R})$ looks like a curve.

An elliptic curve is an Abelian group – points can be added and subtracted. This basic feature is essential for cryptography. In order to give a consistent and complete definition of the addition of points on an elliptic curve, it is not sufficient to consider the points of the affine curve, i.e., the solutions of the defining equation in $\mathbb{A}^2(\overline{k})$. We have to complete the curve with a "point at infinity". In our example curve, the component on the right-hand side stretches to infinity. It is completed to a closed loop by adding one point at infinity, where the two unbounded branches meet.

To introduce the concept of points at infinity in a coherent way, we have to define the projective plane. In the affine plane $\mathbb{A}^2(k)$, parallel lines do not intersect. The projective plane is obtained from the affine plane by adding additional points such that any two distinct lines intersect in exactly one point. Parallel lines without an intersection point no longer exist.

Definition A.95. The *projective plane* over a field k is the set of directions in k^3:
$$\mathbb{P}^2(k) = k^3 \setminus \{(0,0,0)\}/\sim .$$
Here, $\sim$ is an equivalence relation on the vectors (x,y,z): $(x,y,z) \sim (x',y',z')$ if and only if there is a $\lambda \in k^*$ such that $(x,y,z) = \lambda \cdot (x',y',z')$. The equivalence class of (x,y,z) is denoted by $[x,y,z]$, and is called a *system of homogeneous coordinates* of a point in $\mathbb{P}^2(k)$.

Note that a direction in k^3 can be described by a non-zero vector (x,y,z), and two vectors define the same direction if and only if they are equivalent. We have $[x,y,z] = [\lambda x, \lambda y, \lambda z]$ for $\lambda \in k^*$.

The above definition can be generalized to the projective space $\mathbb{P}^n(k)$ of homogeneous $(n+1)$-tuples $(n \geq 1)$. The *projective line* $\mathbb{P}^1(k)$ consists of the directions in k^2.

Via the injective map $\mathbb{A}^2(k) \longrightarrow \mathbb{P}^2(k)$, $(x, y) \longmapsto [x, y, 1]$, the affine plane $\mathbb{A}^2(k)$ becomes a part of the projective plane. The image of this map is $\{[x, y, z] \in \mathbb{P}^2(k) \mid z \neq 0\}$, which is the complement of the line $Z = 0$.

A line $L \subset \mathbb{P}^2(k)$ is the set of solutions $[x, y, z] \in \mathbb{P}^2(k)$ of an equation $aX + bY + cZ = 0$, with $a, b, c \in k$, not all zero. Note that the validity of $ax + by + cz = 0$ is independent of the chosen representative of $[x, y, z]$.

The point $[-b, a, 0]$ is the only point $[x, y, z]$ on L with $z = 0$. A point $P = [x, y, z] \in L$ with $z \neq 0$ has an equivalent representation $[x/z, y/z, 1]$. We see that P is a point on the affine line L_a defined by $aX + bY + c = 0$, and L and L_a differ by exactly one point, i.e., the affine line L_a is completed to L by one additional point.

Let $[\alpha, \beta, 1]$ be a point on L_a. $(-b, a)$ is a direction vector of L_a and we get all of the points on L_a from $[\alpha - t \cdot b, \beta + t \cdot a, 1], t \in k$. If $k = \mathbb{R}$, we can consider the limit

$$\lim_{t \to \pm\infty} [\alpha - tb, \beta + ta, 1] = \lim_{t \to \pm\infty} [\alpha/t - b, \beta/t + a, 1/t] = [-b, a, 0].$$

The point $[-b, a, 0]$ is called *the point at infinity* of the affine line L_a. The direction of L_a, given by $(-b, a)$, can be regarded as a point $[-b, a] \in \mathbb{P}^1(k)$. Affine lines are parallel if and only if they have the same direction. Thus, parallel lines intersect in their common point at infinity. Two lines which are not parallel in $\mathbb{A}^2(k)$ intersect in $\mathbb{A}^2(k)$ and have different points at infinity. The points at infinity of all of the lines in $\mathbb{A}^2(k)$ form a line L_∞, called the *line at infinity*. This is defined by the equation $Z = 0$. L_∞ intersects each of the other lines in a single point. We see that any two distinct lines in $\mathbb{P}^2(k)$ intersect in exactly one point.

$\mathbb{A}^2(k)$ can also be embedded as the complement of the line $X = 0$ by $\mathbb{A}^2(k) \longrightarrow \mathbb{P}^2(k)$, $(y, z) \longmapsto [1, y, z]$, or as the complement of the line $Y = 0$ by $\mathbb{A}^2(k) \longrightarrow \mathbb{P}^2(k)$, $(x, z) \longmapsto [x, 1, z]$. Thus, $\mathbb{P}^2(k)$ is covered by three affine planes.

A map $\varphi : \mathbb{P}^2(k) \longrightarrow \mathbb{P}^2(k)$ is called a *(projective) coordinate transformation* if $\varphi([x, y, z]) = [(x, y, z) \cdot A]$ with an invertible matrix A. Obviously A and λA, $\lambda \in k^*$, define the same map φ, and up to a multiplicative constant A is uniquely determined by φ.

Every point $[x, y, z] \in \mathbb{P}^2(k)$ can be mapped to $[0, 0, 1]$ by using a coordinate transformation. $[0, 0, 1]$ corresponds to the origin $(0, 0)$ in the affine plane.

Definition A.96. Let $F(X, Y, Z) \in k[X, Y, Z]$ be a polynomial. $F(X, Y, Z)$ is *homogeneous* of degree m if

$$F(X, Y, Z) = \sum_{\substack{i,j,k: \\ i+j+k=m}} a_{i,j,k} X^i Y^j Z^k,$$

i.e., F has coefficients $\neq 0$ only if $i + j + k = m$.

Projective curves are solutions of homogeneous polynomials.

Definition A.97. A subset $C \subset \mathbb{P}^2(\overline{k})$ is called a *(plane) projective curve* defined over k if there exists a non-constant homogeneous polynomial $F(X, Y, Z) \in k[X, Y, Z]$ such that C is the set of solutions of F:

$$C = \{[x, y, z] \in \mathbb{P}^2(\overline{k}) \mid F(x, y, z) = 0\}.$$

We say that $F = 0$ (or F for short) is a *(defining) equation* for C. An equation of minimal degree defining C is called a *minimal polynomial* of C. The *degree* of C, $\deg(C)$ for short, is the degree of a minimal polynomial of C. The subset $C(k)$ of points $[x, y, z] \in C$ whose coordinates are elements of k,

$$C(k) := \{[x, y, z] \in C \mid x, y, z \in k\} = \{[x, y, z] \in \mathbb{P}^2(k) \mid F(x, y, z) = 0\},$$

is called the set of *k-rational points* of C.

Since $F(X, Y, Z)$ is a homogeneous polynomial, the validity of $F(x, y, z) = 0$ does not depend on the chosen representative of $[x, y, z]$. The notions of a plane curve C and its degree are independent of the coordinate system: if $\varphi : \mathbb{P}^2(k) \longrightarrow \mathbb{P}^2(k)$ is a coordinate transformation with matrix A, then $F((X, Y, Z)A^{-1})$ is a defining equation for $\varphi(C)$ and is of the same degree as F.

Definition A.98. Let $F(X, Y)$ be a polynomial of degree d. The homogeneous polynomial

$$F_{\mathrm{h}}(X, Y, Z) := Z^d F\left(\frac{X}{Z}, \frac{Y}{Z}\right)$$

is called the *homogenization* of F.

Conversely, let $F = F(X, Y, Z)$ be a homogeneous polynomial. $F_{\mathrm{a}} := F(X, Y, 1)$ is called the *dehomogenization* of F (with respect to the variable Z).

By the mapping

$$\mathbb{A}^2(\overline{k}) \longrightarrow \mathbb{P}^2(\overline{k}), \ (x, y) \longmapsto [x, y, 1]$$

we identify the affine plane with the complement of the line L_∞ defined by $Z = 0$ in $\mathbb{P}^2(\overline{k})$. There is a one-to-one correspondence between the affine curves in $\mathbb{A}^2(\overline{k})$ and the projective curves in $\mathbb{P}^2(\overline{k})$ that do not contain the line L_∞:

1. Let $C \subset \mathbb{P}^2(\overline{k})$ be a projective curve which does not contain the line L_∞. Let $F \in k[X, Y, Z]$ be the minimal polynomial of C. Then $C_{\mathrm{a}} = C \cap \mathbb{A}^2(\overline{k})$ is an affine curve with minimal polynomial $F_{\mathrm{a}}(X, Y)$. C_{a} is called the *affine part* of C with respect to L_∞.

2. Let $C \subset \mathbb{A}^2(\overline{k})$ be an affine curve defined by the minimal polynomial $F(X,Y) = \sum_{i,j} a_{i,j} X^i Y^j \in k[X,Y]$ of degree d, and let $F_{\mathrm{h}}(X,Y,Z)$ be the homogenization of $F(X,Y)$. The curve $\tilde{C} \subset \mathbb{P}^2(\overline{k})$ defined by F_{h} is called the *projective closure* of C.

The finitely many points $\tilde{C} \setminus C$ are called the *points at infinity* of C. They correspond to the roots of the *degree form* $F_d = \sum_{i,j:\, i+j=d} a_{i,j} X^i Y^j$ of F:

$$\tilde{C} \setminus C = \{[x,y,0] \mid F_d(x,y) = 0\}.$$

The operations 1 – taking the affine part of a projective curve – and 2 – taking the projective closure of an affine curve – are inverse to each other. This follows since $(F_{\mathrm{h}})_{\mathrm{a}} = F$ for $F \in k[X,Y]$ and $(\tilde{F}_{\mathrm{a}})_{\mathrm{h}} = \tilde{F}$ if $\tilde{F}(X,Y,Z)$ is a homogeneous polynomial that is not divided by Z.

Intersection of Curves with Lines. A curve and a line may intersect at infinity. So, we need to use homogeneous coordinates. We will see that, in general, a curve C and a line have $\deg(C)$ intersection points. If, for a specific line, the number of intersection points is less than $\deg(C)$, then some of the intersection points are counted with multiplicity. The multiplicity of an intersection point can be defined by using homogeneous polynomials in two variables. Recall that a polynomial $F(X,Y)$ of degree m is homogeneous if and only if

$$F(X,Y) = \sum_{i,j:\, i+j=m} a_{i,j} X^i Y^j.$$

Lemma A.99. *Let $F(X,Y) \in k[X,Y]$ be a homogeneous polynomial of degree $m > 0$. Then, over $\overline{k}$, $F(X,Y)$ splits into linear factors, i.e.,*

$$F(X,Y) = \prod_{i=1}^{m} (a_i X - b_i Y),$$

with $a_i, b_i \in \overline{k}$.

Proof. Let $F(X,Y)$ be a homogeneous polynomial of degree m. Then $F(X,Y)/Y^m = F\left(X/Y, 1\right)$ is a polynomial in one variable, which splits into linear factors. Hence,

$$\frac{F(X,Y)}{Y^m} = \prod_{i=1}^{\tilde{m}} \left(a_i \frac{X}{Y} - b_i\right),$$

with $a_i, b_i \in \overline{k}$. Multiplying the equation by Y^m gives the formula. $\qquad \square$

The linear factors $(bX - aY)$ of a homogeneous polynomial $F(X,Y)$ correspond to the zeros $[a,b]$ of the polynomial. Note that a zero (a,b) is unique up to a factor $\lambda \neq 0$. Therefore, we consider zeros as points in $\mathbb{P}^1(\overline{k})$. If $i \geq 0$

is the largest exponent such that $(bX - aY)^i$ is a factor of $F(X, Y)$, we say that $[a, b]$ is a zero of *order i*.

Now, let $L \subset \mathbb{P}^2(\overline{k})$ be a line defined by $aX + bY + cZ = 0$. Let $P_1 = [x_1, y_1, z_1]$ and $P_2 = [x_2, y_2, z_2] \in L$, $P_1 \neq P_2$. Then

$$\gamma : \mathbb{P}^1(\overline{k}) \longrightarrow L, \ [u, v] \longmapsto uP_1 + vP_2 = [ux_1 + vx_2, uy_1 + vy_2, uz_1 + vz_2]$$

is a bijective map from the projective line $\mathbb{P}^1(\overline{k})$ to L.

We intersect L with the projective curve C that is defined by $F(X, Y, Z) = 0$. The point $\gamma([u, v])$ is on C if and only if $F(ux_1 + vx_2, uy_1 + vy_2, uz_1 + vz_2) = 0$. Therefore, the intersection points of L and C correspond to the zeros of the homogeneous polynomial $G(U, V) = F(Ux_1 + Vx_2, Uy_1 + Vy_2, Uz_1 + Vz_2)$. We assume that L is not a part of C. Then $G(U, V)$ is not the zero polynomial.

Definition A.100. The *intersection multiplicity* $i_P(C, L)$ of C and L at point $P = uP_1 + vP_2$ is the order of the zero $[u, v]$ of $G(U, V)$.

The intersection multiplicity does not depend on the coordinate system, nor on the choice of the points P_1 and P_2 on L.

Proposition A.101. *Let C be a projective curve of degree m and let L be a projective line, $L \not\subseteq C$. Then*

$$\sum_{P \in C \cap L} i_P(C, L) = m.$$

A curve of degree m and a line intersect in m points, if properly counted.

Proof. The polynomial $G(U, V)$ is homogeneous of degree m. It has – counted with multiplicities – m zeros. □

Regular and Singular Points. Most of the points on a projective curve are regular points. Singular points are the exception. We define singular points by using the partial derivatives $\partial F / \partial X$, $\partial F / \partial Y$ and $\partial F / \partial Z$ of a minimal polynomial $F \in k[X, Y, Z]$, which we denote by F_X, F_Y and F_Z.

Definition A.102. Let $C \subset \mathbb{P}^2(\overline{k})$ be a projective curve with minimal polynomial $F(X, Y, Z)$ and let $P \in C$. C is *singular* at P (or P is a *singular point* of C) if

$$F_X(P) = F_Y(P) = F_Z(P) = 0.$$

If P is not a singular point of C, then P is a *regular point* of C. The curve C is called *regular* if all points on C are regular points.

Example. Let C be defined by $F(X, Y, Z) = Y^2Z - X^3 - X^2Z$. Then

$$F_X(X, Y, Z) = -3X^2 - 2XZ,$$
$$F_Y(X, Y, Z) = 2YZ \text{ and}$$
$$F_Z(X, Y, Z) = Y^2 - X^2.$$

C is singular at $[0,0,1]$, as shown in Figure A.2.

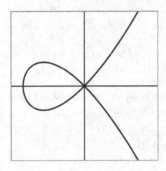

Fig. A.2: The origin is a singular point.

Let φ be a coordinate transformation with matrix A, let $P = [x, y, z]$ and let $\tilde{P} = \varphi(P) = [(x, y, z) \cdot A]$. Then $\tilde{F}(X, Y, Z) = F((X, Y, Z) \cdot A^{-1})$ is an equation for $\varphi(C)$. Applying the chain rule, we obtain

$$(\tilde{F}_X(\tilde{P}), \tilde{F}_Y(\tilde{P}), \tilde{F}_Z(\tilde{P})) = (F_X(P), F_Y(P), F_Z(P)) \cdot A^{-1}.$$

Thus, $F_X(P) = F_Y(P) = F_Z(P) = 0$ if and only if $\tilde{F}_X(\tilde{P}) = \tilde{F}_Y(\tilde{P}) = \tilde{F}_Z(\tilde{P}) = 0$. We see that the definition of singular points does not depend on the coordinates chosen.

Remark. Let C be a projective curve with equation F and let $P = [x, y, 1] \in C$. Let C' be the affine part of C with respect to the line $Z = 0$, and let $G(X, Y) = F(X, Y, 1)$. The following statements are equivalent:

1. C is singular at P.
2. $G_X(x, y) = G_Y(x, y) = 0$.

We have $G_X(x, y) = F_X(x, y, 1)$ and $G_Y(x, y) = F_Y(x, y, 1)$. Thus, statement 2 follows from statement 1. From the formula

$$\deg(F) \cdot F(X, Y, Z) = X \cdot F_X(X, Y, Z) + Y \cdot F_Y(X, Y, Z) + Z \cdot F_Z(X, Y, Z),$$

we conclude that $F_Z(x, y, 1) = 0$ if $F_X(x, y, 1) = F_Y(x, y, 1) = 0$. Therefore, statement 2 implies statement 1.

Definition A.103. Let C be a projective curve, P be a regular point of C and L be a line through P.

1. L is called a *tangent* to C at P if $i_P(C, L) \geq 2$.
2. P is called an *inflection point* of C and L is called the *flex tangent* to C at P if $i_P(C, L) \geq 3$.

Example. Let C be defined by $F(X, Y, Z) = YZ^2 - X^3 + XZ^2$. Then the origin is an inflection point of C, as shown in Figure A.3.

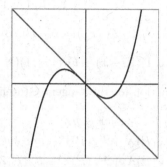

Fig. A.3: The origin is an inflection point.

Proposition A.104. *Let* $C \subset \mathbb{P}^2(\overline{k})$ *be a projective curve with minimal polynomial* F *and let* $P = [x, y, z]$ *be a regular point of* C. *Then there is exactly one tangent to* C *at* P.

1. In the affine plane (i.e., if $z \neq 0$*), the tangent is defined by*

$$F_X(x, y, 1)(X - x) + F_Y(x, y, 1)(Y - y) = 0.$$

2. In the projective plane, the tangent is defined by

$$F_X(P)X + F_Y(P)Y + F_Z(P)Z = 0.$$

Proof. Let $m = \deg(F)$. First, we prove statement 1. Transforming coordinates, we may assume that $P = [0, 0, 1]$. Then, F has the form

$$F(X, Y, Z) = A_1(X, Y)Z^{m-1} + \ldots + A_m(X, Y),$$

where $A_i(X, Y)$ is homogeneous in X, Y of degree i. Consider the Taylor series at P

$$F(X, Y, Z) = (\alpha X + \beta Y)Z^{m-1} + \text{ terms of higher degree in } X \text{ and } Y,$$

with $\alpha = F_X(P)$ and $\beta = F_Y(P)$. Let L be a projective line through P defined by $aX + bY + cZ$. Since $P = [0, 0, 1]$, necessarily $c = 0$. $Q = [-b, a, 0]$ is the point at infinity of L. A point not equal to Q on the line through P and Q is given by $[uP + vQ] = [P + \frac{v}{u}Q]$ (see above). The intersection multiplicity $i_P(C, L)$ is the order of the root $T = 0$ of

$$F(P + TQ) = F(-bT, aT, 1) = (\beta a - \alpha b)T + \ldots .$$

Thus, $i_P(C, L) \geq 2$ if and only if $\beta a - \alpha b = 0$. The last equation holds if and only if L is also defined by $F_X(P)X + F_Y(P)Y = 0$. This proves statement 1.

Next, we derive statement 2 from statement 1. Transforming coordinates, we may assume that $P = [x, y, 1]$ is a point in the affine plane. From

$$\deg(F) \cdot F(X, Y, Z) = X \cdot F_X(X, Y, Z) + Y \cdot F_Y(X, Y, Z) + Z \cdot F_Z(X, Y, Z),$$

we conclude that $-(xF_X(P) + yF_Y(P)) = F_Z(P)$. By homogenizing

$$F_X(x, y, 1)(X - x) + F_Y(x, y, 1)(Y - y),$$

which is the defining equation for the tangent (statement 1), we get

$$
\begin{aligned}
F_X(P)(X - xZ) &+ F_Y(P)(Y - yZ) \\
&= F_X(P)X + F_Y(P)Y - (xF_X(P) + yF_Y(P)) Z \\
&= F_X(P)X + F_Y(P)Y + F_Z(P)Z,
\end{aligned}
$$

which proves statement 2. $\qquad\square$

A.11.2 Normal Forms of Elliptic Curves

A regular projective curve of degree 3 is called an *elliptic curve*. It can be shown that every elliptic curve can be defined by a Weierstrass form, by using a suitable coordinate system. This enables us to define elliptic curves by using Weierstrass forms.

As always, k is a field and $\overline{k}$ is its algebraic closure.

Definition A.105. An equation

$$Y^2 + a_1 XY + a_3 Y = X^3 + a_2 X^2 + a_4 X + a_6$$

is called a *Weierstrass form*.[8] An *elliptic curve* $E \subset \mathbb{A}^2(\overline{k})$ defined over k is a regular curve that is defined by a Weierstrass form with coefficients in k.

Remark. The degree form of a Weierstrass form is X^3. Thus, the only point at infinity is $[0, 1, 0]$ (see page 442). Proposition A.101 implies that the intersection multiplicity of E and the line at infinity at the point $[0, 1, 0]$ is 3. This means that $[0, 1, 0]$ is an inflection point of E and the line at infinity is a flex tangent to E. If $F(X, Y, Z)$ is the homogenization of a Weierstrass form, then $F_Z(0, 1, 0) = 1$. Hence, the projective closure of E is also regular at $[0, 1, 0]$.

In the following, our arguments are often dependent on the characteristic of the base field k.

---

[8] Karl Weierstrass (1815–1897) was a German mathematician who studied these equations comprehensively.

Definition A.106. Let k be a field. The *characteristic* $\text{char}(k)$ of k is defined as

$$\text{char}(k) := \min\{r \in \mathbb{N} \mid r \cdot 1 = \underbrace{1 + 1 + \ldots + 1}_{r} = 0\}$$

if there exists an $r \in \mathbb{N}$ with $r \cdot 1 = 0$ in k, and $\text{char}(k) := 0$ otherwise.

Remark. The characteristic $\text{char}(k)$ of a field is either 0 or a prime number. Namely, if $(r \cdot s) \cdot 1 = (r \cdot 1) \cdot (s \cdot 1) = 0$ in a field, then either $r \cdot 1 = 0$ or $s \cdot 1 = 0$.

Examples:

1. $\text{char}(\mathbb{Q}) = \text{char}(\mathbb{R}) = \text{char}(\mathbb{C}) = 0$.
2. In cryptography, we are interested in elliptic curves that are defined either over a prime field $\mathbb{F}_p = \mathbb{Z}_p$, where p is a large prime, or over a binary field $\mathbb{F}_{2^l}$. Obviously, $\text{char}(\mathbb{Z}_p) = p$ and $\text{char}(\mathbb{F}_{2^l}) = 2$.
3. Let $\mathbb{F}_q$ be a finite field and let $p := \text{char}(\mathbb{F}_q)$, which is a prime number. Then $q = p^l$ for some $l \in \mathbb{N}$.

Remark. In a field k of characteristic 2, then the identities $(a+b)^2 = a^2 + b^2$ and $-a = +a$ are valid, since $2 \cdot 1 = 1 + 1 = 0$. In the following, we will often apply these identities.

Proposition A.107. *Let E be an elliptic curve defined over k.*

1. *If $\text{char}(k) = p$, where p is a prime > 3, then the Weierstrass form can be simplified to $Y^2 = X^3 + bX + c$.*
2. *If $\text{char}(k) = 2$, then the Weierstrass form can be simplified to $Y^2 + XY = X^3 + aX^2 + c$ if $a_1 \neq 0$, and to $Y^2 + aY = X^3 + bX + c$ if $a_1 = 0$.*

Proof. 1. Since $2 \neq 0$ and hence 2 is a unit in k, we can complete the square: the transformation $Y = Y - \frac{1}{2}(a_1 X + a_3)$ yields $Y^2 = X^3 + \tilde{a}X^2 + \tilde{b}X + \tilde{c}$. By substituting $X = X - \frac{1}{3}\tilde{a}$ (note that 3 is also a unit), we get the desired equation

$$Y^2 = \left(X - \frac{\tilde{a}}{3}\right)^3 + \tilde{a}\left(X - \frac{\tilde{a}}{3}\right)^2 + \tilde{b}\left(X - \frac{\tilde{a}}{3}\right) + \tilde{c}$$
$$= X^3 + \bar{b}X + \bar{c}.$$

2. If $a_1 \neq 0$, the substitution

$$X = a_1^2 X + \frac{a_3}{a_1}, \quad Y = a_1^3 Y + \frac{a_1^2 a_4 + a_3^2}{a_1^3}$$

and, if $a_1 = 0$, the substitution

$$X = X + a_2$$

yield the desired equations. □

Remark. We will refer to the simplified equations of Proposition A.107 as *short Weierstrass forms.*

By definition, all points on an elliptic curve are regular. For a curve E that is defined by a Weierstrass form over k, there is a simple criterion for the regularity of all points. This is based on the coefficients of the defining equation and can easily be checked. If $\operatorname{char}(k) = p > 3$, the discriminant

$$D(f) := 4b^3 + 27c^2$$

of a cubic polynomial $f(X) = X^3 + bX + c \in k[X]$ is used.

Lemma A.108. *Let* $\operatorname{char}(k) > 3$ *and let* $f(X) = X^3 + bX + c \in k[X]$. *The following statements are equivalent:*

1. $f(X)$ *and* df/dX *have a common root in* $\overline{k}$.
2. $f(X)$ *has a multiple root in* $\overline{k}$.
3. $D(f) = 0$.

Proof. Let x be a root of f in k. Then $f(X) = (X - x)g(X)$ and $df/dX(X) = g(X) + (X - x)dg/dX(X)$. Hence, $df/dX(x) = 0$ if and only if $g(x) = 0$. This, in turn, is equivalent to $g(X) = (X - x)h(X)$. The equivalence of statements 1 and 2 follows.

To prove the equivalence of statements 2 and 3, assume that $f(X)$ has a multiple root. Then

$$f(X) = X^3 + bX + c = (X - x_1)(X - x_2)^2,$$

where $x_1 + 2x_2 = 0$, $2x_1x_2 + x_2^2 = b$ and $x_1x_2^2 = -c$. Then $b = -3x_2^2$ and $c = 2x_2^3$, and $D(f) = 4b^3 + 27c^2 = 0$ follows. Conversely, assume $D(f) = 0$. If $b = 0$, then $c = 0$ and hence $X = 0$ is a multiple root of $f(X)$. If $b \neq 0$, then $x_1 = 3c/b$ is a simple root and $x_2 = -3c/2b$ is a root of order 2 of $f(X)$. $\square$

Definition A.109. Let

$$Y^2 + a_1XY + a_3Y = X^3 + a_2X^2 + a_4X + a_6$$

be a Weierstrass form. The *discriminant* Δ of the equation is

$$\Delta := -b_2^2b_8 - 8b_4^3 - 27b_6^2 + 9b_2b_4b_6,$$

where

$$b_2 := a_1^2 + 4a_2,$$
$$b_4 := 2a_4 + a_1a_3,$$
$$b_6 := a_3^2 + 4a_6,$$
$$b_8 := a_1^2a_6 + 4a_2a_6 - a_1a_3a_4 + a_2a_3^2 - a_4^2.$$

Proposition A.110. *Let C be defined by*

$$Y^2 + a_1 XY + a_3 Y = X^3 + a_2 X^2 + a_4 X + a_6$$

over k, and let char$(k) \neq 3$. *Then C is regular if and only if* $\Delta \neq 0$.

Remark. Let C be defined by the short Weierstrass form $Y^2 = X^3 + bX + c$ if char$(k) = p > 3$, or $Y^2 + XY = X^3 + aX^2 + c$ if char$(k) = 2$. As we will see in the proof, the discriminant condition means that C is regular if and only if $4b^3 + 27c^2 \neq 0$ or $c \neq 0$, respectively.

Proof (Proposition A.110). If char$(k) = p > 3$, we may assume, by Proposition A.107, that the defining equation is

$$F(X,Y) = Y^2 - (X^3 + bX + c).$$

Then Δ coincides up to a constant factor with the discriminant $D(f) = 4b^3 + 27c^2$ of $f(X) = X^3 + bX + c$. The polynomials $F(X,Y), F_X(X,Y) = -(3X^2 + b)$ and $F_Y(X,Y) = 2Y$ have a common zero if and only if $f(X)$ and $df/dX(X)$ have a common zero, which in turn is equivalent to $\Delta = D(f) = 0$ (Lemma A.108).

Now, let char$(k) = 2$. In our proof, we consider only the case $a_1 \neq 0$ (the case $a_1 = 0$ is not relevant to cryptography). We may assume, by Proposition A.107, that the defining equation is

$$Y^2 + XY = X^3 + aX^2 + c.$$

Then the discriminant reduces to $\Delta = c$ (note that $2 = 0$). A point (x,y) is singular if and only if (x,y) is a common zero of $F(X,Y) = Y^2 + XY + (X^3 + aX^2 + c)$, $F_X(X,Y) = Y + X^2$ and $F_Y(X,Y) = X$. Now, F_X and F_Y have the only common zero, $(0,0)$, and $F(0,0) = c$. Thus, $(0,0)$ is a singular point if and only if $\Delta = c = 0$. $\square$

A.11.3 Point Addition on Elliptic Curves

Projective elliptic curves are Abelian groups – any two points P, Q on the curve can be added to give $P + Q$. We define the addition of points geometrically, applying the *chord-and-tangent method*. Point addition relies on the fact that a line L intersects a projective curve E of degree 3 in three points (Proposition A.101). The points must be counted properly, with multiplicities. If L is a tangent or a flex tangent, two or three of these points coincide.

Let $E \subset \mathbb{P}^2(\overline{k})$ be a (projective) elliptic curve defined over k. We now define the addition $+$ of points on E.

Definition A.111. We choose any point on E as the neutral element $\mathcal{O}$. To add $P, Q \in E$, we first draw a line L joining P and Q (we take the tangent to E at P, if $P = Q$). L intersects E at a third point, which is denoted by $P * Q$. Let $\tilde{L}$ be the line through $\mathcal{O}$ and $P * Q$ (we take the tangent at $\mathcal{O}$ if $P * Q = \mathcal{O}$). The line $\tilde{L}$ intersects E at a third point, say R. We define

$$P + Q = R.$$

Example. Let E be defined by $Y^2 = X^3 - X + 6$. We choose $\mathcal{O} = [0, 1, 0]$. The lines in the (x, y) plane that are parallel to the y-axis intersect E in $[0, 1, 0]$. Figure A.4 shows (for $k = \mathbb{R}$) the addition of two points in the affine part of E.

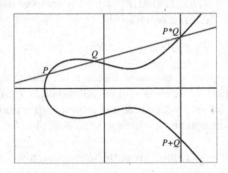

Fig. A.4: Addition of two points.

Theorem A.112. $(E, +)$ *is an Abelian group. In particular:*

1. $P + \mathcal{O} = P$ *for all* $P \in E$ *(identity).*
2. $P + Q = Q + P$ *for* $P, Q \in E$ *(commutative law).*
3. *Let* $P \in E$*. There exists* $-P$ *on* E *with* $P + (-P) = \mathcal{O}$ *(inverse).*
4. $(P + Q) + R = P + (Q + R)$ *for all* $P, Q, R \in E$ *(associative law).*

Proof.

1. Let L be the line joining P and $\mathcal{O}$. Then the line $\tilde{L}$ through $\mathcal{O} * P$ and $\mathcal{O}$ is also L. The third intersection point of $\tilde{L} = L$ and E is again P.
2. This statement is obvious.
3. Let $\tilde{L}$ be the tangent at $\mathcal{O}$ and let S be the third intersection point of $\tilde{L}$ and E. Let L be the line through P and S. Then $-P$ is the third intersection point of L and E.
4. There are many different proofs of the associative law. A proof can be found, for example, in [Kunz05] or [Silverman86]. The proofs use sophisticated methods and are outside the scope of this text.

   Today, by using the formulae for the addition of points (Proposition A.114 and A.115), computer algebra systems are able to verify the associativity explicitly.

□

Remark. From item 3 in the proof, we see that if $\mathcal{O}$ is an inflection point of E, then $S = \mathcal{O}$. Hence, $-P$ is the third intersection point of the line through P and $\mathcal{O}$ with the curve E, and $P + Q = -P * Q$. If $\mathcal{O} = [0, 1, 0]$ is chosen as the point at infinity, then $\mathcal{O}$ is an inflection point, as we have seen above (see page 446).

The addition of points also works in the subset $E(k)$ of k-rational points. The reason is that if two intersection points of E with a line L have coefficients in k, then the third does also. More precisely, we can state the following proposition.

Proposition A.113. *If $\mathcal{O}$ is a k-rational point, then $E(k)$ is a subgroup of E.*

Proof. Let $P, Q \in E(k)$. The polynomial $G(U, V)$ that we used to compute the third intersection point of the line L through P and Q with E has coefficients in k (see page 443). The intersection points correspond to the zeros of $G(U, V)$. If two zeros of $G(U, V)$ have coordinates in k, then the third zero also has coordinates in k. Thus, if two intersection points of a line L with E are k-rational, then the third intersection point is also k-rational. Thus, $P + Q$ and $-P$ are k-rational points. □

Example. Let E be an elliptic curve defined over a finite field $\mathbb{F}_q$, and let $\mathcal{O} = [0, 1, 0]$. Then $E(\mathbb{F}_q)$ is a subgroup of E. In particular, $E(\mathbb{F}_q)$ is an Abelian group.

Formulae for the Addition of Points. From now on, we shall always assume that the zero element of the Abelian group E is the point at infinity, i.e., $\mathcal{O} = [0, 1, 0]$.

Using the short Weierstrass forms, we can derive explicit formulae for the addition of points on E.

Proposition A.114. *Let E be an elliptic curve defined by the equation $Y^2 = X^3 + aX + b$ over a field k with $\mathrm{char}(k) = p > 2$.*

1. *Let $P = (x, y) \in E$. Then $-P = (x, -y)$.*

Let $P_1 = (x_1, y_1) \in E$ and $P_2 = (x_2, y_2) \in E$.

2. *If $P_1 = -P_2$, then $P_1 + P_2 = \mathcal{O}$.*
3. *If $P_1 \neq -P_2$, then $P_3 = P_1 + P_2$ is given by*

$$P_3 = (x_3, y_3) = (\lambda^2 - x_1 - x_2, \lambda(x_1 - x_3) - y_1),$$

where $\lambda = \dfrac{y_2 - y_1}{x_2 - x_1}$, *if $P_1 \neq P_2$ and* $\lambda = \dfrac{3x_1^2 + a}{2y_1}$, *if $P_1 = P_2$.*

Note that $x_2 - x_1 \neq 0$ if $P_1 \neq \pm P_2$, and $y_1 \neq 0$ if $P_1 = P_2$ and $P_1 \neq -P_2$.

Proof.

1. $-P$ is the third intersection point of the line L through $P = (x, y)$ and $\mathcal{O}$ with E (see the remark after the proof of Theorem A.112). The line L is given by $X = x$ and $(-y)^2 = y^2$. Hence $-P = (x, -y)$.
2. This statement is clear.
3. The assumption $P_1 \neq -P_2$ means that the line L through P_1 and P_2 (or, if $P_1 = P_2$, the tangent L at P_1) is not vertical. Therefore, L has an equation $Y = \lambda X + \nu$. Counted with multiplicities, L intersects E in three points, P_1, P_2 and a third point (x_3, y_3'). The third point is also in the affine plane, because $P_1 \neq -P_2$, i.e., $P_1 + P_2 \neq \mathcal{O}$. Substituting $Y = \lambda X + \nu$ for L in the equation of E, we get

$$-X^3 - aX - b + (\lambda X + \nu)^2 = -(X - x_1)(X - x_2)(X - x_3).$$

Comparing the coefficients of X^2 gives

$$x_1 + x_2 + x_3 = \lambda^2,$$

which implies $x_3 = \lambda^2 - x_1 - x_2$. Using the equation for L, we obtain the y-coordinate y_3' of the third intersection point

$$y_3' = \lambda x_3 + \nu = \lambda x_3 + y_1 - \lambda x_1,$$

and then (by part 1)

$$(x_3, y_3) = -(x_3, y_3') = (x_3, -y_3') = (\lambda^2 - x_1 - x_2, \lambda(x_1 - x_3) - y_1).$$

We still have to compute λ. If $P_1 \neq P_2$,

$$\lambda = \frac{y_2 - y_1}{x_2 - x_1}.$$

If $P_1 = P_2 =: P = (x, y)$, the tangent L is

$$(3x^2 + a)(X - x) - 2y(Y - y) = 0.$$

Solving this equation for Y, we get

$$\lambda = \frac{3x^2 + a}{2y}.$$

If $y = 0$, then the tangent is vertical, and hence $P = -P$ and we are in case 2.

$\square$

Proposition A.115. *Let E be an elliptic curve defined by the equation $Y^2 + XY = X^3 + aX^2 + b$ over a field k with $\mathrm{char}(k) = 2$.*

1. *If $P = (x, y) \in E$, then $-P = (x, x + y)$.*

Let $P_1 = (x_1, y_1) \in E$ and $P_2 = (x_2, y_2) \in E$.

2. *If $P_1 = -P_2$, then $P_1 + P_2 = \mathcal{O}$.*
3. *If $P_1 \neq -P_2$, then $P_1 + P_2 = P_3 = (x_3, y_3)$, with*

$$x_3 = \lambda^2 + \lambda + a + x_1 + x_2,$$
$$y_3 = (x_1 + x_3)\lambda + x_3 + y_1$$

and

$$\lambda = \begin{cases} \dfrac{y_2 + y_1}{x_2 + x_1} & \text{if } x_1 \neq x_2, \\[2ex] x_1 + \dfrac{y_1}{x_1} & \text{if } x_1 = x_2. \end{cases}$$

Note that we never have $x_1 = x_2 = 0$ if $P_1 \neq -P_2$.

Proof.

1. $-P$ is the third intersection point of the line L through P and $\mathcal{O}$ with E (see the remark after the proof of Theorem A.112). L is defined by $X = x$, and y is a root of $Y^2 + xY = x^3 + ax^2 + b$. Since $(x+y)^2 + x(x+y) = y^2 + xy$, the other root is $x + y$ and hence $-P = (x, x + y)$.
2. This statement is clear.
3. The assumption $P_1 \neq -P_2$ means that the line L through P_1 and P_2 (or, if $P_1 = P_2$, the tangent L at P_1) is not vertical. Therefore, L has an equation $Y = \lambda X + \nu$. Counted with multiplicities, L intersects E in three points, P_1, P_2 and a third point (x_3, y_3'). The third point is also in the affine plane, because $P_1 \neq -P_2$, i.e., $P_1 + P_2 \neq \mathcal{O}$. Substituting $Y = \lambda X + \nu$ for L in the equation of E, we get

$$(\lambda X + \nu)^2 + X(\lambda X + \nu) + X^3 + aX^2 + b = (X + x_1)(X + x_2)(X + x_3).$$

Comparing the coefficients of X^2 gives

$$x_1 + x_2 + x_3 = \lambda^2 + \lambda + a \text{ or, equivalently, } x_3 = \lambda^2 + \lambda + a + x_1 + x_2.$$

The third intersection point (x_3, y_3') of L and E is $P_1 * P_2$, and $P_1 + P_2 = -P_1 * P_2 = -(x_3, y_3') = (x_3, x_3 + y_3')$. The equation of L yields $y_3' = \lambda x_3 + \nu = \lambda x_3 + y_1 + \lambda x_1 = (x_1 + x_3)\lambda + y_1$. Hence

$$y_3 = y_3' + x_3 = (x_1 + x_3)\lambda + x_3 + y_1.$$

We still have to compute λ.

   i. If $x_1 \neq x_2$ then $\lambda = (y_2 + y_1)/(x_2 + x_1)$.

ii. If $x_1 = x_2 =: x$, then y_1 and $y_1 + x$ are the two roots of $Y^2 + xY = x^3 + ax^2 + b$ (see part 1). Since $P_2 \neq -P_1$, $y_2 \neq y_1 + x$ and hence $y_2 = y_1 =: y$, i.e., $P_1 = P_2 =: P$ and L is the tangent at P. The tangent is given by

$$(x^2 + y)(X + x) + x(Y + y) = 0 \text{ or, equivalently, } xY = (x^2 + y)X + x^3.$$

We see that $\lambda = (x^2 + y)/x = x + y/x$, if $x \neq 0$. If $x = 0$, then the tangent is vertical, and hence $P = -P$ and we are in case 2.

$\square$

Remark. In the case of non-binary fields, Edwards introduced a (singular) model for elliptic curves ([Edwards07]). These curves, in *Edwards form*, are defined by an equation

$$X^2 + Y^2 = a^2(1 + X^2Y^2).$$

The group operation can be expressed by a simple, uniform addition law

$$((x_1, y_1), (x_2, y_2)) \longmapsto \left(\frac{x_1y_2 + y_1x_2}{a(1 + x_1x_2y_1y_2)}, \frac{y_1y_2 - x_1x_2}{a(1 - x_1x_2y_1y_2)} \right),$$

without any exceptions. The zero element is $(0, a)$, and the inverse of $P = (x, y)$ is $(-x, y)$. In [BerLan07], the defining equation is slightly modified, which leads to a bigger class of curves, and the group operations of curves in Edwards form are comprehensively studied. There are faster algorithms for curves in Edwards form. A curve C in Edwards form is not an elliptic curve in the sense of Definition A.105. C has degree 4 and is regular in the affine plane. The points at infinity, $[1, 0, 0]$ and $[0, 1, 0]$, are singular points of C. Every elliptic curve over a non-binary field can be transformed (more precisely, it is birationally equivalent) to a curve in Edwards form over an extension field (in many cases, this is the original field), and the group operations can be carried out there.

Multiplication of a Scalar and a Point. Repeatedly adding the same point P on an elliptic curve E can be viewed as multiplication of P by a scalar $r \in \mathbb{N}$. Let P be a point on an elliptic curve and let $r \in \mathbb{N}$. We set

$$r \cdot P := \underbrace{P + P + \ldots + P}_{r} \text{ for } r > 0, \text{ and } 0 \cdot P := \mathcal{O}.$$

Obviously, $(r + s)P = rP + sP$ and $r(-P) = -rP$.

Scalar–point multiplication rP in the Abelian group E is the analogue of the exponentiation g^r in a prime residue class group $\mathbb{Z}_p^*$. In cryptography, it is used as a one-way function and enables the analogous implementation on elliptic curves of cryptographic schemes that are based on the discrete logarithm problem (Section 3.4).

Exponentiation can be executed very efficiently by applying the square-and-multiply algorithm (see the description of fast exponentiation in Section A.2). Scalar–point multiplication can be executed efficiently by using the additive version of fast exponentiation, the double-and-add algorithm.

Let $r = r_{n-1} \ldots r_1 r_0$ be the binary representation of a number $r \in \mathbb{N}$, $r_{n-1} = 1$. Then

$$
\begin{aligned}
rP &= (2^{n-1} \cdot r_{n-1} + \ldots + 2 \cdot r_1 + r_0)P \\
&= (2^{n-1} \cdot r_{n-1} + \ldots + 2 \cdot r_1)P + r_0 P \\
&= 2((2^{n-2} \cdot r_{n-1} + \ldots + r_1)P) + r_0 P \\
&\;\;\vdots \\
&= 2(\ldots(2P + r_{n-2}P) + \ldots + r_1 P) + r_0 P.
\end{aligned}
$$

The following double-and-add algorithm computes rP:

Algorithm A.116.
    int *Multiply*(point P, bitString $r_{n-1} \ldots r_0$)
    1   $Q \leftarrow P$
    2   for $i \leftarrow n - 2$ down to 0 do
    3       $Q \leftarrow 2Q + r_i P$
    4   return Q

The multiplication rP can be done by l point doublings and l point additions, where $l = \lfloor \log_2 r \rfloor$.

A.11.4 Group Order and Group Structure of Elliptic Curves

Let E be an elliptic curve that is defined over a finite field $\mathbb{F}_q$. Determining the group order $|E(\mathbb{F}_q)|$, i.e., determining the number of points in $E(\mathbb{F}_q)$, is essential if E is to be used in cryptography. Algorithms for computing $|E(\mathbb{F}_q)|$ in a reasonable time are necessary.

Let E be defined by

$$
Y^2 + a_1 XY + a_3 Y = X^3 + a_2 X^2 + a_4 X + a_6.
$$

For every x, this equation has at most two solutions. This yields – including the point at infinity – the weak bound

$$
|E(\mathbb{F}_q)| \leq 2q + 1.
$$

If $\mathbb{F}_q = \mathbb{F}_{2^l}$ is a binary field, then we consider curves that are defined by an equation
$$
Y^2 + XY = X^3 + aX^2 + c.
$$

The point $(0, \sqrt{c}) \in E$ is the only one with $x = 0$. If $\tilde{x} \neq 0$, we divide by x^2 and get

$$W^2 + W = x + a + \frac{c}{x^2},$$

where $W = Y/x$. This equation has a solution w if and only if $\mathrm{Tr}(x+a+c/x^2) = 0$ (see Section A.6), and in this case $w + 1$ is a second solution. This shows that $|E(\mathbb{F}_{2^l})|$ is even.

Half of the elements $x \in \mathbb{F}_{2^l}$ have $\mathrm{Tr}(x) = 0$ (see Proposition A.51). If we assume that the values $x + a + c/x^2$ are uniformly distributed among the elements with trace 0 and the elements with trace 1, then we can conclude that $|E(\mathbb{F}_{2^l})|$ is a number near q.

If $\mathbb{F}_q = \mathbb{F}_p$ is a prime field ($p > 3$ a prime), then E can be defined by an equation

$$Y^2 = X^3 + bX + c.$$

The polynomial $f(X) = X^3 + bX + c$ has 0, 1 or 3 roots in k. For each root x of $f(X)$, we get one point on $E(\mathbb{F}_p)$, namely $(x, 0)$. If $f(x) \neq 0$, there are 0 or 2 solutions of $Y^2 = f(x)$. Including the point at infinity, we find that $|E(\mathbb{F}_p)|$ is odd if $f(X)$ has no root and it is even in the other case.

By using the Legendre symbol, which can be efficiently computed, we can find out for every $x \in \mathbb{F}_p$ whether $Y^2 = f(x)$ has a solution (Proposition A.58). In this way, we can compute $|E(\mathbb{F}_p)|$ for small values of p.

By applying this method, we can derive a formula for $|E(\mathbb{F}_p)|$ in special cases. If $c = 0$ and $p \equiv 3 \bmod 4$, then $|E(\mathbb{F}_p)| = p+1$. Namely, the point $(0, 0)$ is always on E. If $-b \in QR_p$, then $(\sqrt{-b}, 0)$ and $(-\sqrt{-b}, 0)$ are $\mathbb{F}_p$-rational points on E. Now, assume $p \equiv 3 \bmod 4$. Then -1 is not a quadratic residue modulo p (Theorem A.59). Thus we get

$$\left(\frac{f(-x)}{p}\right) = \left(\frac{-f(x)}{p}\right) = -\left(\frac{f(x)}{p}\right),$$

which means that either $Y^2 = f(x)$ or $Y^2 = f(-x)$ has (two) solutions, provided $f(x) \neq 0$. Thus, for exactly half of the elements $x \in \mathbb{F}_p$ with $f(x) \neq 0$, the equation $Y^2 = f(x)$ has two solutions. Summarizing, we count $p + 1$ points, including the point at infinity. These curves, however, show weaknesses if used in cryptography.

Again, as in the binary case, we expect $|E(\mathbb{F}_p)|$ to be close to p, assuming that the values $f(x)$ are uniformly distributed among squares and non-squares in $\mathbb{F}_p^*$. Recall that there are as many squares as non-squares (Proposition A.56).

The following theorem gives a bound for $|E(\mathbb{F}_q)|$.

Theorem A.117 *(Hasse). Let E be an elliptic curve over $\mathbb{F}_q$. Then*

$$||E(\mathbb{F}_q)| - (q + 1)| \leq 2\sqrt{q}.$$

For a proof, see [Silverman86].

The first polynomial-time algorithm for computing the order $|E(\mathbb{F}_q)|$ of arbitrary elliptic curves over finite fields was found by Schoof in 1985 ([Schoof85]). Subsequently, the algorithm was improved by Elkies and Atkin. The resulting algorithm, which is called the Schoof–Elkies–Atkin (SEA) algorithm, can be used to compute $|E(\mathbb{F}_q)|$ for all values of q of practical interest. A comprehensive exposition of Schoof's algorithm and its improvements by Elkies and Atkin can be found in [BlaSerSma99].

In 2000, Satoh published an algorithm for computing $|E(\mathbb{F}_{p^k})|$ for elliptic curves defined over $\mathbb{F}_{p^k}$, where p is a small prime ≥ 5 ([Satoh00]). Later, the algorithm was extended to elliptic curves over binary fields (see [FouGauHar00] and [Skjernaa03]). Variants of the Satoh algorithm are very fast for binary fields.

The structure theorem for finite Abelian groups says that every finite Abelian group is isomorphic to a finite product of residue class groups of $\mathbb{Z}$. For an elliptic curve, the number of factors is ≤ 2.

Theorem A.118. *Let E be an elliptic curve defined over $\mathbb{F}_q$. Then*

$$E(\mathbb{F}_q) \cong \mathbb{Z}_{d_1} \times \mathbb{Z}_{d_2},$$

where d_1 and d_2 are uniquely determined, and d_1 divides d_2 and $q - 1$.

If $d_1 = 1$, then $E(\mathbb{F}_q)$ is cyclic. A proof of the theorem can be found in [Enge99].

Finally, we give two examples.

1. We define $\mathbb{F}_8$ by using the polynomial $f(X) = X^3 + X + 1$ (see Section A.5.3). Let ξ be the residue class of X in $\mathbb{F}_2[X]/f(X)$ and $E(\mathbb{F}_8)$ be the elliptic curve $Y^2 + XY = X^3 + (\xi + 1)X^2 + \xi$. We have

$$\begin{aligned}
E(\mathbb{F}_8) = \{&\mathcal{O}, (0, \xi + \xi^2), \\
&(1, 0), (1, 1), (\xi, 1 + \xi + \xi^2), (\xi, 1 + \xi^2), \\
&(1 + \xi, 1), (1 + \xi, \xi), (\xi + \xi^2, \xi), (\xi + \xi^2, \xi^2)\}.
\end{aligned}$$

   Since $|E(\mathbb{F}_8)| = 10 = 2 \cdot 5$ has no multiple factors, we have $d_1 = 1$ and $E(\mathbb{F}_8) \cong \mathbb{Z}_{10}$ is a cyclic group.

2. Let E be the elliptic curve $Y^2 = X^3 + 10X = X(X+1)(X+10)$ over $\mathbb{F}_{11}$. It has $|E(\mathbb{F}_{11})| = 12$ points. According to Theorem A.118, both $\mathbb{Z}_2 \times \mathbb{Z}_6$ and $\mathbb{Z}_{12}$ are possible structures. To select the right one, we compute the orders of the points:

point	$\mathcal{O}$	$(0,0)$	$(1,0)$	$(4, \pm 4)$	$(6, \pm 1)$	$(8, \pm 3)$	$(9, \pm 4)$	$(10, 0)$
order	1	2	2	3	6	6	6	2

   Thus $E(\mathbb{F}_{11}) \cong \mathbb{Z}_2 \times \mathbb{Z}_6$.

B. Probabilities and Information Theory

We review some basic notions and results used in this book concerning probability, probability spaces and random variables. This chapter is also intended to establish notation. There are many textbooks on probability theory, including [Bauer96], [Feller68], [GanYlv67], [Gordon97] and [Rényi70].

B.1 Finite Probability Spaces and Random Variables

We summarize basic concepts and derive a couple of elementary results. They are useful in our computations with probabilities. We consider only finite probability spaces.

Definition B.1.

1. A *probability distribution* (or simply *distribution*) $p = (p_1, \ldots, p_n)$ is a tuple of elements $p_i \in \mathbb{R}, 0 \leq p_i \leq 1$, called *probabilities* such that $\sum_{i=1}^{n} p_i = 1$.
2. A *probability space* (X, p_X) is a finite set $X = \{x_1, \ldots, x_n\}$ equipped with a probability distribution $p_X = (p_1, \ldots, p_n)$. p_i is called the probability of $x_i, 1 \leq i \leq n$. We also write $p_X(x_i) := p_i$ and consider p_X as a map $X \to [0, 1]$, called the *probability measure* on X, associating with $x \in X$ its probability.
3. An *event* $\mathcal{E}$ in a probability space (X, p_X) is a subset $\mathcal{E}$ of X. The probability measure is extended to events:

$$p_X(\mathcal{E}) = \sum_{y \in \mathcal{E}} p_X(y).$$

Example. Let X be a finite set. The *uniform distribution* $p_{X,u}$ is defined by $p_{X,u}(x) := 1/|X|$, for all $x \in X$. All elements of X have the same probability.

Notation. If the probability measure is determined by the context, we often do not specify it explicitly and simply write X instead of (X, p_X) and $\mathrm{prob}(x)$ or $\mathrm{prob}(\mathcal{E})$ instead of $p_X(x)$ or $p_X(\mathcal{E})$. If $\mathcal{E}$ and $\mathcal{F}$ are events, we write $\mathrm{prob}(\mathcal{E}, \mathcal{F})$ instead of $\mathrm{prob}(\mathcal{E} \cap \mathcal{F})$ for the probability that both events $\mathcal{E}$ and $\mathcal{F}$ occur. Separating events by commas means combining them with

AND. The events $\{x\}$ containing a single element are simply denoted by x. For example, $\text{prob}(x, \mathcal{E})$ means the probability of $\{x\} \cap \mathcal{E}$ (which is 0 if $x \notin \mathcal{E}$ and $\text{prob}(x)$ otherwise).

Remark. Let X be a probability space. The following properties of the probability measure are immediate consequences of Definition B.1:

1. $\text{prob}(X) = 1, \; \text{prob}(\emptyset) = 0$.
2. $\text{prob}(\mathcal{A} \cup \mathcal{B}) = \text{prob}(\mathcal{A}) + \text{prob}(\mathcal{B})$, if $\mathcal{A}$ and $\mathcal{B}$ are disjoint events in X.
3. $\text{prob}(X \setminus \mathcal{A}) = 1 - \text{prob}(\mathcal{A})$.

Definition B.2. Let X be a probability space and $\mathcal{A}, \mathcal{B} \subseteq X$ be events, with $\text{prob}(\mathcal{B}) > 0$. The *conditional probability* of $\mathcal{A}$ assuming $\mathcal{B}$ is

$$\text{prob}(\mathcal{A}\,|\,\mathcal{B}) := \frac{\text{prob}(\mathcal{A}, \mathcal{B})}{\text{prob}(\mathcal{B})}.$$

In particular, we have

$$\text{prob}(x\,|\,\mathcal{B}) = \begin{cases} \text{prob}(x)/\text{prob}(\mathcal{B}) & \text{if } x \in \mathcal{B}, \\ 0 & \text{if } x \notin \mathcal{B}. \end{cases}$$

The conditional probabilities $\text{prob}(x\,|\,\mathcal{B}), x \in X$, define a probability distribution on X. They describe the probability of x assuming that event $\mathcal{B}$ occurs.

If $\mathcal{C}$ is a further event, then $\text{prob}(\mathcal{A}\,|\,\mathcal{B}, \mathcal{C})$ is the conditional probability of $\mathcal{A}$ assuming $\mathcal{B} \cap \mathcal{C}$. Separating events by commas in a condition means combining them with AND.

We now derive a couple of elementary results. They are used in our computations with probabilities.

Proposition B.3. *Let X be a finite probability space, and let X be the disjoint union of events $\mathcal{E}_1, \ldots, \mathcal{E}_r \subseteq X$, with $\text{prob}(\mathcal{E}_i) > 0$ for $i = 1 \ldots r$. Then*

$$\text{prob}(\mathcal{A}) = \sum_{i=1}^{r} \text{prob}(\mathcal{E}_i) \cdot \text{prob}(\mathcal{A}\,|\,\mathcal{E}_i)$$

for every event $\mathcal{A} \subseteq X$.

Proof.

$$\text{prob}(\mathcal{A}) = \sum_{i=1}^{r} \text{prob}(\mathcal{A} \cap \mathcal{E}_i) = \sum_{i=1}^{r} \text{prob}(\mathcal{E}_i) \cdot \text{prob}(\mathcal{A}\,|\,\mathcal{E}_i).$$

$\square$

Lemma B.4. *Let $\mathcal{A}, \mathcal{B}, \mathcal{E} \subseteq X$ be events in a probability space X, with $\text{prob}(\mathcal{E}) > 0$. Suppose that $\mathcal{A}$ and $\mathcal{B}$ have the same conditional probability assuming $\mathcal{E}$, i.e., $\text{prob}(\mathcal{A}\,|\,\mathcal{E}) = \text{prob}(\mathcal{B}\,|\,\mathcal{E})$. Then*

$$|\text{prob}(\mathcal{A}) - \text{prob}(\mathcal{B})| \leq \text{prob}(X \setminus \mathcal{E}).$$

Proof. By Proposition B.3, we have

$$\text{prob}(\mathcal{A}) = \text{prob}(\mathcal{E}) \cdot \text{prob}(\mathcal{A}|\mathcal{E}) + \text{prob}(X \setminus \mathcal{E}) \cdot \text{prob}(\mathcal{A}|X \setminus \mathcal{E})$$

and an analogous equality for $\text{prob}(\mathcal{B})$. Subtracting both equalities, we get the desired inequality. ☐

Lemma B.5. *Let $\mathcal{A}, \mathcal{E} \subseteq X$ be events in a probability space X, with $\text{prob}(\mathcal{E}) > 0$. Then*

$$|\text{prob}(\mathcal{A}) - \text{prob}(\mathcal{A}|\mathcal{E})| \leq \text{prob}(X \setminus \mathcal{E}).$$

Proof. By Proposition B.3, we have

$$\text{prob}(\mathcal{A}) = \text{prob}(\mathcal{E}) \cdot \text{prob}(\mathcal{A}|\mathcal{E}) + \text{prob}(X \setminus \mathcal{E}) \cdot \text{prob}(\mathcal{A}|X \setminus \mathcal{E}).$$

Hence,

$$\begin{aligned}
|\text{prob}(\mathcal{A}) &- \text{prob}(\mathcal{A}|\mathcal{E})| \\
&= \text{prob}(X \setminus \mathcal{E}) \cdot |\text{prob}(\mathcal{A}|X \setminus \mathcal{E}) - \text{prob}(\mathcal{A}|\mathcal{E})| \\
&\leq \text{prob}(X \setminus \mathcal{E}),
\end{aligned}$$

as desired. ☐

Random variables are maps that are defined on probability spaces. They induce image distributions.

Definition B.6. Let (X, p_X) be a probability space and Y be a finite set. A map $S : X \longrightarrow Y$ is called a Y-valued *random variable* on X. We call S a *real-valued random variable* if $Y \subset \mathbb{R}$, and a *binary random variable* if $Y = \{0, 1\}$. Binary random variables are also called *Boolean predicates*. The random variable S and the distribution p_X induce a distribution p_S on Y:

$$p_S(y) := \text{prob}(S = y) := p_X(\{x \in X \mid S(x) = y\}).$$

The distribution p_S is called the *image distribution* of p_X under S, and it is also called the *distribution of the random variable S*.

Remarks:

1. The probability distribution p_S is concentrated on the image of S – only the elements y in $S(X)$ can have a probability > 0. The probability of $y \in Y$ is the probability that $S(x) = y$ if x is randomly chosen from X.
2. The concepts of probability spaces and random variables are closely related. Considering a Y-valued random variable S means considering the probability space (Y, p_S). Vice versa, it is sometimes convenient (and common practice in probability theory) to look at a probability space (X, p_X) as if it were a random variable. Namely, we consider (X, p_X)

as a random variable S_X that is defined on some not-further-specified probability space (Ω, p_Ω) and samples over X according to the given distribution:

$$S_X : \Omega \longrightarrow X \text{ and } p_{S_X} = p_X,$$

i.e., $\mathrm{prob}(S_X = x) = p_X(x)$ for every $x \in X$.

Independent events and joint probability spaces are core concepts of probability calculus.

Definition B.7. Let $\mathcal{A}, \mathcal{B} \subseteq X$ be events in a probability space X. $\mathcal{A}$ and $\mathcal{B}$ are called *independent* if and only if $\mathrm{prob}(\mathcal{A}, \mathcal{B}) = \mathrm{prob}(\mathcal{A}) \cdot \mathrm{prob}(\mathcal{B})$. If $\mathrm{prob}(\mathcal{B}) > 0$, then this condition is equivalent to $\mathrm{prob}(\mathcal{A}|\mathcal{B}) = \mathrm{prob}(\mathcal{A})$.

Definition B.8. A probability space (X, p_X) is called a *joint probability space* with factors $(X_1, p_{X_1}), \ldots, (X_r, p_{X_r})$, denoted for short by $X_1 X_2 \ldots X_r$, if:

1. The set X is the Cartesian product of the sets $X_1, \ldots, X_r$:

$$X = X_1 \times X_2 \times \ldots \times X_r.$$

2. The distribution $p_{X_i}, 1 \leq i \leq r$, is the image of p_X under the projection

$$\pi_i : X \longrightarrow X_i, \ (x_1, \ldots, x_r) \longmapsto x_i,$$

which means

$$p_{X_i}(x) = p_X(\pi_i^{-1}(x)), \text{ for } 1 \leq i \leq r \text{ and } x \in X_i.$$

The probability spaces $X_1, \ldots, X_r$ are called *independent* if and only if

$$p_X(x_1, \ldots, x_r) = \prod_{i=1}^{r} p_{X_i}(x_i), \text{ for all } (x_1, \ldots, x_r) \in X$$

(or, equivalently, if the fibers $\pi_i^{-1}(x_i), 1 \leq i \leq r,$[1] are independent events in X, in the sense of Definition B.7). In this case, X is called the *direct product* of the X_i, denoted for short by $X = X_1 \times X_2 \times \ldots \times X_r$.

Notation. Let (XY, p_{XY}) be a joint probability space with factors (X, p_X) and (Y, p_Y). Let $x \in X$ and $y \in Y$. Then we denote by $\mathrm{prob}(y|x)$ the conditional probability $p_{XY}((x, y)|\{x\} \times Y)$ of (x, y), assuming that the first component is x. Thus, $\mathrm{prob}(y|x)$ is the probability that y occurs as the second component if the first component is x.

---

[1] The set $\pi^{-1}(x)$ of pre-images of a single element x under a projection map π is called the *fiber* over x.

The distribution of a random variable may be considered as the distribution of a probability space, and vice versa (see Remark 2 on page 461). In this way, joint probability spaces correspond to jointly distributed random variables.

Definition B.9. Random variables $S_1, \ldots, S_r$ are called *jointly distributed* if there is a joint probability distribution p_S of $S_1, \ldots, S_r$:

$$\mathrm{prob}(S_1 = x_1, S_2 = x_2, \ldots, S_r = x_r) = p_S(x_1, \ldots, x_r).$$

They are called *independent* if and only if

$$\mathrm{prob}(S_1 = x_1, S_2 = x_2, \ldots, S_r = x_r) = \prod_{i=1}^{r} \mathrm{prob}(S_i = x_i),$$

for all $x_1, \ldots, x_r$.

Definition B.10. Let $S : X \longrightarrow Y$ be a real-valued random variable. The probabilistic average

$$\mathrm{E}(S) := \sum_{x \in X} \mathrm{prob}(x) \cdot S(x),$$

which is equal to

$$\sum_{y \in Y} \mathrm{prob}(S = y) \cdot y,$$

is called the *expected value* of S.

Remark. The expected value $\mathrm{E}(S)$ is a weighted average. The weight of each value y is the probability $\mathrm{prob}(S = y)$ that it appears as the value of S.

Proposition B.11. *Let R and S be real-valued random variables:*

1. $\mathrm{E}(R + S) = \mathrm{E}(R) + \mathrm{E}(S)$.
2. *If R and S are independent, then $\mathrm{E}(R \cdot S) = \mathrm{E}(R) \cdot \mathrm{E}(S)$.*

Proof. Let $W_R, W_S \subset \mathbb{R}$ be the images of R and S. Since we consider only finite probability spaces, W_R and W_S are finite sets. We have

$$\mathrm{E}(R + S) = \sum_{x \in W_R, y \in W_S} \mathrm{prob}(R = x, S = y) \cdot (x + y)$$

$$= \sum_{x \in W_R, y \in W_S} \mathrm{prob}(R = x, S = y) \cdot x$$

$$+ \sum_{x \in W_R, y \in W_S} \mathrm{prob}(R = x, S = y) \cdot y$$

$$= \sum_{x \in W_R} \mathrm{prob}(R = x) \cdot x + \sum_{y \in W_S} \mathrm{prob}(S = y) \cdot y$$

$$= \mathrm{E}(R) + \mathrm{E}(S),$$

and if R and S are independent,

$$
\begin{aligned}
\mathrm{E}(R \cdot S) &= \sum_{x \in W_R, y \in W_S} \mathrm{prob}(R = x, S = y) \cdot x \cdot y \\
&= \sum_{x \in W_R, y \in W_S} \mathrm{prob}(R = x) \cdot \mathrm{prob}(S = y) \cdot x \cdot y \\
&= \left(\sum_{x \in W_R} \mathrm{prob}(R = x) \cdot x \right) \cdot \left(\sum_{y \in W_S} \mathrm{prob}(S = y) \cdot y \right) \\
&= \mathrm{E}(R) \cdot \mathrm{E}(S),
\end{aligned}
$$

as desired.   □

A probability space X is a model of a random experiment. n independent repetitions of the random experiment are modeled by the direct product $X^n = X \times \ldots \times X$. The following lemma answers the question as to how often we have to repeat the experiment until a given event is expected to occur.

Lemma B.12. *Let $\mathcal{E}$ be an event in a probability space X, with $\mathrm{prob}(\mathcal{E}) = p > 0$. Repeatedly, we execute the random experiment X independently. Let G be the number of executions of X, until $\mathcal{E}$ occurs the first time. Then the expected value $\mathrm{E}(G)$ of the random variable G is $1/p$.*

Proof. We have $\mathrm{prob}(G = t) = p \cdot (1 - p)^{t-1}$. Hence,

$$
\mathrm{E}(G) = \sum_{t=1}^{\infty} t \cdot p \cdot (1 - p)^{t-1} = -p \cdot \frac{\mathrm{d}}{\mathrm{d}p} \sum_{t=1}^{\infty} (1 - p)^t = -p \cdot \frac{\mathrm{d}}{\mathrm{d}p} \frac{1}{p} = \frac{1}{p}.
$$

□

Remark. G is called the *geometric random variable* with respect to p.

Lemma B.13. *Let R, S and B be jointly distributed random variables with values in $\{0, 1\}$. Assume that B and S are independent and that B is uniformly distributed: $\mathrm{prob}(B = 0) = \mathrm{prob}(B = 1) = 1/2$. Then*

$$
\mathrm{prob}(R = S) = \frac{1}{2} + \mathrm{prob}(R = B \mid S = B) - \mathrm{prob}(R = B).
$$

Proof. We denote by $\overline{B}$ the complementary value $1 - B$ of B. First we observe that $\mathrm{prob}(S = B) = \mathrm{prob}(S = \overline{B}) = 1/2$, since B is uniformly distributed and independent of S:

$\mathrm{prob}(S = B)$

$= \mathrm{prob}(S = 0) \cdot \mathrm{prob}(B = 0 \mid S = 0) + \mathrm{prob}(S = 1) \cdot \mathrm{prob}(B = 1 \mid S = 1)$

$= \mathrm{prob}(S = 0) \cdot \mathrm{prob}(B = 0) + \mathrm{prob}(S = 1) \cdot \mathrm{prob}(B = 1)$

$= (\mathrm{prob}(S = 0) + \mathrm{prob}(S = 1)) \cdot \dfrac{1}{2} = \dfrac{1}{2}.$

Now, we compute

$\text{prob}(R = S)$

$= \dfrac{1}{2} \cdot \text{prob}(R = B \mid S = B) + \dfrac{1}{2} \cdot \text{prob}(R = \overline{B} \mid S = \overline{B})$

$= \dfrac{1}{2} \cdot (\text{prob}(R = B \mid S = B) + 1 - \text{prob}(R = B \mid S = \overline{B}))$

$= \dfrac{1}{2} + \dfrac{1}{2} \cdot (\text{prob}(R = B \mid S = B) - \text{prob}(R = B \mid S = \overline{B}))$

$= \dfrac{1}{2} + \dfrac{1}{2} \cdot \Big(\text{prob}(R = B \mid S = B)$

$\qquad - \dfrac{\text{prob}(R = B) - \text{prob}(S = B) \cdot \text{prob}(R = B \mid S = B)}{\text{prob}(S = \overline{B})} \Big)$

$= \dfrac{1}{2} + \dfrac{1}{2} \cdot (\text{prob}(R = B \mid S = B)$

$\qquad\qquad - (2 \cdot \text{prob}(R = B) - \text{prob}(R = B \mid S = B)))$

$= \dfrac{1}{2} + \text{prob}(R = B \mid S = B) - \text{prob}(R = B),$

and the lemma is proven. □

In this book we often meet joint probability spaces in the following way: a set X and a family $W = (W_x)_{x \in X}$ of sets are given. Then we may join the set X with the family W, to get

$$X \bowtie W := \{(x, w) \mid x \in X, w \in W_x\} = \bigcup_{x \in X} \{x\} \times W_x.$$

The set W_x becomes the fiber over x.

Assume that probability distributions p_X on X and p_{W_x} on $W_x, x \in X$, are given. Then we get a joint probability distribution p_{XW} on $X \bowtie W$ using

$$p_{XW}(x, w) := p_X(x) \cdot p_{W_x}(w).$$

Conversely, given a distribution p_{XW} on $X \bowtie W$, we can project $X \bowtie W$ to X. We get the image distribution p_X on X using

$$p_X(x) := \sum_{w \in W_x} p_{XW}(x, w)$$

and the probability distributions p_{W_x} on $W_x, x \in X$, using

$$p_{W_x}(w) = p_{XW}((x, w) \mid \{x\} \times W_x) = {p_{XW}(x, w)} / {p_X(x)}.$$

The probabilities p_{W_x} are conditional probabilities.
$\text{prob}(w \mid x) := p_{W_x}(w)$ is the conditional probability that w occurs as the second component, assuming that the first component x is given.

We write XW for short for the probability space $(X \bowtie W, p_{XW})$, as before, and we call it a joint probability space. At first glance it does not meet Definition B.8, because the underlying set is not a Cartesian product (except in the case where all sets W_x are equal). However, all sets are assumed to be finite. Therefore, we can easily embed all the sets W_x into one larger set $\tilde{W}$. Then $X \bowtie W \subseteq X \times \tilde{W}$, and p_{XW} may also be considered as a probability distribution on $X \times \tilde{W}$ (extend it by zero, so that all elements outside $X \bowtie W$ have a probability of 0). In this way we get a joint probability space $X\tilde{W}$ in the strict sense of Definition B.8. XW and $X\tilde{W}$ are practically the "same" as probability spaces (the difference has a measure of 0).

As an example, think of the domain of the modular squaring one-way function (Section 3.2) $(n, x) \mapsto \cdot x^2 \bmod n$, where $n \in I_k = \{pq \mid p, q \text{ distinct primes}, |p| = |q| = k\}$ and $x \in \mathbb{Z}_n$, and assume that the moduli n (the keys; see Rabin encryption, Section 3.5.1) and the elements x (the messages) are selected according to some probability distributions (e.g., the uniform ones). Here $X = I_k$ and $W = (\mathbb{Z}_n)_{n \in I_k}$.

The joining may be iterated. Let $X_1 = (X_1, p_1)$ be a probability space and let $X_j = (X_{j,x}, p_{j,x})_{x \in X_1 \bowtie \ldots \bowtie X_{j-1}}, 2 \le j \le r$, be families of probability spaces. Then by iteratively joining the fibers, we get a joint probability space $(X_1 X_2 \ldots X_r, p_{X_1 X_2 \ldots X_r})$.

At the end of this section, we introduce some notation which turns out to be very useful in many situations.

Notation. Let (X, p_X) be a probability space and $B : X \longrightarrow \{0,1\}$ be a Boolean predicate. Then

$$\text{prob}(B(x) = 1 : x \overset{p_X}{\leftarrow} X) := p_X(\{x \in X \mid B(x) = 1\}).$$

The notation suggests that $p_X(\{x \in X \mid B(x) = 1\})$ is the probability for $B(x) = 1$ if x is randomly selected from X according to p_X. If the distribution p_X is clear from the context, we simply write

$$\text{prob}(B(x) = 1 : x \leftarrow X)$$

instead of $\text{prob}(B(x) = 1 : x \overset{p_X}{\leftarrow} X)$. If p_X is the uniform distribution, we write
$$\text{prob}(B(x) = 1 : x \overset{u}{\leftarrow} X).$$

Sometimes, we denote the probability distribution on X by $x \leftarrow X$ and the uniform distribution by $x \overset{u}{\leftarrow} X$. We emphasize in this way that the members of X are chosen randomly.

Assume that p_X is the distribution p_S of an X-valued random variable S (see Definition B.6 above). We then write

$$\text{prob}(B(x) = 1 : x \leftarrow S)$$

instead of $\mathrm{prob}(B(x) = 1 : x \overset{ps}{\leftarrow} X)$. This intuitively suggests that the elements x are randomly produced by the random variable S, and the probability of x is the probability that S outputs x.

Now let (XW, p_{XW}) be a joint probability space, $W = (W_x)_{x \in X}$, as introduced above, and let p_X and $p_{W_x}, x \in X$, be the probability distributions induced on X and the fibers W_x. We write

$$\mathrm{prob}(B(x, w) = 1 : x \overset{p_X}{\leftarrow} X, w \overset{p_{W_x}}{\leftarrow} W_x), \text{ or simply}$$

$$\mathrm{prob}(B(x, w) = 1 : x \leftarrow X, w \leftarrow W_x)$$

instead of $p_{XW}(\{(x, w) \mid B(x, w) = 1\})$. Here, B is a Boolean predicate on $X \bowtie W$. The notation suggests that we mean the probability for $B(x, w) = 1$, if first x is randomly selected and then w is randomly selected from W_x.

More generally, if $(X_1 X_2 \ldots X_r, p_{X_1 X_2 \ldots X_r})$ is formed by iteratively joining fibers (see above; we also use the notation from above), then we write

$$\mathrm{prob}(B(x_1, \ldots, x_r) = 1 : x_1 \leftarrow X_1, x_2 \leftarrow X_{2, x_1},$$

$$x_3 \leftarrow X_{3, x_1 x_2}, \quad \ldots \quad , x_r \leftarrow X_{r, x_1 \ldots x_{r-1}})$$

instead of $p_{X_1 X_2 \ldots X_r}(\{(x_1, \ldots, x_r) \mid B(x_1, \ldots, x_r) = 1\})$. Again we write more precisely $\overset{u}{\leftarrow}$ (or $\overset{p}{\leftarrow}$) instead of $\leftarrow$ if the distribution is the uniform one (or not clear from the context).

The distribution $x_j \leftarrow X_{j, x_1 \ldots x_{j-1}}$ is the conditional distribution of $x_j \in X_{j, x_1 \ldots x_{j-1}}$, assuming $x_1, \ldots, x_{j-1}$; i.e., it gives the conditional probabilities $\mathrm{prob}(x_j \mid x_1, \ldots, x_{j-1})$. We have (for $r = 3$) that

$$\mathrm{prob}(B(x_1, x_2, x_3) = 1 : x_1 \leftarrow X_1, x_2 \leftarrow X_{2, x_1}, x_3 \leftarrow X_{3, x_1 x_2})$$

$$= \sum_{(x_1, x_2, x_3) : B(x_1, x_2, x_3) = 1} \mathrm{prob}(x_1) \cdot \mathrm{prob}(x_2 \mid x_1) \cdot \mathrm{prob}(x_3 \mid x_2, x_1).$$

B.2 Some Useful and Important Inequalities

Here we derive some useful and well-known inequalities. In the first part, our exposition follows [Cachin97, Section 2.2.2].

Lemma B.14. *Let X be a real-valued random variable and $f : \mathbb{R} \longrightarrow \mathbb{R}$ be a function. Let $D \subseteq \mathbb{R}$ be an interval and $c \in \mathbb{R}, c > 0$ such that $f(x) \geq c$ for all $x \notin D$. Let*

$$\chi_D(x) := \begin{cases} 1 & \text{if } x \in D, \\ 0 & \text{if } x \in \mathbb{R} \setminus D \end{cases}$$

be the characteristic function of D. Then

$$c \cdot \mathrm{prob}(X \notin D) + \mathrm{E}(f(X) \cdot \chi_D(X)) \leq \mathrm{E}(f(X)).$$

Proof.

$$f(X) = f(X) \cdot \chi_D(X) + f(X) \cdot \chi_{\mathbb{R} \setminus D}(X) \text{ and } f(X) \cdot \chi_{\mathbb{R} \setminus D}(X) \geq c \cdot \chi_{\mathbb{R} \setminus D}(X)$$

imply

$$\mathrm{E}(f(X)) = \mathrm{E}(f(X) \cdot \chi_D(X)) + \mathrm{E}(f(X) \cdot \chi_{\mathbb{R} \setminus D}(X))$$
$$\geq \mathrm{E}(f(X) \cdot \chi_D(X)) + c \cdot \mathrm{prob}(X \notin D).$$

$\square$

Corollary B.15 *(k-th moment inequality). For every $k \in \mathbb{N}$ and $\varepsilon > 0$,*

$$\mathrm{prob}(|X| \geq \varepsilon) \leq \frac{\mathrm{E}(|X|^k)}{\varepsilon^k}.$$

Proof. Take $f(x) := |x|^k$ and $D :=\,]-\varepsilon, +\varepsilon[$ in Lemma B.14 and observe that $f(x) \geq \varepsilon^k$ for $x \notin D$ and $\mathrm{E}(f(X) \cdot \chi_D(X)) \geq 0$. $\square$

Some special cases of the corollary are well-known inequalities. The case $k = 1$ is

Theorem B.16 *(Markov's inequality). Let X be a real-valued random variable with $X \geq 0$. Then, for every $\varepsilon > 0$,*

$$\mathrm{prob}(X \geq \varepsilon) \leq \frac{\mathrm{E}(X)}{\varepsilon}.$$

The variance of a random variable X measures how much the values of X differ on average from the expected value.

Definition B.17. Let X be a real-valued random variable. The expected value $\mathrm{E}((X - \mathrm{E}(X))^2)$ of $(X - \mathrm{E}(X))^2$ is called the *variance* of X or, for short, $\mathrm{Var}(X)$.

If the variance is small, the probability that the value of X is far from $\mathrm{E}(X)$ is low. This fundamental fact is stated precisely in Chebyshev's inequality, which is case $k = 2$ of Corollary B.15 applied to the random variable $X - \mathrm{E}(X)$:

Theorem B.18 *(Chebyshev's inequality). Let X be a real-valued random variable with variance σ^2. Then for every $\varepsilon > 0$,*

$$\mathrm{prob}(|X - \mathrm{E}(X)| \geq \varepsilon) \leq \frac{\sigma^2}{\varepsilon^2}.$$

We will also need the following implication of Lemma B.14.

Corollary B.19. *Let X be a real-valued random variable and $a, r \in \mathbb{R}, a \geq 1$. Then*

$$\mathrm{prob}(X \geq r) \leq \mathrm{E}\left(a^{X-r}\right).$$

Proof. Take $f(x) := a^{x-r}$ and $D :=]-\infty, r[$ in Lemma B.14 and observe that $f(x) \geq 1$ for $x \notin D$ and $\mathrm{E}(f(X) \cdot \chi_D(X)) \geq 0$. $\qquad\square$

Definition B.20. Let $D \subseteq \mathbb{R}$ be an interval and $f : D \longrightarrow \mathbb{R}$ be a real-valued function. f is called *convex* if and only if

$$f(\lambda x_1 + (1 - \lambda)x_2) \leq \lambda f(x_1) + (1 - \lambda)f(x_2)$$

for all $x_1, x_2 \in D$ and $\lambda \in [0, 1]$. If the inequality is strict for $0 < \lambda < 1$ and $x_1 \neq x_2$, then f is called *strictly convex*.

Example. $-\log_2 : \mathbb{R}_+ \longrightarrow \mathbb{R}$ is a strictly convex function.

Proposition B.21 *(Jensen's inequality).* *Let* $D \subseteq \mathbb{R}$ *be an interval and* $f : D \longrightarrow \mathbb{R}$ *be a convex function. Let* $(p_1, \ldots, p_n)$ *be a probability distribution and* $x_1, x_2, \ldots, x_n \in D$. *Then*

$$f \left(\sum_{i=1}^n p_i x_i \right) \leq \sum_{i=1}^n p_i f(x_i).$$

If f *is strictly convex, then equality holds only if all* x_i *with* $p_i \neq 0$ *are equal, i.e.,* $x_i = x$ *for every* $i, 1 \leq i \leq n$, *with* $p_i \neq 0$, *for some fixed* $x \in \mathbb{R}$.

Proof. Without loss of generality, we assume that $p_i \neq 0, 1 \leq i \leq n$. We use induction on n. For $n = 2$, the statement follows immediately from the convexity of f:

$$\begin{aligned}
f(p_1 x_1 + p_2 x_2) &= f(p_1 x_1 + (1 - p_1)x_2) \\
&\leq p_1 f(x_1) + (1 - p_1)f(x_2) = p_1 f(x_1) + p_2 f(x_2).
\end{aligned}$$

If f is strictly convex, then we have $<$ for $x_1 \neq x_2$.

Suppose the statement is true for n. Then it is also true for $n + 1$, as the following computation shows:

$$f \left(\sum_{i=1}^{n+1} p_i x_i \right) = f \left((1 - p_{n+1}) \left(\sum_{i=1}^n \frac{p_i}{1 - p_{n+1}} x_i \right) + p_{n+1} x_{n+1} \right)$$

$$\leq (1 - p_{n+1}) f \left(\sum_{i=1}^n \frac{p_i}{1 - p_{n+1}} x_i \right) + p_{n+1} f(x_{n+1})$$

$$\leq (1 - p_{n+1}) \left(\sum_{i=1}^n \frac{p_i}{1 - p_{n+1}} f(x_i) \right) + p_{n+1} f(x_{n+1}) = \sum_{i=1}^{n+1} p_i f(x_i).$$

If f is strictly convex and equality holds, the induction hypothesis says that

$$x_i = x, 1 \leq i \leq n, \text{ for some } x \in \mathbb{R} \text{ and } x_{n+1} = \sum_{i=1}^n \frac{p_i}{1 - p_{n+1}} x_i,$$

which implies

$$x_{n+1} = \frac{\sum_{i=1}^{n} p_i}{1 - p_{n+1}} x = x.$$

$\square$

Corollary B.22. *Let* $f : D \longrightarrow \mathbb{R}$ *be a convex function and* X *be a random variable with values in* D. *Then*

$$E(f(X)) \geq f(E(X)).$$

If f *is strictly convex, then equality holds if and only if* X *is constant, i.e.,* $X = E(X)$ *with probability* 1.

Proof. This is an immediate consequence of Proposition B.21. Let $x_1, \ldots, x_n$ be the values of X and take $p_i := \mathrm{prob}(X = x_i)$. $\square$

Gibb's inequality is also an implication of Jensen's inequality.

Corollary B.23 *(Gibb's inequality).* *Let* $(p_1, \ldots, p_n)$ *and* $(q_1, \ldots, q_n)$ *be probability distributions. Assume that* $p_i \neq 0$ *and* $q_i \neq 0$ *for* $1 \leq i \leq n$. *Then*

$$-\sum_{i=1}^{n} p_i \log_2(p_i) \leq -\sum_{i=1}^{n} p_i \log_2(q_i),$$

with equality if and only if $p_i = q_i, 1 \leq i \leq n$.

Proof. We apply Proposition B.21 with $x_i := q_i/p_i, 1 \leq i \leq n$, and the strictly convex function $-\log_2$:

$$0 = -\log_2(1) = -\log_2\left(\sum_{i=1}^{n} q_i\right) = -\log_2\left(\sum_{i=1}^{n} p_i\left(\frac{q_i}{p_i}\right)\right)$$

$$\leq -\sum_{i=1}^{n} p_i \log_2\left(\frac{q_i}{p_i}\right) = -\sum_{i=1}^{n} p_i(\log_2(q_i) - \log_2(p_i)),$$

with equality if and only if $q_i = x p_i, 1 \leq i \leq n$, for some fixed x. Since $\sum_{i=1}^{n} p_i = \sum_{i=1}^{n} q_i = 1$, necessarily $x = 1$. $\square$

B.3 The Weak Law of Large Numbers

Proposition B.24 *(Weak Law of Large Numbers).* *Let* $S_1, \ldots, S_t$ *be pairwise-independent real-valued random variables, with a common expected value and a common variance:* $E(S_i) = \alpha$ *and* $\mathrm{Var}(S_i) = \sigma^2, i = 1, \ldots, t$. *Then, for every* $\varepsilon > 0$,

$$\mathrm{prob}\left(\left|\frac{1}{t}\sum_{i=1}^{t} S_i - \alpha\right| < \varepsilon\right) \geq 1 - \frac{\sigma^2}{t\varepsilon^2}.$$

Proof. Let $Z = \frac{1}{t} \sum_{i=1}^{t} S_i$. Then we have $E(Z) = \alpha$ and

$$\operatorname{Var}(Z) = E((Z - \alpha)^2) = E\left(\left(\frac{1}{t} \sum_{i=1}^{t} (S_i - \alpha)\right)^2\right)$$

$$= \frac{1}{t^2} E\left(\sum_{i=1}^{t} (S_i - \alpha)^2 + \sum_{i \neq j} (S_i - \alpha) \cdot (S_j - \alpha)\right)$$

$$= \frac{1}{t^2} \sum_{i=1}^{t} E((S_i - \alpha)^2) + \sum_{i \neq j} E((S_i - \alpha) \cdot (S_j - \alpha))$$

$$= \frac{1}{t^2} \sigma^2 t = \frac{\sigma^2}{t}.$$

Here observe that for $i \neq j$, $E((S_i - \alpha) \cdot (S_j - \alpha)) = E(S_i - \alpha) \cdot E(S_j - \alpha)$, since S_i and S_j are independent, and $E(S_i - \alpha) = E(S_i) - \alpha = 0$ (Proposition B.11).

From Chebyshev's inequality (Theorem B.18), we conclude that

$$\operatorname{prob}(|Z - \alpha| \geq \varepsilon) \leq \frac{\sigma^2}{t \varepsilon^2},$$

and the Weak Law of Large Numbers is proven. □

Remark. In particular, the Weak Law of Large Numbers may be applied to t executions of the random experiment underlying a random variable S. It says that the mean of the observed values of S comes close to the expected value of S if the executions are pairwise-independent (or even independent) and its number t is sufficiently large.

If S is a binary random variable observing whether an event occurs or not, there is an upper estimate for the variance, and we get the following corollary.

Corollary B.25. *Let $S_1, \ldots, S_t$ be pairwise-independent binary random variables, with a common expected value and a common variance: $E(S_i) = \alpha$ and $\operatorname{Var}(S_i) = \sigma^2, i = 1, \ldots, t$. Then for every $\varepsilon > 0$,*

$$\operatorname{prob}\left(\left|\frac{1}{t} \sum_{i=1}^{t} S_i - \alpha\right| < \varepsilon\right) \geq 1 - \frac{1}{4t\varepsilon^2}.$$

Proof. Since S_i is binary, we have $S_i = S_i^2$ and get

$$\operatorname{Var}(S_i) = E((S_i - \alpha)^2) = E(S_i^2 - 2\alpha S_i + \alpha^2)$$

$$= E(S_i^2) - 2\alpha E(S_i) + \alpha^2 = \alpha - 2\alpha^2 + \alpha^2 = \alpha(1 - \alpha) \leq \frac{1}{4}.$$

Now apply Proposition B.24. □

Corollary B.26. *Let $S_1, \ldots, S_t$ be pairwise-independent binary random variables, with a common expected value and a common variance:* $\mathrm{E}(S_i) = \alpha$ *and* $\mathrm{Var}(S_i) = \sigma^2, i = 1, \ldots, t$. *Assume* $\mathrm{E}(S_i) = \alpha = 1/2 + \varepsilon, \varepsilon > 0$. *Then*

$$\mathrm{prob}\left(\sum_{i=1}^{t} S_i > \frac{t}{2}\right) \geq 1 - \frac{1}{4t\varepsilon^2}.$$

Proof.

$$\text{If } \left|\frac{1}{t}\sum_{i=1}^{t} S_i - \alpha\right| < \varepsilon, \text{ then } \frac{1}{t}\sum_{i=1}^{t} S_i > \frac{1}{2}.$$

We conclude from Corollary B.25 that

$$\mathrm{prob}\left(\sum_{i=1}^{t} S_i > \frac{t}{2}\right) \geq \mathrm{prob}\left(\left|\frac{1}{t}\sum_{i=1}^{t} S_i - \alpha\right| < \varepsilon\right) \geq 1 - \frac{1}{4t\varepsilon^2},$$

and the corollary is proven.   □

B.4 Distance Measures

We define the distance between probability distributions. Further, we prove some statements on the behavior of negligible probabilities when the probability distribution is varied a little.

Definition B.27. Let p and $\tilde{p}$ be probability distributions on a finite set X. The *statistical distance* between p and $\tilde{p}$ is

$$\mathrm{dist}(p, \tilde{p}) := \frac{1}{2}\sum_{x \in X} |p(x) - \tilde{p}(x)|.$$

Remark. The statistical distance defines a metric on the set of distributions on X.

Lemma B.28. *The statistical distance between probability distributions p and $\tilde{p}$ on a finite set X is the maximum distance between the probabilities of events in X, i.e.,*

$$\mathrm{dist}(p, \tilde{p}) = \max_{\mathcal{E} \subseteq X} |p(\mathcal{E}) - \tilde{p}(\mathcal{E})|.$$

Proof. Recall that the events in X are the subsets of X. Let

$$\mathcal{E}_1 := \{x \in X \mid p(x) > \tilde{p}(x)\}, \quad \mathcal{E}_2 := \{x \in X \mid p(x) < \tilde{p}(x)\},$$

$$\mathcal{E}_3 := \{x \in X \mid p(x) = \tilde{p}(x)\}.$$

Then

$$0 = p(X) - \tilde{p}(X) = \sum_{i=1}^{3} p(\mathcal{E}_i) - \tilde{p}(\mathcal{E}_i),$$

and $p(\mathcal{E}_3) - \tilde{p}(\mathcal{E}_3) = 0$. Hence, $p(\mathcal{E}_2) - \tilde{p}(\mathcal{E}_2) = -(p(\mathcal{E}_1) - \tilde{p}(\mathcal{E}_1))$, and we obviously get

$$\max_{\mathcal{E} \subseteq X} |p(\mathcal{E}) - \tilde{p}(\mathcal{E})| = p(\mathcal{E}_1) - \tilde{p}(\mathcal{E}_1) = -(p(\mathcal{E}_2) - \tilde{p}(\mathcal{E}_2)).$$

We compute

$$\text{dist}(p, \tilde{p}) = \frac{1}{2} \sum_{x \in X} |p(x) - \tilde{p}(x)|$$

$$= \frac{1}{2} \left(\sum_{x \in \mathcal{E}_1} (p(x) - \tilde{p}(x)) - \sum_{x \in \mathcal{E}_2} (p(x) - \tilde{p}(x)) \right)$$

$$= \frac{1}{2} (p(\mathcal{E}_1) - \tilde{p}(\mathcal{E}_1) - (p(\mathcal{E}_2) - \tilde{p}(\mathcal{E}_2)))$$

$$= \max_{\mathcal{E} \subseteq X} |p(\mathcal{E}) - \tilde{p}(\mathcal{E})|.$$

The lemma follows. $\qquad \square$

Lemma B.29. *Let (XW, p_{XW}) be a joint probability space (see Section B.1, p. 462 and p. 465). Let p_X be the induced distribution on X, and $\text{prob}(w|x)$ be the conditional probability of w, assuming x ($x \in X, w \in W_x$). Let $\tilde{p}_X$ be another distribution on X. Setting $\text{prob}(x, w) := \tilde{p}_X(x) \cdot \text{prob}(w|x)$, we get another probability distribution $\tilde{p}_{XW}$ on XW (see Section B.1). Then*

$$\text{dist}(p_{XW}, \tilde{p}_{XW}) \leq \text{dist}(p_X, \tilde{p}_X).$$

Proof. We have

$$|p_{XW}(x, w) - \tilde{p}_{XW}(x, w)| = |(p_X(x) - \tilde{p}_X(x)) \cdot \text{prob}(w|x)| \leq |p_X(x) - \tilde{p}_X(x)|,$$

and the lemma follows immediately from Definition B.27. $\qquad \square$

Throughout the book we consider families of sets, probability distributions on these sets (often the uniform one) and events concerning maps and probabilistic algorithms between these sets. The index sets J are partitioned index sets with security parameter k, $J = \bigcup_{k \in \mathbb{N}} J_k$, usually written as $J = (J_k)_{k \in \mathbb{N}}$ (see Definition 6.2). The indexes $j \in J$ are assumed to be binarily encoded, and k is a measure for the binary length $|j|$ of an index j. Recall that, by definition, there is an $m \in \mathbb{N}$, with $k^{1/m} \leq |j| \leq k^m$ for $j \in J_k$. As an example, think of the family of RSA functions (see Chapters 3 and 6):

$$(\text{RSA}_{n,e} : \mathbb{Z}_n^* \longrightarrow \mathbb{Z}_n^*, \ x \longmapsto x^e)_{(n,e) \in I},$$

where $I_k = \{(n, e) \mid n = pq, \ p \neq q \text{ primes}, \ |p| = |q| = k, e \text{ prime to } \varphi(n)\}$. We are often interested in asymptotic statements, i.e., statements holding for sufficiently large k.

Definition B.30. Let $J = (J_k)_{k \in \mathbb{N}}$ be an index set with security parameter k, and let $(X_j)_{j \in J}$ be a family of sets. Let $p = (p_j)_{j \in J}$ and $\tilde{p} = (\tilde{p}_j)_{j \in J}$ be families of probability distributions on $(X_j)_{j \in J}$.

p and $\tilde{p}$ are called *polynomially close*,[2] if for every positive polynomial P there is a $k_0 \in \mathbb{N}$ such that for all $k \geq k_0$ and $j \in J_k$

$$\text{dist}(p_j, \tilde{p}_j) \leq \frac{1}{P(k)}.$$

Remarks:

1. "Polynomially close" defines an equivalence relation between distributions.
2. Polynomially close distributions cannot be distinguished by a statistical test implemented as a probabilistic polynomial algorithm (see [Luby96], Lecture 7). We do not consider probabilistic polynomial algorithms in this appendix. Probabilistic polynomial statistical tests for pseudorandom sequences are studied in Chapter 8 (see, e.g., Definition 8.2).

The following lemma gives an example of polynomially close distributions.

Lemma B.31. *Let* $J_k := \{n \mid n = rs, r, s \text{ primes}, |r| = |s| = k, r \neq s\}$ [3] *and* $J := \bigcup_{k \in \mathbb{N}} J_k$. *The distributions* $x \xleftarrow{u} \mathbb{Z}_n$ *and* $x \xleftarrow{u} \mathbb{Z}_n^*$ *are polynomially close. In other words, uniformly choosing any* x *from* $\mathbb{Z}_n$ *is polynomially close to choosing only units.*

Proof. Let p_n be the uniform distribution on $\mathbb{Z}_n$ and let $\tilde{p}_n$ be the distribution $x \xleftarrow{u} \mathbb{Z}_n^*$. Then $p_n(x) = 1/n$ for all $x \in \mathbb{Z}_n$, $\tilde{p}_n(x) = 1/\varphi(n)$ if $x \in \mathbb{Z}_n^*$, and $\tilde{p}_n(x) = 0$ if $x \in \mathbb{Z}_n \setminus \mathbb{Z}_n^*$. We have $|\mathbb{Z}_n^*| = \varphi(n) = n \prod_{p|n} \frac{p-1}{p}$, where the product is taken over the distinct primes p dividing n (Corollary A.31). Hence,

$$\text{dist}(p_n, \tilde{p}_n) = \frac{1}{2} \sum_{x \in \mathbb{Z}_n} |p_n(x) - \tilde{p}_n(x)|$$

$$= \frac{1}{2} \left(\sum_{x \in \mathbb{Z}_n^*} \left(\frac{1}{\varphi(n)} - \frac{1}{n} \right) + \sum_{x \in \mathbb{Z}_n \setminus \mathbb{Z}_n^*} \frac{1}{n} \right) = 1 - \frac{\varphi(n)}{n} = 1 - \prod_{p|n} \frac{p-1}{p}.$$

If $n = rs \in J_k$, then

$$\text{dist}(p_n, \tilde{p}_n) = 1 - \frac{(r-1)(s-1)}{rs} = \frac{1}{r} + \frac{1}{s} - \frac{1}{rs} \leq 2 \cdot \frac{1}{2^{k-1}} = \frac{1}{2^{k-2}},$$

and the lemma follows. $\square$

---

[2] Also called (ε)-*statistically indistinguishable* (see, e.g., [Luby96]) or *statistically close* (see, e.g., [Goldreich01]).

[3] If $r \in \mathbb{N}$, we denote, as usual, the binary length of r by $|r|$.

Remark. The lemma does not hold for arbitrary n, i.e., with $\tilde{J}_k := \{n \mid |n| = k\}$ instead of J_k. Namely, if n has a small prime factor q, then $1 - \prod_{p|n} \frac{p-1}{p} \geq 1 - \frac{q-1}{q}$, which is not close to 0.

Example. The distributions $x \xleftarrow{u} \mathbb{Z}_n$ and $x \xleftarrow{u} \mathbb{Z}_n \cap \text{Primes}$ are not polynomially close.[4] Their statistical distance is almost 1 (for large n).

Namely, let $k = |n|$, and let p_1 and p_2 be the uniform distributions on $\mathbb{Z}_n$ and $\mathbb{Z}_n \cap \text{Primes}$. As usual, we extend p_2 by 0 to a distribution on $\mathbb{Z}_n$. The number $\pi(x)$ of primes $\leq x$ is approximately $x/\ln(x)$ (Theorem A.82). Thus, we get (with $c = \ln(2)$)

$$
\begin{aligned}
\text{dist}(p_1, p_2) &= \frac{1}{2} \sum_{x \in \mathbb{Z}_n} |p_1(x) - p_2(x)| \\
&= \frac{1}{2} \left(\pi(n) \left(\frac{1}{\pi(n)} - \frac{1}{n} \right) + (n - \pi(n)) \frac{1}{n} \right) \\
&= \left(1 - \frac{\pi(n)}{n} \right) \approx 1 - \frac{1}{\ln(n)} \\
&= 1 - \frac{1}{c \log_2(n)} \geq 1 - \frac{1}{c(k-1)},
\end{aligned}
$$

and see that the statistical distance is close to 1 for large k.

Lemma B.32. *Let $J = (J_k)_{k \in \mathbb{N}}$ be an index set with security parameter k, and let $(X_j)_{j \in J}$ be a family of sets. Let $p = (p_j)_{j \in J}$ and $\tilde{p} = (\tilde{p}_j)_{j \in J}$ be families of probability distributions on $(X_j)_{j \in J}$ which are polynomially close. Let $(\mathcal{E}_j)_{j \in J}$ be a family of events $\mathcal{E}_j \subseteq X_j$.*
Then for every positive polynomial P, there is a $k_0 \in \mathbb{N}$ such that for all $k \geq k_0$

$$
|p_j(\mathcal{E}_j) - \tilde{p}_j(\mathcal{E}_j)| \leq \frac{1}{P(k)},
$$

for all $j \in J_k$.

Proof. This is an immediate consequence of Lemma B.28. $\qquad\square$

Definition B.33. *Let $J = (J_k)_{k \in \mathbb{N}}$ be an index set with security parameter k, and let $(X_j)_{j \in J}$ be a family of sets. Let $p = (p_j)_{j \in J}$ and $\tilde{p} = (\tilde{p}_j)_{j \in J}$ be families of probability distributions on $(X_j)_{j \in J}$.*
$\tilde{p}$ is polynomially bounded by p if there is a positive polynomial Q such that $p_j(x)Q(k) \geq \tilde{p}_j(x)$ for all $k \in \mathbb{N}, j \in J_k$ and $x \in X_j$.

Examples. In both examples, let $J = (J_k)_{k \in \mathbb{N}}$ be an index set with security parameter k:

---

[4] As always, we consider $\mathbb{Z}_n$ as the set $\{0, \ldots, n-1\}$.

1. Let $(X_j)_{j \in J}$ and $(Y_j)_{j \in J}$ be families of sets with $Y_j \subseteq X_j, j \in J$. Assume there is a polynomial Q such that $|Y_j|Q(k) \geq |X_j|$ for all $k, j \in J_k$. Then the image of the uniform distributions on $(Y_j)_{j \in J}$ under the inclusions $Y_j \subseteq X_j$ is polynomially bounded by the uniform distributions on $(X_j)_{j \in J}$. This is obvious, since $1/|Y_j| \leq Q(k)/|X_j|$ by assumption.

   For example, $x \xleftarrow{u} \mathbb{Z}_n \cap \text{Primes}$ is polynomially bounded by $x \xleftarrow{u} \mathbb{Z}_n$, because the number of primes $\leq n$ is of the order n/k, with $k = |n|$ being the binary length of n (by the Prime Number Theorem, Theorem A.82).

2. Let $(X_j)_{j \in J}$ and $(Y_j)_{j \in J}$ be families of sets. Let $f = (f_j : Y_j \longrightarrow X_j)_{j \in J}$ be a family of surjective maps, and assume that for $j \in J_k$, each $x \in X_j$ has at most $Q(k)$ pre-images, Q a polynomial. Then the image of the uniform distributions on $(Y_j)_{j \in J}$ under f is polynomially bounded by the uniform distributions on $(X_j)_{j \in J}$.

   Namely, let p_u be the uniform probability distribution and p_f be the image distribution. We have for $j \in J_k$ and $x \in X_j$ that

$$p_f(x) \leq \frac{Q(k)}{|Y_j|} \leq \frac{Q(k)}{|X_j|} = Q(k)p_u(x).$$

Proposition B.34. *Let $J = (J_k)_{k \in \mathbb{N}}$ be an index set with security parameter k. Let $(X_j)_{j \in J}$ be a family of sets. Let $p = (p_j)_{j \in J}$ and $\tilde{p} = (\tilde{p}_j)_{j \in J}$ be families of probability distributions on $(X_j)_{j \in J}$. Assume that $\tilde{p}$ is polynomially bounded by p. Let $(\mathcal{E}_j)_{j \in J}$ be a family of events $\mathcal{E}_j \subseteq X_j$, whose probability is negligible with respect to p; i.e., for every positive polynomial P there is a $k_0 \in \mathbb{N}$ such that $p_j(\mathcal{E}_j) \leq 1/P(k)$ for $k \geq k_0$ and $j \in J_k$.*
Then the events $(\mathcal{E}_j)_{j \in J}$ have negligible probability also with respect to $\tilde{p}$.

Proof. There is a polynomial Q such that $\tilde{p}_j \leq Q(k) \cdot p_j$ for $j \in J_k$. Now let R be a positive polynomial. Then there is some k_0 such that for $k \geq k_0$ and $j \in J_k$

$$p_j(\mathcal{E}_j) \leq \frac{1}{R(k)Q(k)} \quad \text{and hence} \quad \tilde{p}_j(\mathcal{E}_j) \leq Q(k) \cdot p_j(\mathcal{E}_j) \leq \frac{1}{R(k)},$$

and the proposition follows. $\qquad\qquad\qquad\qquad\qquad\qquad\qquad\qquad \square$

B.5 Basic Concepts of Information Theory

Information theory and the classical notion of provable security for encryption algorithms go back to Shannon and his famous papers [Shannon48] and [Shannon49].

   We give a short introduction to some basic concepts and facts from information theory that are needed in this book. Textbooks on the subject include [Ash65], [Hamming86], [CovTho92] and [GolPeiSch94].

Changing the point of view from probability spaces to random variables (see Section B.1), all the following definitions and results can be formulated and are valid for (jointly distributed) random variables as well.

Definition B.35. Let X be a finite probability space. The *entropy* or *uncertainty* of X is defined by

$$H(X) := \sum_{x \in X, prob(x) \neq 0} prob(x) \cdot \log_2 \left(\frac{1}{prob(x)} \right)$$

$$= - \sum_{x \in X, prob(x) \neq 0} prob(x) \cdot \log_2(prob(x)).$$

Remark. The probability space X models a random experiment. The possible outcomes of the experiments are the elements $x \in X$. If we execute the experiment and observe the event x, we gain information. The amount of information we obtain with the occurrence of x (or, equivalently, our uncertainty whether x will occur) – measured in bits – is given by

$$\log_2 \left(\frac{1}{prob(x)} \right) = -\log_2 (prob(x)).$$

The lower the probability of x, the higher the uncertainty. For example, tossing a fair coin we have $prob(heads) = prob(tails) = 1/2$. Thus, the amount of information obtained with the outcome *heads* (or *tails*) is 1 bit. If you throw a fair die, each outcome has probability $1/6$. Therefore, the amount of information associated with each outcome is $\log_2(6) \approx 2.6$ bits.

The entropy $H(X)$ measures the average (i.e., the expected) amount of information arising from executing the experiment X. For example, toss a coin which is unfair, say $prob(heads) = 3/4$ and $prob(tails) = 1/4$. Then we obtain $\log_2(4/3) \approx 0.4$ bits of information with outcome *heads*, and 2 bits of information with outcome *tails*; so the average amount of information resulting from tossing this coin – the entropy – is $3/4 \cdot \log_2(4/3) + 1/4 \cdot 2 \approx 0.8$ bits.

Proposition B.36. Let X be a finite probability space which contains n elements, $X = \{x_1, \ldots, x_n\}$:

1. $0 \leq H(X) \leq \log_2(n)$.
2. $H(X) = 0$ if and only if there is some $x \in X$ with $prob(x) = 1$ (and hence all other elements in X have a probability of 0).
3. $H(X) = \log_2(n)$ if and only if the distribution on X is uniform.

Proof. Since $-prob(x) \cdot \log_2(prob(x)) \geq 0$ for every $x \in X$, the first inequality in statement 1, $H(X) \geq 0$, and statement 2 are immediate consequences of the definition of $H(X)$.

To prove statements 1 and 3, set

$$p_k := \text{prob}(x_k), q_k := \frac{1}{n}, 1 \leq k \leq n.$$

Applying Corollary B.23 we get

$$\sum_{k=1}^{n} p_k \log_2 \left(\frac{1}{p_k} \right) \leq \sum_{k=1}^{n} p_k \log_2 \left(\frac{1}{q_k} \right) = \sum_{k=1}^{n} p_k \log_2(n) = \log_2(n).$$

Equality holds instead of $\leq$ if and only if

$$p_k = q_k = \frac{1}{n}, k = 1, \ldots, n,$$

see Corollary B.23. $\square$

In the following we assume without loss of generality that all elements of probability spaces have a probability > 0.

We consider joint probability spaces XY and will see how to specify the amount of information gathered about X when learning Y. We will often use the intuitive notation $\text{prob}(y|x)$, for $x \in X$ and $y \in Y$. Recall that $\text{prob}(y|x)$ is the probability that y occurs as the second component, if the first component is x (see Section B.1, p. 462).

Definition B.37. Let X and Y be finite probability spaces with joint distribution XY.
The *joint entropy* $\text{H}(XY)$ of X and Y is the entropy of the joint distribution XY of X and Y:

$$\text{H}(XY) := - \sum_{x \in X, y \in Y} \text{prob}(x, y) \cdot \log_2\left(\text{prob}(x, y)\right).$$

Conditioning X over $y \in Y$, we define

$$\text{H}(X|y) := - \sum_{x \in X} \text{prob}(x|y) \cdot \log_2(\text{prob}(x|y)).$$

The *conditional entropy* (or *conditional uncertainty*) of X assuming Y is

$$\text{H}(X|Y) := \sum_{y \in Y} \text{prob}(y) \cdot \text{H}(X|y).$$

The *mutual information* of X and Y is the reduction of the uncertainty of X when Y is learned:
$$\text{I}(X;Y) = \text{H}(X) - \text{H}(X|Y).$$

$\text{H}(XY)$ measures the average amount of information gathered by observing both X and Y. $\text{H}(X|Y)$ measures the average amount of information arising from (an execution of) the experiment X, knowing the result of experiment Y. $\text{I}(X;Y)$ measures the amount of information about X obtained by learning Y.

Proposition B.38. *Let* X *and* Y *be finite probability spaces with joint distribution* XY. *Then:*

1. $H(X|Y) \geq 0$.
2. $H(XY) = H(X) + H(Y|X)$.
3. $H(XY) \leq H(X) + H(Y)$.
4. $H(Y) \geq H(Y|X)$.
5. $I(X;Y) = I(Y;X) = H(X) + H(Y) - H(XY)$.
6. $I(X;Y) \geq 0$.

Proof. Statement 1 is true since $H(X|y) \geq 0$ by Proposition B.36. The other statements are a special case of the more general Proposition B.43. □

Proposition B.39. *Let* X *and* Y *be finite probability spaces with joint distribution* XY. *The following statements are equivalent:*

1. X *and* Y *are independent.*
2. $\text{prob}(y|x) = \text{prob}(y)$, *for* $x \in X$ *and* $y \in Y$.
3. $\text{prob}(x|y) = \text{prob}(x)$, *for* $x \in X$ *and* $y \in Y$.
4. $\text{prob}(x|y) = \text{prob}(x|y')$, *for* $x \in X$ *and* $y, y' \in Y$.
5. $H(XY) = H(X) + H(Y)$.
6. $H(Y) = H(Y|X)$.
7. $I(X;Y) = 0$.

Proof. The equivalence of statements 1, 2 and 3 is an immediate consequence of the definitions of independence and conditional probabilities (see Definition B.7). Statement 3 obviously implies statement 4. Conversely, statement 3 follows from statement 4 using

$$\text{prob}(x) = \sum_{y \in Y} \text{prob}(y) \cdot \text{prob}(x|y).$$

The equivalence of the latter statements follows as a special case from Proposition B.44. □

Next we study the mutual information of two probability spaces conditional on a third one.

Definition B.40. *Let* X, Y *and* Z *be finite probability spaces with joint distribution* XYZ. *The* conditional mutual information $I(X;Y|Z)$ *is given by*

$$I(X;Y|Z) := H(X|Z) - H(X|YZ).$$

The conditional mutual information $I(X;Y|Z)$ is the average amount of information about X obtained by learning Y, assuming that Z is known.

When studying entropies and mutual informations for jointly distributed finite probability spaces X, Y and Z, it is sometimes useful to consider the conditional situation where $z \in Z$ is fixed. Therefore we need the following definition.

Definition B.41. Let X, Y and Z be finite probability spaces with joint distribution XYZ and $z \in Z$.

$$H(X \mid Y, z) := \sum_{y \in Y} \text{prob}(y \mid z) \cdot H(X \mid y, z),$$

$$I(X; Y \mid z) := H(X \mid z) - H(X \mid Y, z).$$

(See Definition B.37 for the definition of $H(X \mid y, z)$.)

Proposition B.42. *Let X, Y and Z be finite probability spaces with joint distribution XYZ and $z \in Z$:*

1. $H(X \mid YZ) = \sum_{z \in Z} \text{prob}(z) \cdot H(X \mid Y, z)$.
2. $I(X; Y \mid Z) = \sum_{z \in Z} \text{prob}(z) \cdot I(X; Y \mid z)$.

Proof.

$$H(X \mid YZ) = \sum_{y,z} \text{prob}(y, z) \cdot H(X \mid y, z) \text{ (by Definition B.37)}$$

$$= \sum_{z} \text{prob}(z) \sum_{y} \text{prob}(y \mid z) \cdot H(X \mid y, z) = \sum_{z} \text{prob}(z) \cdot H(X \mid Y, z).$$

This proves statement 1.

$$I(X; Y \mid Z) = H(X \mid Z) - H(X \mid YZ)$$

$$= \sum_{z} \text{prob}(z) \cdot H(X \mid z) - \sum_{z} \text{prob}(z) \cdot H(X \mid Y, z) \text{ (by Def. B.37 and 1.)}$$

$$= \sum_{z} \text{prob}(z) \cdot I(X; Y \mid z).$$

This proves statement 2.   □

Proposition B.43. *Let X, Y and Z be finite probability spaces with joint distribution XYZ.*

1. $H(XY \mid Z) = H(X \mid Z) + H(Y \mid XZ)$.
2. $H(XY \mid Z) \leq H(X \mid Z) + H(Y \mid Z)$.
3. $H(Y \mid Z) \geq H(Y \mid XZ)$.
4. $I(X; Y \mid Z) = I(Y; X \mid Z) = H(X \mid Z) + H(Y \mid Z) - H(XY \mid Z)$.
5. $I(X; Y \mid Z) \geq 0$.
6. $I(X; YZ) = I(X; Z) + I(X; Y \mid Z)$.
7. $I(X; YZ) \geq I(X; Z)$.

Remark. We get Proposition B.38 from Proposition B.43 by taking $Z := \{z_0\}$ with $\text{prob}(z_0) := 1$ and $XYZ := XY \times Z$.

Proof.

1. We compute

$$\text{H}(X\,|\,Z) + \text{H}(Y\,|\,XZ)$$

$$= -\sum_{z\in Z} \text{prob}(z) \sum_{x\in X} \text{prob}(x\,|\,z) \cdot \log_2(\text{prob}(x\,|\,z))$$

$$\quad - \sum_{x\in X, z\in Z} \text{prob}(x,z) \sum_{y\in Y} \text{prob}(y\,|\,x,z) \cdot \log_2(\text{prob}(y\,|\,x,z))$$

$$= -\sum_{x,y,z} \text{prob}(z)\text{prob}(x,y\,|\,z) \cdot \log_2(\text{prob}(x\,|\,z))$$

$$\quad - \sum_{x,y,z} \text{prob}(x,z)\text{prob}(y\,|\,x,z) \cdot \log_2(\text{prob}(y\,|\,x,z))$$

$$= -\sum_{x,y,z} \text{prob}(x,y,z) \cdot \left(\log_2(\text{prob}(x\,|\,z)) + \log_2\left(\frac{\text{prob}(x,y\,|\,z)}{\text{prob}(x\,|\,z)} \right) \right)$$

$$= -\sum_{x,y,z} \text{prob}(x,y,z) \cdot \log_2(\text{prob}(x,y\,|\,z))$$

$$= -\sum_{z} \text{prob}(z) \sum_{x,y} \text{prob}(x,y\,|\,z) \cdot \log_2(\text{prob}(x,y\,|\,z))$$

$$= \text{H}(XY\,|\,Z).$$

2. We have

$$\text{H}(X\,|\,Z) = -\sum_{z\in Z} \text{prob}(z) \sum_{x\in X} \text{prob}(x\,|\,z) \cdot \log_2(\text{prob}(x\,|\,z))$$

$$= -\sum_{z\in Z} \text{prob}(z) \sum_{y\in Y} \sum_{x\in X} \text{prob}(x,y\,|\,z) \cdot \log_2(\text{prob}(x\,|\,z)),$$

$$\text{H}(Y\,|\,Z) = -\sum_{z\in Z} \text{prob}(z) \sum_{y\in Y} \text{prob}(y\,|\,z) \cdot \log_2(\text{prob}(y\,|\,z))$$

$$= -\sum_{z\in Z} \text{prob}(z) \sum_{x\in X} \sum_{y\in Y} \text{prob}(x,y\,|\,z) \cdot \log_2(\text{prob}(y\,|\,z)).$$

Hence,

$$\text{H}(X\,|\,Z) + \text{H}(Y\,|\,Z)$$

$$= -\sum_{z\in Z} \text{prob}(z) \sum_{x,y} \text{prob}(x,y\,|\,z) \cdot \log_2(\text{prob}(x\,|\,z)\text{prob}(y\,|\,z)).$$

By definition,

$$\text{H}(XY\,|\,Z) = -\sum_{z\in Z} \text{prob}(z) \sum_{x,y} \text{prob}(x,y\,|\,z) \cdot \log_2(\text{prob}(x,y\,|\,z)).$$

Since $(\text{prob}(x,y\,|\,z))_{(x,y)\in XY}$ and $(\text{prob}(x\,|\,z) \cdot \text{prob}(y\,|\,z))_{(x,y)\in XY}$ are probability distributions, the inequality follows from Corollary B.23.

3. Follows from statements 1 and 2.
4. Follows from the definition of the mutual information, since $H(X|YZ) = H(XY|Z) - H(Y|Z)$ by statement 1.
5. Follows from statements 2 and 4.
6. $I(X;Z) + I(X;Y|Z) = H(X) - H(X|Z) + H(X|Z) - H(X|YZ) = I(X;YZ)$.
7. Follows from statements 5 and 6.

The proof of Proposition B.43 is finished.   □

Proposition B.44. *Let X, Y and Z be finite probability spaces with joint distribution XYZ. The following statements are equivalent:*

1. *X and Y are independent assuming z, i.e.,*
   $$\text{prob}(x,y|z) = \text{prob}(x|z) \cdot \text{prob}(y|z), \text{ for all } (x,y,z) \in XYZ.$$
2. *$H(XY|Z) = H(X|Z) + H(Y|Z)$.*
3. *$H(Y|Z) = H(Y|XZ)$.*
4. *$I(X;Y|Z) = 0$.*

Remark. We get the last three statements in Proposition B.39 from Proposition B.44 by taking $Z := \{z_0\}$ with $\text{prob}(z_0) := 1$ and $XYZ := XY \times Z$.

Proof. The equivalence of statements 1 and 2 follows from the computation in the proof of Proposition B.43, statement 2, by Corollary B.23. The equivalence of statements 2 and 3 is immediately derived from Proposition B.43, statement 1. The equivalence of statements 2 and 4 follows from Proposition B.43, statement 4.   □

Remark. All entropies considered above may in addition be conditional on some event $\mathcal{E}$. We make use of this fact in our overview about some results on "unconditional security" in Chapter 10. There, mutual information of the form $I(X;Y|Z,\mathcal{E})$ appears. $\mathcal{E}$ is an event in XYZ (i.e., a subset of XYZ) and the mutual information is defined as usual, but the conditional probabilities assuming $\mathcal{E}$ have to be used. More precisely:

$$H(X|YZ,\mathcal{E}) := - \sum_{(x,y,z)\in\mathcal{E}} \text{prob}(y,z|\mathcal{E}) \cdot \text{prob}(x|y,z,\mathcal{E}) \cdot \log_2(\text{prob}(x|y,z,\mathcal{E})),$$

$$H(X|Z,\mathcal{E}) := - \sum_{(x,y,z)\in\mathcal{E}} \text{prob}(z|\mathcal{E}) \cdot \text{prob}(x|z,\mathcal{E}) \cdot \log_2(\text{prob}(x|z,\mathcal{E})),$$

$$I(X;Y|Z,\mathcal{E}) := H(X|Z,\mathcal{E}) - H(X|YZ,\mathcal{E}).$$

Here recall that, for example, $\text{prob}(y,z|\mathcal{E}) = \text{prob}(X \times \{y\} \times \{z\}|\mathcal{E})$ and $\text{prob}(x|y,z,\mathcal{E}) = \text{prob}(\{x\} \times Y \times Z | X \times \{y\} \times \{z\}, \mathcal{E})$ The results on entropies from above remain valid if they are, in addition, conditional on $\mathcal{E}$. The same proofs apply, but the conditional probabilities assuming $\mathcal{E}$ have to be used.

References

Textbooks

[Ash65] R.B. Ash: Information Theory. New York: John Wiley & Sons, 1965.

[Assche06] G. van Assche: Quantum Cryptography and Secret-Key Distillation. Cambridge, UK: Cambridge University Press, 2006.

[BalDiaGab95] J.L. Balcázar, J. Díaz, J. Gabarró: Structural Complexity I. Berlin, Heidelberg, New York: Springer-Verlag, 1995.

[Bauer07] F.L. Bauer: Decrypted Secrets – Methods and Maxims of Cryptology. 4th ed. Berlin, Heidelberg, New York: Springer-Verlag, 2007.

[Bauer96] H. Bauer: Probability Theory. Berlin: de Gruyter, 1996.

[BerPer85] J. Berstel, D. Perrin: Theory of Codes. Orlando: Academic Press, 1985.

[Billingsley12] P. Billingsley: Probability and Measure. Anniversary Edition, New York: John Wiley & Sons, 2012.

[BlaSerSma99] I. Blake, G. Seroussi, N. Smart: Elliptic Curves in Cryptography. Cambridge: Cambridge University Press, 1999.

[BroSemMusMüh07] I.N. Bronshtein, K.A. Semendyayev, G. Musiol, H. Mühlig: Handbook of Mathematics. 5th ed., Berlin, Heidelberg, New York: Springer-Verlag, 2007.

[Buchmann00] J.A. Buchmann: Introduction to Cryptography. Berlin, Heidelberg, New York: Springer-Verlag, 2000.

[Cohen95] H. Cohen: A Course in Computational Algebraic Number Theory. Berlin, Heidelberg, New York: Springer-Verlag, 1995.

[CohFre06] H. Cohen, G. Frey (eds.): Handbook of Elliptic and Hyperelliptic Curve Cryptography. Boca Raton, London, New York, Singapore: Chapman & Hall/CRC, 2006.

[CohDiuLal77] C. Cohen-Tannoudji, B. Diu, F. Laloë: Quantum Mechanics. New York: John Wiley & Sons, 1977.

[CovTho92] T.M. Cover, J.A. Thomas: Elements of Information Theory. New York: John Wiley & Sons, 1992.

[DaeRij02] J. Daemen, V. Rijmen: The Design of Rijndael – AES – The Advanced Encryption Standard. Berlin, Heidelberg, New York: Springer-Verlag, 2002.

[DevBer12] J. L. Devore, K.N. Berk: Modern Mathematical Statistics with Applications. 2nd edition, Berlin, Heidelberg, New York: Springer-Verlag, 2012.

[Enge99] A. Enge: Elliptic Curves and Their Applications to Cryptography. Boston, Dordrecht, London: Kluwer Academic Publishers, 1999.

[Feller68] W. Feller: An Introduction to Probability Theory and Its Applications. 3rd ed. New York: John Wiley & Sons, 1968.

[Forster96] O. Forster: Algorithmische Zahlentheorie. Braunschweig, Wiesbaden: Vieweg, 1996.

[GanYlv67] R.A. Gangolli, D. Ylvisaker: Discrete Probability. New York: Harcourt, Brace & World, 1967.

484   References

[Goldreich99] O. Goldreich: Modern Cryptography, Probabilistic Proofs and Pseudorandomness. Berlin, Heidelberg, New York: Springer-Verlag, 1999.
[Goldreich01] O. Goldreich: Foundations of Cryptography – Basic Tools. Cam-
bridge University Press, 2001.
[Goldreich04] O. Goldreich: Foundations of Cryptography. Volume II Basic Appli-
cations. Cambridge University Press, 2004.
[GolPeiSch94] S.W. Golomb, R.E. Peile, R.A. Scholtz: Basic Concepts in Informa-
tion Theory and Coding. New York: Plenum Press, 1994.
[Gordon97] H. Gordon: Discrete Probability. Berlin, Heidelberg, New York:
Springer-Verlag, 1997.
[Hamming86] R.W. Hamming: Coding and Information Theory. 2nd ed. Englewood
Cliffs, NJ: Prentice Hall, 1986.
[HanMenVan04] H. Hankerson, A. Menezes, S. Vanstone: Guide to Elliptic Curve
Cryptography. Berlin, Heidelberg, New York: Springer-Verlag, 2004.
[HarWri79] G.H. Hardy, E.M. Wright: An Introduction to the Theory of Numbers.
5th ed. Oxford: Oxford University Press, 1979.
[HopUll79] J. Hopcroft, J. Ullman: Introduction to Automata Theory, Languages
and Computation. Reading, MA: Addison-Wesley Publishing Company, 1979.
[IreRos82] K. Ireland, M.I. Rosen: A Classical Introduction to Modern Number
Theory. Berlin, Heidelberg, New York: Springer-Verlag, 1982.
[Kahn67] D. Kahn: The Codebreakers: The Story of Secret Writing. New York:
Macmillan Publishing Co. 1967.
[Knuth98] D.E. Knuth: The Art of Computer Programming. 3rd ed. Volume
2/Seminumerical Algorithms. Reading, MA: Addison-Wesley Publishing Com-
pany, 1998.
[Koblitz94] N. Koblitz: A Course in Number Theory and Cryptography. 2nd ed.
Berlin, Heidelberg, New York: Springer-Verlag, 1994.
[Kunz05] E. Kunz: Introduction to Plane Algebraic Curves. Basel: Birkhäuser,
2005.
[Lang05] S. Lang: Algebra. 3rd ed. Berlin, Heidelberg, New York: Springer-Verlag,
2005.
[Luby96] M. Luby: Pseudorandomness and Cryptographic Applications. Princeton,
NJ: Princeton University Press, 1996.
[MenOorVan96] A. Menezes, P.C. van Oorschot, S.A. Vanstone: Handbook of Ap-
plied Cryptography. Boca Raton, New York, London, Tokyo: CRC Press, 1996.
[MotRag95] R. Motwani, P. Raghavan: Randomized Algorithms. Cambridge, UK:
Cambridge University Press, 1995.
[NieXin09] H. Niederreiter, C. Xing: Algebraic Geometry in Coding Theory and
Cryptography. Princeton, NJ: Princeton University Press, 2009.
[NieChu00] M.A. Nielsen, I.L. Chuang: Quantum Computation and Quantum In-
formation. Cambridge, UK: Cambridge University Press, 2000.
[Osen74] L.M. Osen: Women in Mathematics. Cambridge, MA: MIT, 1974.
[Papadimitriou94] C.H. Papadimitriou: Computational Complexity. Reading, MA:
Addison-Wesley Publishing Company, 1994.
[Rényi70] A. Rényi: Probability Theory. Amsterdam: North-Holland, 1970. Repub-
lication: Mineola, NY: Dover Publications, 2007.
[Riesel94] H. Riesel: Prime Numbers and Computer Methods for Factorization.
Boston, Basel: Birkhäuser, 1994.
[Rose94] H.E. Rose: A Course in Number Theory. 2nd ed. Oxford: Clarendon Press,
1994.
[Rosen00] K.H. Rosen: Elementary Number Theory and Its Applications. 4th ed.
Reading, MA: Addison-Wesley Publishing Company, 2000.

[Salomaa90] A. Salomaa: Public-Key Cryptography. Berlin, Heidelberg, New York: Springer-Verlag, 1990.

[Schneier96] B. Schneier: Applied Cryptography. New York: John Wiley & Sons, 1996.

[Silverman86] J.H. Silverman: The Arithmetic of Elliptic Curves. Berlin, Heidelberg, New York: Springer-Verlag, 1986.

[Simmons92] G.J. Simmons (ed.): Contemporary Cryptology. Piscataway, NJ: IEEE Press, 1992.

[Stinson95] D.R. Stinson: Cryptography – Theory and Practice. Boca Raton, New York, London, Tokyo: CRC Press, 1995.

[TilJaj11] H.C.A. van Tilborg, S. Jajodia (eds.): Encyclopedia of Cryptography and Security. 2nd ed. Berlin, Heidelberg, New York: Springer-Verlag, 2011.

Papers

[AleChoGolSch88] W.B. Alexi, B. Chor, O. Goldreich, C.P. Schnorr: RSA/Rabin functions: certain parts are as hard as the whole. SIAM Journal on Computing, 17(2): 194–209, April 1988.

[AgrKaySax04] M. Agrawal, N. Kayal, N. Saxena: PRIMES is in P. Ann. of Math., 160: 781 - 793, 2004.

[AraFouTra07] R. Araújo, S. Foulle, J. Traoré: A practical and secure coercion-resistant scheme for remote elections. Frontiers of Electronic Voting 2007 (FEE 2007), Schloss Dagstuhl, Germany, 2007.

[AraFouTra10] R. Araújo, S. Foulle, J. Traoré: A practical and secure coercion-resistant scheme for remote elections. In: D. Chaum, M. Jakobsson, R. L. Rivest, P.Y.A. Ryan, J. Benaloh, M. Kutylowski, B. Adida (eds.): Towards Trustworthy Elections - New Directions in Electronic Voting. Lecture Notes in Computer Science, 2951: 330–342, Springer-Verlag, 2010.

[AtiSti96] M. Atici, D. R. Stinson: Universal hashing and multiple authentication. Advances in Cryptology - CRYPTO '96, Lecture Notes in Computer Science, 1109: 16–30, Springer-Verlag, 1996.

[AumDinRab02] Y. Aumann, Y.Z. Ding, M.O. Rabin: Everlasting security in the bounded storage model. IEEE Transactions on Information Theory, 48(6): 1668-1680, 2002..

[AumRab99] Y. Aumann, M.O. Rabin: Information-theoretically secure communication in the limited storage space model. Advances in Cryptology - CRYPTO '99, Lecture Notes in Computer Science, 1666: 65–79, Springer-Verlag, 1999.

[Bach88] E. Bach: How to generate factored random numbers. SIAM Journal on Computing, 17(2): 179–193, April 1988.

[BalKob98] R. Balsubramanian, N. Koblitz: The improbability that an elliptic curve has subexponential discrete log problem under the Menezes-Okamoto-Vanstone algorithm. Journal of Cryptology, 11(2): 141 – 145, 1998.

[BaFoPoiPouSt01] O. Baudron, P.A. Fouque, D. Pointcheval, G. Poupard, J. Stern: Practical Multi-Candidate Election System. 20th ACM Symposium on Principles of Distributed Computing (PODC '01): 274 – 283, New York: ACM Press, 2001.

[BarPfi97] N. Barić, B. Pfitzmann: Collision-free accumulators and fail-stop signature schemes without trees. Advances in Cryptology - EUROCRYPT '97, Lecture Notes in Computer Science, 1233: 480–494, Springer-Verlag, 1997.

[BatHor62] P. Bateman, R. Horn: A heuristic formula concerning the distribution of prime numbers. Mathematics of Computation, 16: 363–367, 1962.

[BatHor65] P. Bateman, R. Horn: Primes represented by irreducible polynomials in one variable. Proc. Symp. Pure Math., 8: 119–135, 1965.

[BayGro12] S. Bayer, J. Groth: Efficient Zero-Knowledge Argument for Correctness of a Shuffle. Advances in Cryptology - EUROCRYPT 2012, Lecture Notes in Computer Science, 7237: 263–280, Springer-Verlag, 2012.

[BeGrGwHåKiMiRo88] M. Ben-Or, O. Goldreich, S. Goldwasser, J. Håstad, J. Kilian, S. Micali, P. Rogaway: Everything provable is provable in zero-knowledge. Advances in Cryptology - CRYPTO '88, Lecture Notes in Computer Science, 403: 37–56, Springer-Verlag, 1990.

[Bellare99] M. Bellare: Practice oriented provable security. Lectures on Data Security. Lecture Notes in Computer Science, 1561: 1–15, Springer-Verlag, 1999.

[BelRog93] M. Bellare, P. Rogaway: Random oracles are practical: a paradigm for designing efficient protocols, Proc. First Annual Conf. Computer and Communications Security, ACM, New York, 1993:6273, 1993.

[BelRog94] M. Bellare, P. Rogaway: Optimal asymmetric encryption. Advances in Cryptology - EUROCRYPT '94, Lecture Notes in Computer Science, 950: 92–111, Springer-Verlag, 1995.

[BelRog96] M. Bellare, P. Rogaway: The exact security of digital signatures, how to sign with RSA and Rabin. Advances in Cryptology - EUROCRYPT '96, Lecture Notes in Computer Science, 1070: 399–416, Springer-Verlag, 1996.

[BelRog97] M. Bellare, P. Rogaway: Collision-resistant hashing: towards making UOWHF practical. Advances in Cryptology - CRYPTO '97, Lecture Notes in Computer Science, 1294: 470–484, Springer-Verlag, 1997.

[BenBra84] C. H. Bennett, G. Brassard: Quantum Cryptography: Public key distribution and coin tossing. Proceedings of the IEEE International Conference on Computers, Systems, and Signal Processing, Bangalore, India: 175–179, 1984.

[BenBraCréMau95] C. H. Bennett, G. Brassard, C. Crépeau, U. Maurer: Generalized privacy amplification. IEEE Transactions on Information Theory, 41 (6): 1915–1923, 1995..

[BenBesBraSalSmo92] C. H. Bennett, F. Bessette, G. Brassard, L. Salvail, J. Smolin: Experimental quantum cryptography. Journal of Cryptology, 5: 3–28, 1992.

[BenBraRob88] C. H. Bennett, G. Brassard, J.M. Robert: Privacy amplification by public discussion. SIAM Journal on Computing, 17(2): 1988, 210–229.

[BerLan07] D.J. Bernstein, T. Lange: Faster addition and doubling on elliptic curves. ASIACRYPT 2007, Lecture Notes in Computer Science, 4833: 29 – 50, Springer-Verlag, 2007.

[BiBoBoMoRo06] E. Biham, M. Boyer, P. O. Boykin, T. Mor, V. Roychowdhury: A proof of the security of quantum key distribution. Journal of Cryptology, 19(4): 2006, 381–439.

[Bleichenbacher96] D. Bleichenbacher: Generating ElGamal signatures without knowing the secret key. Advances in Cryptology - EUROCRYPT '96, Lecture Notes in Computer Science, 1070: 10–18, Springer-Verlag, 1996.

[Bleichenbacher98] D. Bleichenbacher: A chosen ciphertext attack against protocols based on the RSA encryption standard PKCS #1. Advances in Cryptology - CRYPTO '98, Lecture Notes in Computer Science, 1462: 1–12, Springer-Verlag, 1998.

[BluBluShu86] L. Blum, M. Blum, M. Shub: A simple unpredictable pseudorandom number generator. SIAM Journal on Computing, 15(2): 364–383, 1986.

[BluGol85] M. Blum, S. Goldwasser: An efficient probabilistic public-key encryption scheme which hides all partial information. Advances in Cryptology - Proceedings of CRYPTO '84, Lecture Notes in Computer Science, 196: 289–299, Springer-Verlag, 1985.

[Blum82] M. Blum: Coin flipping by telephone: a protocol for solving impossible problems. Proceedings of the 24th IEEE Computer Conference, San Francisco, Calif., February 22–25, 1982: 133–137, 1982.

[Blum84] M. Blum: Independent unbiased coin flips from a correlated biased source. Proceedings of the IEEE 25th Annual Symposium on Foundations of Computer Science, Singer Island, Fla., October 24–26, 1984: 425–433, 1984.

[BluMic84] M. Blum, S. Micali: How to generate cryptographically strong sequences of pseudorandom bits. SIAM Journal on Computing, 13(4): 850–863, November 1984.

[Boer88] B. Den Boer: Diffie–Hellman is as strong as discrete log for certain primes. Advances in Cryptology - CRYPTO '88, Lecture Notes in Computer Science, 403: 530–539, Springer-Verlag, 1990.

[BoeBos93] B. den Boer, A. Bosselaers: Collisions for the compression function of MD5. Advances in Cryptology - EUROCRYPT '93, Lecture Notes in Computer Science, 765: 293–304, Springer-Verlag, 1994.

[Boneh01] D. Boneh: Simplified OAEP for the RSA and Rabin functions. Advances in Cryptology - CRYPTO 2001, Lecture Notes in Computer Science, 2139: 275–291, Springer-Verlag, 2001.

[BonDur00] D. Boneh, G. Durfee: Cryptanalysis of RSA with private key d less than $N^{0.292}$. IEEE Transactions on Information Theory, 46(4): 1339–1349, 2000.

[BonGol02] D. Boneh, P. Golle: Almost entirely correct mixing with applications to voting. ACM Conference on Computer and Communications Security 2002: 68–77, 2002.

[BonVen96] D. Boneh, R. Venkatesan: Hardness of computing the most significant bits of secret keys in Diffie–Hellman and related schemes. Advances in Cryptology - CRYPTO '96, Lecture Notes in Computer Science, 1109: 129–142, Springer-Verlag, 1996.

[BonVen98] D. Boneh, R. Venkatesan: Breaking RSA may not be equivalent to factoring. Advances in Cryptology - EUROCRYPT '98, Lecture Notes in Computer Science, 1403: 59–71, Springer-Verlag, 1998.

[Brands93] S. Brands: An efficient off-line electronic cash system based on the representation problem. Technical Report CS-R9323. Amsterdam, NL: Centrum voor Wiskunde en Informatica (CWI), 1993.

[BraCre96] G. Brassard, C. Crépeau: 25 years of quantum cryptography. SIGACT News 27(3): 13–24, 1996.

[BraSal93] G. Brassard, L. Salvail: Secret-key reconciliation by public discussion. Advances in Cryptology - EUROCRYPT '93, Lecture Notes in Computer Science, 765: 410–423, Springer-Verlag, 1994.

[BuLaToNiDoPe03] W. T. Buttler, S. K. Lamoreaux, J. R. Torgerson, G. H. Nickel, C. H. Donahue, C. G. Peterson: Fast, efficient error reconciliation for quantum cryptography. arXiv:quant-ph/0203096, 2003. http://arxiv.org/abs/quant-ph/0203096v2 .

[Cachin97] C. Cachin: Entropy measures and unconditional security in cryptography. ETH Series in Information Security and Cryptography, vol. 1 (Reprint of Ph.D. dissertation No. 12187, Swiss Federal Institute of Technology (ETH), Zürich). Konstanz: Hartung-Gorre Verlag, 1997.

[CachMau97] C. Cachin, U.M. Maurer: Unconditional security against memory-bounded adversaries. Advances in Cryptology - CRYPTO '97, Lecture Notes in Computer Science, 1294: 292–306, Springer-Verlag, 1997.

[CamMauSta96] J. Camenisch, U.M. Maurer, M. Stadler: Digital payment systems with passive anonymity revoking trustees. Proceedings of ESORICS '96, Lecture Notes in Computer Science, 1146: 33–43, Springer-Verlag, 1996.

[CamWie92] K.W. Campbell, M.J. Wiener: DES is not a group. Advances in Cryptology - CRYPTO '92, Lecture Notes in Computer Science, 740: 512–520, Springer-Verlag, 1993.

[CamPivSta94] J.L. Camenisch, J.M. Piveteau, M.A. Stadler: Blind signatures based on the discrete logarithm problem. Advances in Cryptology - EUROCRYPT '94, Lecture Notes in Computer Science, 950: 428–432, Springer-Verlag, 1995.

[CanGolHal98] R. Canetti, O. Goldreich, S. Halevi: The random oracle methodology, revisited. STOC'98, Dallas, Texas: 209–218, New York, NY: ACM, 1998.

[CanGolHal04] R. Canetti, O. Goldreich, S. Halevi: On the random-oracle methodology as applied to length-restricted signature schemes. First Theory of Cryptography Conference, TCC 2004, Lecture Notes in Computer Science, 2951: 40–57, Springer-Verlag, 2004.

[CarWeg79] J.L. Carter, M.N. Wegman: Universal classes of hash functions. Journal of Computer and System Sciences, 18: 143–154, 1979.

[ChaPed92] D. Chaum, T. Pedersen: Wallet databases with observers. Advances in Cryptology - CRYPTO '92, Lecture Notes in Computer Science, 740: 89–105, Springer-Verlag, 1993.

[Chaum81] D. Chaum: Untraceable electronic mail, return addresses, and digital pseudonyms. Communications of the ACM, 24(2): 84–90, 1981.

[Chaum82] D. Chaum: Blind signatures for untraceable payments. Advances in Cryptology - Proceedings of CRYPTO '82: 199–203, Plenum Press 1983.

[ChePaiPoi06] B. Chevallier-Mames, P. Paillier, D. Pointcheval: Encoding-Free ElGamal Encryption Without Random Oracles. Proceedings Theory and Practice in Public Key Cryptography - PKC 2006, Lecture Notes in Computer Science, 3958: 91–104, Springer-Verlag, 2006.

[ClaChoMye07] M.R. Clarkson, S. Chong, A.C. Myers: Civitas: A Secure Remote Voting System. Cornell University Computing and Information Science Technical Report TR2007-2081. Frontiers of Electronic Voting 2007 (FEE 2007), Schloss Dagstuhl, Germany, 2007.

[Coppersmith97] D. Coppersmith: Small solutions to polynomial equations and low exponent RSA vulnerabilities. Journal of Cryptology, 10(4): 233–260, 1997.

[CorMay07] J.-S. Coron, A. May: Deterministic polynomial-time equivalence of computing the RSA secret key and factoring. Journal of Cryptology, 20(1): 39–50, 2007.

[CraDamSch94] R. Cramer, I. Damgård, B. Schoenmakers: Proofs of partial knowledge and simplified design of witness hiding protocols. Advances in Cryptology - CRYPTO '94, Lecture Notes in Computer Science, 839: 174–187, Springer-Verlag, 1994.

[CraDam96] R. Cramer, I. Damgård: New generation of secure and practical RSA-based signatures. Advances in Cryptology - CRYPTO '96, Lecture Notes in Computer Science, 1109: 173–185, Springer-Verlag, 1996.

[CraFraSchYun96] R. Cramer, M.K. Franklin, B. Schoenmakers, M. Yung: Multi-authority secret-ballot elections with linear work. Advances in Cryptology - EUROCRYPT '96, Lecture Notes in Computer Science, 1070: 72–83, Springer-Verlag, 1996.

[CraGenSch97] R. Cramer, R. Gennaro, B. Schoenmakers: A secure and optimally efficient multi-authority election scheme. Advances in Cryptology - EUROCRYPT '97, Lecture Notes in Computer Science, 1233: 103–118, Springer-Verlag, 1997.

[CraSho98] R. Cramer, V. Shoup: A practical public key cryptosystem provably secure against adaptive chosen ciphertext attack. Advances in Cryptology -

CRYPTO '98, Lecture Notes in Computer Science, 1462: 13–25, Springer-Verlag, 1998.

[CraSho00] R. Cramer, V. Shoup: Signature schemes based on the strong RSA assumption. ACM Transactions on Information and System Security, 3(3): 161–185, 2000.

[Damgård87] I.B. Damgård: Collision-free hash functions and public-key signature schemes. Advances in Cryptology - EUROCRYPT '87, Lecture Notes in Computer Science, 304: 203–216, Springer-Verlag, 1988.

[DamJur01] I. Damgård, M. Jurik: A generalisation, a simplification and some applications of Paillier's probabilistic public-key system. Proceedings Theory and Practice in Public Key Cryptography - PKC 2001, Lecture Notes in Computer Science, 1992: 119–136, Springer-Verlag, 2001.

[DamKop01] I. Damgård, M. Koprowski: Practical threshold RSA signatures without a trusted dealer. Advances in Cryptology - EUROCRYPT 2001, Lecture Notes in Computer Science, 2045: 152–165, Springer-Verlag, 2001.

[DanGol09] G. Danezis, I. Goldberg: Sphinx: A compact and provably secure mix format. IEEE Symposium on Security and Privacy 2009: 269–282.

[DanDinMat03] G. Danezis, R. Dingledine, N. Mathewson: Mixminion: Design of a type III anonymous remailer Protocol. IEEE Symposium on Security and Privacy 2003: 2–15.

[Diffie88] W. Diffie: The first ten years of public key cryptology. In: G.J. Simmons (ed.): Contemporary Cryptology, 135–175, Piscataway, NJ: IEEE Press, 1992.

[DifHel76] W. Diffie, M.E. Hellman: New directions in cryptography. IEEE Transactions on Information Theory, IT-22: 644–654, 1976.

[DifHel77] W. Diffie, M. E. Hellman: Exhaustive cryptanalysis of the NBS data encryption standard. Computer, 10: 74–84, 1977.

[Ding01] Y.Z. Ding: Provable everlasting security in the bounded storage model. PhD Thesis, Cambridge, MA: Harvard University, May 2001.

[DinRab02] Y.Z. Ding, M.O. Rabin: Hyper-encryption and everlasting security. 19th Annual Symposium on Theoretical Aspects of Computer Science (STACS) 2002, Lecture Notes in Computer Science, 2285: 1–26, Springer-Verlag, 2002.

[Dobbertin96] H. Dobbertin: Welche Hash-Funktionen sind für digitale Signaturen geeignet? In: P. Horster (ed.): Digitale Signaturen, 81–92, Braunschweig, Wiesbaden: Vieweg 1996.

[Dobbertin96a] H. Dobbertin: Cryptanalysis of MD5. Presented at the rump session, Advances in Cryptology - EUROCRYPT '96.

[DusLütHen06] M. Dušek, N. Lütkenhaus, M. Hendrych: Quantum Cryptography. In: E. Wolf (ed.): Progress in Optics 49: 381–454, Amsterdam: Elsevier 2006.

[DwoNao94] C. Dwork, M. Naor: An efficient unforgeable signature scheme and its applications. Advances in Cryptology - CRYPTO '94, Lecture Notes in Computer Science, 839: 234–246, Springer-Verlag, 1994.

[DziMau02] S. Dziembowski, U. Maurer: Tight security proofs for the bounded-storage model. Proceedings of the 34th Annual ACM Symposium on Theory of Computing, Montréal, Québec, Canada, May 19–21, 2002: 341–350, 2002.

[DziMau04a] S. Dziembowski, U. Maurer: Optimal randomizer efficiency in the bounded-storage model. Journal of Cryptology, 17(1): 5–26, January 2004.

[DziMau04b] S. Dziembowski, U. Maurer: On generating the initial key in the bounded-storage model. Advances in Cryptology - EUROCRYPT 2004, Lecture Notes in Computer Science, 3027: 126–137, Springer-Verlag, 2004.

[Edwards07] H.M. Edwards: A normal form for elliptic curves. Bulletin of the American Mathematical Society 44: 393 – 422, 2007.

[ElGamal84] T. ElGamal: A public key cryptosystem and a signature scheme based on discrete logarithms. Advances in Cryptology - Proceedings of CRYPTO '84, Lecture Notes in Computer Science, 196: 10–18, Springer-Verlag, 1985.

[FiaSha86] A. Fiat, A. Shamir: How to prove yourself: practical solutions to identification and signature problems. Advances in Cryptology - CRYPTO '86, Lecture Notes in Computer Science, 263: 186–194, Springer-Verlag, 1987.

[FisCarShe06] K. Fisher, R. Carback, A.T. Sherman: Punchscan - introduction and system definition of a high-integrity election system. IAVoSS Workshop On Trustworthy Elections (WOTE 2006), Cambridge, UK, 2006.

[FIPS46 1977] FIPS46: Data Encryption Standard. Federal Information Processing Standards Publication 46, U.S. Department of Commerce/National Bureau of Standards, National Technical Information Service, Springfield, Virginia, 1977.

[FIPS 113] FIPS 113: Computer data authentication. Federal Information Processing Standards Publication 113, U.S. Department of Commerce/National Bureau of Standards, http://www.itl.nist.gov/fipspubs/, 1985.

[FIPS 180-2] FIPS 180-2: Secure hash signature standard. Federal Information Processing Standards Publication 180-2, U.S. Department of Commerce/National Bureau of Standards, http://www.itl.nist.gov/fipspubs/, 2002.

[FIPS 186-4] FIPS 186-4: Digital Signature Standard (DSS). Federal Information Processing Standards Publication 186-4, National Institute of Standards and Technology, 2013.

[FIPS 198] FIPS 198: The keyed-hash message authentication code (HMAC). Federal Information Processing Standards Publication 198, U.S. Department of Commerce/National Bureau of Standards, http://www.itl.nist.gov/fipspubs/, 2002.

[FisSch00] R. Fischlin, C.P. Schnorr: Stronger security proofs for RSA and Rabin bits. Journal of Cryptology, 13(2): 221–244, 2000.

[FraMaKYun98] Y. Frankel, P.D. MacKenzie, M. Yung: Robust efficient distributed RSA-key generation. The 30th Annual ACM Symposium on Theory of Computing – STOC '98, 663–672, 1998.

[FouGauHar00] M. Fouquet, P. Gaudry, R. Harley: An extension of Satoh's algorithm and its implementation. Journal of the Ramanujan Mathematical Society, 15: 281–318, 2000

[FouSte01] P.A. Fouque, J. Stern: Fully distributed threshold RSA under standard assumptions. ASIACRYPT 2001, Lecture Notes in Computer Science, 2248: 310-330, Springer-Verlag, 2001.

[Frey98] G. Frey: How to disguise an elliptic curve. Talk at ECC 98, Waterloo.

[FreRue94] G. Frey, H. Rück: A remark concerning m-divisibility and the discrete logarithm in the divisor class group of curves. Math. Comp., 62: 865–874, 1994.

[FujOkaPoiSte01] E. Fujisaki, T. Okamoto, D. Pointcheval, J. Stern: RSA-OAEP is secure under the RSA assumption. Advances in Cryptology - CRYPTO 2001, Lecture Notes in Computer Science, 2139: 260–274, Springer-Verlag, 2001.

[FurSak01] J. Furukawa, K. Sako: An efficient scheme for proving a shuffle. Advances in Cryptology - CRYPTO 2001, Lecture Notes in Computer Science, 2139: 368–387, Springer-Verlag, 2001.

[Furukawa04] J. Furukawa: Efficient, verifiable shuffle decryption and its requirement of unlinkability. Proceedings Public Key Cryptography - PKC 2004, Lecture Notes in Computer Science, 2947: 319–332, Springer-Verlag, 2004.

[FurYam01] E. Furukawa, K. Yamazaki: Application of existing perfect code to secret key reconciliation. Proceedings International Symposium on Communications and Information Technologies ISCIT 2001: 397–400, 2001.

[GalMck00] S.D. Galbraith, J. McKee: The probability that the number of points on an elliptic curve over a finite field is prime. Journal of the London Math. Soc. 62(3): 671 – 684, 2000.

[GauHesSma02] P. Gaudry, F. Hess, N.P. Smart: Constructive and destructive facets of Weil descent on elliptic curves. Journal of Cryptology, 15(1): 19–46, 2002.

[GenHalRab99] R. Gennaro, S. Halevi, T. Rabin: Secure hash-and-sign signatures without the random oracle. Advances in Cryptology - EUROCRYPT '99, Lecture Notes in Computer Science, 1592: 123–139, Springer-Verlag, 1999.

[GenJarKraRab99] R. Gennaro, S. Jarecki, H. Krawczyk, T. Rabin: Secure distributed key generation for discrete-log based cryptosystems. Advances in Cryptology - EUROCRYPT '99, Lecture Notes in Computer Science, 1592: 295–310, Springer-Verlag, 1999.

[Gill77] J. Gill: Computational complexity of probabilistic Turing machines. SIAM Journal on Computing, 6(4): 675–695, December 1977.

[GolLev89] O. Goldreich, L. Levin: A hard-core predicate for all one-way functions. Proceedings of the 21st Annual ACM Symposium on Theory of Computing, Seattle, Wash., May 15–17, 1989: 25–32, 1989.

[GolMic84] S. Goldwasser, S. Micali: Probabilistic encryption. Journal of Computer and System Sciences, 28(2): 270–299, 1984.

[GolMicRac89] S. Goldwasser, S. Micali, C. Rackoff: The knowledge complexity of interactive proof systems. SIAM Journal on Computing, 18: 185–208, 1989.

[GolMicRiv88] S. Goldwasser, S. Micali, R. Rivest: A digital signature scheme secure against chosen message attacks. SIAM Journal on Computing, 17(2): 281–308, 1988.

[GolMicTon82] S. Goldwasser, S. Micali, P. Tong: Why and how to establish a private code on a public network. Proceedings of the IEEE 23rd Annual Symposium on Foundations of Computer Science, Chicago, Ill., November 3–5, 1982: 134–144, 1982.

[GolMicWid86] O. Goldreich, S. Micali, A. Wigderson: Proofs that yield nothing but their validity and a methodology of cryptographic protocol design. Proceedings of the IEEE 27th Annual Symposium on Foundations of Computer Science, Toronto, October 27–29, 1986: 174–187, 1986.

[GolTau03] S. Goldwasser, Y. Tauman: On the (in)security of the Fiat-Shamir paradigm. Proceedings of the IEEE 44th Annual Symposium on Foundations of Computer Science, Cambridge, MA, USA, October 11–14, 2003: 102–113, 2003.

[GolTau03a] S. Goldwasser, Y. Tauman: On the (in)security of the Fiat-Shamir paradigm. Cryptology ePrint Archive, http://eprint.iacr.org, Report 034, 2003.

[Gordon84] J.A. Gordon: Strong primes are easy to find. Advances in Cryptology - EUROCRYPT '84, Lecture Notes in Computer Science, 209: 216–223, Springer-Verlag, 1985.

[Groth03] J. Groth: A verifiable secret shuffle of homomorphic encryptions. Proceedings Theory and Practice in Public Key Cryptography - PKC 2003, Lecture Notes in Computer Science, 2567: 145–160, Springer-Verlag, 2003.

[GroLu07] J. Groth, S. Lu: Verifiable shuffle of large size ciphertexts. Proceedings Theory and Practice in Public Key Cryptography - PKC 2007, Lecture Notes in Computer Science, 4450: 377–392, Springer-Verlag, 2007.

[Hirt01] M. Hirt: Multi-Party Computation – Efficient Protocols, General Adversaries, and Voting. PhD dissertation no. 14376, Swiss Federal Institute of Technology (ETH), Zürich, 2001.

[Hirt10] M. Hirt: Receipt-free k-out-of-l voting based on ElGamal encryption. In: D. Chaum, M. Jakobsson, R. L. Rivest, P.Y.A. Ryan, J. Benaloh, M. Kutylowski,

B. Adida (eds.): Towards Trustworthy Elections - New Directions in Electronic Voting. Lecture Notes in Computer Science, 2951: 64–82, Springer-Verlag, 2010.

[HåsNäs98] J. Håstad, M. Näslund: The security of all RSA and Discrete Log bits. Proceedings of the IEEE 39th Annual Symposium on Foundations of Computer Science, Palo Alto, CA, November 8–11, 1998: 510–519, 1998.

[HåsNäs99] J. Håstad, M. Näslund: The security of all RSA and Discrete Log bits. Electronic Colloquium on Computational Complexity, http://eccc.hpi-web.de, ECCC Report TR99-037, 1999.

[ImpLevLub89] R. Impagliazzo, L. Levin, M. Luby: Pseudo-random number generation from one-way functions. Proceedings 21st ACM Symposium on Theory of Computing, STOC '89: 12–24, New York, NY: ACM, 1989.

[ISO/IEC 9594-8] ISO/IEC 9594-8: Information technology - Open Systems Interconnection - The Directory: Authentication framework. International Organization for Standardization, Geneva, Switzerland, http://www.iso.org, 1995.

[ISO/IEC 9797-1] ISO/IEC 9797-1: Message Authentication Codes (MACs) – Part 1: Mechanisms using a block cipher. International Organization for Standardization, Geneva, Switzerland, http://www.iso.org, 1999.

[ISO/IEC 9797-2] ISO/IEC 9797-2: Message Authentication Codes (MACs) – Part 2: Mechanisms using a dedicated hash-function. International Organization for Standardization, Geneva, Switzerland, http://www.iso.org, 2002.

[ISO/IEC 10116] ISO/IEC 10116: Information processing - Modes of operation for an n-bit block cipher algorithm. International Organization for Standardization, Geneva, Switzerland, http://www.iso.org, 1991.

[ISO/IEC 10118-2] ISO/IEC 10118-2: Information technology - Security techniques - Hash-functions - Part 2: Hash-functions using an n-bit block cipher algorithm. International Organization for Standardization, Geneva, Switzerland, http://www.iso.org, 1994.

[JakJue00] M. Jakobsson, A. Juels: Mix and match: Secure function evaluation via ciphertexts. ASIACRYPT 2000, Lecture Notes in Computer Science, 1976: 162–177, Springer-Verlag, 2000.

[JakJueRiv02] M. Jakobsson, A. Juels, R. L. Rivest: Making mix nets robust for electronic voting by randomized partial checking. Proceedings of the 11th USENIX Security Symposium, Dan Boneh (ed.). USENIX Association, Berkeley, CA, USA, 339–353, 2002.

[JakSakImp96] M. Jakobsson, K. Sako, R. Impagliazzo: Designated verifier proofs and their applications. Advances in Cryptology - EUROCRYPT '96, Lecture Notes in Computer Science, 1070: 143–154, Springer-Verlag, 1996.

[JueCatJak05] A. Juels, D. Catalano, M. Jakobsson: Coercion-resistant electronic elections. Proceedings of the 2005 ACM Workshop on Privacy in the Electronic Society (WPES 2005) : 61–70, ACM, 2005.

[JueCatJak10] A. Juels, D. Catalano, M. Jakobsson: Coercion-resistant electronic elections. In: D. Chaum, M. Jakobsson, R. L. Rivest, P.Y.A. Ryan, J. Benaloh, M. Kutylowski, B. Adida (eds.): Towards Trustworthy Elections - New Directions in Electronic Voting. Lecture Notes in Computer Science, 2951: 37–63, Springer-Verlag, 2010.

[Koblitz88] N. Koblitz: Primality of the number of points on an elliptic curve over a finite field. Pacific Journal of Mathematics, 131(1): 157–165, 1988.

[KobMen05] N. Koblitz, A.J. Menezes: Another look at "provable security". Journal of Cryptology, Online First: OF1–OF35, November 2005.

[Klima06] V. Klima: Tunnels in Hash Functions: MD5 Collisions Within a Minute. Cryptology ePrint Archive, http://eprint.iacr.org, Report 105, 2006.

[KoeMauRen04] R. König, U. Maurer, R. Renner: Privacy amplification secure against an adversary with selectable knowledge. IEEE International Symposium on Information Theory ISIT 2004: p. 231, 2004.

[KoeMauRen05] R. König, U. Maurer, R. Renner: On the power of quantum memory. IEEE Transactions on Information Theory, 51(7): 2391–2401, 2005.

[KoeRenBarMau07] R. König, R. Renner, A. Bariska, U. Maurer: Small accessible quantum information does not imply security. Physical Review Letters, vol. 98, no. 140502, 2007.

[LeeKim03] B. Lee, K. Kim: Receipt-free electronic voting scheme with a tamper-resistant randomizer. Information Security and Cryptology - ICISC 2002: 5th International Conference, Seoul, Korea, November 28-29, 2002, Lecture Notes in Computer Science, 2587: 389 – 406, Springer-Verlag, 2003.

[LeeMooShaSha55] K. de Leeuw, E.F. Moore, C.E. Shannon, N. Shapiro: Computability by probabilistic machines. In: C.E. Shannon, J. McCarthy (eds.): Automata Studies, 183–212, Princeton, NJ: Princeton University Press, 1955.

[Lo03] H.-K. Lo: Method for decoupling error correction from privacy amplification. New J. Phys. 5 36, 2003.

[Lu02] C. Lu: Hyper-encryption against space-bounded adversaries from on-line strong extractors. Advances in Cryptology - CRYPTO 2002, Lecture Notes in Computer Science, 2442: 257–271, Springer-Verlag, 2002.

[LyWiWiElSkMa10] L. Lydersen, C. Wiechers, C. Wittmann, D. Elser, J. Skaar, V. Makarov: Hacking commercial quantum cryptography systems by tailored bright illumination. Nature Photonics 4: 686–689, 2010.

[MagBurChr01] E. Magkos, M. Burmester, V. Chrissikopoulos: Receipt-freeness in large-scale elections without untappable channels. In: 1st IFIP Conference on E-Commerce / E-Business/ E-Government (IFIP / SEC 01), Kluwer Academic Publishers, pp. 683–693, 2001.

[Makarov09] V. Makarov: Controlling passively quenched single photon detectors by bright light, New J. Phys. 11, 065003, 2009.

[MatMeyOse85] S.M. Matyas, C.H. Meyer, J. Oseas: Generating strong one way functions with cryptographic algorithm. IBM Techn. Disclosure Bull., 27(10A), 1985.

[MatTakIma86] T. Matsumoto, Y. Takashima, H. Imai: On seeking smart public-key-distribution systems. The Transactions of the IECE of Japan, E69: 99–106, 1986.

[Maurer92] U.M. Maurer: Conditionally-perfect secrecy and a provably-secure randomized cipher. Journal of Cryptology, 5(1): 53–66, 1992.

[Maurer93] U.M. Maurer: Protocols for secret-key agreement based on common information. IEEE Transactions on Information Theory, 39(3): 733–742, 1993.

[Maurer94] U.M. Maurer: Towards the equivalence of breaking the Diffie–Hellman protocol and computing discrete logarithms. Advances in Cryptology - CRYPTO '94, Lecture Notes in Computer Science, 839: 271–281, Springer-Verlag, 1994.

[Maurer95] U.M. Maurer: Fast generation of prime numbers and secure public-key cryptographic parameters. Journal of Cryptology, 8: 123–155, 1995.

[Maurer97] U.M. Maurer: Information-theoretically secure secret-key agreement by not authenticated public discussion. Advances in Cryptology - EUROCRYPT '92, Lecture Notes in Computer Science, 658: 209–225, Springer-Verlag, 1993.

[Maurer99] U.M. Maurer: Information-theoretic cryptography. Advances in Cryptology - CRYPTO '99, Lecture Notes in Computer Science, 1666: 47–65, Springer-Verlag, 1999.

[MauRenHol04] U. Maurer, R. Renner, C. Holenstein: Indifferentiability, impossibility results on reductions, and applications to the Random Oracle methodology. First Theory of Cryptography Conference, TCC 2004, Lecture Notes in Computer Science, 2951: 21–39, Springer-Verlag, 2004.

[MauWol96] U.M. Maurer, S. Wolf: Diffie–Hellman oracles. Advances in Cryptology - CRYPTO '96, Lecture Notes in Computer Science, 1109: 268–282, Springer-Verlag, 1996.

[MauWol97] U.M. Maurer, S. Wolf: Privacy amplification secure against active adversaries. Advances in Cryptology - CRYPTO '96, Lecture Notes in Computer Science, 1109: 307–321, Springer-Verlag, 1996.

[MauWol98] U.M. Maurer, S. Wolf: Diffie–Hellman, decision Diffie–Hellman, and discrete logarithms. Proceedings of ISIT '98, Cambridge, MA, August 16–21, 1998, IEEE Information Theory Society: 327, 1998.

[MauWol00] U.M. Maurer, S. Wolf: The Diffie–Hellman protocol. Designs, Codes, and Cryptography, Special Issue Public Key Cryptography, 19: 147–171, Kluwer Academic Publishers, 2000.

[MauWol00a] U.M. Maurer, S. Wolf: Information-theoretic key agreement: From weak to strong secrecy for free. Advances in Cryptology - EUROCRYPT 2000, Lecture Notes in Computer Science, 1807: 351–368, Springer-Verlag, 2000.

[Mayers96] D. Mayers: Quantum key distribution and string oblivous transfer in noisy channels. Advances in Cryptology - CRYPTO '96, Lecture Notes in Computer Science, 1109: 343–357, Springer-Verlag, 1996.

[MenOkaVan93] A. Menezes, T. Okamoto, S. Vanstone: Reducing elliptic curve logarithms to logarithms in a finite field. IEEE Transactions on Information Theory, 39: 1639–1646, 1993.

[MenQu01] A. Menezes, M. Qu: Analysis of the Weil descent attack of Gaudry, Hess and Smart. Topics in Cryptology - CT-RSA 2001, Lecture Notes in Computer Science 2020: 308–318, Springer-Verlag, 2001.

[Merali09] Z. Merali: Hackers blind quantum cryptographers. Nature News, DOI:10.1038/news.2010.436, 2009.

[Moeller03] B. Möller: Provably secure public-key encryption for length-preserving Chaumian mixes. M. Joye (ed.): Topics in Cryptology - CT-RSA 2003, Lecture Notes in Computer Science, 2612: 2003, Springer-Verlag, 244–262.

[MicRacSlo88] S. Micali, C. Rackoff, B. Sloan: The notion of security for probabilistic cryptosystems. SIAM Journal on Computing, 17: 412–426, 1988.

[MiKiGrWePeEuSc13] R. Mingesz, L.B. Kish, Z. Gingl, C.-G. Granqvist, H. Wen, F. Peper, T. Eubanks, G. Schmera: Unconditional security by the laws of classical physics. Metrol. Meas. Syst. XX (1): 3–16, 2013.

[NaoYun89] M. Naor, M. Yung: Universal one-way hash functions and their cryptographic applications. Proceedings of the 21st Annual ACM Symposium on Theory of Computing, Seattle, Wash., May 15–17, 1989: 33–43, 1989.

[NaoYun90] M. Naor, M. Yung: Public-key cryptosystems provably secure against chosen ciphertext attack. Proceedings of the 22nd Annual ACM Symposium on Theory of Computing, Baltimore, MD, May 14–16, 1990: 427–437, 1990.

[Neff01] C. A. Neff: A verifiable secret shuffle and its application to e-voting. ACM conference on Computer and Communications Security CCS '01: 116–125, 2001.

[NeeSch78] R.M. Needham, M.D. Schroeder: Using encryption for authentication in large networks of computers. Communications of the ACM, 21: 993–999, 1978.

[von Neumann63] J. von Neumann: Various techniques for use in connection with random digits. In: von Neumann's Collected Works, 768–770. New York: Pergamon, 1963.

[NeuTs'o94] B.C. Neuman, T. Ts'o: Kerberos: an authentication service for computer networks. IEEE Communications Magazine, 32: 33–38, 1994.

[Newman80] D.J. Newman: Simple analytic proof of the prime number theorem. Am. Math. Monthly 87: 693–696, 1980.

[Nguyen02] K.-C. Nguyen: Extension des protocoles de réconciliation en cryptographie quantique. Master's thesis, Université libre de Bruxelles, 2002.

[NisZuc96] N. Nisan, D. Zuckerman: Randomness is linear in space. Journal of Computer and System Sciences, 52(1): 43–52, 1996.

[NIST94] National Institute of Standards and Technology, NIST FIPS PUB 186, Digital Signature Standard, U.S. Department of Commerce, 1994.

[NIST12] National Institute of Standards and Technology: Recommendation for Key Management Part 1: General (Revision 3). NIST Special Publication 800–57, U.S. Department of Commerce, 2012.

[Okamoto92] T. Okamoto: Provably secure and practical identification schemes and corresponding signature schemes. Advances in Cryptology - CRYPTO '92, Lecture Notes in Computer Science, 740: 31–53, Springer-Verlag, 1993.

[OkaOht92] T. Okamoto, K. Ohta: Universal electronic cash. Advances in Cryptology - CRYPTO '91, Lecture Notes in Computer Science, 576: 324–337, Springer-Verlag, 1992.

[Okamoto97] T. Okamoto: Receipt-free electronic voting schemes for large scale elections. Security Protocols, 5th International Workshop, Paris, France, April 7–9, 1997, Proceedings, Lecture Notes in Computer Science, 1361: 25–35, Springer-Verlag, 1997.

[OngSchSha84] H. Ong, C.P. Schnorr, A. Shamir: Efficient signature schemes based on quadratic equations. Advances in Cryptology - Proceedings of CRYPTO '84, Lecture Notes in Computer Science, 196: 37–46, Springer-Verlag, 1985.

[Paillier99] P. Paillier: Public-key cryptosystems based on composite degree residuosity classes. Advances in Cryptology - EUROCRYPT '99, Lecture Notes in Computer Science, 1592: 223–238, Springer-Verlag, 1999.

[Pedersen91] T. Pedersen: A threshold cryptosystem without a trusted party. Advances in Cryptology - EUROCRYPT '91, Lecture Notes in Computer Science, 547: 522–526, Springer-Verlag, 1991.

[Peralta92] R. Peralta: On the distribution of quadratic residues and nonresidues modulo a prime number. Mathematics of Computation, 58(197): 433–440, 1992.

[Pfitzmann96] B. Pfitzmann: Digital signature schemes - general framework and fail-stop signatures. Lecture Notes in Computer Science, 1100, Springer-Verlag, 1996.

[PohHel78] S.C. Pohlig, M.E. Hellman: An improved algorithm for computing logarithms over GF(p) and its cryptographic significance. IEEE Transactions on Information Theory, IT24: 106–110, January 1978.

[PoiSte00] D. Pointcheval, J. Stern: Security arguments for digital signatures and blind signatures. Journal of Cryptology, 13(3): 361–396, 2000.

[PolSch87] J.M. Pollard, C.P. Schnorr: An efficient solution of the congruence $x^2 + ky^2 = m(\mathrm{mod}\,n)$. IEEE Transactions on Information Theory, 33(5): 702–709, 1987.

[Rabin63] M.O. Rabin: Probabilistic automata. Information and Control, 6: 230–245, 1963.

[Rabin79] M.O. Rabin: Digitalized signatures and public key functions as intractable as factorization. MIT/LCS/TR-212, MIT Laboratory for Computer Science, 1979.

[RacSim91] C. Rackoff, D.R. Simon: Non-interactive zero-knowledge proof of knowledge and chosen ciphertext attack. Advances in Cryptology - CRYPTO '91, Lecture Notes in Computer Science, 576: 433–444, Springer-Verlag, 1992.

[RasKol10] S. Rass, C. Kollmitzer: Adaptive Cascade. In: C. Kollmitzer, M. Pivk (eds.): Applied Quantum Cryptography. Lecture Notes in Physics 797: 49–69, Berlin, Heidelberg, New York: Springer-Verlag, 2010.

[Renner05] R. Renner: Security of Quantum Key Distribution. PhD Thesis, Swiss Federal Institute of Technology (ETH), Zürich, 2005.

[Renner12] R. Renner: Reply to recent scepticism about the foundations of quantum cryptography. arXiv:1209.2423 [quant-ph], 2012.

[RenKoe05] R. Renner, R. König: Universally composable privacy amplification against quantum adversaries. Second Theory of Cryptography Conference, TCC 2005, Lecture Notes in Computer Science, 2951: 407–425, Springer-Verlag, 2005.

[Rényi61] A. Rényi: On measures of entropy and information. Proc. 4th Berkeley Symposium on Mathematical Statistics and Probability, vol. 1: 547–561, Berkeley: University of California Press, 1961.

[RFC 1510] J. Kohl, C. Neuman: The Kerberos network authentication service (V5). Internet Request for Comments 1510 (RFC 1510), http://www.ietf.org, 1993.

[RFC 2104] H. Krawczyk, M. Bellare, R. Canetti: HMAC: Keyed-hashing for message authentication. Internet Request for Comments 2104 (RFC 2104), http://www.ietf.org, 1997.

[RFC 2313] B. Kaliski: PKCS#1: RSA encryption, Version 1.5. Internet Request for Comments 2313 (RFC 2313), http://www.ietf.org, 1998.

[RFC 2409] The Internet Key Exchange (IKE). Internet Request for Comments 2409 (RFC 2409), http://www.ietf.org, 1998.

[RFC 3174] US Secure Hash Algorithm 1 (SHA1). Internet Request for Comments 3174 (RFC 3174), http://www.ietf.org, 2001.

[RFC 3369] R. Housley: Cryptographic Message Syntax (CMS). Internet Request for Comments 3369 (RFC 3369), http://www.ietf.org, 2002.

[RFC 3447] J. Jonsson, B. Kaliski: Public-Key Cryptography Standards (PKCS) #1: RSA cryptography specifications, Version 2.1. Internet Request for Comments 3447 (RFC 3447), http://www.ietf.org, 2003.

[RFC 4492] S. Blake-Wilson, N. Bolyard, V. Gupta, C. Hawk, B. Moeller: Elliptic curve cryptography (ECC) cipher suites for transport layer security (TLS). Internet Request for Comments 4492 (RFC 4492), http://www.ietf.org, 2006.

[RFC 5246] T. Dierks, E. Rescorla: The Transport Layer Security (TLS) protocol, Version 1.2. Internet Request for Comments 5246 (RFC 5246), http://www.ietf.org, 2008.

[Rivest90] R. Rivest: The MD4 message digest algorithm. Advances in Cryptology - CRYPTO '90, Lecture Notes in Computer Science, 537: 303–311, Springer-Verlag, 1991.

[RivShaAdl78] R. Rivest, A. Shamir, and L.M. Adleman: A method for obtaining digital signatures and public key cryptosystems. Communications of the ACM, 21(2): 120–126, 1978.

[RosSch62] J. Rosser, L. Schoenfield: Approximate formulas for some functions of prime numbers. Illinois J. Math. 6: 64–94, 1962.

[Santos69] E.S. Santos: Probabilistic Turing machines and computability. Proc. Amer. Math. Soc. 22: 704–710, 1969.

[Satoh00] T. Satoh: The canonical lift of an ordinary elliptic curve over a prime field and its point counting. Journal of the Ramanujan Mathematical Society, 15: 247–270, 2000.

[SatAra98] T. Satoh, K. Araki: Fermat quotients and the polynomial time discrete log algorithm for anomalous elliptic curves. Commentarii Mathematici Universitatis Sancti Pauli, 47: 81–92, 1998.

[Schauer10] S. Schauer: Attack strategies on QKD protocols. In: C. Kollmitzer, M. Pivk: Applied Quantum Cryptography. Lecture Notes in Physics 797. Berlin, Heidelberg, New York: Springer-Verlag, 2010.

[Schoof85] R. Schoof: Elliptic curves over finite fields and the computation of square roots mod p. Math. Comp., 44(170): 483–494, 1985.

[SchnAle84] C.P. Schnorr, W. Alexi: RSA-bits are 0.5 + epsilon secure. Advances in Cryptology - EUROCRYPT '84, Lecture Notes in Computer Science, 209: 113–126, Springer-Verlag, 1985.

[Schweisgut07] J. Schweisgut: Elektronische Wahlen unter dem Einsatz kryptografischer Observer. Dissertation, Universität Giessen, Germany, http://geb.uni-giessen.de/geb/volltexte/2007/4817/, 2007.

[Semaev98] I. Semaev: Evaluation of discrete logarithms in a group of p-torsion points of an elliptic curve in characteristic p. Mathematics of Computation, 67: 353–356, 1998.

[Shannon48] C.E. Shannon: A mathematical theory of communication. Bell Systems Journal, 27: 379–423, 623–656, 1948.

[Shannon49] C.E. Shannon: Communication theory of secrecy systems. Bell Systems Journal, 28: 656–715, 1949.

[Shor94] P.W. Shor: Algorithms for quantum computation: discrete log and factoring. Proceedings of the IEEE 35th Annual Symposium on Foundations of Computer Science, Santa Fe, New Mexico, November 20–22, 1994: 124–134, 1994.

[ShoPre00] P.W. Shor, J. Preskill: Simple proof of security of the BB84 quantum key distribution protocol. Phys. Rev. Lett. 85: 441–444, 2000.

[Shoup01] V. Shoup: OAEP reconsidered. Advances in Cryptology - CRYPTO 2001, Lecture Notes in Computer Science, 2139: 239–259, Springer-Verlag, 2001.

[Skjernaa03] B. Skjernaa: Satoh's algorithm in characteristic 2. Mathematics of Computation, 72: 477–487, 2003.

[Smart99] N. Smart: The discrete logarithm problem on elliptic curves of trace one. Journal of Cryptology, 12(3): 193–196, 1999.

[Smith05] W. D. Smith: New cryptographic election protocol with best-known theoretical properties. Frontiers in Electronic Elections (FEE 2005) – Workshop at ESORICS 2005, Milan, Italy, 2005.

[Stinson94] D.R. Stinson: Universal hashing and authentication codes. Designs, Codes and Cryptography 4 (4): 369–380, 1994.

[TerWik10] B. Terelius, D. Wikström: Proofs of restricted shuffles. AFRICACRYPT 2010, Lecture Notes in Computer Science, 6055: 100–113, Springer-Verlag, 2010.

[TsiYun98] Y. Tsiounis, M. Yung: On the security of ElGamal based encryption. Proceedings Theory and Practice in Public Key Cryptography - PKC 1998, Lecture Notes in Computer Science, 1431: 117–134, Springer-Verlag, 1998.

[Vadhan03] S. Vadhan: On constructing locally computable extractors and cryptosystems in the bounded storage model. Advances in Cryptology - CRYPTO 2003, Lecture Notes in Computer Science, 2729: 61–77, Springer-Verlag, 2003.

[Vazirani85] U.V. Vazirani: Towards a strong communication complexity, or generating quasi-random sequences from slightly random sources. Proceedings of the 17th Annual ACM Symposium on Theory of Computing, Providence, RI, May 6–8, 1985: 366–378, 1985.

[VazVaz84] U.V. Vazirani, V.V. Vazirani: Efficient and secure pseudorandom number generation. Proceedings of the IEEE 25th Annual Symposium on Foundations of Computer Science, Singer Island, Fla., October 24–26, 1984: 458–463, 1984.

[Vernam19] G.S. Vernam: Secret signaling system. U.S. Patent #1, 310, 719, 1919.

[Vernam26] G.S. Vernam: Cipher printing telegraph systems for secret wire and radio telegraphic communications. Journal of American Institute for Electrical Engineers, 45: 109–115, 1926.

[WaiPfi89] M. Waidner, B. Pfitzmann: The dining cryptographers in the disco: unconditional sender and recipient untraceability with computationally secure serviceability. Advances in Cryptology - EUROCRYPT '89, Lecture Notes in Computer Science, 434: 690, Springer-Verlag, 1990.

[WanFenLaiYu04] X. Wang, D. Feng, X. Lai, H. Yu: Collisions for hash functions MD4, MD5, HAVAL-128, RIPEMD. Rump Session, Advances in Cryptology - CRYPTO 2004. Cryptology ePrint Archive, http://eprint.iacr.org, Report 199, 2004.

[WanYinYu05] X. Wang, Y.L. Yin, H. Yu: Finding Collisions in the Full SHA-1. Advances in Cryptology - CRYPTO 2005, Lecture Notes in Computer Science, 3621: 17–36, Springer-Verlag, 2005.

[WatShiIma03] Y. Watanabe, J. Shikata, H. Imai: Equivalence between semantic security and indistinguishability against chosen ciphertext attacks. Proceedings Theory and Practice in Public Key Cryptography - PKC 2003, Lecture Notes in Computer Science, 2567: 71–84, Springer-Verlag, 2003.

[WebAraBuc07] S. Weber, R. Araújo, J. Buchmann: On coercion-resistant electronic elections with linear work. Second International Conference on Availability, Reliability and Security (ARES 2007): 908 – 916, 2007.

[WegCar81] M.N. Wegman, J.L. Carter: New hash functions and their use in authentication and set equality. Journal of Computer and System Sciences, 22: 265–279, 1981.

[Wiener90] M.J. Wiener: Cryptanalysis of short RSA secret exponents. IEEE Transactions on Information Theory, 36: 553–558, 1990.

[Wikström05] D. Wikström: A sender verifiable mix-net and a new proof of a shuffle. ASIACRYPT 2005, Lecture Notes in Computer Science, 3788: 273–292, Springer-Verlag, 2005.

[Wikström09] D. Wikström: A commitment-consistent proof of a shuffle. Proceedings of the 14th Australasian Conference on Information Security and Privacy ACISP '09, Lecture Notes in Computer Science, 5594: 407–421, Springer-Verlag, 2009.

[Wolf98] S. Wolf: Unconditional security in cryptography. In: I. Damgård (ed.): Lectures on Data Security. Lecture Notes in Computer Science, 1561: 217–250, Springer-Verlag, 1998.

[Yuen12] H.P. Yuen: Fundamental and practical problems of QKD security - The actual and the perceived situation. arXiv:1109.1066v4, 2012.

Internet

[BelDesJokRog97] M. Bellare, A. Desai, E. Jokipii, P. Rogaway: A concrete security treatment of symmetric encryption: analysis of the DES modes of operation. http://www-cse.ucsd.edu/users/mihir/papers/sym-enc.html, 1997.

[BerDaePeeAss11] G. Bertoni, J. Daemen, M. Peeters, G. Van Assche: Cryptographic sponge functions, Version 0.1. http://sponge.noekeon.org/CSF-0.1.pdf, 2011.

[BerDaePeeAss11a] G. Bertoni, J. Daemen, M. Peeters, G. Van Assche: The Keccak reference, Version 3.0. http://keccak.noekeon.org/Keccak-reference-3.0.pdf, 2011.

[BerDaePeeAss11b] G. Bertoni, J. Daemen, M. Peeters, G. Van Assche: The Keccak SHA-3 submission, Version 3.0. http://keccak.noekeon.org/Keccak-submission-3.pdf, 2011.

[GolBel01] S. Goldwasser, M. Bellare: Lecture notes on cryptography. http://www-cse.ucsd.edu/users/mihir/papers/gb.html, 2001.

[DistributedNet] Distributed Net. http://www.distributed.net.

[ID Quantique] ID Quantique. http://www.idquantique.com.

[MagiQ] MagiQ Technologies. http://www.magiqtech.com.

[NIST00] National Institute of Standards and Technology. Advanced Encryption algorithm (AES) development effort. http://www.nist.gov/aes.

[NIST13] National Institute of Standards and Technology: Cryptographic hash and SHA-3 Standard development.
http://csrc.nist.gov/groups/ST/hash/index.html, 2013.

[RSALabs] RSA Laboratories. http://www.rsasecurity.com/rsalabs/.

[TCG] Trusted Computing Group: Trusted Platform Module Specifications. https://www.trustedcomputinggroup.org/specs/TPM/.

Index

admissible key generator, 234, 269
advanced encryption standard, 19
advantage distillation, 327
AES, see advanced encryption standard
affine cipher, 318
affine plane, 438
algebraic closure, 438
algebraic curve
- affine curve, 438
- flex tangent, 444
- inflection point, 444
- k-rational point, 441
- points at infinity, 442
- projective curve, 441
- regular point, 443
- singular point, 443
- tangent, 444
algorithm
- coin-tossing, 203
- deterministic, 203
- deterministic extension, 205
- distinguishing, 290, 294
- efficient, 7
- Euclid, 52, 399, 414
- extended Euclid, 400, 414
- Las Vegas, 208
- Lucifer, 18
- Miller–Rabin, 436
- Monte Carlo, 208
- oracle, 245
- polynomial algorithm, 207
- probabilistic, see probabilistic
  algorithm
- randomized encryption, 27, 284
- Rijndael, 20
- sampling, 223
- Solovay–Strassen, 436
assumption
- composite residuosity, 89
- decisional composite residuosity, 89
- decision Diffie–Hellman, 198, 313

- Diffie–Hellman, 112
- discrete logarithm, 218
- factoring, 229
- quadratic residuosity, 231
- RSA, 227
- strong RSA, 382
attack
- adaptively-chosen-ciphertext, 5, 65,
  302
- adaptively-chosen-message, 374
- adaptively-chosen-plaintext, 5
- against encryption, 4
- against signatures, 373
- birthday, 38
- Bleichenbacher's 1-Million-Chosen-
  Ciphertext, 64
- chosen-message, 374
- chosen-plaintext, 5
- ciphertext-only, 5
- common-modulus, 62
- encryption and signing with RSA, 71
- key-only, 374
- known-plaintext, 5
- known-signature, 374
- low-encryption-exponent, 63
- partial chosen-ciphertext, 65
- replay, 108
- RSA, 62
- small-message-space, 63
attacks and levels of security, 373
authentication, 2

big-O, 402
binary
- encoding of $x \in \mathbb{Z}$, 401
- encoding of $x \in \mathbb{Z}_n$, 403
- field 437
- random variable, 461
- symmetric channel, 348
binding property, 127, 129
bit-security

– Exp family, 243
– RSA family, 250
– Square family, 258
Bleichenbacher's 1-Million-Chosen-
    Ciphertext Attack, 64
blind signatures
– Nyberg–Rueppel, 202
– RSA, 201
– Schnorr, 189
blindly issued proofs, 186
block cipher, 15
Blum–Blum–Shub generator, 271, 281
Blum–Goldwasser probabilistic
    encryption, 298
Blum–Micali generator, 270
Boolean predicate, 216, 461
bounded storage model, 322

Caesar's shift cipher, 1
cardinality, 404
Carmichael number, 433
certificate, 115
certification authority, 115
challenge-response, 6, 108, 113
characteristic, 447
Chebyshev's Inequality, 468
Chernoff bound, 213
cipher-block chaining, 26
cipher feedback, 27
ciphertext, 1, 12
ciphertext-indistinguishable, 295, 303
claw, 376
claw-free pair of one-way permutations,
    376
coin, see electronic coin
coin tossing by telephone, 126
collision, 31
– entropy, 329, 337
– resistant hash function, see hash
    function
– probability, 329
commitment, 126
– binding property, 127
– discrete logarithm, 128
– generalized Pedersen, 155
– hiding property, 127
– homomorphic, 130
– Pedersen, 128
– quadratic residue, 127
completeness, 118, 199
complexity class
– $\mathcal{BPP}$, 212
– $\mathcal{IP}$, 200

– $\mathcal{NP}$, 212
– $\mathcal{RP}$, 212
– $\mathcal{ZPP}$, 213
composite, 52, 402
compression function, 32, 39
computationally perfect pseudorandom
    generator, 268
conditional
– entropy, 478
– mutual information, 479
– probability, 460
– uncertainty, 478
confidentiality, 1
congruent modulo n, 53, 403
creation of a certificate, 116
cryptanalysis, 4
cryptogram, 12
cryptographic protocol, 5, 107, see also
    protocol
cryptographically secure, 268
cryptography, 1
cryptology, 4
cyclic group, 410

data encryption standard, 16
data integrity, 2
data origin authentication, 2
decision Diffie–Hellman problem, 198,
    313
decryption, 1
– key, 1
– threshold, 136
DES, see data encryption standard
deterministic extension, 205
Diffie–Hellman
– key agreement, 111
– problem, 78, 111
digital
– digital signature algorithm, 80
– digital signature standard, 80
– fingerprint, 42
– signature scheme, 373
– signatures, 3, see also signatures
digital cash, 184
– coin and owner tracing, 193, 196
– customer anonymity, 193
– deposit, 196
– electronic coin, 184
– fair payment systems, 185, 192
– offline, 194
– online, 193
– opening an account, 193
– owner tracing, 192

– payment and deposit protocol, 193
– payment protocol, 195
– security, 196
– withdrawal protocol, 193, 194
direct product, 462
discrete
– exponential function, 55, 217
– exponential generator, 270
– logarithm function, 55, 218
distance measures, 472
division with remainder, 52, 398
divisor, 52
DSA, see digital signature algorithm
DSS, see digital signature standard

electronic cash, see digital cash
electronic codebook mode, 26
electronic election, 135
– authority's proof, 138
– coercion-resistant, 168
– communication model, 135
– forced-abstention attack, 177
– multi-way, 143
– randomization attack, 177
– receipt-free, 168
– re-encryption of votes, 169
– setting up the scheme, 135
– simulation attack, 177
– tally computing, 137
– threshold decryption, 136
– trusted center, 144
– vote casting, 137
– vote duplication, 201
– voter's proof, 141
ElGamal encryption, 77
– in a prime-order subgroup, 82
– with elliptic curves, 100
ElGamal signatures, 78
elliptic curve, 446
– chord-and-tangent method, 449
– Edwards form, 454
– formulas for the addition of points, 451
– group order and group structure, 455
– normal forms, 446
– point addition, 449
– scalar point multiplication, 454
– Weierstrass form, 446
elliptic curve cryptography, 90
– curve B-163, 98
– curve P-192, 98
– Diffie–Hellman key agreement, 99
– discrete logarithm assumption, 91

– discrete logarithm problem, 91
– domain parameters, 93
– ElGamal encryption, 100
– Elliptic Curve Digital Signature Algorithm, 102
– encoding of plaintexts, 101
– insecure curves, 93
– Menezes–Vanstone variant, 102
– nearly prime order, 93
– prime-order subgroup, 93
– selecting the curve and the base point, 93
encoding of messages, 84, 101
encryption, see also public-key encryption
– ciphertext-indistinguishable, 295, 303
– Cramer–Shoup, 313
– methods, 1
– malleable, 152
– OAEP, 67
– perfectly secret, 285
– provably secure, 283
– randomized, 284
– SAEP, 304
– symmetric, 2
– symmetric-key, 11
– Vernam's one-time pad, 7, 13, 284
entity authentication, 3, 107, 108, 112
entropy, 477
– collision, 329, 337
– conditional, 478
– smoothing, 328, 340
– joint, 478
– Rényi, 329, 338
existential forger, 375
existentially forged, 70
Exp family, 218
extraction of a secret key, 341

fail-stop signature schemes, 376
family of
– claw-free one-way permutations, 376
– hard-core predicates, 235
– one-way functions, 231
– one-way permutations, 232
– trapdoor functions, 233
fast modular exponentiation, 406
feasible, 7
Feistel cipher, 18
Fibonacci numbers, 400
field
– binary 437
– finite, 413

– prime, 413
forward secrecy, 114
function
– compression, 32, 39
– discrete exponential, 55, 217
– discrete logarithm, 55, 218
– Euler phi or totient, 54, 405
– hash, see hash function
– one-way, see one-way function
– polynomial function, 268
– RSA, 55, 227
– square, 56, 229, 230
– stretch function, 268

Galois field, 417
geometric random variable, 464
GMR signature scheme, 380
Golden Ratio, 400
Goldwasser–Micali probabilistic
  encryption, 294
greatest common divisor, 397, 414

half-trace, 420
hard-core predicate, 235, 280
hash function, 30
– birthday attack, 38
– collision resistant, free, 31, 376, 377
– compression function, 32
– construction, 32
– digital fingerprint, 42
– ε-almost strongly universal class, 333
– ε-universal class, 333
– real hash functions, 40
– second pre-image resistant, 31
– strongly collision resistant, 31
– strongly universal class, 329
– universal class, 329
– universal one-way family, 385
– weakly collision resistant, 31
hash-then-decrypt paradigm, 72
hiding property, 127, 129
HMAC, 43
homomorphic encryption algorithms,
  87
homomorphism, 408
honest-verifier zero-knowledge, 140

identification schemes, 117
image distribution, 461
independent events, 462
index set with security parameter, 220
information
– reconciliation, 332, 352, 362
– theoretic security, 284

– theory, 476
inner product bit, 235
integer quotient, 398
integers modulo n, 53
interactive proof
– completeness, 118, 199
– designated verifier, 172
– diverted, 174
– move, 117, 119, 120
– of knowledge, 117
– prover, 117, 118
– prover's secret, 117
– re-encryption, 172
– rounds, 117
– soundness, 118, 200
– special soundness, 140
– system, 117
– system for a language, 199
– verifier, 117, 118
– zero-knowledge, 121, 200
isomorphism, 408
iterated cipher, 18

Jacobi symbol, 423

Keccak, 36
Kerberos, 108
– authentication protocol, 109
– authentication server, 108
– credential, 109
– ticket, 109
Kerckhoffs' Principle, 4
key, 12
– exchange, 107, 112
– generator, 231, 268
– set with security parameter, 220
– stream, 13
knowledge
– completeness, 118
– soundness, 118
– zero-knowledge, 121, 200

Lagrange interpolation formula, 132
least common multiple, 397
Legendre symbol, 422
Log family, 218

malleable, 152
master keys, 107
MD5, 40
Merkle–Damgård construction, 32
Merkle's meta method, 32
message, 1
message authentication code, 3, 43

– unconditionally secure, 333, 335
message digest, 42
mix net, 146
– decryption, 147
– re-encryption, 150
– source routing, 146
modes of operation, 25
– cipher-block chaining, 26
– cipher feedback, 27
– electronic codebook mode, 26
– output feedback mode, 28
modification detection codes, 42
modular
– arithmetic, 51
– powers, 55, 226
– Square Roots, 426
– Squares, 56
– Squaring, 85, 229
multiplicative inverse, 404
mutual authentication, 112
mutual information, 478

natural representatives, 403
negligible, 219
next-bit predictor, 275
next-bit test, 275
NIST, 19
noisy channel model, 332
non-interactive proofs of knowledge, 142
non-repudiation, 3
Nyberg–Rueppel signatures, 201

OAEP, 67
one-way function, 2, 50, 215, 218, 231
– bijective one-way function, 232
– bit security of, 243
– definition of, 231
– Exp family, 218
– pseudorandomness, 267
– RSA family, 227
– square family, 230
– weak one-way function, 240
one-way permutation, 227, 232
OR-combination of Σ-proofs, 141
order of x, 409
order of a set, 404
output feedback mode, 28

Paillier encryption, 88
password scheme, 118
perfectly secret, 7, 285
permutation, 15
plaintext, 1, 12

plaintext equivalence, 178
plaintext equivalence test, 179
polynomial
– degree, 413
– irreducible, 415
– prime to, 414
– reducible, 415
polynomial-time indistinguishability, 8, 296
polynomial-time reduction, 219
polynomially bounded, 475
polynomially close, 474
positive polynomial, 209
primality test, 432
prime number, 52, 402
prime residue class group, 54, 404
prime to, 52, 397
primitive element, 55
primitive root, 410
principal square root, 244
privacy amplification, 328, 340
probabilistic
– public-key encryption, 294
– primality test, 436
– signature scheme, 75
probabilistic algorithm, 203
– coin-tossing algorithm, 203
– distinguishing algorithm, 290, 294
– Las Vegas, 208
– Miller–Rabin primality test, 436
– Monte Carlo, 208
– polynomial algorithm, 207
– Solovay–Strassen primality test, 436
probability, 459
– distribution, 459
– image distribution, 461
– independent events, 462,
– joint space, 462
– measure, 459
– notation for, 216, 466
– of success, 289, 375
– space, 459
projective line, 440
projective plane, 439
proof of knowledge, see also interactive proof
– interactive, 117
– of a plaintext, 154
– of a secret key, 171
– non-interactive, 142
– of a logarithm, 186
– of a representation, 202
– of a square root, 119, 123

- of the equality of two logarithms, 138
- one of two pairs, 141
protocol,
- Diffie–Hellman key agreement, 111
- digital cash, 192
- electronic elections, 135
- Fiat–Shamir identification, 119, 123
- honest-verifier zero-knowledge protocol, 140
- Kerberos authentication, 109
- public-coin, 139
- Schnorr's identification, 186
- station-to-station, 114
- special honest-verifier zero-knowledge protocol, 141
- strong three-way authentication, 113
- two-way authentication, 113
- zero-knowledge, 121, 200
provable security, 6
provably secure digital signature, 373
provably secure encryption, 283
pseudorandom
- function model, 29
- one-time pad induced by G, 292
- permutation model, 29
pseudorandom bit generator, 268
- Blum–Blum–Shub generator, 271, 281
- Blum–Micali generator, 270
- discrete exponential generator, 270
- induced by f, B and Q, 270
- RSA generator, 270
- x^2 mod n generator, 271
PSS, see probabilistic signature scheme
public key, 49
- cryptography, 2, 49
- management techniques, 115
public-key encryption, 2, 49
- asymmetric encryption, 2
- Blum–Goldwasser probabilistic encryption, 298
- ciphertext-indistinguishable, 295
- ElGamal encryption, 77
- OAEP, 67
- probabilistic public-key encryption, 294
- provably secure encryption, 283
- public-key one-time pad, 293
- Rabin's encryption, 85
- randomized encryption, 284
- RSA, 58

quadratic

- equations in binary fields, 419
- non-residue, 200, 421
- residue, 422
quadratic residuosity
- family, 231
- property, 230
quantum key distribution, 343
- BB84 protocol, 350
- binary symmetric channel, 348
- Breidbart basis, 347
- Cascade, 362
- circular basis, 346
- circularly polarized photons, 347
- conjugate bases, 350
- diagonal basis, 346
- elliptically polarized photons, 347
- encoding of quantum bits, 347, 351
- estimation of the error rate, 352
- Hermitian inner product, 344
- Hilbert space of a quantum bit, 345
- individual attacks, 354
- information reconciliation, 352, 362
- intercept-and-resend attack, 354
- linearly polarized photons, 347
- measurements, 347
- phase factors, 348
- privacy amplification, 353
- quantum bit, 345
- quantum channel, 348
- raw key, 352
- rectilinear basis, 346
- sending a bit through a quantum channel, 348
- sifted key, 352
- superpositions, 345
- transmission of a key string, 360
- transmission of a single key bit, 355
- unitary matrix, 346

Rabin's encryption, 85
Rabin's signature scheme, 86
random
- function, 9
- oracle model, 73, 304, 312
- self-reducible, 222
random variable, 461
- binary, 461
- distribution, 461
- expected value, 463
- geometric, 464
- independent, 463
- jointly distributed, 463
- real-valued, 461

– variance, 468
randomized algorithm, see probabilistic algorithm
rational approximation, 251
real-valued random variable, 461
re-encryption of ciphertexts, 89
relatively prime to, 52, 397
Rényi entropy, 329, 338
representation problem, 197
residue, 403
– residue class ring, 403
– residue class, 403
retrieval of secret keys, 374
revocation of a certificate, 116
Rijndael cipher, 19
RIPEMD-160, 40
RSA, 58
– assumption, 227
– attack on encryption and signing, 71
– attacks, 62
– digital signatures, 70
– encryption and decryption, 59
– encryption and decryption exponent, 58
– existentially forged, 70
– family, 227
– function, 55
– generator, 270
– key generation, 58, 228
– low-encryption-exponent attack, 63
– probabilistic encryption, 67
– security, 60
– speeding up encryption and decryption, 62

Σ-proof, 141
SAEP, 304
sampling algorithm, 223
Schnorr's blind signature, 189
Schnorr's identification protocol, 186
second pre-image resistance, 31
secret key, 49
secret key encryption, see encryption
secret sharing, 130
secure
– against adaptively-chosen-ciphertext attacks, 303
– against adaptively-chosen-message attacks, 376
– bit, 243
– cryptographically, 268
– indistinguishability-secure, 303
security

– everlasting, 324
– information-theoretic, 284
– in the random oracle model, 73, 304
– levels of, 373
– modes of operation, 29
– proof by reduction, 74
– RSA, 60
– unconditional, 321
– under standard assumptions, 313
seed, 267
selective forgery, 374
semantically secure, 296
session keys, 107
SHA-1, 40
SHA-3, 41
Shamir's threshold scheme, 131
shuffle, 146
– argument, 156
– exponentiation argument, 161
– product argument, 164
– proofs, 154
– re-encryption, 151
signature
– authentication-tree-based signatures, 379
– blind, see blind signature
– Cramer–Shoup signature scheme, 383
– Digital Signature Algorithm, 80
– ElGamal signatures, 78
– fail-stop, 376
– Fiat–Shamir, 125
– full-domain-hash RSA, 72
– GMR signature scheme, 380
– one-time signature scheme, 379, 394
– probabilistic, see probabilistic signature scheme
– provably secure digital, 373
– Rabin's signature scheme, 86
– RSA signature scheme, 70
– signed message, 70
– state-free, 381
simulator, 122
simultaneously secure bits, 280
single-length MDCs, 40
Sophie Germain prime, 381
soundness, 118, 200
special soundness, 140
sponge construction, 34
square family, 230
square root, 56, 426
squares, 422
statistical distance, 472
statistically close, 474

stream ciphers, 12
strong primes, 61

Theorem
- Blum and Micali, 274
- Chinese Remainder, 407
- Euler's Theorem, 54, 406
- Fermat's Little Theorem, 54, 406
- Fundamental Theorem of Arithmetic,
  53
- Hasse, 456
- Prime Number, 432
- Shannon, 286
- Yao, 275
threshold scheme, 130
tight reduction, 308
trace map, 419
transcript, 121
trapdoor
- function, 50, 215, 233
- information, 50, 227
- permutation, 227

uncertainty, 477
uniform distribution, 459
uniform sampling algorithm, 224
unit, 404
universal forgery, 374
untappable channel, 174

variance, 468
Vernam's one-time pad, 7, 13, 284
virtually uniform, 223
voting, see electronic election

Weak Law of Large Numbers, 470
witness, 436

zero divisor, 54
zero-knowledge, 121, 200
- non-interactive zero-knowledge
  proofs, 143
- proofs of shuffles, 154
- (special) honest-verifier zero-
  knowledge protocol, 140, 141

Printed in the United States
By Bookmasters